Creative Construction Conference (CCC 2021)

IOP Conference Series: Materials Science and Engineering
Volume 1218

Budapest, Hungary
28 - 30 June 2021

Editors:

Miroslaw Skibniewski
Miklos Hajdu
Attila Varga
Gergely Szakats

ISBN: 978-1-7138-4677-2
ISSN: 1757-8981

Printed from e-media with permission by:

Curran Associates, Inc.
57 Morehouse Lane
Red Hook, NY 12571

Some format issues inherent in the e-media version may also appear in this print version.

Institute of Physics
Dirac House, Temple Back
Bristol BS1 6BE UK

Phone: 44 1 17 929 7481
Fax: 44 1 17 920 0979

techtracking@iop.org

Additional copies of this publication are available from:

Curran Associates, Inc.
57 Morehouse Lane
Red Hook, NY 12571 USA
Phone: 845-758-0400
Fax: 845-758-2633
Email: curran@proceedings.com
Web: www.proceedings.com

TABLE OF CONTENTS

IOP Conf. Series: Materials Science and Engineering 1218 (2022) 011001 doi:10.1088/1757-899X/1218/1/011001

PREFACE

Creation is an inspiring endeavor. This is especially true when one creates objects of a size that only a few people can. Cathedrals or skyscrapers, bridges or aqueducts, highways, or dams are not only fascinating due to their size but also because they significantly affect the lives of all people in their proximity – for decades or even centuries. The act of construction is an act of "true" creation. Growing demand for a "better" built environment – even though the meaning of "better" is often hard to define in this context – requires increasing levels of diverse knowledge. As time progresses, it is getting harder to live up to the ever-growing societal expectations: trains begin to compete with the speed of airplanes, buildings are often expected to produce rather than to consume energy, and in the not-so-distant future, we build settlements in outer space, most likely with advanced robotics.

Creative Construction Conferences are organized to provide an annual forum for ambitious researchers and practitioners from around the world who have realized that the construction industry is rapidly changing and who accept that building materials, construction technologies, and project management will quickly and creatively evolve in the future.
The 10th Creative Construction Conference (CCC2021) – due to COVID 19 - was held online on 28-30 June 2021. Academics from 40 countries and all continents attended CCC2021 organized across six tracks:

- Creative management;
- Creative construction technologies and materials;
- Advanced automation and robotics for construction;
- Creative scheduling;
- Sustainable construction, health, and safety;
- Recent advances in Building Information Modeling, visualization, and virtual reality.

Original papers have been published in this Proceedings, followed by a rigorous peer-review process.

We are grateful to all the authors, speakers, poster presenters, members of the International Advisory Board, the Organizing and Scientific Committees, reviewers, and the session chairs for contributing to the success of CCC2021. Their invaluable contribution helped us become the largest construction engineering and management conference in Europe.

Prof. Miroslaw E. Skibniewski
University of Maryland, USA
Chair of International Advisory Board
Chair of International Scientific Committee

Prof. Miklós Hajdu
Budapest University of Technology, Hungay
Chair of International Organizing Committee
Co-Chair of International Advisory Board

IOP Conf. Series: Materials Science and Engineering 1218 (2022) 011001 doi:10.1088/1757-899X/1218/1/011001

International Scientific Committee *(permanent)*

IOP Conf. Series: Materials Science and Engineering 1218 (2022) 011001 doi:10.1088/1757-899X/1218/1/011001

Editors:

Mirosław Skibniewski
University of Maryland, USA
E-mail: mirek@umd.edu

Miklós Hajdu
Budapest University of Technology and Economics, Hungary
E-mail: miklos.hajdu63@gmail.com

Attila Varga
Diamond Congress Ltd., Hungary
E-mail: diamond@diamond-congress.hu

Gergely Szakáts
Diamond Congress Ltd., Hungary
E-mail: gergely@diamond-congress.hu

Peer review declaration

All papers published in this volume of *IOP Conference Series: Materials Science and Engineering* have been peer reviewed through processes administered by the Editors. Reviews were conducted by expert referees to the professional and scientific standards expected of a proceedings journal published by IOP Publishing.

- **Type of peer review: Double-blind**

- **Conference submission management system:**
  DPCS3 – Diamond Professional Congress System

- **Number of submissions received:** 110

- **Number of submissions sent for review:** 110

- **Number of submissions accepted:** 72

- **Acceptance Rate (Number of Submissions Accepted / Number of Submissions Received X 100):** 65%

- **Average number of reviews per paper: minimum 2**
  The 4 reviewers had to judge the continuously arriving materials from November until June. Quite a few did not meet the requirements of the conference, they were rejected.

- **Total number of reviewers involved: 4**
  Miklós Hajdu - miklos.hajdu63@gmail.com
  Mirosław J. Skibniewski
  Levente Mályusz
  Zoltán Vattai

- **Any additional info on review process:**
  Not only scientific but also professional and formal requirements were scored.

- Detailed grading:
  01. The references are representative and correct.
  02. The abstract and keywords describe the content.
  03. The paper makes a contribution to construction research.
  04. The title reflects the content and purpose of the abstract.
  05. The English is satisfactory.
  06. The paper is in the template and the size is appropriate.
  07. The quality of the illustrations, figures and tables is satisfactory.
  08. The topic of the paper is relevant to the constituent themes of the conference.
  09. The list of references contains the DOI nr. of the referred material.
  10. The paper is of sufficient quality for CCC2021.
  11. I recommend this version of the paper to be published in the Conference Proceedings

- **Contact person for queries:**

Gergely Szakáts, Diamond Congress Ltd. gergely@diamond-congress.hu

IOP Conf. Series: Materials Science and Engineering 1218 (2022) 012001 doi:10.1088/1757-899X/1218/1/012001

A Systematic Literature Review of Influencing Factors and Strategies of Artificial Intelligence Adoption in the Construction Industry

M M Tjebane[1*], I Musonda[1] and C S Okoro[2]

1. Department of Construction Management and Quantity Surveying, University of Johannesburg, South Africa
2. Department of Finance and Investment Management, College of Business and Economics, University of Johannesburg, South Africa

*motheometa@gmail.com

Abstract. The construction industry has seen an increase in Artificial Intelligence(AI) in recent years, a paradigm shift in many industries. It puts under pressure for technological advancement. Therefore, AI is under great attention in the construction industry as a new strategic paver. This paper adopts a systematic literature review (SLR) approach and followed the Preferred Reporting Items for Systematic Reviews and Meta-Analyses (PRISMA) guidelines to observe and understand the influencing factors and strategies for AI adoption. The SLR for AI-related research carried out between 2010 and 2020. Data was collected from ASCE Journals, Emerald Insight, Elsevier ScienceDirect, Engineering Village, Google Scholar, ICE virtual library, IOPscience, IEEE Xplore, ScienceDirect, Scopus, SpringerLink and Taylor & Francis. The paper identifies and classifies the new developments in AI research, making its implementation and adoption a reality in the construction industry. This review has the potential for construction industry stakeholders, especially those in developing countries, to utilise the accumulated evidence from selected systematic reviews to enable the usage of AI for infrastructure development

1. Introduction

The construction industry of countries plays a vital role in growing the economy [1]. More significantly in developing countries. It contributes to job creation, infrastructure development, and GDP growth [2]. However, it is lagging when it comes to technology usage [3]. This is compared to most industries that have instilled digitalisation and innovation. As a result, those industries have seen improvement and productivity gains. The adoption of new-age technologies such as Artificial Intelligence (AI) bridges the gap with other industries and developed countries [4].

AI refers to technology that can think independently of humans. It is a constructed, artificial, or machine intelligence, distinguished from natural intelligence by its very definition [5]. AI technologies are as important as those that provoked socially, politically, and environmentally in previous centuries [6]. Some of the construction industry's AI technologies include Building Information Modelling (BIM), sensor-based technologies, robotics, and wearable technologies [7]. This study aims to identify the influencing factors and strategies of AI adoption using a systematic literature review. This is to improve and assist in adopting AI technologies in the construction industry of developing countries.

2. Methodology

This study adopts a methodological approach to conducting a systematic literature review of influencing factors and AI adoption strategies in the construction industry. Systematic literature review (SLR) is a type of literature review that follows a specific review protocol and quality procedures to select relevant (primary) studies, extract and analyse appropriate information from the selected studies to answer specific research questions [8]. This approach can be used to draw logical inferences systematically based on the collected literature. Compared to other approaches, content analysis has the advantage of performing both qualitative and quantitative analyses of AI factors and strategies in the construction industry [9].

Various approaches and steps are taken when doing a systematic literature review with the same outcome [10]. This study took steps performed in accordance with [11].

2.1. Search Strategy

A systematic search was performed to identify the relevant AI articles published in the various databases from 2010 until 2020. Data was collected from ASCE Journals, Emerald Insight, Elsevier ScienceDirect, Engineering Village, Google Scholar, ICE virtual library, IOPscience, IEEE Xplore, Elsevier ScienceDirect, Elsevier Scopus, SpringerLink and Taylor & Francis. When searching for the relevant documents, various search terms were used. Furthermore, Boolean and database-specific operators such as AND, OR and special characters such as truncation (*) or (?) were associated with the search terms were used[12]. The search terms used were *Adoption, Artificial Intelligence, Construction industry, Construction Management, Factors, Influencing Factors, Strategies,*

2.2. Inclusion and Exclusion Criteria

This review included empirical studies based on the relevant research topic, AI adoption in the construction industry, written in English and published between 2010 and 2020. Exclusion criteria are any empirical which pieces were excluded if they focused on any other than the construction industry.

A total of 837 articles was extracted from the various databases. Thirty-one duplicate publications were removed. Four hundred and twenty records were excluded as they were either not written in English, show no peer review process, did not focus on AI and the construction industry, leaving 386 studies. These 386 articles abstracts were read, excluding 350 records as they did not look at influencing factors or strategies of AI adoption in the construction industry, and 36 papers were selected for the final classification and analysis. From those papers, 20 papers focused on influencing factors and 16 on strategies. Fig. 1 shows the approach used during the selection of the studies.

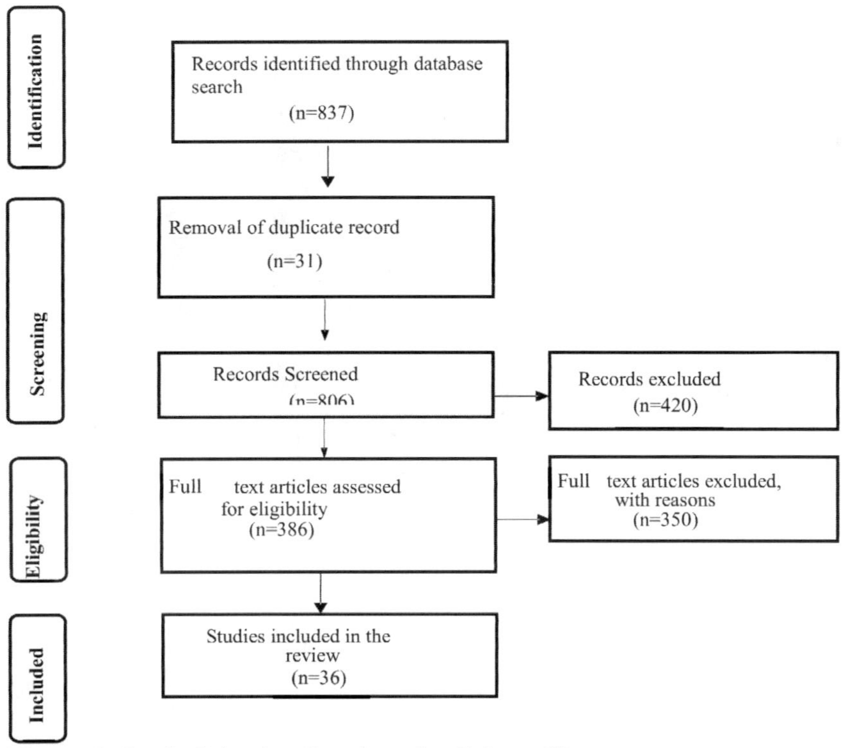

Figure 1: Study Selection Results using Prisma Chart

3. Results

3.1. Descriptive analysis of the SLR findings

Figure 2 shows the 36 publications that identified AI influencing factors and strategies within the construction industry that has grown from 2010 to 2020, implying that AI is an emerging topic. 2019 has the highest number of publications of 12, while 2005 and 2006 each have one publication.

Figure 3 represents the publications per country. China leads in AI research in the construction industry accounting for eight publications, approximately 23% for articles used in the SLR. Portugal, Singapore, Saudi Arabia and Oman each have one publication

Figure 4 shows the type of publications that used the SLR. Of the 36 publications used, 26 are journal articles making up approximately 73% of the study.

Figure 5 shows the research method of the publications. `

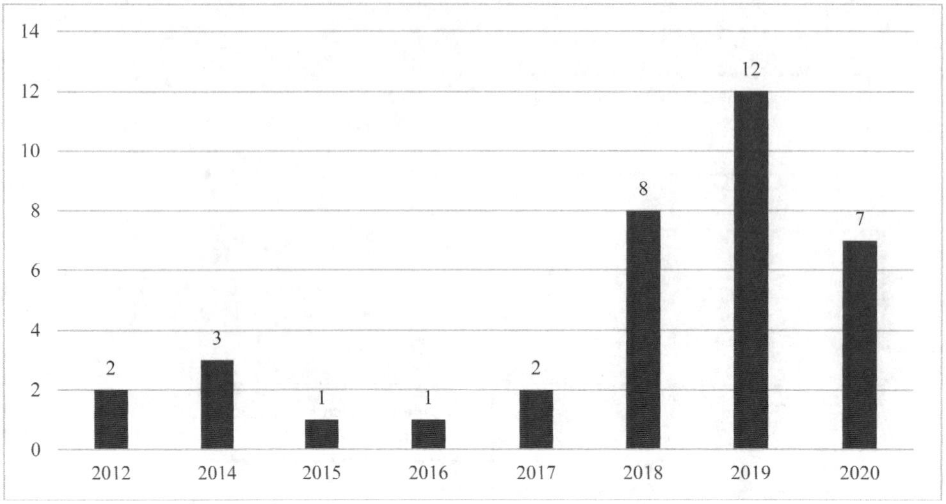

Figure 2: Number of publications per year between 2010 and 2020

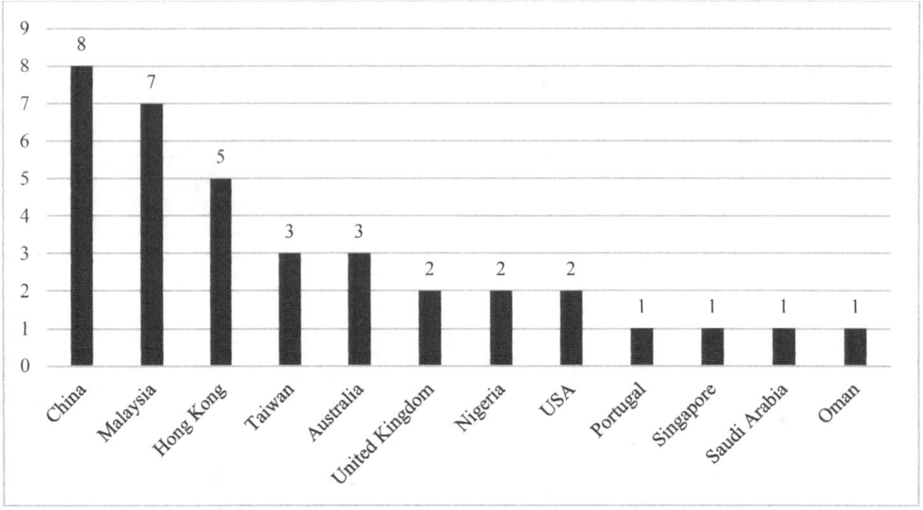

Figure 3: Country publications of studies

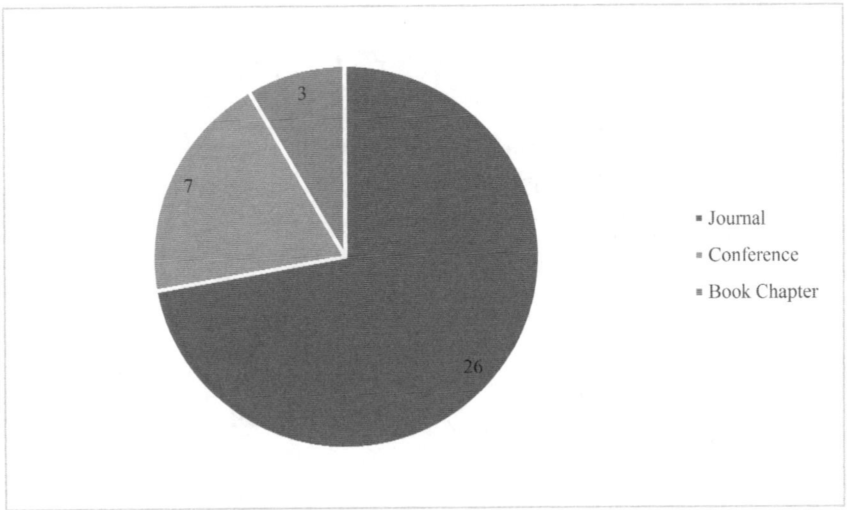

Figure 4: Publication Type

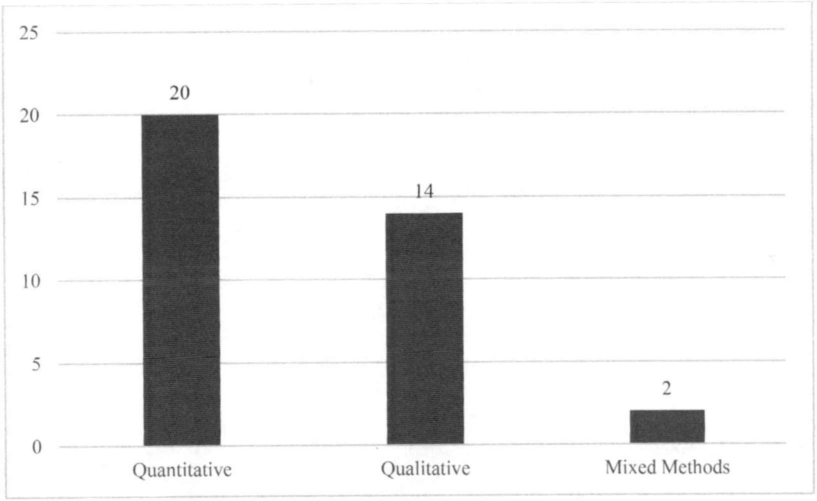

Figure 5: Publications Research Method

3.2. Factors Influencing the adoption of AI in the Construction Industry

Table 1 indicates factors influencing the adoption of AI in the construction industry. The most influential factor is the cost of adopting AI with 15 sources. This is closely followed by Skills and expert knowledge with 13 sources. Fragmented industry, improve work culture, internet services and perform technical tasks are the least influential factors. This is together with power supply, professional bodies support and project documentation with one source each.

Table 1: Factors influencing AI adoption

Influencing Factors	Source
3D detailing	[13], [15],[20],[28]
Clash Detection	[13],[20],[28]
Collaboration/communication	[15] ,[19],[22],[24],[25],[27],[28]
Comparability	[14],[17],[18],[23],[30]
Complexity	[14],[17],[18]
Cost of adopting AI	[14],[15],[16],[18],[19],[20],[21],[22],[23],[24],[25],[26],[28],[29],[31]
Costing and estimating	[13],[28]
Define project roles and responsibilities	[13],[28]
Documentation	[13],[28]
Facilities Management	[13],[28]
Fragmented industry	[15]
Goal Setting	[13],[22]
Government support	[15],[16],[17],[19],[20],[21],[22],[23],[24],[25],[28],[29]
Improve work culture	[15]
Internet Services	[23]
Legal Security	[25],[26],[27]
Market Pressure and Demand	[14],[15],[16],[17],[19], [20],[21],[22],[24],[26],[29]
Organisational Readiness	[14],[16],[17], [18],[29]
Organisational Size	[14],[17]
Perform Technical tasks	[13]
Power Supply	[23]
Professional bodies support	[19]
Project documentation	[13]
project effectiveness and productivity	[13], [15],[28]
project life cycling	[13],[20]
Reducing risk factors on project	[13],[15],[18],[21],[22],[24]
Relative Advantage	[14],[16],[17],[29]
Resistance to Change	[21],[26],[31],[32]
Scheduling	[20],[24],[28]
Site Layout Planning	[13],[28]
Skills and expert knowledge	[15], [18], [19],[20],[21],[22],[23],[24],[25],[28],[30],[31],[32]
Standardisation	[15], [19],[20],[21],[23],[24],[25],[26],[27],[28]
Successful delivery of project on time, within cost and expected quality	[13], [15],[24],[25],[26],[27],[29]

Supply Chain	[22],[23]
Top management support	[13], [14], [15],[16],[17], [18],[22],[23],[28],[29],[30],[32]
Trading Partners Support	[14], [29]

3.3. Strategies Influencing the adoption of AI in the Construction Industry

Strategies of AI adoption are indicated in table 2 below. First, training and skills acquisition of personnel cited as a critical strategy in adopting AI in the construction industry. This is followed by AI academia, research and development, and the development of government policies. Next, evaluating Project delivery method, example projects, industry leadership are the least critical strategies of AI adoption. This is followed by measured benchmarks, merger and realistic timescale for adoption.

Table 2: Strategies of AI adoption

Strategies	Sources
AI academia, research and development	[36],[37],[38],[39],[40],[43],[46],[47],[48]
Benchmarking of best practice	[37],[39]
Change Management	[34],[43]
Communicating benefits	[34],[38],[39],[40],[47]
Communicating with vendors	[42],[47],[48]
Development of government policies	[35],[37],[38],[40],[43],[45],[46],[47],[48]
Evaluating Project Delivery Method	[44]
Example projects	[37]
Financial incentives for AI adoption	[36],[45]
Industry leadership	[38]
Legal contracts and property rights	[38],[41],[43],[48]
Low-cost loans and subsidies from government and financial institutions	[36],[37],[42],[44],[45],[46],[47],[48]
Mandatory AI policies and regulations	[36],[37],[38],[45],[46],[47]
Measured benchmarks	[34]
Merger	[33]
Provision of trail usage	[40],[42],[45],[46]
Public awareness and initiative	[36],[37],[42],[44],[45],[46],[47]
Realistic timescale for adoption	[34]
Senior Leadership Commitment	[34],[36],[44],[48]
Sharing of knowledge of AI technologies	[35],[46]
Standardisation of AI practice in the construction industry	[33],[35],[37],[38],[39],[43],[46],[47]
Training and skills acquisition of personnel	[33],[34],[35],[37],[39],[40],[43],[44],[46],[47],[48]

4. Conclusion and Future work

This research presented the influencing factors and strategies of AI adoption in the construction industry. This is based on publications made between 2010 and 2020 through a systematic literature review. A total of 36 publications were analysed in the review. It was determined that the number of publications was highest in 2019. However, only seven publications were made in 2017. This implies a decreasing interest in AI in the construction domain.

Geographically, the Asian countries are leading with 29 publications in total, while Nigeria is the only African country with only two publications. Also, it could tell us, Africa lacks in AI research. This could d be due to a lot of reasons such as lack of facilities, collaboration and expertise. Thus, focus on expertise and collaboration must be encouraged while the government aids institutions with facilities. From the publications, the cost of adopting AI was found to be the most influential factor. This implies that the construction industry considers the transactional cost which goes with adopting AI. When first adopting AI, the infrastructure and equipment, in the long run, save money time for the organisations.

Training and skills acquisition of personnel was found to be the critical strategy. Stakeholders in various construction organisations are questioning whether AI will increase productivity. They are also curious if AI will eventually replace employees' existing employment. Organisations are constantly changing. This is accomplished by transitioning to modern technology and software systems to help with a more dynamic approach to employee training and development.

Future studies should combine the literature review knowledge and acquire primary data through interviews and questionnaires. Future research should also look at developing a roadmap of AI adoption in the construction industry, focusing on developing countries.

References

[1] S. B. Rane, P. R. Potdar, and S. Rane, "Development of Project Risk Management framework based on Industry 4.0 technologies," Benchmarking: An International Journal, vol. ahead-of-print, no. ahead-of-print, Oct. 2019, DOI: 10.1108/bij-03-2019-0123.

[2] T. Haupt and N. Harinarain, "The image of the construction industry and its employment attractiveness," *Acta Structilia 23(2)*, vol. 23, no. 1, Feb. 2017, doi: 10.18820/24150487/as23i2.4.

[3] F. Craveiro, J. P. Duarte, H. Bartolo, and P. J. Bartolo, "Additive manufacturing as an enabling technology for digital construction: A perspective on Construction 4.0," *Automation in Construction*, vol. 103, pp. 251–267, Jul. 2019, DOI: 10.1016/j.autcon.2019.03.011.

[4] S. Girginkaya Akdag and U. Maqsood, "A roadmap for BIM adoption and implementation in developing countries: the Pakistan case," *Archnet-IJAR: International Journal of Architectural Research*, vol. 14, no. 1, pp. 112–132, Sep. 2019, DOI: 10.1108/arch-04-2019-0081.

[5] M. Ryan, "In AI We Trust: Ethics, Artificial Intelligence, and Reliability," *Science and Engineering Ethics*, Jun. 2020, DOI: 10.1007/s11948-020-00228-y.

[6] F. Fogliano, F. Fabbrini, A. Souza, G. Fidélio, J. Machado, and R. Sarra, "Edgard, the Chatbot: Questioning Ethics in the Usage of Artificial Intelligence Through Interaction Design and Electronic Literature," *Digital Human Modeling and Applications in Health, Safety, Ergonomics and Risk Management. Healthcare Applications*, pp. 325–341, 2019, DOI: 10.1007/978-3-030-22219-2_25.

[7] J. M. D. Delgado *et al.*, "Robotics and automated systems in construction: Understanding industry-specific challenges for adoption," *Journal of Building Engineering*, vol. 26, p. 100868,

Nov. 2019, DOI: 10.1016/j.jobe.2019.100868.

[8] M. Cubric, "Drivers, barriers and social considerations for AI adoption in business and management: A tertiary study," *Technology in Society*, p. 101257, Apr. 2020, DOI: 10.1016/j.techsoc.2020.101257.

[9] H. Yan, N. Yang, Y. Peng, and Y. Ren, "Data mining in the construction industry: Present status, opportunities, and future trends," *Automation in Construction*, vol. 119, p. 103331, Nov. 2020, DOI: 10.1016/j.autcon.2020.103331.

[10] W. ten Ham-Baloyi and P. Jordan, "Systematic review as a research method in post-graduate nursing education," *Health SA Gesondheid*, vol. 21, pp. 120–128, Dec. 2016, DOI: 10.1016/j.hsag.2015.08.002.

[11] D. Moher, A. Liberati, J. Tetzlaff, and D. G. Altman, "Preferred Reporting Items for Systematic Reviews and Meta-Analyses: The PRISMA Statement," *PLoS Medicine*, vol. 6, no. 7, p. e1000097, Jul. 2009, DOI: 10.1371/journal.pmed.1000097.

[12] S. Madigan, M. Wade, G. Tarabulsy, J. M. Jenkins, and M. Shouldice, "Association Between Abuse History and Adolescent Pregnancy: A Meta-analysis," *Journal of Adolescent Health*, vol. 55, no. 2, pp. 151–159, Aug. 2014, DOI: 10.1016/j.jadohealth.2014.05.002.

[13] A. Chegu Badrinath and S.-H. Hsieh, "Empirical Approach to Identify Operational Critical Success Factors for BIM Projects," *Journal of Construction Engineering and Management*, vol. 145, no. 3, p. 04018140, Mar. 2019, DOI: 10.1061/(asce)co.1943-7862.0001607.

[14] M. Pan and W. Pan, "Understanding the Determinants of Construction Robot Adoption: Perspective of Building Contractors," *Journal of Construction Engineering and Management*, vol. 146, no. 5, p. 04020040, May 2020, DOI: 10.1061/(asce)co.1943-7862.0001821.

[15] J. M. D. Delgado, L. Oyedele, T. Beach, and P. Demian, "Augmented and Virtual Reality in Construction: Drivers and Limitations for Industry Adoption," *Journal of Construction Engineering and Management*, vol. 146, no. 7, p. 04020079, Jul. 2020, DOI: 10.1061/(asce)co.1943-7862.0001844.

[16] C.-C. Yeh and Y.-F. Chen, "Critical success factors for adoption of 3D printing," *Technological Forecasting and Social Change*, vol. 132, pp. 209–216, Jul. 2018, DOI: 10.1016/j.techfore.2018.02.003.

[17] Y. Chen, Y. Yin, G. J. Browne, and D. Li, "Adoption of building information modeling in Chinese construction industry," *Engineering, Construction and Architectural Management*, vol. 26, no. 9, pp. 1878–1898, Oct. 2019, DOI: 10.1108/ecam-11-2017-0246.

[18] H. M. F. Shehzad, R. B. Ibrahim, A. F. Yusof, and K. A. M. Khaidzir, "Building Information Modeling: Factors Affecting the Adoption in the AEC Industry," *2019 6th International Conference on Research and Innovation in Information Systems (ICRIIS)*, Dec. 2019, DOI: 10.1109/icriis48246.2019.9073581.

[19] M. Abubakar, Y. M. Ibrahim, D. Kado, and K. Bala, "Contractors' Perception of the Factors Affecting Building Information Modelling (BIM) Adoption in the Nigerian Construction Industry," *Computing in Civil and Building Engineering (2014)*, Jun. 2014, DOI: 10.1061/9780784413616.022.

[20] Y. Le, X. Zhang, and M. Liu, "Study on Influencing Factors of BIM Adoption in China's Construction Industry Based on FA-SEM Model," *ICCREM 2019*, Aug. 2019, DOI: 10.1061/9780784482308.010.

[21] O. Aljobaly and A. Banawi, "Evaluation of the Saudi Construction Industry for Adoption of Building Information Modelling," *Advances in Intelligent Systems and Computing*, pp. 488–498, Jun. 2019, DOI: 10.1007/978-3-030-20454-9_49.

[22] S. Akmam Syed Zakaria, T. Gajendran, M. Skitmore, and G. Brewer, "Key factors influencing the decision to adopt industrialised building systems technology in the Malaysian construction industry: an inter-project perspective," *Architectural Engineering and Design Management*, vol. 14, no. 1–2, pp. 27–45, Mar. 2017, DOI: 10.1080/17452007.2017.1298512.

[23] A. Afolabi, E. Ibem, E. Aduwo, P. Tunji-Olayeni, and O. Oluwunmi, "Critical Success Factors

(CSFs) for e-Procurement Adoption in the Nigerian Construction Industry," *Buildings*, vol. 9, no. 2, p. 47, Feb. 2019, DOI: 10.3390/buildings9020047.
[24] M. N. A. Rashid, M. R. Abdullah, and D. Ismail, "Critical Success Factors CSFs to Automation and Robotics in Industrialized Building System IBS," *International Journal of Academic Research in Business and Social Sciences*, vol. 8, no. 12, Jan. 2019, DOI: 10.6007/ijarbss/v8-i12/5432.
[25] T. O. Olawumi and D. W. M. Chan, "Critical success factors for implementing building information modeling and sustainability practices in construction projects: A Delphi survey," *Sustainable Development*, Jan. 2019, DOI: 10.1002/sd.1925.
[26] P. Li, S. Zheng, H. Si, and K. Xu, "Critical Challenges for BIM Adoption in Small and Medium-Sized Enterprises: Evidence from China," *Advances in Civil Engineering*, vol. 2019, pp. 1–14, Feb. 2019, DOI: 10.1155/2019/9482350.
[27] Y. Dou, X. Xue, Z. Zhao, and Y. Jiang, "Measuring the Factors that Influence the Diffusion of Prefabricated Construction Technology Innovation," *KSCE Journal of Civil Engineering*, vol. 23, no. 9, pp. 3737–3752, Aug. 2019, DOI: 10.1007/s12205-019-2029-3.
[28] M. Attarzadeh, R. L. K. Tiong, and T. Nath, "Identifying key factors for building information modelling adoption in Singapore," *Proceedings of the ICE - Management, Procurement and Law*, vol. 168, no. 5, pp. 220–231, Oct. 2015, DOI: 10.1680/mpal.15.00030.
[29] T. Umar, "Key factors influencing the implementation of three-dimensional printing in construction," *Proceedings of the Institution of Civil Engineers - Management, Procurement and Law*, pp. 1–14, Apr. 2020, DOI: 10.1680/jmapl.19.00029.
[30] W. Wang, S. Zhang, Y. Su, and X. Deng, "Key Factors to Green Building Technologies Adoption in Developing Countries: The Perspective of Chinese Designers," *Sustainability*, vol. 10, no. 11, p. 4135, Nov. 2018, DOI: 10.3390/su10114135.
[31] I. D. Mohd. Aripin, E. M. A. Zawawi, and Z. Ismail, "Factors Influencing the Implementation of Technologies Behind Industry 4.0 in the Malaysian Construction Industry," *MATEC Web of Conferences*, vol. 266, p. 01006, 2019, DOI: 10.1051/matecconf/201926601006.
[32] K. Sargent, P. Hyland, and S. Sawang, "Factors influencing the adoption of information technology in a construction business," *Construction Economics and Building*, vol. 12, no. 2, p. 86, Jun. 2012, DOI: 10.5130/ajceb.v12i2.2448.
[33] Y. Tan, L. Shen, and C. Langston, "Competition Environment, Strategy, and Performance in the Hong Kong Construction Industry," *Journal of Construction Engineering and Management*, vol. 138, no. 3, pp. 352–360, Mar. 2012, DOI: 10.1061/(asce)co.1943-7862.0000407.
[34] B. C. Lines and P. K. Reddy Vardireddy, "Drivers of Organizational Change within the AEC Industry: Linking Change Management Practices with Successful Change Adoption," *Journal of Management in Engineering*, vol. 33, no. 6, p. 04017031, Nov. 2017, DOI: 10.1061/(asce)me.1943-5479.0000548.
[35] W. Pan, L. Chen, and W. Zhan, "PESTEL Analysis of Construction Productivity Enhancement Strategies: A Case Study of Three Economies," *Journal of Management in Engineering*, vol. 35, no. 1, p. 05018013, Jan. 2019, DOI: 10.1061/(asce)me.1943-5479.0000662.
[36] A. Darko and A. P. C. Chan, "Strategies to promote green building technologies adoption in developing countries: The case of Ghana," *Building and Environment*, vol. 130, pp. 74–84, Feb. 2018, DOI: 10.1016/j.buildenv.2017.12.022.
[37] I. Y. Wuni and G. Q. Shen, "Barriers to the adoption of modular integrated construction: Systematic review and meta-analysis, integrated conceptual framework, and strategies," *Journal of Cleaner Production*, vol. 249, p. 119347, Nov. 2019, DOI: 10.1016/j.jclepro.2019.119347.
[38] P. Smith, "BIM Implementation – Global Strategies," *Procedia Engineering*, vol. 85, pp. 482–492, 2014, DOI: 10.1016/j.proeng.2014.10.575.
[39] T. O. Olawumi and D. W. M. Chan, "Development of a benchmarking model for BIM implementation in developing countries," *Benchmarking: An International Journal*, vol. 26, no. 4, pp. 1210–1232, May 2019, DOI: 10.1108/bij-05-2018-0138.

[40] A. H. Memon, I. A. Rahman, I. Memon, and N. I. A. Azman, "BIM in Malaysian Construction Industry: Status, Advantages, Barriers and Strategies to Enhance the Implementation Level," *Research Journal of Applied Sciences, Engineering and Technology*, vol. 8, no. 5, pp. 606–614, Aug. 2014, DOI: 10.19026/rjaset.8.1012.
[41] T.-N. Dao, T.-Q. Nguyen, and P.-H. Chen, "BIM Adoption in Construction Projects Funded with State-managed Capital in Vietnam: Legal Issues and Proposed Solutions," *Lecture Notes in Civil Engineering*, pp. 1211–1216, Oct. 2019, DOI: 10.1007/978-981-15-0802-8_194.
[42] S. M. E. Sepasgozar and S. Davis, "Digital Construction Technology and Job-site Equipment Demonstration: Modelling Relationship Strategies for Technology Adoption," *Buildings*, vol. 9, no. 7, p. 158, Jun. 2019, DOI: 10.3390/buildings9070158.
[43] K. O. Ayinla and Z. Adamu, "Bridging the digital divide gap in BIM technology adoption," *Engineering, Construction and Architectural Management*, vol. 25, no. 10, pp. 1398–1416, Nov. 2018, DOI: 10.1108/ecam-05-2017-0091.
[44] W. Ferron and Y. Turkan, "Toward a Roadmap for BIM Adoption and Implementation by Small-Sized Construction Companies," *Advances in Informatics and Computing in Civil and Construction Engineering*, pp. 873–879, Oct. 2018, DOI: 10.1007/978-3-030-00220-6_105.
[45] N. A. A. Mahat, S. N. A. S. Alwee, H. Adnan, and A. A. Hassan, "Propelling Green Building Technologies Adoption in Malaysia Construction Industry," *IOP Conference Series: Earth and Environmental Science*, vol. 233, p. 022032, Feb. 2019, DOI: 10.1088/1755-1315/233/2/022032.
[46] M. J. F. Silva, F. Salvado, P. Couto, and Á. V. e Azevedo, "Roadmap Proposal for Implementing Building Information Modelling (BIM) in Portugal," *Open Journal of Civil Engineering*, vol. 06, no. 03, pp. 475–481, 2016, DOI: 10.4236/ojce.2016.63040.
[47] A. P. C. Chan, A. Darko, and E. E. Ameyaw, "Strategies for Promoting Green Building Technologies Adoption in the Construction Industry—An International Study," *Sustainability*, vol. 9, no. 6, p. 969, Jun. 2017, DOI: 10.3390/su9060969.
[48] A. A. Zaini, A. W. Razali, H. C. Gui, N. Zaini, and S. D. Tamjehi, "Assessing Strategies of Building Information Modeling (BIM) Implementation in Sarawak Construction Industry," *IOP Conference Series: Earth and Environmental Science*, vol. 498, p. 012086, Jun. 2020, DOI: 10.1088/1755-1315/498/1/012086.

A Systematic Review of the Benefits of Automation Inspection Tools for Quality Housing Delivery

T D Nena[1*], I Musonda[1] and C S Okoro[2]

1. Department of Construction Management and Quantity Surveying, University of Johannesburg, South Africa
2. Department of Finance and Investment Management, College of Business and Economics, University of Johannesburg, South Africa

*tholangdavid@yahoo.co.uk

Abstract. Efficient and effective inspection is vital for the quality delivery of housing construction. Houses are being inspected less frequently due to incapacitated inspectorates, high cost, and the long duration of the current inspection process. Furthermore, traditional inspection processes often contribute to poor quality housing delivery are inspector dependent and complex and unsafe working environment is exposed to inspectors. Automated inspection technologies play a vital role in the inspection of housing construction. Automation-assisted technologies such as laser scanners, drones and Augmented Reality (AR)-based inspection equipped with multiple imaging and sensing systems have been developed to mitigate some of these issues with the traditional method of inspection processes. This paper reviews relevant peer-reviewed articles to explore automation inspection process tools and technologies and highlight their benefits in the housing construction process. Following the systematic literature assessment methodology, five (5) databases were queried, based on the specific search string. Specifically, 26 journal articles satisfying a set of inclusion criteria were reviewed, its results highlighting the benefits of adopting an automated inspection process for quality housing delivery. The review of the journal articles revealed that the automation inspection process is available with several benefits demonstrating the potential to improve the quality housing delivery with the aid of tools and technologies like Augmented Reality (AR)-based inspection, 3D laser scanning, Drones, and BIM-based inspection. Benefits such as time-saving, limited human intervention during the construction process, inspection accuracy, reduced resource requirements, and improved collaboration between stakeholders. This study also revealed that while new technologies are evolving with benefits to improve the inspection process, there is still a gap in the practical application of these automated inspection process technologies during housing construction. The outcome of this study would benefit both the housing owners (clients), building inspection entities, building construction researchers, and the industry professionals to realise the potential of the automation inspection process and understand the requirements and benefits for their successful implementation.

1. Introduction

Construction industry productivity has declined over the last decades compared to the manufacturing industry which doubled its productivity over the same period [1]. The benefits of automation in the manufacturing industry could be the reasons for high productivity. In manufacturing, it has been observed that, automation helps reduce the rate of production errors, does not compromise production quality for speed, drastically reduces production time, and increased workers' safety [2]. Furthermore, automation recorded a sustainable reduced cost on labour spending, and increased labour productivity also cut down manual tasks [3]. These immense benefits have led to the emergence of research in applying automation to construction processes. Concurrent with this is the need to automate the processes vital to enabling rapid delivery of quality housing infrastructure [4]. Principal to these processes is the housing inspection process which enables the compliance of construction projects with acceptable government standards and guidelines.

The use of automation for housing inspection is increasingly being adopted in the construction industry. The use of an automated inspection process offers the benefit of carrying out inspections of houses without being on the actual construction site [4]. For instance, automated technologies such as drones, laser scanners improve upon traditional methods of inspecting which are human reliant [5]. However, while research in automation inspection is an emerging area, it is imperative to establish an outlook of relevant previous studies and their impact in furthering research in automation during housing construction inspection. The review is also pivotal in identifying areas that are lacking and areas that need further research to avail a seamless automation inspection adoption across the AEC sector.

Accordingly, the systematic review reported here aims to explore the benefits of the automated inspection process for construction of houses following the Preferred Reporting Items for Systematic Review and Meta-Analysis (PRISMA) statement for systematic review reporting format [6,7]. To explore the recent automated inspection processes with their benefits, peer-reviewed journal articles published from 2012 to 2020 were reviewed. The specific objectives were to 1) determine the tools and technologies used for the automation-enabled inspection process, 2) to determine the benefits of the automation inspection process for quality housing construction, and finally, 3) to identify the gap in the literature and propose future research agenda.

2. Methodology

Systematic reviews are important in summarising the empirical evidence related to technology or method that helps in suggesting feasibilities for further research in the same area. Sometimes, contradiction or support of the hypothesis can be achieved through a systematic review. First, a protocol to select studies and extract necessary information from the data set along with the proper justifiable need for the review needs to be identified and later take efforts to synthesise the selected data and report the overall review properly [8]. Hence, the studies related to automation in the housing construction domain were systematically reviewed to explore the benefits of the automation inspection process. For the development of theory, a systematic literature review plays a vital role in the development of theory and also helps in fulfilling the possible gaps in research. Moreover, it highlights the areas where further research is needed within the available literature [8]. The main objective of the paper was to provide a systematic review and the benefits of the automation inspection process in housing construction within the last nine years. For that purpose, a systematic review was done following the PRISMA template presented below [6].

2.1 Search strategy

A systematic search strategy was used to identify relevant literature available on the automated inspection process methods for housing construction. This strategy was tailored to five databases: Scopus, Ebscohost, Google scholar, African Journals (Formerly: SA ePublications), American Society of Civil Engineers (ASCE), and the terms used as search string were the following: "inspection process" OR "inspection systems", "inspection technology" AND benefits, "housing" OR "building infrastructure" AND delivery, as advised by the Boolean operator [9]. Keywords used were automation, inspection, technology, housing construction, benefits, and the search span was from the year 2012 to 2020.

2.2 Quality assessment

The study is based on peer-reviewed journal articles. A thorough check of duplicated articles was carried out on Microsoft excel [10] to maintain the quality of the review and 4 duplicate articles were removed. Also, the abstract and conclusion of articles were thoroughly checked to ensure the quality and relevance of academic literature included in the review process [11] at the later stage, the careful evaluation of each research paper was carried out. The next was to consider exclusion and inclusion criteria where the study was only limited to English-written articles. For exclusion criterion, only four (4) articles were written in non- English language and were excluded from the study. The literature search is also explained in the PRISMA flowchart below.

2.3 Studies included in the qualitative synthesis
The final 26 journal articles are used after assessing each article based on the above-mentioned exclusion and inclusion criteria. These articles are used for the final process based on different steps. In the first step, the corresponding of Microsoft excel to a descriptive analysis of the literature published on automation inspection process in the field of housing construction like the distribution of year, distribution of countries, distribution of the citations, and journal area identifications. In the other step, the content analysis was done to identify and analyse the main direction and research done by the researchers in the year 2012 to 2020 as well as mentioning the benefits and reporting on the future opportunities to research [10].

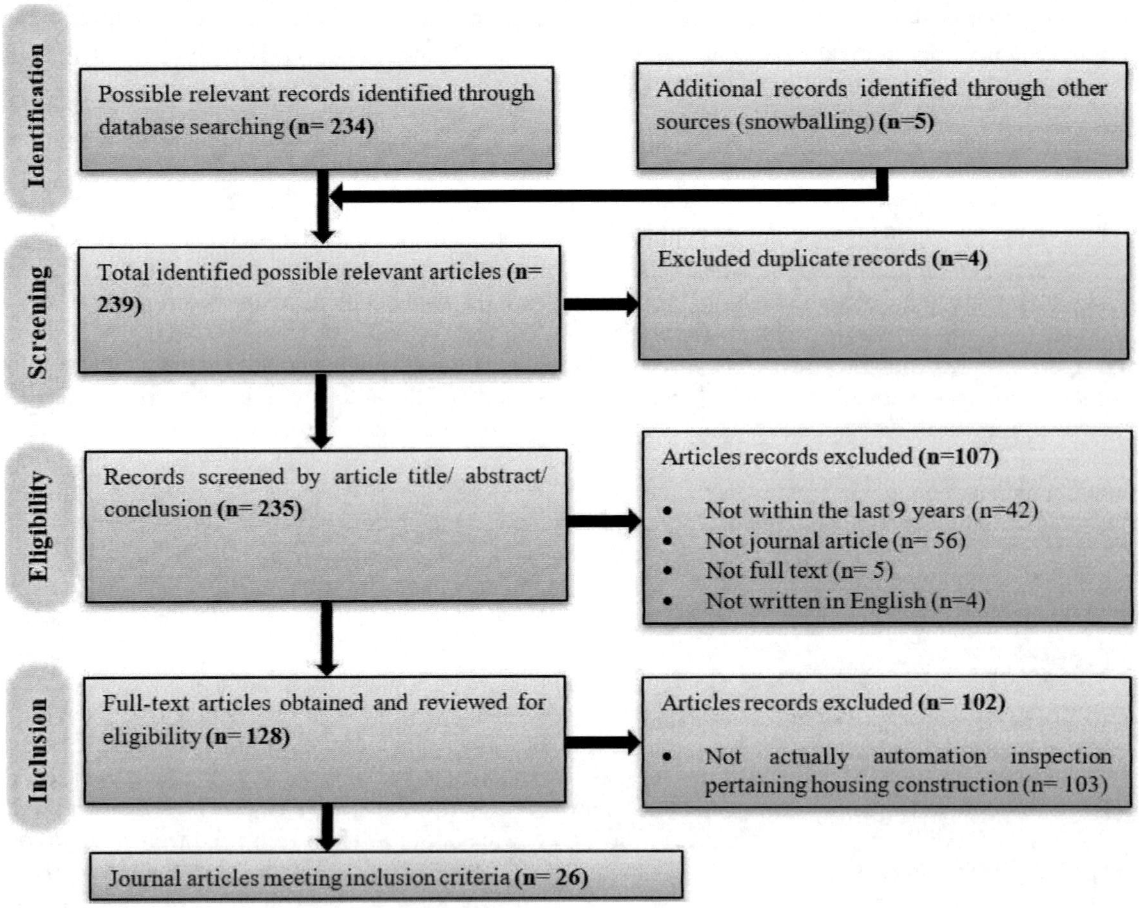

Figure 1: Study selection results using PRISMA chart

3. Results

3.1 Journal article year-wise publications
The journal year-wise publications focusing only on automation-enabled inspection process are shown in figure 2 below with 2012 and 2015 selected papers being the lowest with one (1) paper respectively, followed by 2013 with only two (2) papers. The year 2016, 2017, 2019, and 2020 were high with three

(3) papers, respectively. 2014 selected publications were higher with four (4) papers while 2018 was the highest number of selected papers with six (6) papers. The Journal article year is imperative in gauging the evolution of research in the field of discourse [12]. The low number of publications in 2019 and 2020 could be due to the Coronavirus pandemic which shut down numerous countries across the globe and therefore prevented, amongst other activities, research, and academics. However, the high number of publications gotten from 2018 signifies a degree of enlarging awareness amongst researchers in the area. The finding also instructs that the emergence is slow, and much effort is needed between researchers to advance this important discourse.

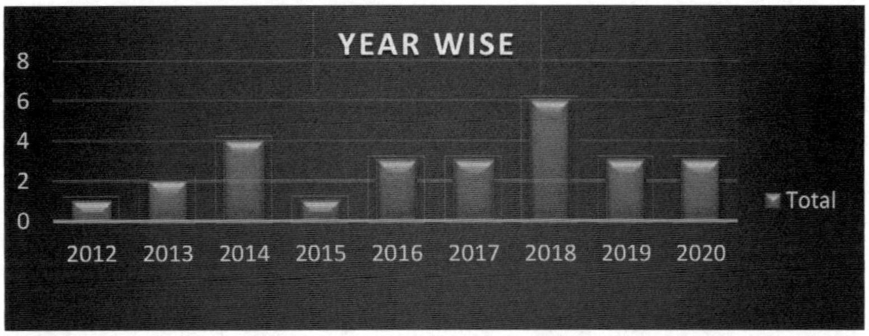

Figure 2: Year-wise publications per year between 2012 and 2020

3.2 Journal based publications
The other important assessment that was made for the study is article journal area identification of the study. Most of the papers were selected from the automation in construction with eleven (11) papers. The second journal area identification is the journal of infrastructure systems and Proceedings of the Creative Construction Conference with two (2) papers, respectively. And finally, one (1) paper was selected from each remaining journal area identification contributed to the study. The results are shown in figure 3 below. The high rate of publications from Automation in Construction could be due to the focus of the Journal on Industry 4.0 technologies; however, it also raises the need for more Journals reviewing and accepting publications more importantly as research in Industry 4.0 currently dominates the academic discourse. As the publication is pivotal to research drive, journals as stakeholders in the research process must encourage more publication in this discourse to further expand the body of knowledge on automation of the inspection process.

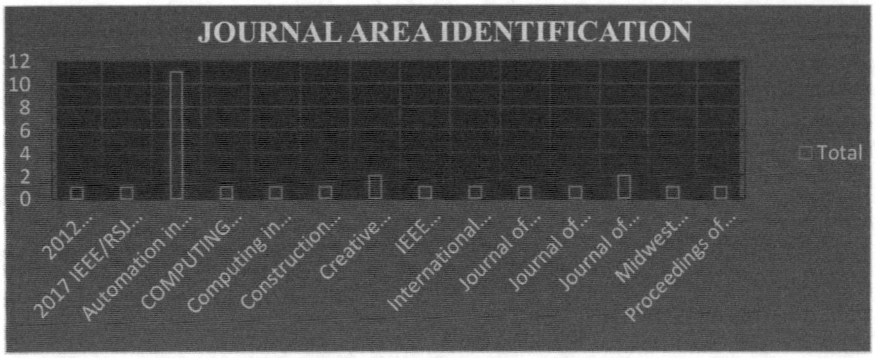

Figure. 3: Journal based publication

4. Classification of literature

4.1 Automated inspection tools application and their benefits
The application of automation in inspection processes is wide-ranging. As stated by [13], UAV/Drones could help in accurate inspections, improved time efficiency, reduced paperwork, and enabling safety of labour on site. While with robotics, they could aid with the flexibility of the inspection process [14].

Furthermore, they can be integrated with BIM to automatically generate inspection lots, improve quality supervision of projects, and also avail a safe, efficient, and cost-effective platform. Other applications and benefits include; Automated document management & data collection [15,16] in Augmented reality platforms and 3D laser scanning, [17,18].

Table 1: Table of classifications

ID	Automated inspection tools and technology	Benefits of the automated inspection process	Sources
1.	UAV/Drone-based inspection	• Effectiveness and more robust in automated navigation and visual data collection • accurate inspections • Improve time efficiency • reduce paperwork	[5,13,19,20,21,22]
2.	Robotic – Based inspection	• Cost-effective, robust, and flexible enough to be employed in a variety of inspection scenarios.	[3,14,23,24,25]
3.	BIM-based inspection	• Automatically generate inspection lots, items, and points. • The system makes it easier for inspectors to learn the inspection object intuitively and comprehensively before the construction phase • More effective to fill out the customized forms without worrying about omission and negligence • More convenient to input the data in the original records into computers without second input to reduce huge workload and data falsification.	[26,27,28,29,30,31,32]
4.	Augmented Reality (AR)	• Automated document management & data collection, • Minimal set-up required • No need for training to operate the tool	[15,16,33,34]
5.	3D laser scanning	• Automated document management & data analysis • Improved accuracy	[17,18,35,36]

4.2 Time saving

Time-saving is one of the benefits of the automation inspection process where more houses are being inspected effectively and efficiently. According to [14] robotic inspection system was proven to being an effective and efficient tool for inspection without time-consuming where defects during the construction process are detected without the involvement of human inspectors. Also, the study conducted by [16] developed an inspection system that does not save time only but also reduces construction reworks related costs on site. These studies present the possibility of improving the inspection process by automation in the construction of housing.

4.3 Reduced resource requirements

Daily site visits to conduct inspection can cause fatigue to humans, as a result, more errors could be encountered during decision making. The study conducted by [20] emphasised that utilising and implementing new technologies such as drones can be useful in improving efficiency, reduce paperwork and automate the entire process of inspection process. The study found that the ability for drones to use the on-board processor to receive the location information of the scheduled inspection visits makes it easy for inspectorates not to drive around searching for a location and reduce the unnecessary site visits, with the use of images and data collected with the help of drone from site to detect any non-compliance then inspectorates use the mobile application to issue out non-compliance if detected during the inspection process. This is supported by [14] who suggest the use of robots to assist with housing inspection with the argument that it reduces labor while improving efficiency and effectiveness during the inspection process. Another study by [3] designed the robotic inspection tool to assist human inspectors to reduce the need for more labor to conduct an inspection task of housing construction. These studies present the possibility of adopting the use of automated tools for housing inspection under construction where it was found to be more beneficial than the use of traditional tools of inspecting.

4.4 Improved collaboration

The collaboration of all construction stakeholders plays a vital role in ensuring quality housing delivery [37]. Currently, structural damage results from compromised construction quality standards [38]. For effective and efficient inspections, builders and inspectors should work together for a better quality of construction work [39]. Also, transforming construction into a technology-driven industry could improve quality, collaboration between stakeholders, productivity, and efficiency [40]. According to [27], this technology could be Building Information Model (BIM) which will improve inspection efficiency where collaboration work among inspectors and builders is improved leading to improved quality housing delivery. For better collaboration, multiple mobile robots can also work simultaneously and send the inspection data to a cloud server and the results can be reviewed by multiple inspectors retrieved from the cloud server at the same time [41]. The BIM-based inspection makes it easier for housing inspectors to collaborate during the inspection process.

5. Conclusion and future work

Research trends indicate that in recent years automation in housing construction is receiving more attention. An automation inspection process is a key to efficient and effective quality housing delivery. A systematic search and the literature screening procedures in this research were used following the PRISMA statement template to identify relevant research studies on the topic. After the literature selection procedure, 26 articles were selected for further analysis. The results show that more of the research activities are done in developed countries where the benefits of the automation inspection process are realised. The most proposed tools and technologies used to enable the automation inspection process are 3D laser scanners, Drones, BIM-based inspection, Robotic–based inspection, and Augmented Reality (AR). While some of these tools and technologies are evolving and tested mostly in laboratories with mock-ups, the main existing gap is the implementation of these tools and technologies to the actual housing construction and monitor their benefits. The existing research shows that the main challenge is to adopt and implement these automation tools and technologies which can collect data effectively and efficiently. Automation tools and technologies are usually complicated to use and expensive to acquire. In addition to the use of these tools and technologies, experienced inspectors are required. Future studies should test the use of automated inspection tools and technologies on the actual construction of houses. Future research should also look at developing an automated inspection process framework for housing construction projects focusing on developing countries.

6. References

[1] Oesterreich T D and Teuteberg F 2016 Understanding the implications of digitisation and automation in the context of Industry 4.0: A triangulation approach and elements of a research agenda for the construction industry *Computers in industry* **83** pp 121-139

[2] Gregori F, Papetti A, Pandolfi M, Peruzzini M and Germani M 2017 Digital manufacturing systems: a framework to improve social sustainability of a production site. *Procedia CIRP*, **63** pp 436-442

[3] Yan R J, Kayacan E, Chen I M, Tiong L K and Wu J 2018 QuicaBot: Quality inspection and assessment robot *IEEE Transactions on Automation Science and Engineering* **16** pp 506-517
[4] Elshafey A, Saar C C, Aminudin E B, Gheisari M and Usmani A 2020 Technology acceptance model for Augmented Reality and Building Information Modeling integration in the construction industry *J. Inf. Technol. Constr.* **25** pp 161-172
[5] Rakha T and Gorodetsky A 2018 Review of Unmanned Aerial System (UAS) applications in the built environment: Towards automated building inspection procedures using drones *Automation in Construction* **93** pp 252-264
[6] Moher D, Liberati A, Tetzlaff J and Altman D G 2009 Preferred Reporting Items for Systematic Reviews and Meta-Analyses: The PRISMA Statement *PLoS Med* **6** pp 1-6
[7] Albeaino G, Gheisari M and Franz B W 2019 A systematic review of unmanned aerial vehicle application areas and technologies in the AEC domain *J. Inf. Technol. Constr.* **24** pp 381-405
[8] Abideen A Z, Mohamad F B and Fernando Y 2020 Lean simulations in production and operations management–a systematic literature review and bibliometric analysis. *Journal of Modelling in Management*.
[9] Aghajafari F, Nagulesapillai T, Ronksley P E, Tough S C, O'Beirne M and Rabi D M 2013 Association between maternal serum 25-hydroxyvitamin D level and pregnancy and neonatal outcomes: systematic review and meta-analysis of observational studies. *Bmj*, **346**
[10] Khan N and Qureshi M 2020 A systematic literature review on online medical services in Malaysia
[11] Olawumi T O, Chan D W and Wong J K 2017 Evolution in the intellectual structure of BIM research: A bibliometric analysis *Journal of Civil Engineering and Management*, **23** pp 1060-1081
[12] Saka A B and Chan D W 2019 A scientometric review and metasynthesis of building information modelling (BIM) research in Africa *Buildings* **9** p 85
[13] Greenwood W W, Lynch J P and Zekkos D 2019 Applications of UAVs in civil infrastructure *Journal of infrastructure systems*, **25** p 04019002
[14] Wang J and Luo C 2019 Automatic Wall Defect Detection Using an Autonomous Robot: A Focus on Data Collection In *Computing in Civil Engineering 2019: Data, Sensing, and Analytics* pp 312-319
[15] Kopsida M, Brilakis I and Vela P A 2015 A review of automated construction progress monitoring and inspection methods In *Proc. of the 32nd CIB W78 Conference 2015* pp 421-431
[16] Kwon O S, Park C S and Lim C R 2014 A defect management system for reinforced concrete work utilizing BIM, image-matching and augmented reality *Automation in construction* **46** pp 74-81
[17] Maalek R, Ruwanpura J and Ranaweera K 2014 Evaluation of the state-of-the-art automated construction progress monitoring and control systems In *Construction Research Congress 2014: Construction in a Global Network* pp 1023-1032
[18] Liu D, Lu W and Niu Y 2018 Extended technology-acceptance model to make smart construction systems successful *Journal of Construction Engineering and Management* **144** p 04018035
[19] Freimuth H and König M 2018 Planning and executing construction inspections with unmanned aerial vehicles *Automation in Construction* **96** pp 540-553
[20] Ashour R, Taha T, Mohamed F, Hableel E, Kheil Y A, Elsalamouny M, Kadadha M, Rangan K, Dias J, Seneviratne L and Cai G 2016 Site inspection drone: A solution for inspecting and regulating construction sites In *2016 IEEE 59th International Midwest Symposium on Circuits and Systems (MWSCAS)* pp 1-4
[21] Asadi K, Suresh A K, Ender A, Gotad S, Maniyar S, Anand S, Noghabaei M, Han K, Lobaton E and Wu T 2020 An integrated UGV-UAV system for construction site data collection *Automation in Construction* **112** p 103068

[22] Musonda I and Pillay N 2019 Using UAV's and BIM integration to improve infrastructure delivery–A case of Gauteng department of Infrastructure Development, South Africa In *Creative Construction Conference 2019* pp 922-928
[23] Lattanzi D and Miller G 2017 Review of robotic infrastructure inspection systems Journal of Infrastructure Systems **23** p 04017004
[24] Yan R J, Kayacan E, Chen I M and Tiong L K 2017 A novel building post-construction quality assessment robot: Design and prototyping In *2017 IEEE/RSJ International Conference on Intelligent Robots and Systems (IROS)* pp 6020-6023
[25] Cai S, Ma Z, Skibniewski M J, Bao S and Wang H 2020 Construction Automation and Robotics for High-Rise Buildings: Development Priorities and Key Challenges *Journal of Construction Engineering and Management* **146** p 04020096
[26] Ma Z, Cai S, Mao N, Yang Q, Feng J and Wang P 2018 Construction quality management based on a collaborative system using BIM and indoor positioning. *Automation in Construction* **92** pp 35-45
[27] Ma Z, Mao N and Yang Q 2016 A BIM based approach for quality supervision of construction projects In *Creative Construction Conference, 25-28 June 2016, Budapest, Hungary* pp 644- 649
[28] Hamledari H, Davari S, Azar E R, McCabe B, Flager F and Fischer M 2017 UAV-enabled site-to-BIM automation: Aerial robotic-and computer vision-based development of as-built/as-is BIMs and quality control In *Construction research congress* pp 336-346
[29] Ding L, Li K, Zhou Y and Love P E 2017 An IFC-inspection process model for infrastructure projects: Enabling real-time quality monitoring and control *Automation in Construction* **84** pp 96-110
[30] Tsai Y H, Hsieh S H and Kang S C 2014 A BIM-enabled approach for construction inspection In *Computing in Civil and Building Engineering (2014)* pp 721-728
[31] Tsai Y H and Hsieh S H 2016 Process modeling of a BIM-enabled construction inspection approach with BPMN In *Proceedings of the International Conference on Innovative Production and Construction, Darwin, Australia* pp 3-5
[32] Chen L and Luo H 2014 A BIM-based construction quality management model and its applications *Automation in construction* **46** pp 64-73
[33] Lee J Y, Kwon O S, Choi J S and Park C S 2012 A study on construction defect management using augmented reality technology In *2012 International Conference on Information Science and Applications* pp 1-6
[34] Park C S, Lee, D Y, Kwon O S and Wang X 2013 A framework for proactive construction defect management using BIM, augmented reality and ontology-based data collection template *Automation in construction* **33** pp 61-71
[35] Anil E B, Tang P, Akinci B and Huber D 2013 Deviation analysis method for the assessment of the quality of the as-is Building Information Models generated from point cloud data *Automation in construction* **35** pp 507-516
[36] Kim M K, Thedja J P P and Wang Q 2020 Automated dimensional quality assessment for formwork and rebar of reinforced concrete components using 3D point cloud data *Automation in Construction* **112** p 103077
[37] Akinyede I J, Fapohunda J A and Haldwang R 2017 Evaluation of Critical Factors for Achieving Quality Housing Delivery within Budget In *International Conference on Sustainable Infrastructure 2017* pp 153-164
[38] Adenuga O A 2013 Factors affecting quality in the delivery of public housing projects in Lagos State, Nigeria
[39] Fore S 2015 An Analysis of Factors Influencing the Quality of Housing Construction Projects in the Western Cape, South Africa *International Journal of Science and Technology* **1** pp 195-211
[40] Aripin I D M, Zawawi E M A and Ismail Z 2019 Factors influencing the implementation of technologies behind industry 4.0 in the Malaysian construction industry In *Matec Web of Conferences* **266** p 01006

[41] Liu L, Chen I M, Kayacan E, Tiong L K and Maruvanchery 2015 Automated construction quality assessment: A review In *2015 10th International Symposium on Mechatronics and its Applications (ISMA)* pp 1-6

IOP Conf. Series: Materials Science and Engineering 1218 (2022) 012003 doi:10.1088/1757-899X/1218/1/012003

A Theoretical Assessment on the Implementation of Artificial Intelligence (AI) for an Improved Learning Curve on Construction in South Africa

M G Phaladi[1,6], **X N Mashwama**[2], **W D Thwala**[3], **C O Aigbavboa**[4]
[1,2,3,4] Department of Construction Management and Quantity Surveying, University of Johannesburg, 25 Louisa Street, Doornfontein, 2028, South Africa

Email: phaladimg@gmail.com

Abstract. Generally, the construction sector is lacking behind with the usage of technological software. Artificial Intelligence(AI) technology is viewed as a technological tool that can improve the timeous delivery of construction project(s). Therefore, the study is a literature review on the assessment of the implementation of AI technology on South African construction projects. The qualitative approach and purposive sampling technique were adopted in this study. The primary findings emanating from this study reveals that the past other underground empirical studies have identified several important impacts of AI technology and how it can transform the construction industry. Furthermore, the study findings revealed that if AI technology is fully exposed and exploited on construction projects especially to both developed and developing countries it would certainly improve productivity, enhance efficiency, and performance issues. On the other hand, the level of implementation of AI technology in South Africa is still very low. Therefore, the study would contribute to the body of existing knowledge of AI technology on construction projects, especially in South Africa.

1. Introduction

The construction sector is one of the less digitalized industries in the world when compared to other industries. Adopting the latest technologies can be daunting for the construction teams. However, the construction sector is still lacking behind with the usage of digital technologies (Barbosa *et al.,* 2017). In the past years, researches have been conducted regarding the possible methods that can be adopted to increase construction project productivity. On the contrary, the introduction of the Fourth Industrial Revolution (4IR) features to our daily lives, has forced the construction industry to implement those features to improve their level of competitiveness on the project(s). One of the recommended features is Artificial Intelligence(AI) technology. The majority of the developed countries have adopted the AI technology where its benefits have been realized when compared to the developing countries especially African countries. Furthermore, companies have been advised to adopt digital technologies to their construction activities to achieve ultimate efficiency. Gotthardt *et al.* (2019), justified that the implementation of AI technology increases efficiency and eliminates cost overruns. Previous studies have indicated that the adoption of AI on construction projects has increased over the past years. The application of AI on construction projects has benefits such as enhanced productivity growth (Jesuthasan, 2018), changes to the approach of work (Jariwala, 2015), improves decision making, improves profitability, reduces construction design errors, and eliminates accidents on site. Despite the positive influence AI has on construction projects, they are concerns about future employment opportunities especially for construction workers such as certain skills would be obsolete whilst the adoption of digital technologies might require experienced personnel who knows how to use AI systems. This brings uncertainty about whether the adoption of AI would be successful or not (Nickerson, 2019). This paper aims to assess the role of implementing AI on construction projects, and also to examine how AI technology can be used to enhance construction project productivity. This paper will further present a literature review, methodology, summary, conclusion, and recommendations for future studies.

2. Literature review

2.1 South Africa and the Construction industry

In 2020, the world has and still experiencing the disease called coronavirus pandemic which was firstly discovered in 2019 in China. Hui *et al.* (2020), further said that the disease is caused by "severe acute respiratory syndrome". However, construction companies in both developing and developed countries were drastically affected by the coronavirus pandemic where most of the firms were forced to shut down particularly small and medium enterprises. The situation has resulted in companies facing commercial and legal difficulties (Woertz, 2020). Most of the industries have experienced economic downfalls and recessions. According to Fernandes (2020), further stated that the effect of the pandemic on the economy has resulted in to decrease in Gross Domestic Product (GDP) by approximately 2-6% in some of the nations. Most of the workers were forced to operate remotely as a remedy for easing coronavirus infections. However, considering the type of activities performed on construction projects, it is always difficult to work remotely as a result of activities that have to be inspected and monitored by the project team. One of the critical challenges during the pandemic was that the majority of the people lost their employment. As for the construction sector, human laborers on-site lost their jobs as a result of lockdown (KHL, 2020). The coronavirus pandemic has affected the construction sector in various ways which include; shortage of supplies, high cost of materials, and late deliveries (Mordor Intelligence, 2020). Trowers and Hamlins (2020), added that a decrease in supply and demand resulted in to decrease in revenues where firms are forced with the tough decision of reducing the wages and dismissing the workers. That has led to construction projects being completed longer than anticipated. During the coronavirus pandemic, the rate of unemployment has increased in both developed and developing countries. Generally, the construction sector does not have a high number of the skilled and experienced workforce (Al-Amri and Marey-Perez, 2020).

On the other hand, the construction industry is one of the biggest sectors that create employment opportunities and contributes to the national economic growth in South Africa (Oyewobi *et al.*, 2015). The construction industry in South Africa is not growing as expected due to various factors that hinder its progress. Some of the factors include the poor implementation of the latest technology on a construction project (Kapoor, 2011). Windapo and Cattell (2013), further stated that the South African construction industry is still lacking behind with the usage of technologies when compared to other developed nations. Again, the South African construction sector is still suffering from poor productivity. Factors affecting productivity growth include; lack of skilled manpower, inexperienced contractors, shortage of materials on-site, insufficient finances to proceed with construction works, and design errors, etc. Rust *et al.* (2013) opined that if productivity growth is poor on construction projects, the performance would massively be affected as well. Modares and Sepehri (2009), further postulated that the reasons for South Africa to experience productivity issues is because of the corruption, fraud, and lack of proper construction methods performed. Moreover, some of the contractors cannot afford to pay and maintain experienced personnel such as permanent supervisors and skilled laborers during the project life span (Ofori, 2013). Windapo (2016), stated that South African contractors have a problem paying experienced laborers, thus resulting in opting for inexperienced laborers to perform work on-site which has often led to defective works. For construction organizations to increase economic growth and to improve their construction projects, the government institutions have to take center stage by encouraging and financially supporting them to adopt the latest technologies to their projects.

2.2 The role of Artificial Intelligence on construction projects

Artificial Intelligence(AI) is defined “as the ability of a machine to mimic intelligent human behavior, this seeking to use human-inspired algorithms for approximating conventionally challenging problems” (Salehi and Burgueno, 2018). The implementation of AI technology on construction projects differs depending on each firm’s requirements. According to Marr (2019), said companies can adopt AI to enhance efficiency. Previous studies conducted found out that AI will enhance productivity (Jariwala, 2015), quality of activities performed would be improved (Lin and Hazelbaker, 2019), and creates employment opportunities (Greenman, 2017). Construction companies can use AI to achieve the client's objectives such as timeous delivery of the projects. However, the usage of AI in the construction sector is very incomplete (Blanco *et al.,* 2018) compared to the other industries. The construction sector has lost its good reputation due to poor productivity, performance, and low usage of technological software. Moreover, AI is seen as a technological tool that can eliminate the industry's greatest predicaments such as safety issues, additional cost, and time (Blanco *et al.,* 2018). Generally, the usage of AI technology on South African construction projects is very low since it is still at the development stage. The reason behind slow adoption is due to poor leadership and lack of faith by company owners. Besides, the high cost associated with the implementation of AI proves to be a major stumbling block for the majority of construction firms, particularly small and medium firms. Luo *et al.* (2018), claim that AI systems cannot produce the expected results. In contrast, AI technology would certainly improve the lack of productivity and efficiency on construction projects in the following ways;

2.2.1 Monitoring of construction activities

Construction projects often require supervision of activities where supervisors/construction project managers need to deliver the project on time, within the estimated budget, of good quality, and recommended safety to meet the client objectives. Artificial Intelligence technology can be used to monitor the activities happening on site such as workers who are not wearing personal protective clothing and also to store important information about the construction site. According to Strukova and Liska (2012), added that construction project managers can be able to access stored information.

2.2.2 Estimation of quantities and planning

With the usage of Artificial Intelligence systems project teams can be able to plan their tasks and estimate accurate costs on the project. The AI software will break down and plan the activities automatically which makes the work of construction project managers easier (CastroLacouture, 2009).

2.2.3 During the design phase

The application of Artificial Intelligence(AI) technology on construction projects would allow the contractors and project team to view designed structures in Three dimensions which reduces the design errors on-site (Castro-Lacouture, 2009). Engineers and Architects can leverage AI technology to eliminate unnecessary mistakes done by the contractors as a result of missing information on designed structures. Furthermore, the usage of smart technology can show a clear connection of pipework to avoid any future leakages of the pipes.

2.2.4 Controlling quality

The majority of the construction projects have been affected by a lack of quality on work done especially after few months of handing over the project to a client(s). This is caused by the low usage of technologies on construction projects to detect faults. However, the introduction of Artificial Intelligence(AI) technology brings the quality that has been lacking over the years on construction projects. The adoption of AI technology can be used to inspect the work if it is of good quality (Balaguer and Abderrahim, 2008).

2.2.5 Overcoming shortages of experienced labors

Construction projects are often facing the challenge of keeping experienced laborers because most of the contractors not willing to meet their wage demands. The investment in Artificial Intelligence(AI) machines by firm owners would replace the headache of maintaining experienced laborers. This is because the implemented intelligent machines on-site can be able to perform the tasks which were performed by human beings (Boyd and Holton, 2017) which in return decreases the design errors caused by human labor.

3. Research Methodology

The study is a theoretical literature review that was conducted with the usage of previously published conference papers and journal articles. The main research study is still in progress. However, a qualitative research approach was selected. The qualitative data would be collected through in-depth interviews with the respondents to gain knowledge of AI technology directly from them. The preferred respondents are built environment professionals such as Architects, Quantity Surveying, Construction Project Managers, and Engineers. Purposive sampling was chosen as one of the sampling techniques. The respondents would be contacted and selected from the database list provided by professional regulatory bodies such as ECSA, SACPCMP, SACAP, and SACQSP.

4. Lesson learnt

The literature revealed that the construction sector plays a massive role in increasing economic growth and Gross Domestic Product. The implementation of the latest technological software such as Artificial Intelligence technology on South African construction projects is still lacking due to poor leadership display by government leaders, lack of improper infrastructure development, etc. Reviewed literature has outlined facts leading to low productivity and performance such as shortage of materials, design errors, lack of finances to proceed with construction activities. Most importantly, the timeous delivery of the project is the most rated factor experienced in developing and undeveloped countries. Modares and Sepehri (2009), pointed out that fraud and corruption in construction projects have become a norm to project teams. Hence, productivity and efficiency have not been easily achieved lately. The literature revealed has further stated that for construction projects to achieve maximum productivity, the latest technological software offered by the Fourth Industrial Revolution such as Artificial Intelligence technology has to be adopted to improve project timeous delivery. Also, the study has indicated that the usage of AI in South Africa is still very low as a result of a lack of faith in digital technology. Literature has revealed the benefits of using AI such as improved quality of work, enhanced productivity, elimination of design errors, and improved safety on construction projects. Moreover, the literature reviewed has also indicated how AI technology can be used on a construction project(s) for enhancing efficiency and productivity such as workers controlling the quality, during the design and construction phase. Literature has further revealed the impact of coronavirus pandemic on construction projects such as the majority of the workers losing their employment.

5. Conclusion and Recommendation

The study has successfully revealed the potential benefits of implementing AI technology on construction projects such as timeous delivery of the projects, enhanced productivity, elimination of cost and schedule overruns, etc. On the other hand, they are challenges of implementing AI technology such as loss of jobs by human labors, high cost associated with the adoption of AI which the study has successfully revealed. Again, the study has revealed how AI technology can be used on construction projects. The study recommends that the government institutions and professional regulatory bodies must encourage the company owners to use the Fourth Industrial Revolution features and also to financially support them particularly Small and Medium organizations. further research has to be conducted on the stages of the project lifecycles on where AI technology can mostly be used to enlighten the construction firm owners on how to use AI technology from inception to completion of the project for improvement of productivity.

6. References

[1] Barbosa F, Woetzel J, Mischke J, Riberirinho M J, Sridhar M, Pearsons M, Bertram N and Brown S 2017 Reinventing construction: A route to higher productivity, Reinventing construction: A route to higher productivity (Accessed 25 March 2021).

[2] Blanco J L, Fuchs S, Parsons M and Ribeirinho M J 2018 Artificial intelligence: Construction technology's next frontier Artificial Intelligence Construction technology's next frontier.

[3] Greenman C 2017 Exploring the impact of artificial intelligence on the accounting profession. *Journal of Research in Business, Economics, and Management* **8(3)** 14-51. Retrieved from:

http://scitecresearch.com/journals/index.php/jrbem/article/view/1063. (Accessed 20 March 2021).

[4] Gotthardt M, Koivulaakso D, Paksoy O, Saramo C, Martikainen M and Lehner O M 2019 Current state and challenges in the implementation of robotic process automation and artificial intelligence in accounting and auditing. *ACRN Oxford Journal of Finance & Risk Perspectives* **8** 31-46. Retrieved from: http://www.acrn-journals.eu/resources/SI08_2019c.pdf (Accessed 08 January 2021).

[5] Jariwala B 2015 Exploring AI and the accountancy profession: Opportunity, Threat, both, neither? Retrieved from: https://www.ifac.org/knowledge-gateway/developingaccountancy/profession/discussion/exploring-artificial-intelligence (Accessed 07 February 2021).

[6] Jesuthasan R 2018 Future of work – reinventing jobs: A 4-step approach for applying automation to jobs Retrieved from https://www.willistowerswatson.com/en-GB/insights/2018/08/future-ofwork-reinventing-jobs (Accessed 12 February 2021).

[7] Nickerson M A 2019 AI New risks and rewards Retrieved from https://sfmagazine.com/postentry/april-2019-ai-new-risks-and-rewards/ (Accessed 20 February 2021).

[8] Decanio S J 2016 Robots and humans–complements or substitutes? *Journal of Macroeconomics* **49** 280-291.

[9] Marr B 2019 Why every company needs an artificial intelligence (AI) strategy for 2019 Retrieved from https://www.forbes.com/sites/bernardmarr/2019/03/21/why-every-companyneeds-an-artificial-intelligence-ai-strategy-for-2019/#52cf3eab6 (Accessed 23 February 2021).

[10] Ofori G 2013 Construction in developing countries Construction Management and Economics **25(1)** 1–19 Available at https://doi.org/10.1080/01446190601114134 (Accessed 06 March 2021).

[11] Oyewobi L O, Windapo A O and James R O B 2015 An empirical analysis of construction organizations competitive strategies and performance Built Environment Project and Asset Management **5(4)** 417–431. Available at: http://doi. org/10.1108/BEPAM-10-2013-0045 (Accessed 23 January 2021).

[12] Rust F C, Botha C, Van Wyk L, Steyn W, Du Plessis C and Landman K 2013 *South African construction industry technology foresight study:* Summary report of a desktop study of CSIR Technical Report CSIR, Pretoria.

[13] Windapo A O 2016 Skilled labour supply in the South African construction industry: The nexus between certification quality of work output and shortages SA *Journal of Human Resource Management/SA* **14(1)** 1–8.

[14] Windapo A O and Cattell K 2013 The South African construction industry: Perceptions of key challenges facing its performance development and growth. Journal of construction in developing countries **18(2)** 65-79.

[15] Salehi H and Burgueño R 2018 Emerging artificial intelligence methods in structural engineering *journal of Engineering Structures* **171** 170–189.

[16] Luo J, Meng Q and Cai Y 2018 Analysis of the impact of AI application on the development of accounting industry *Open Journal of Business and Management* **6(4)** 850-856 Available from https://doi.org/10.4236/ojbm.2018.64063 (Accessed 27 February 2021).

[17] Al-Amri T and Marey-Pérez M 2020 Towards a sustainable construction industry: Delays and cost overrun causes in construction projects of Oman *Journal of Project Management* **5(2)** 87102.

[18] KHL 2020 Europe's Construction Industry. Retrieved from: https://www.khl.com/constructioneurope/86-of-sites-now-open-in-england-and-wales/144207.article (Accessed 27 March 2021).

[19] Trowers G and Hamlins N 2020 Covid-19: Employment update –Issues to consider in Oman Retrieved from https://www.trowers.com/insights/2020/april/covid-19---employment-update--issues-to-consider-in-Oman (Accessed 23 March 2021).

[20] Woertz E 2020 COVID-19 in the Middle East and North Africa: Reactions Vulnerabilities Prospects.
[21] Hui D S I, Azhar E, Madani T A, Ntoumi F, Kock R, Dar O and Zumla A 2020 The continuing 2019-nCoV epidemic threat of novel coronaviruses to global health-The latest 2019 novel coronavirus outbreak in Wuhan, China *International Journal of Infectious Diseases* **91** 264-266.
[22] Fernandes N 2020 Economic effects of coronavirus outbreak (COVID-19) on the world economy. Available at: www. SSRN 3557504 (Accessed 21 March 2021).
[23] Mordor Intelligence 2020 Oman Construction Market Growth, Trends, and Forecast Retrieved from https://www.mordorintelligence.com/industry-reports/oman-construction-market (Accessed 27 March 2021).
[24] Balaguer C and Abderrahim M 2008 Trends in robotics and automation in construction.
[25] Struková Z and Líška M 2012 Application of automation and robotics in construction work execution AD ALTA *Journal of Interdisciplinary Research* **2** 121-5.
[26] Castro-Lacouture D 2009 Construction automation. In Springer handbook of automation (pp. 1063-1078) Springer Berlin Heidelberg.
[27] Boyd R and Holton R J 2017 Technology, innovation, employment, and power: Does robotics and artificial intelligence really mean social transformation? *Journal of Sociology* **1** 1–15.

Acknowledgements
Thanks to the National Research Foundation (NRF) for financially supporting me throughout the study and also to my supervisors for their expect guidance throughout the compilation of the research paper.

IOP Conf. Series: Materials Science and Engineering 1218 (2022) 012004 doi:10.1088/1757-899X/1218/1/012004

Analyses of Automated Bricklaying Workflow Regarding Time and Arrangement

I Vidovszky[1] and A Pém[2]

[1] Department of Construction Technology and Management, Budapest University of Technology and Economics, Budapest, Hungary

[2] Department of Construction Technology and Management, Budapest University of
Technology and Economics, Budapest, Hungary vidovszky.istvan@epk.bme.hu

Abstract. Bricklaying has been one of the determinant technologies of the construction industry for several thousand years. Even in our modern times, it is one of the most commonly used fields of building construction technologies. There are several reasons, like the lack of human resources on the construction markets or the increasing demands on both reasonably short construction time and high construction quality, the combination of which leads to consideration of applying innovative, automated technologies in more and more fields. With the spread of these advanced techniques, changes and further development on the construction site can be predicted. One of these developing fields will be presumably the field of small, solid construction units, the innovation of which would result in changes regarding the elements and the technology too. In our research, automated bricklaying construction situations are modeled by applying a robotic arm equipped with a linear rail. Some new features are analyzed at the work of using automated systems. Namely, different brick wall construction methods are tested, measured, and analyzed to find optimal solutions for the cases of various sets of influencing factors, like element placement or palette arrangement.

1. Introduction

There have been many attempts regarding the automation of construction in the last sixty years, which means applying electromechanical robots in the first place. In the construction industry, and particularly in bricklaying, the two main directions of applying robotic technology are the use of 4 or 6 axis robotic arms and large-scale robotic structures, mainly external lifting frames above the manipulation area. According to some researchers, flying robotic systems can be regarded as the third category (Sousa et al., 2016). It is a developing field with great future potentials; however, its application is far more experimental than the other two.

The theoretical background of construction automation was evaluated in the 1980ies and 1990ies already. The initiation of the ROD (Robotic Oriented Design) (Bock, 1988) has been a new approach for planning and executing buildings. The ABC (Automated Building Construction) emerged as a new way of thinking about construction automation, which involves the organization, the machinery, and the logistics of the supply chain (Skibniewski-Wooldridge, 1992). This approach leads us to the evaluations of the last decade when large-scale robotics-based constructions are considered as integrated automated construction sites (Bock-Langerberg, 2014).

As far as there are many highly repetitive tasks in bricklaying, it is one of the filed that can be automatized with success. There were also many projects focusing on the automation of bricklaying in the last decades. The first attempts for such automation were maybe the early patents for bricklaying machines in the 1960ies, like the so-called “Motor Mason”, filmed in Essex (Britsh Pathé, 1967). It was followed by some further attempts in the 1970ies 1980ies too.

In the 1990ies, the theoretical basics of bricklaying robotics (Pritschow, 1993), was followed by the European ESPRIT II ROCCO (Robot Assembly System for Computer Integrated Construction) (Andres et al. 1994; Gambao et al., 1997) and BRONCO (Bricklaying Robot for Use on the Construction Site)

project (Pritschow et al., 1996). These researches made several basic statements about the details of bricklaying robots' possible equipment and working methods.

In our time, there are some industrial bricklaying robots already, like the MULE, or the SAM (SemiAutomated Mason) developed by the American *Construction Robotics* company, or the Hadrian X developed by the Australian *Fastbrick Robotics*. While MULE is a semi-automated cooperative system helping the bricklayers to lift heavy construction blocks, SAM is an automated bricklaying technology developed for large-scale monotonous bricklaying activities, like long straight walls (Petters and Belden 2014). The most complex system of the ones mentioned above is Hadrian X, which erects the entire wall structure of a minor scale building autonomously.

About the unique forms and shaping of the walls with the help of robotic arms, the research in the ETH Zürich can be mentioned as an example (Bärtschi et al., 2010). Apart from the spectacular attempts of free-shaping of the bricklaying structures, they also developed a new brick bond for their twisted wall project. It was followed by many bricklaying projects with curved walls (Helm et al., 2012) or even with vaults (Parascho et al., 2020). An extension of these works is the research of bricklaying in a previously established, existing environment, where self-navigating mobile robots construct walls (Dörfler et al., 2016).

The research and technologies mentioned above basically use robotic arms for the executions, but there are some attempts to apply large frame structures too. Such an example is applying a cable-driven system for bricklaying (Bruckmann et al., 2016).

There have been some attempts with bricklaying drones as well. An example is published by Wilmann et al. (2012). However, the application of flying robots on the construction site is seemingly more complicating than other solutions. Apart from the technical challenges like the effect of the wind in outer environmental circumstances or the safety questions while working in the same area with human workers, flying is a strictly overregulated traffic issue in several countries, especially above a certain weight. Not forgetting its possible advantages, we can state that it is currently the less easily applicable one among the mentioned technologies.

In traditional construction, the allocation of construction objects on site is a partial task of the construction site layout planning (CSLP). In practice, the construction object allocation is carried out routinely (Sanad et al. 2008) based on human judgment using the first-come-first-served method (Zouein and Tommelein, 1999) or using the construction manager's experience (Cheng and O'Connor, 1996). There are two basic methods to deal with CSLP problem (Moore, 1980). The first method is placing every kind of resource everywhere or composing a couple of arrangement combinations and selecting the best from these. The other is bringing objects of one type at a time only in a specific order and calculating the optimal arrangement after each step (Hegazy and Elbeltagi, 1999).

Due to the number of factors involved in the CSLP, computer-aided systems were applied lately to solve the related problems. These systems were either CAD-based (Sadeghpour et al., 2004), or applied AI techniques (Tommelein et al. 1992) or used genetic algorithms (Osman et al. 2003; Li and Love, 1998; Hegazy and Elbeltagi, 1999; Mawdesley et al. 2002; Tam and Tong, 2003, and others). Most developed models identify the number and size of the temporary facilities serving the construction site and search for optimal arrangement by minimizing the total transportation costs between the facilities or between the facility and the structure to be built (Sadeghpour et al., 2004).

Besides the allocation of the objects, there are other aspects to take into consideration. The wellknown time and motion studies of Frank and Lillian Gilbreth in the early 1900s illuminated the problem of unnecessary motions, and they invented new ideas to arrange bricks and pallets to improve productivity.

Bricklaying with robotics and the physical constraints of the bricklaying technology establish a similar set of tasks with CSLP and motion studies.

Despite the numerous researches mentioned above, there are many open questions in this field too. In this paper, some bricklaying methods and related pallet arrangements are investigated.

2. The applied methods

2.1 Tools and layout

This research aims to analyze the connection between the possible ways of bricklaying in combination with the pallet layouts. We were looking for the answers to real construction site problems, at the analyses of which we applied a modeled construction situation. For this purpose, a robotic arm and drystacked 30x15x9mm ceramic bricks were used. Our sample constructions were six full brick long and four brick high stretch bond walls.

We used Dobot Magician 4 axis robotic arm equipped with a sliding rail and a pneumatic gripper end-effector (Fig 1). It was initially programmed by the built-in application of the Dobot Studio; however, for the more complicated final tasks, we applied Python codes.

In this examination, we used a model environment with the following differences to the reality which have to be considered at the evaluation:

- no weather effects were considered – in a real environment, the weather may have some impact on the construction works,
- the proportion of the gripper to the brick is not realistic; the gripper was oversized, which enabled less smooth work and less freedom of the placement than in a real environment,
- the bricks were dry-stacked – no mortar was applied
- the pallet heights and the arrangements were limited by the model robotic environment.

Figure 1. The arrangement of the robotic environment

2.2 The bricklaying and the palleting

We applied different bricklaying and palleting methods to test out what is optimal in our model.

We considered two types of bricklaying methods. The first one is the traditional way of bricklaying, in which case the wall is constructed course by course (Fig 2a). The second one is a stair-wise method (Fig 2b) that would not come up too often in traditional bricklaying performed by human resources because it neither allows easy aligning at laying of the bricks nor enables the rapid work with the application of string lines. However, under certain circumstances, this second method might allow faster bricklaying in the case of automated technologies.

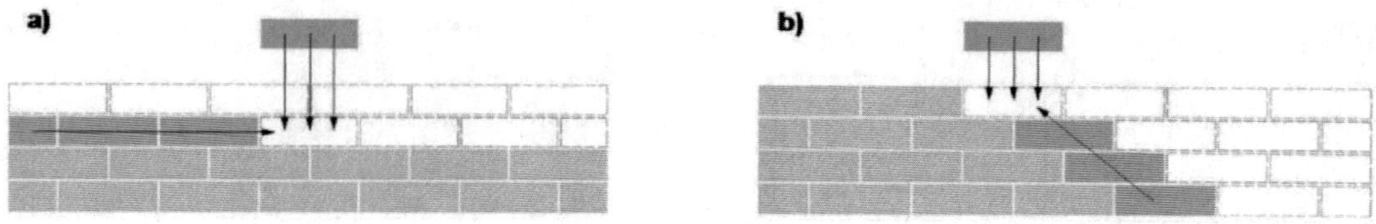

Figure 2. Bricklaying modes: a) traditional bricklaying b) stair-wise bricklaying.

According to the pallet arrangement, we applied two different conceptions. In the first case, the pallets were at one end of the wall symmetrically arranged (the wall was in the axis of the pallet configuration). We called it *"head" palleting arrangement*. (Fig 3a) The other type, when we placed the pallets parallel to one side of the wall. This one we called *parallel palleting arrangement* (Fig 3b).

As a start, we made a typical linear layout constructed course by course with traditional bricklayer method. We started with half a brick in every second course to establish a proper stretch-bond wall. As a second step, we applied the same bond type with the stair-wise bricklaying method.

Both bricklaying was tested by "head" and by "parallel" palleting too. The results of the measurements are summarized in *Table 1*.

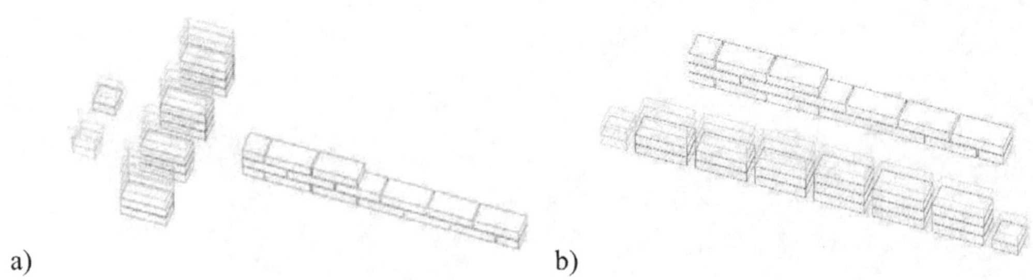

Figure 3. Pallet arrangements: a) "head" palleting b) parallel palleting

3. Results and discussion

The optimal way of using pallets could be when the total delivery path of all built-in elements would be the shortest possible (Pém & Malyusz, 2008). However, it would require the analyses of many combinations, most of which are irrelevant or are not applicable because of some physical barrier, like the physical extension of the gripper head.

For this reason, as a first step, we considered some possible cases by the calculation of the shortest route from the available (top) elements of the pallets at each brick placement. This method resulted in an unbalanced pallet consumption and seemingly long placing routes after some pallets were emptied. Therefore, we decided to apply additional rules. Our modified python code forced the differences between the pallets to be less than half of an element from the average value of the pallets in height, which resulted in a more balanced consumption of the pallets. However, it caused a longer execution time in our case. (Table 1, 5-6)

Finally, we applied a somewhat unconventional arrangement as well (Fig 4). In traditional construction circumstances, pallets have to be placed outside of the marked area of the walls, which is quite logical in any common cases. In the case of working with robotic equipment, however, it can be reconsidered.

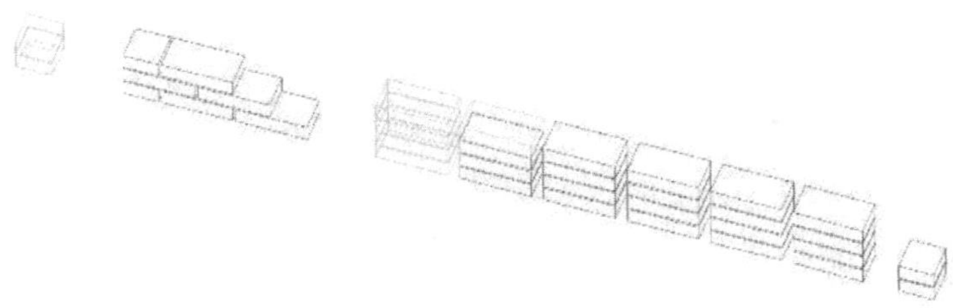

Figure 4. Placing the pallets in the marked out area of the wall

The stair-wise bricklaying method enables the placement of the pallets in the marked-out area of the wall, right where the wall will be constructed later, resulting in the absolute shortest routes and time to the final positions of the wall bricks. (Table 1, 7)

Table 1. The measure times of the tests

	Pallet arrangement	Bricklaying mode	Time(h/m/s)
1	"Head" pallet arrangement	course by course	00:10:46
		stair-wise	00:10:46
2		course by course	00:10:06
3	Parallel pallet arrangement	stair-wise	00:10:03
4		course by course	00:10:24
5	Parallel pallet arrangement with limitations	stair-wise	00:10:21

6		stair-wise	00:09:20
7	Pallets in the marked out area of the wall		

4. Summary

In this research, we made some examinations to get better times at automatized bricklaying in the case of straight solid walls by two means: establishing better arrangements of pallets and applying an unconventional bricklaying work method. Finally, the combination of the two, namely applying stairwise bricklaying and placing the pallets in the marked-out area of the wall to be constructed, resulted in the shortest of the times of our attempts.

5. Possible extension of the research

All the examined cases were bricklaying of solid, straight walls, without geometrical turns or any openings. It means that our observations are validated to the simplest layouts only. More complex cases, such as bricklaying tasks at brick walls with geometrical turns and openings, more complex pallet arrangements, and the application of a more convenient gripper, are planned to be investigated.

Acknowledgment

The research reported in this paper was supported by the BME Artificial Intelligence TKP2020 IE grant of NKFIH Hungary (BME IE-MI-SC TKP2020).

References

[1] Andres J, Bock T, Gebhart F, Steck W 1994 First results of the development of the masonry robot system ROCCO: a fault tolerant assembly tool. 11th International Symposium on Automation and Robotics in Construction (ISARC) pp 87-93

[2] Bärtschi R, Knauss M, Bonwetsch T, Gramazio F, Kohler M 2010 Wiggled brick bond Advances (Architectural Geometry, 2010) pp 137-148

[3] Bock T 1988 Robot-Oriented Design. Proc. The 5th International Symposium on Robotics in Construction, June 6-8, Tokyo, Japan pp 135-144

[4] Bock T. 2015 The future of construction automation: Technological disruption and the upcoming ubiquity of robotics (Automation in Construction Vol.59) pp 113-121

[5] British Pathé https://www.britishpathe.com/video/mechanical-bricklayer/query/Motor+mason 2021-05-26

[6] Bruckmanna T, Matternb H, Spenglerc A, Reicherta C, Malkwitzc A König M 2016 Automated Construction of Masonry Buildings using Cable-Driven Parallel Robots Proc. 33rd International Symposium on Automation and Robotics in Construction (ISARC) pp.565-568.

[7] Cheng M Y, O'Connor J T, 1996 ArcSite: Enhanced GIS for construction site layout (Journal of Construction Engineering and Management, Vol.122) pp 329-336

[8] Dörfler K, Sandy T, Giftthaler M, Gramazio F, Kohler M, Buchlin J 2016 Mobile Robotic Brickwork. Automation of a Discrete Robotic Fabrication Process Using an Autonomous Mobile Robot In: Robotic Fabrication in Architecture, Art and Design ed Willmann J, Block, P, Hutter M, Byrne K, Schork T. (Springer, 2016) pp 204-217

[9] Gambao E, Balaguer C, Barrientos A, Saltaren R, Puente E A 1997 Robot assembly system for the construction process automation. In Proceedings of International Conference on Robotics and Automation IEEE. (Vol.1) pp 46-51

[10] Sanad H M, Ammar M A, Ibrahim M E 2008 Optimal Construction Site Layout Considering Safety and Environmental Aspects (Journal of Construction Engineering and Management, Vol.134) pp. 536-544

[11] Hegazy T, Elbeltagi E 1999 EvoSite: Evolution-Based Model for Site Layout Planning (Journal of Computing in Civil Engineering Vol.13) pp 198-206

[12] Helm V, Ercan S, Gramazio F, Kohler M 2012 In-Situ Robotic Construction: Extending the Digital Fabrication Chain in Architecture, in Synthetic Digital Ecologies Proc. 32nd Annual Conference of the Association for Computer Aided Design in Architecture (ACADIA) San Francisco pp 169–176

[13] Osman H M, Georgy M E, Ibrahim M E, 2003 A hybrid CAD-based construction site layout planning system using genetic algorithms (Automation in Construction, Vol.12) pp 749-764

[14] Li H, Love P E D 1998 Site-level facilities layout using genetic algorithms (Journal of Computing in Civil Engineering Vol.12) pp 227-231.

[15] Mawdesley M J, Al-jibouri S H, Yang H 2002 Genetic algorithm for construction site layout in project planning (Journal of Construction Engineering and Management, Vol. 128.5) pp 418426

[16] Moore J 1980 Computer method in facility layout (Industrial Engineering. Vol. 12.9) pp 82-93 [17] Parascho S, Han I X, Walker S, Beghini A, Bruun E P, Adriaenssens S 2020 Robotic vault: a cooperative robotic assembly method for brick vault construction. (Construction Robotics, Vol. 4) pp 117-126

[18] Pém A., Mályusz L. 2008 Optimisation of deposit location at a building site as a "knapsack problem" Proc. International Conference - Organization, Technology and Management in Construction, Umag, Horvátország, Proceedings pp 1-6

[19] Pritschow G. et al. 1996 Technological aspects in the development of a mobile bricklaying robot. (Automation in Construction Vol 5) pp 3-13.

[20] Pritschow G, Dalacker M, Kurz J 1993 Configurable Control System of a Mobile Robot for OnSite Construction of Masonry In: ed G.ll. Watson, R.L. Tucker, and J.K. Walters 1993 Elsevier Science Publishers R.V. pp 85-92

[21] Sadeghpour F, Moselhi O, Alkass T S 2004 A CAD-based model for site planning (Automation in Construction Vol.13) pp 701-715

[22] Skibniewski M J, Wooldridge S C 1992 Robotic materials handling for automated building construction technology (Automation in Construction Vol.1) pp 251-266

[23] Sousa J P, Gassó Palop C, Moreira E, Pinto A M, Lima J, Costa P, Costa P, Veiga G, Moreira A P 2016 The SIDERobot: A Cabel Robot System for On-site Construction in Architecture In: ed Reinhardt D, Saunders R, Burry J Robotic Fabrication in Architecture, Art and Design 2016 Springer pp 231-239

[24] Tam C M, Tong T K L, 2003 GA-ANN model for optimizing the locations of tower crane and supply points for high-rise public housing construction (Construction Management and Economics Vol.21) pp 257–266

[25] Tommelein I D, Lewitt R E, and Hayes-Roth B 1992 SightPlan model for site layout (Journal of Construction Engineering and Management, Vol. 118) pp 749-766.

[26] Zouein P P, Tommelein I D 1999 Dynamic layout planning using a hybrid incremental solution method (Journal of Construction Engineering and Management, Vol.125) pp 400-408

IOP Conf. Series: Materials Science and Engineering 1218 (2022) 012005 doi:10.1088/1757-899X/1218/1/012005

Exploratory Research towards Automated Masonry Construction using UAVs

B Elkhapery and E P Small

Department of Civil & Environmental Engineering, University of Delaware, Newark, Delaware, USA

bkhapery@udel.edu

Abstract. Unmanned autonomous vehicles (UAVs), also known as drones, are pervasive in modern society for both recreational and business purposes. Construction sites routinely employ UAVs to capture imagery used for photo documentation, progress reporting, surveying, and inspection. Drones can also carry infrared cameras and LIDAR, increasing opportunities for their use on construction sites. They are not limited, however, to remote sensing and photogrammetric applications. Research is pushing the boundaries to explore other UAV opportunities, such as material deliveries and physical construction. This paper reports on exploratory research to investigate current efforts in automated construction of masonry walls using drones and drone swarms. Masonry was chosen due to the repetitive nature of the processes, which increases the probability of successful implementation. The results of a literature review are first presented to establish the state of the art in this area. The proposed approach will be evaluated based on the construction quality in accordance with standards, entailing both accuracy and precision requirements in the laying process. Approaches for continuing research in drone-based masonry construction automation will be outlined and discussed in the conclusion.

1. Introduction

The construction industry global market is expected to be a total worth of $15 trillion by 2025, a projected increase of 70% compared to its worth in 2013 [1]. Investments in the construction sector are projected to continue to increase steadily in the near future beyond 2025 as global economic growth expands. In Western nations, this growth is hindered by challenges with labor availability. To advance the industry capabilities, professional societies have been proactive in recruiting and training. Material suppliers introduce products with less labor demand for installation, such as self-consolidating concrete. Prefabrication and off-site manufacturing further reduce on-site labor requirements. Accordingly, there is a need for continuous innovation to support and enable the industry to match the rising demand.

Furthermore, current construction practices depict a rising demand for skilled laborers. With the construction industry shift towards high-rise structure development, workers' safety and health risk have undoubtedly escalated. Consequently, construction workers have a 1 in 200 chance to die or get severely injured from the job on the field over a 45-year career span period [2]. Robotics and the use of technologies, such as UAVs, for physical construction, can provide further positive benefits by eliminating risks associated with workers' safety. This is a further additional benefit complementing projected improvements in quality and productivity.

The incorporation of off-site manufacturing and just-in-time delivery has enabled construction innovators in China to build commercial structures in days as opposed to months. Robotics and automation promise the ability to bring techniques commonplace in manufacturing to the construction site. Significant research in this area has been done at the Technical University of Munich and throughout Japan and Korea [1]. Moreover, these techniques tie to the broader field of construction automation (CA), which is considered an integral technological advancement to the development of construction 4.0 practices – practices that promise to revolutionize the industry.

Building construction includes many disparate activities, such as site surveying, excavation, bricklaying, concrete casting, tiling, painting. Construction robotics can incorporate the latest technologies, such as sensors, cameras, mapping, planning tools, to allow enhanced control and efficient task execution. CA and robotic systems development are seen to be able to highly contribute to the development of safer construction, with lower cost, better quality, and reduced duration [3]. In regard to autonomous on-site construction, three families of robotics can be recognized: 3D printers, mobile robots, and unmanned aerial vehicles (UAVs). Construction 4.0 incorporates all three of these families; however, each has limitations that must still be overcome by research and development. 3D printers, for instance, are restricted because they need to be on a gigantic scale and are often larger than the structure to be constructed. Mobile robots are restricted in their movement space, which results in construction progress disruption. Finally, UAVs are limited to their payload capacity and the required accuracy imprecision in the elements' placement.

It is beyond the scope of this paper to investigate the current state of the art for all three of these areas. Thus, this research explores how to un-tether an automated process using UAVs for construction activities and assistance in construction activity completion. UAVs are becoming more common worldwide and frequently used for surveying, documentation, communication, and recording site conditions [4]. Each iteration of drone advancement incorporates more sensors allowing for safer flight coupled with software advances that allow for ease of use and autonomy [4]. Furthermore, drone technology has advanced to allow for interchangeable and increasing payloads, supporting several construction applications. This research summarizes activities completed and published to date and develops a conceptual framework for furthering knowledge in this area through additional research.

2. Objectives

Significant research involving UAVs has been identified in the literature in recent years. Some of this research involves using drones for the delivery of the actual construction for a given activity. Where this has been examined, the approaches were tested and validated under controlled environmental conditions and generally focused on the use of single drones. Accuracy limitations were observed and documented in the assembly process in most research activities. This research recognizes these challenges but also recognizes that tools and techniques continue to advance, which will address the precision in future iterations and with the coupling of drones with alternative location systems, such as Vicon and Optitrack - motion capture systems and Real-Time Kinetic (RTK) systems.

This research will build upon the state of the art as identified in the literature to investigate how UAV-based drones and drone swarms can be utilized for construction. Furthermore, the research aims to develop a new construction process by building further promising studies in this field. Specifically, the research will focus on masonry blocks and brick walls, which involve significant task repetition. UAVs facilitate enhanced construction efficiency with lower costs and allow new shapes and materials to be realized with the new way of thought resulting from such technological adoption [5]. The research will address drone technology and UAV development, the construction elements, and the construction process in terms of sequencing. The proposed approach will be evaluated based on the construction quality according to standards specified for masonry construction. Accordingly, tolerances and sequencing are critical facets relating to the elements and the process, which will be addressed in this research.

Furthermore, the research hypothesizes the cooperation of several UAVs or a swarm of UAVs acting in parallel to attest to its presumed increase in overall construction efficiency in terms of cost and time. Experimentation will first be performed in controlled, indoor conditions. Then, research results will be extended to outdoor environments to investigate the role of wind and other environmental factors on the resulting techniques.

3. Literature review

UAVs ability to physically interact with the real world; allows them to contribute to construction directly and not just perform inspection applications. Seminal research in this area was carried out by a joint effort between MIT and UCL [6]. The team explored using structural elements of different shapes and sizes, including conical-shaped bricks developed to build circular columns and rectangular bricks as typically used in current practice. Ultimately, the team developed "Droxels" reimagined masonry units that can overcome mortar use and accommodate complex and intricate architectural designs. However, while MIT and UCL team reimagined the concrete masonry unit, introducing such units in the marketplace has significant barriers to implementation. This includes code compliance and building code accommodation, as well as the development of market forces. For these reasons, this research will be performed using traditional masonry units and shapes currently available in the market. The rest of this section reviews three major studies in this area (Table 1) and highlights significant findings and achievements.

Table 1. Related research methodologies and aspects

Research team	Reference	Structure constructed	Task sequence	Guidance system	Grasping Mechanism	Payload
Kohler and Raffaello d'Andrea	[7]	Polystyrene foam brick tower	-	FMA - motion capture system	Electromagnetic gripper	-
MIT and UCL	[8] [9] [6]	Prototypes using dricks and droxels	CAD to UAV instructions	Image recognition, laser, GPS/RTK, and GPS/automatic theodolite	Specially designed grippers	-
CTU, UPenn, and NYU	[10]	Dropped bricks on a wall	Special algorithm	RGB-D detection of plate, GPS & down-pointing Garmin LiDAR lite v3 sensor	Electromagnetic gripper	1kg

3.1. The flying machine arena platform

The flying machine arena (FMA) platform is a vision-based autonomous system and is considered the benchmark for future innovations and designs in UAV construction applications. In research led by R. D'Andrea, the FMA platform was also demonstrated in 2011 at Orleans, France exhibition (Figure 1), where 4 UAVs were utilized to construct a 6-meter-high composed of 1500 polystyrene foam brick tower. This experiment is considered the first architectural construction assembly with UAVs. Other demonstrations for the FMA platform were carried in 2012 Zurich - Hannover Messe in San Francisco and in 2013 at Edinburgh TED Global events.

The platform is equipped with 19 Vicon T40 cameras to cover the arena flight space capacity of almost 720 m^3. The reported system setup time for the platform is between four and six hours [7]. The designed system can record measurements at a rate of at least 300 Hz (200 Hz typically required). Measurements inside the FMA are taken with respect to a user-defined coordinate frame. The core system components include global sensing (motion capture system), on-board sensors, off-board control/estimation modules, wireless command interface, state machine, and actuator. The object's positions (x, y, z) are tracked with the help of the cameras imagining sensor/optics capabilities along with near-infrared strobes and 3–4cm retro-reflective markers. Moreover, a calibration algorithm is utilized to take account of the cameras and other benchmark positions. The data is then reused by the camera simulation tool to calculate required adjustments from the current motion capture coordinate and accurately mirror the arena.

It is typically mandated to simulate the exact flight plan multiple times; however, it is challenging to realize this in an asynchronous simulation [7]. The simulation was performed mathematically with nominal equations of motion, using input from the on-board control/estimation and state machine. Parameters that can alter the UAV simulation's flight control include disturbances (wind), sensor inaccuracy, and simulation speed. To consider the average deviance in control of the UAV flight, simulation calculations are performed on the expected flight state.

Figure 1. Orleans, France Exhibition 2011 - Polystyrene foam brick tower [7]

3.2. MIT & UCL collaboration

In aims to present an alternative to additive manufacturing building scale limitation. A research collaboration between MIT and UCL; investigated the feasibility of constructing real building scale structures using UAVs. Traditional masonry wall construction requires accuracy in the brick assembly of less than 1 cm [8]. Therefore, the team explored several guidance systems: image recognition and color tags, laser system mounted on the UAV, GPS coupled with RTK, and GPS coupled with an automatic theodolite instrument (Figure 2). Nevertheless, to overcome UAVs' flight placement imprecision, the study took a different direction by modifying the building's structural components to better fit UAVs' construction limitations. The research team came up with four families of dricks (a portmanteau of "drone" and "brick") and droxels (a portmanteau of "drone" and "voxel"); geometrically modified bricks with defined grooves to enable easy sliding and alignment. Furthermore, they enable the design of complex 3D structures such as curved walls and cantilever. Both bricks advancements, i.e., designed indentations, guide the bricks to slide in place and allow a placement position inaccuracy of almost 5cm.

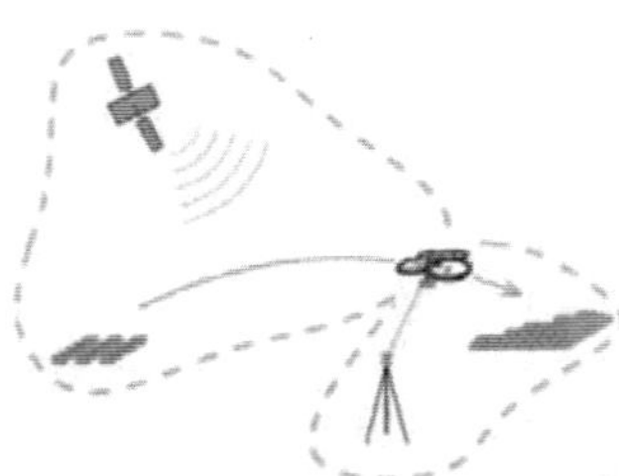

Figure 2. GPS and automatic theodolite navigation system [8]

Additionally, the team demonstrated CAD benefits and its feasibility in being utilized to formulate UAV flight instructions. The CAD drawing is transformed into a ".dxf file" extension compatible with the C++ subroutine, which translates each brick on the line into x-y coordinates. A ".txt file" is then accordingly generated, holding the UAV flight instructions between the loading bay and the required position of the brick's placement (Figure 3).

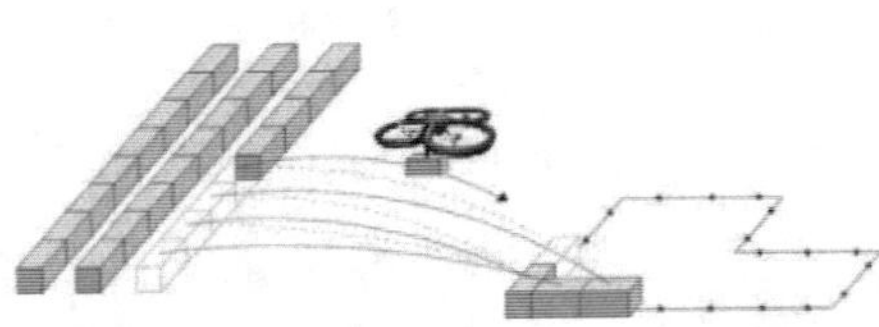

Figure 3. Path planning between the loading bay and planned wall [8]

Further renovations were presented by the research group a year later [9]. The authors confirmed the feasibility of constructing a real scale structure, i.e., concrete precast column using UAVs (Figure 4). A large-scale drone testing was performed using a custom-built UAV with a 40kg payload capacity. The study implemented manual piloting of the UAVs as a proof of concept pertained by the allowed position inaccuracy due to the introduced bricks. The column was constructed by stacking 15-20kg cylindrical bricks (with a metallic plate on top) designed with an allowable position inaccuracy of 6.5cm. The UAV used in this experiment was equipped with an electromagnetic guidance cone for enhanced grip and placement mechanism.

Figure 4. UCL drone assembled precast concrete column [6]

Among the introduced developments are new brick designs, i.e., Drick60. Nevertheless, the introduced bricks still had drawbacks, including required complex formwork and weakness of the brick's thin sides. Similarly, other dricks introduced had other complex issues, i.e., requiring different types and sizes of bricks based on their application (walls and corners). Correspondingly, to ease the complexity of the several dricks introduced, the authors designed dronicks software that automatically portrays the required dricks types at each location of the wall structure.

3.3. CTU, UPenn, and NYU collaboration

The discussed system is a joint research between CTU, UPenn, and NYU, designed for Mohamed Bin Zayed International Robotics Challenge (MZIRC) 2020. The challenge consisted of picking and placing bricks by a UAV team on a highlighted wall pattern. The presented system was the best performer in the competition, with correctly dropping ten bricks. The designed system can scan an area to locate the loading bay and placement location, precisely grasp the bricks, and drop them in position. Accordingly, one of the UAVs in the swarm is responsible for site scanning and creating a topological map (highlighting the brick stack and the wall planned to be constructed). Subsequently, the map is shared with the rest of the UAV swarm. Three UAVs were employed for the joint wall construction; however, only two UAVs were allowed to fly simultaneously to mitigate collisions. Additionally, predicted trajectories were utilized to ensure the UAVs do not cross pass within their planned routes.

Among the research challenges was establishing proper navigation to effectively grasp the bricks from the stack and precisely place them on the wall structure. The study utilizes visual surveying techniques instead of a motion capture system. The experiment guidance system combines Red-Green-Blue-Depth (RGB-D) cameras, GPS, and Light Detection and Ranging (LiDAR) sensors. Robust and prompt brick detection is fundamental for effective grasping [10]. The bricks need to be grasped from their center of mass; to prevent the additional torque from the additional mass and allow for stable flight control and accurate placement. RGB-D cameras can precisely allocate the brick position and the constructed wall while hovering over the site area. Moreover, the bricks are amended with ferromagnetic plates on their top; to allow for better grip using UAV. Nonetheless, it should be noted that the established accuracy for brick placement is not appropriate to constructing a real-wall structure. Blocks were dropped onto any position on the laid-out wall structure and used no glue or mortar to secure the brick in place. Another challenge with the process is that the grasped bricks slightly impact the wall detection and placement process where the brick covers the sensors during its travel.

4. Planned research methodology

The research will investigate possible assembly techniques in laying out the bricks and establishing rules and scenarios to guide the UAVs through their flight assembly (Figure 5); to allow for a better efficient construction process. A scaled-down experiment "small-scale testing" of the proposed methodology will be initially applied on Styrofoam-shaped blocks conducted under a controlled environment. After that, testing will be carried for the outdoor area to ensure system efficiency in constructing an actual wall. The initial stage of testing aims to determine the appropriate guidance system and establish the path planning algorithms that the UAVs will use. Furthermore, the authors will investigate the coupling and use of traditional cementitious mortars and specialized polymer adhesive to ensure completed construction within the required load-carrying capacity and code constraints. Lastly, swarm drone applications will be explored to establish the work distribution for each drone and discover methods to prevent clash detection, i.e., simulating real-time coordinates of each UAV.

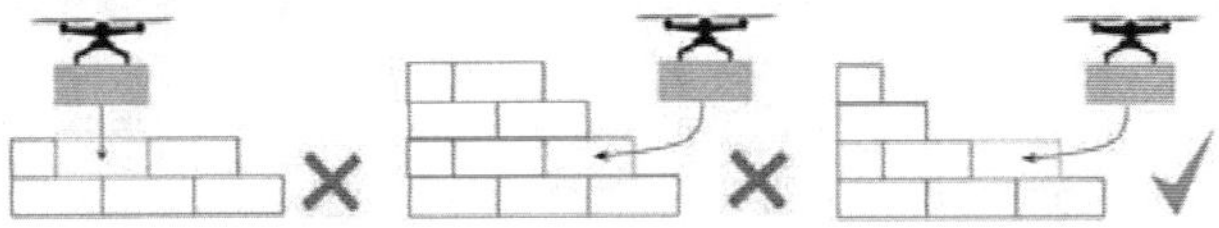

Figure 5. UAV rules for laying out bricks [6]

4.1. *UAV construction process*

Current technology can transform the bricklaying process and wall building to a completely autonomous approach. First, the process is initiated by scanning the site area and identifying the coordinates for the loading bay and the planned wall structure. Next, each UAV is assigned a brick to grasp from the stack, and an algorithm creates a plan for each UAV routes to place the brick in place. The process is repeated until the wall is constructed as per plan or until the UAV battery is depleted. Intermediate steps include hovering above the loading bay and the planned wall structure to locate the bricks' center while loading and placing.

4.2. *Guidance control*

This study's planned UAV guidance control is a motion capture system, i.e., Optitrack cameras. The system will utilize reflective markers to compute the UAVs position 100-200times/sec, allowing precise navigation within the system's frame sight. Furthermore, on-board sensors and GPS could also enhance UAV guidance and act as a backup when the reflective markers fall out of sight. The projected placement accuracy from this system is in order of 3-5mm.

4.3. *Grasping mechanism and path planning*

Designing an efficient gripper is highly important in the proposed application's success. The gripper directly relates to the successful grasping of the brick's center of gravity and placement accuracy. The most promising grasping mechanisms discovered in the literature include electromagnet couplers, special transportation grips, and micro-spine grippers. Each will be considered and evaluated with performed experimentation. As for the path planning, Dijkstra's algorithm will be deployed to find the optimum route, i.e., the shortest path between two coordinates (loading bay & planned wall) while avoiding known obstacles.

4.4. *Modeling and simulation*

Robotics are designed to enable physical world interaction using lines of coding to perform a specific task. The coding varies based on the required task that needs to be executed by the robot. Therefore, the modeling process is vigorous, and effort is required to develop a code that works with the robot size and components and executes tasks depending on the predefined work settings. Construction simulation is a process of replicating reality to extract specific parameters to help refine construction works efficiency. Simulation models seek to refine the defined workflow process by recognizing and eliminating activities that do not add value to the process [11]. Although the simulation process might add more time to the project planning phase, the payoffs are considered rather rewarding with the reduced cost of making unnecessary decisions and their corresponding time. Furthermore, simulation can help identify activities' duration and the optimum resources needed for a particular construction task. These results can be integrated into construction planning and management to maximize efficiency and reflect better project schedules and budgets.

4.5. UAV swarm coordination
A UAV swarm are a set of aerial robots that collaborate to establish a certain user-defined goal. Swarm utilization enables completing tasks in parallel and therefore results in a more productive, efficient process. Other benefits to multi-UAV systems are that a single drone out of control will not stop the entire process since; other UAVs will operate ordinarily. Designing a good communication architecture will enable efficient coordination between the UAVs in completing the assigned task, help prevent clashes in UAV flights, and classify UAVs' work distribution. However, the number of control computations required to navigate the UAV does not scale with more deployed UAVs. The form of communication deployed between the UAVs and the control center is determined based on the number of UAVs deployed and the required coverage area. Two main communication approaches were discovered in the literature: centralized and decentralized communication. A centralized communication approach is a system designed with a central station/node "infrastructure" as the only communicator providing control commands to every UAV in the swarm. On the other hand, the decentralized communication is a more complex approach recommended for huge UAV swarms, where the dependence on one central station is eliminated with real-time interactive communication [12]. Both forms of communication are planned to be explored to determine which approach fits the proposed process.

5. Conclusion
The research explores construction applications that take advantage of UAV capabilities in forming an automated construction environment with increased construction value and reduced risks. The anticipated primary benefit that we could obtain from the research is automating the brick construction process and achieving higher productivity and associated cost savings. Significant background work has been performed by researchers from MIT, UCL, ETHZ, Penn State, NYU, and elsewhere with notable efforts summarized herein. These efforts demonstrated the progress that has been made towards the automation of construction activities using Unmanned Aerial Vehicles (UAVs, i.e., drones) and drone swarms. A methodology for using drones to construct masonry walls was outlined, with salient elements of the approach discussed. The approach utilized construction quality as the metric for success, with swarm elements added for productivity and cost elements. The research effort seeks to advance research knowledge towards the ultimate successful implementation of drones for masonry construction on commercial construction sites.

References

[1] A.-M. Văduva-Şahhanoğlu, C. M. Mădălina Xenia and S. Smid, “Automated and Robotic Construction-a Solution for the Social Challenges of the Construction Sector,” February 2019.

[2] E. Dustin, “25 Construction Safety Statistics and Trends for 2020,” 9 March 2020. [Online]. Available: https://www.bigrentz.com/blog/construction-safety-statistics.

[3] P. Mi, L. Thomas, P. Wei, C. Huimin and B. Thomas, “Structuring the context for construction robot development through integrated scenario approach,” Automation in Construction, Vols. 114, doi: 10.1016/j.autcon.2020.103174, June 2020.

[4] E. P. Small, “Opportunities for UAV’s in Construction Planning, Performance and Contract Close-Out,” in Creative Construction Conference 2018, Ljubljana, 2018.

[5] S. Eloy and N. Pereira Da Silva, “Will drones have a role in building construction?,” in Proceedings of the 4th International Symposium Formal Methods in Architecture, 2018.

[6] S. Goessens, C. Mueller and P. Latteu, “Feasibility study for drone-based masonry construction of real-scale structures,” Automation in Construction, vol. 94, pp. 458-480. doi:10.1016/j.autcon.2018.06.015, 2018.

[7] S. Lupashin, M. Hehn, M. W. Mueller, A. P. Schoellig, M. Sherback and R. D'Andrea, "A platform for aerial robotics research and demonstration: The Flying Machine Arena," Mechatronics, pp. 41-54. doi:10.1016/j.mechatronics.2013.11.006, 2014.

[8] P. Latteur, S. Goessens, J.-S. Breton, J. Leplat, Z. Ma and C. Mueller, "Drone-Based Additive Manufacturing of Architectural Structures," in Proceedings of the International Association for Shell and Spatial Structures (IASS) Symposium, Amsterdam, 2015.

[9] P. Latteur, S. Goessens and C. Mueller, "Masonry construction with drones," in Proceedings of the IASS Annual Symposium, 2016.

[10] T. Baca, R. Penicka, P. Stepan, M. Petrlik, V. Spurny, D. Hert and M. Saska, "Autonomous Cooperative Wall Building by a Team of Unmanned Aerial Vehicles in the MBZIRC 2020 Competition," Robotics and Autonomous Systems, pp. 1-21, 14 December 2020.

[11] A. G. Cabrera, H. P. Martínez and Y. A. Vargas, "Digital simulation as a tool for transforming the construction industry," in VII Elage, Bogotá, 2016.

[12] X. Chen, J. Tang and S. Lao, "Review of Unmanned Aerial Vehicle Swarm Communication Architectures and Routing Protocols," Applied Sciences, pp. pp. 1-23. doi: 10.3390/app10103661, 2020.

IOP Conf. Series: Materials Science and Engineering 1218 (2022) 012006 doi:10.1088/1757-899X/1218/1/012006

Integrated Prescriptive Maintenance System (PREMSYS) for Construction Equipment Based on Productivity

K Petroutsatou and I Ladopoulos

Transport and Project Management Division, Department of Civil Engineering, School of Engineering Aristotle University of Thessaloniki (AUTh), Thessaloniki, Greece

kpetrout@civil.auth.gr

Abstract. It is now imperative to create smart systems that prevent mechanical damage through timely preventive maintenance, particularly in construction projects with strict time schedules and budgets. Construction equipment is the largest capital investment for construction companies. Proper maintenance is of major importance for efficiency, productivity, minimization of equipment costs, and environmental management. The aim of this study is to propose an integrated smart system that will monitor the condition based on productivity of the equipment and will provide diagnostic data, helping to optimize the production process, achieve timely maintenance and increasing the expected "economic" life of the equipment. At the same time, it will positively provide the concept for sustainable or green construction with the minimization and elimination of harmful effects on the environment and on the human. The system will include sensors, placed on specific construction equipment components, and will collect measurements for their use and condition, through real-time data export. These data will then be sent using wireless networks to a main server. Extraction of performance measurements and machine learning (ML) data processing will determine when the equipment needs predictive maintenance and repair. Proper, timely and prescriptive maintenance of construction equipment will reduce their environmental footprint and any human harmful emissions, while saving energy during their operating phase and optimizing production processes through monitoring "dead" time.

1. Introduction

This study aims to cover the need for timely and accurate prediction for construction equipment maintenance and productivity monitoring, in large and costly projects. In the European Union, it is a common practice to execute most of the public works through co-founded financial projects. Those projects are characterized by tight budgets and strict timetables. Therefore, it is imperative for the project managers to adopt a coherent strategy for reducing the total cost and the project's completion time, to eliminate any financial problems during its execution. Additionally, current construction industry faces an imbalance between the vast amount of data, extracted originally by the equipment, and the lack of appropriate domain structures for storing, filtering, processing and exploiting the above data.

The main goal of this study is to propose an integrated smart system, able to constantly monitor the real time condition and productivity of the equipment. This system will provide real time diagnostic data, contributing to the equipment's profitability optimization, through timely maintenance and increase its economic life expectancy. At the same time, it will contribute to the adaptation of a sustainable or "green" environment.

The successful completion and implementation of the proposed system will provide prescriptive guidance to the equipment owners regarding the most efficient type of maintenance needed, while it will offer constant awareness about the equipment's economic life. Namely, it will recognize if the equipment is still profitable or not, and needs to be replaced. The smart system will be trained by three types of data: a. historical equipment's data, regarding its operability and maintenance, gathered from the equipment owners, b. equipment's standard data, provided by the Original Equipment Manufacturers (OEMs) preinstalled sensors and indicators, and c. data coming from the sensors installed on specific equipment's parts or components, in order to collect additional data from their current use. Those specially designed sensors will collect and analyze those data in real time, into a central server (cloud or standard), transmitted through a chosen type of wireless network. The data transmission will be done in certain forms, allowing much faster and efficient processing.

Fundamental subject of this study is to develop targeted industrial research actions for monitoring, rendering, analyzing and evaluating the most significant factors affecting the equipment's efficiency. At the same time, environmental polluting factors (gas emissions, noise pollution) will be restricted, with beneficial effects also for the equipment's operators (less vibrations, increased safety through constant monitoring and timely treatment of potential failures).

The proposed system includes nine scientific and technological challenges: a. Design, development and installation of a prototype life-cycle management system and its comparative evaluation to demonstrate its proof of concept, b. Comprehensive approach of the equipment's efficiency and behavior, through real-time status recording by specially designed sensors and data transmission through modern wireless networks, c. Classification and ranking of the observed equipment's metrics for each type of equipment, d. Development and installation of specially designed sensors, where their proper operation and adequacy will be scientifically documented, e. Design and development of smart systems able to provide guidance for operational decision-making regarding the equipment's maintenance and strategic decision-making for the holistic approach regarding the equipment's efficiency, behavior and economic life, f. Feed the smart system and its subsystems with real-time collected data, followed by their constant improvement and comparative evaluation through Machine Learning models, able to monitor the equipment's current status, by simultaneously knowing their operational and maintenance history, g. Investigate the commercial utilization of the final product or some of its parts, by developing a comprehensive business plan, in order to quantify its business oriented results, and h. Construction equipment market investigation about the application of fleet management software, in order to integrate the results of this study to their functionality. The last-mentioned investigation has been completed and its results are presented in CCC 2021.

The innovation of this study is based on the development of a coherent, multi-parametric approach, for the optimization of a comprehensive construction equipment management. Predictive maintenance, productivity and operational analysis of construction equipment will be integrated into a smart information system, giving constant real-time results. The use of sensors, with IoT capabilities will provide scalable abilities to the initial system infrastructure with low cost.

2. Literature Review

Marinelli and Lambropoulos (2012) investigated the cost implications that the earth loading and hauling trade-off involves for an earthmoving fleet and introduced a method for the selection of the most expedient loading practice.

Alshibani and Moselhi (2012) presented a newly developed optimization simulation model for fleet selection for earthmoving operations. Global positioning system (GPS) data was used to build and update in near real time the developed model. The model is designed to assist contractors in selecting equipment fleet configurations for earthmoving operations; taking into consideration: (1) uncertainties associated with a set of quantitative variables that represent loading, hauling, and dumping duration, as well as project direct and indirect cost; (2) availability of resources to contractors; (3) project cost and (or) time constraints; (4) project indirect cost; and (5) scope of work.

Marinelli and Lambropoulos (2013) managed to optimize scraper's load-time by applying an algorithmic method that incorporated the golden section search and the bisection algorithm.

Memarzadeh et. al. (2013) presented a computer vision-based algorithm for automated 2D detection of construction workers and equipment from site video streams. Their research proposed semi-automated detection methods for tracking of construction workers and equipment. Unlike other state-of-the-art algorithms in automated resource tracking, their method particularly detected idle resources and did not needed manual or semi-automated initialization of the resource locations in 2D video frames.

Aziz et. al. (2014) developed a model which incorporates the basic concepts of Critical Path Method ''CPM'' with a multi-objective Genetic Algorithm ''GA'' simultaneously. This model suggested a practical support for compound horizontally and vertically mega construction planners who needed to optimize resource utilization in order to minimize project duration and its cost with maximizing its quality simultaneously.

Marinelli et. al. (2014) presented an artificial neural network (ANN) model that predicts earthmoving trucks condition level using simple predictors; the model's performance is compared to the respective predictive accuracy of the statistical method of discriminant analysis (DA). Their data processing identified a particularly strong connection of kilometers travelled and maintenance level with the earthmoving trucks condition level. Moreover, the validation process revealed that the predictive efficiency of the proposed ANN model was very high. Similar findings emerged from the application of DA to the same data set using the same predictors.

Said et. al. (2014) developed a novel telematics-based computational methodology to support two major equipment fleet management tasks: fleet use assessment and equipment health monitoring.

Yip et. al. (2014) presented a comparative study on the applications of general regression neural network (GRNN) models and conventional Box–Jenkins time series models to predict the maintenance cost of construction equipment. Both GRNN and Box–Jenkins time series models described the behaviour and predicted the maintenance costs of different equipment categories and fleets with an acceptable level of accuracy. Forecasting with multivariate GRNN models was improved significantly after incorporating parallel fuel consumption data as an explanatory time series.

Heidari and Marr (2015) employed a portable emission measurement system (PEMS) for real-time measurement of carbon dioxide (CO2), nitrogen oxides (NOx), hydrocarbon, and carbon monoxide (CO) emissions from construction equipment to derive emission rates (mass of pollutant emitted per unit time) and emission factors (mass of pollutant emitted per unit volume of fuel consumed) under real-world operating conditions.

Kim et. al. (2016) presented an on-site safety-assessment system for monitoring struck by accidents with moving entities based on computer vision and fuzzy inference. Next, fuzzy inference was used to assess the proper safety levels of each entity using spatial information. The proposed system provided valuable information regarding worker safety represented as a numerical value.

Naskoudakis and Petroutsatou (2016) conducted a meta-analysis of the latest journal papers dedicated to construction machinery, in order to systematically review research themes related with optimization, maintenance/downtime, productivity, robotics and automation, operator's competence, innovation and environment. In the second part of their review, they provided useful pointers regarding the optimum selection of fleet equipment as a key factor for the success of any construction project.

Vanraj et. al. (2016) presented a state-of-the-art review of various Condition Monitoring methods and signal processing techniques used for the Predictive Maintenance of mechanical systems. Conclusively, vibration analysis has proven to be an important criterion for fault diagnosis in manufacturing processes and maintenance scheduling for various manufacturing equipment.

Azar (2017) presented an automated system to analyse the content of heavy-construction videos, including videos with moving viewpoints, and to index them based on midlevel (i.e., objects) and high-level (i.e., operations) semantic content.

Burgler et. al. (2017) presented a method for estimating the productivity of soil removal by combining two technologies based on computer vision: photogrammetry and video analysis. The automated generation of progress and activity statistics from both measurement methods supported interpretation of the productivity estimates. Comparison to annotated ground truth for the tracking and activity monitoring method highlighted the reliability of the extracted information.

Hamledari et. al. (2017) presented a computer vision-based algorithm that automatically detects the components of an interior partition and infers its current state using 2D digital images. The method's high accuracy rates, its fast performance, and applicability to different contexts such as automated robotic inspection are indicative of its promising performance.

Holt and Edwards (2017) conducted a numeric, theoretical analysis using the Caterpillar® hydraulic excavator productivity model to estimate excavator production, given: first, variance in modifying factors based on derived maximum and minimum values; and second, variance resulting from linear calculations based on excavator operator competence. Their findings dictated that excavator productivity resulting from incremental variance of modifying factors in isolation is shown to be linear except, in the case of bucket payload.

Park et. al. (2017) presented the Stochastic Dozer Productivity Estimation (SDPE) method, which integrates dozer production estimating curves and equipment specification obtained from manufacturers, defines the probability density functions (PDFs) of job condition correction factors, executes the dozer productivity estimating format (DPEF) for the user-defined number of iterations, and estimates the best-fit PDFs of productivity and that of total owning and operating (O&O) cost.

In order to strengthen the supervision of construction workers to avoid accidents, Fang et. al. (2018) proposed the use of a high precision, high speed and widely applicable faster Region Based Convolutional Neural Networks (R-CNN) method to detect construction workers' NHU, with data collected from far-field surveillance videos of different construction sites.

Gwak et. al (2018) presented a computational method called Optimal cut-fill Pairing and Sequencing (OPS), which identified the most economical earth allocation planning (EAP). They identified the optimal cut-fill pairs and their sequence which minimizes the total earthwork cost by hybridizing the mixed integer linear programming (MILP) and evolutionary algorithm.

Kim et. al. (2018) investigated the feasibility of measuring machinery cycle times by using inertial measurement units (IMUs) embedded in a smartphone. To enhance the recognition of the equipment's mixed activities and translate the results into reliable cycle time measurements, a dynamic time warping (DTW) algorithm was applied and the DTW distances of IMU signals were used as additional features in activity classification.

Lee et. al. (2018) developed a GPS only-based fleet telematics system for heavy earthwork equipment which could analyse time log information of each equipment using GPS location data without utilizing any other on-board sensors and CAN bus data. Algorithms to analyse the utilization time of earthwork equipment using GPS location information was designed and their reliability was verified through field tests.

Petroutsatou and Marinelli (2018) published their book about the operational analysis of construction equipment and highlighted the necessity of predictive maintenance and proper equipment management when coming to enhance their productivity.

Kim et. al. (2019) presented a non-intrusive earthmoving productivity analysis method using imaging and simulation, to analyse jobsite contexts and the necessity of a complex multi-camera surveillance system or workers' privacy issues. An algorithm for license plate detection and recognition in an uncontrolled environment was developed to automatically produce the site access log, by leveraging video deinterlacing, a deep convolutional network, and rule-based postprocessing. Based on the site access log, simulation-based productivity analysis was conducted to produce a daily productivity report, which can provide the basis for earthmoving resource planning.

Kisi et. al. (2019) proposed a two-prong strategy for estimating optimal labour productivity by quantifying systematic and operational inefficiencies. In that way, they confirmed the feasibility of applying this approach to complex operations involving multiple workers or sequential and/or parallel tasks and actions. Their study augmented the two-prong strategy's methodology to apply the approach to a complex, multi-worker operation necessitating both sequential and parallel tasks and actions.

Petroutsatou and Giannoulis (2020) analysed the Greek construction equipment market sales and its consumers from 2000 to 2015 and identified the factors that affect a machine's value and the most important equipment characteristics. This analysis determined the equipment to be used to develop the proposed integrated system of this research.

3. Theoretical foundation and concept development

This paper proposes a conceptual model for the holistic integration of four major research components: a. Construction equipment life cycle management system development, b. Development of sub-systems of data information and their integration into a smart system concerning the equipment's life cycle management, c. Development, installation, and pilot operation of sensors, and d. Pilot operation and system evaluation.

3.1. Construction Equipment Life Cycle Management System

3.1.1 Literature review and market research

In this stage, a scientometric analysis will take place, to investigate the scientific and the enterprises interest on the research, with the use of VOSViewer and RapidMiner Studio software. It will ensure the adequacy of the collected data, on factors related to construction equipment's work load. A structured questionnaire has already been disseminated among construction equipment OEMs, owners, practitioners and other experts, to identify the type, frequency and weight of importance for the most common construction equipment's failures and damages, during their operation. Six types of equipment will be used: a. Dozers, b. Graders, c. Front Loaders, d. Backhoe Loaders, e. Excavators, and f. Dump Tracks. The evaluators were asked for each type of failure: a. To report the frequency of each failure that occurs yearly, b. How serious and how risky is each failure, regarding productivity, downtime and safety, and c. The type and number of sensors needed to constantly measure certain indicators, to avoid each failure. The questionnaire results are summarized in Table 1. F stands for Flow, H for Humidity, L for Length, P for Pressure, S for Sound, Tem for Temperature, Ten for Tension and V for Vibration.

Table 1. Most Common Construction Equipment Failures.

S/N	Equipment Type	Failure Type	Failure Frequency (occurrence/ year)	Average interval time between failures (months)	Failure weight (scale 1-3)	Failure Risk (scale 1-3)	Number of Required Sensors	Type of Measurement
1	Crawler Dozer	Filter - Gas Pump	2	6	3	1	1	F
2		Gearbox	1	12	3	3	2	V, S
3		Main Front Blade Hydraulic System x Axis Pistons Oil Leak	3	4	2	2	3	L, P, H
4		Main Front Blade Hydraulic System y Axis Oil Leak	3	4	2	2	3	L, P, H
5		Main Front Blade Ware	3	4	3	1	1	L
6		Crawler Tension Mechanism (spring)	3	4	1	2	1	Ten
7		Fuel Consumption & Working Hours					1+1	OEM
		TOTAL					**13**	
1	Grader	Main Leveling Blade Ware	3	4	3	1	1	L
2		Main Leveling Blade Rotating System Stabilizers Ware	1	12	3	1	1	V
3		Main Leveling Blade Mounting System Hydraulic Lifting Pistons Oil Leak	2	6	2	2	3	L, P, H
4		Gearbox	1	12	3	3	2	V, S
5		Filter - Gas Pump	2	6	3	1	1	F
6		Tire Replacement	0.4	30	3	3	2	P, Tem
7		Fuel Consumption & Working Hours					1+1	OEM
		TOTAL					**12**	

1	**Wheeled Backhoe Loader**	Front Bucket Hydraulic Lifting Pistons Oil Leak	2	6	2	2	3	L, P, H
2		Front bucket hydraulic pistons oil leak	2	6	2	2	3	L, P, H
3		Boom arm hydraulic pistons oil leak	2	6	2	2	3	L, P, H
4		Backhoe hydraulic pistons oil leak	2	6	2	2	3	L, P, H
5		Gearbox	1	12	3	3	2	V, S
6		Filter - Gas Pump	3	4	3	1	1	F
7		Front/rear bucket teeth ware	3	4	1	1	1	L
8		Tire Replacement	0.4	30	2	3	2	P, Tem
9		Fuel Consumption & Working Hours					1+1	OEM
		TOTAL					**20**	
1	**Wheeled or Crawled Excavator**	Bucket Teeth Ware	3	4	1	1	1	L
2		Boom Arm Hydraulic Pistons Oil Leak	2	6	2	2	3	L, P, H
3		Bucket Hydraulic Pistons Oil Leak	2	6	2	2	3	L, P, H
4		Filter - Hydraulic Oil Pump	3	4	3	1	1	F
5		Filter - Gas Pump	3	4	3	1	1	F
6		Tire Replacement	0.4	30	1	2	2	P, Tem
7		Crawler Tension Mechanism (spring)	3	4	1	2	1	Ten
8		Fuel Consumption & Working Hours					1+1	OEM
		TOTAL					**14**	
1	**Wheeled or Crawled Loader**	Gearbox	1	1 year	3	3	2	V, S
2		Bucket arm hydraulic lifting pistons oil leak	3	4 months	2	2	3	L, P, H
3		Bucket hydraulic pistons oil leak	3	4 months	2	2	3	L, P, H
4		Bucket Teeth Ware	3	4 months	1	1	1	L
5		Filter - Gas Pump	3	4 months	3	1	1	F
6		Tire Replacement	0.4	2,5 years	2	3	2	P, Tem
7		Fuel Consumption & Working Hours					1+1	OEM
		TOTAL					**14**	
1	**Dump Truck**	Gearbox	1	1 year			2	V, S
2		Dumper lifting pistons oil leak	3	4 months			3	L, P, H
3		Wheel Suspension piston leak/ware	0.5	2 years			3	L, P, H
4		Wheel Suspension Spring tension ware	0.5	2 years			1	Ten
5		Filter - Gas Pump	3	4 months			1	F
6		Filter - Hydraulic Oil Pump	3	4 months			1	F
7		Tire Replacement	0.4	2,5 years			2	P, Tem
8		Fuel Consumption & Working Hours					1+1	OEM
		TOTAL					**13**	

3.1.2 Data Collection, Evaluation and Life Cycle Management Database Development

In this stage, historical operating and maintenance data is collected and evaluated. The data concerns each equipment's operating and maintenance logbooks; these records are being kept by the equipment owners or users. For the purpose of this research, those records will be categorized, classified and verified accordingly, in order to form a homogeneous database. Specialized metrics will be extracted, regarding operational equipment's trends, equipment operators' behaviour, failure frequencies, equipment's downtime, maintenance strategy and policy application, etc.

3.1.3 Collection and Evaluation of the Sensor's Data

The measurements being derived from the sensors, installed on specific equipment's components and parts, will supplement the above "historical" database, to test the existing machine learning model and to lead to more accurate results.

3.2. Informational Sub-System Development - Integration to a Smart Equipment's Life Cycle Management System

3.2.1. Mathematical Modelling for Predicting the Equipment's "Economic Life" and Decision-Making Modelling

At this stage, an artificial intelligence integrated system will be developed for predicting the equipment's "economic life", by providing instructions for a new form of guided maintenance and decision making (prescriptive maintenance). This smart system will be the integration result of two sub-systems. The first one will accept the historical data information, while the second will analyse the data coming from the specifically installed sensors on certain equipment's parts or components, which are recording in real time the equipment's condition and efficiency on the field.

3.2.2. Development of the Software's Final Version
After the completion of the pilot equipment's operation and the system's evaluation, the final version of the management software will follow.

3.3. Construction, Installation, and Pilot Operation of Sensors

3.3.1. Sensors Construction
Those specially constructed sensors will utilize the most developed communication networks, to transfer the collected raw data into the main database. Those networks will allow the sensors to communicate with the project's base stations under severe electromagnetic conditions, such as being next to the equipment's motor engine, inside underground constructions or in distant and inaccessible working areas. The sensors will include three main components: a. The sensing elements, integrated into the electronic circuit, to transform the physical measurements into manageable electronic data, b. A special extremely low-energy microcontroller, which will receive the data from the main processing unit in real time, and c. The communication component, able to transmit great amount of data.

3.3.2. Installation and Pilot Operation
The sensors will be placed on specific equipment's parts or components. During this pilot phase the sensors will be charged directly from the equipment's battery. The aim is to record the necessary physical measurements (e.g. temperature, vibrations, pressures, fluid flow), which will constitute a rich database, sufficient to feed the smart system and train it.

3.4. Pilot Operation and System Evaluation

3.4.1. Pilot Operation Design
The selection of the main operating elements of the system, which will constitute the pilot's system testing elements, is crucial for the overall design of the evaluation process. This process also includes the operational design for the sensors installation and the system's implementation, to be able to operate in real-time. In addition, expedient interfaces will be developed, to monitor the pilot system operability.

3.4.2. Pilot Operation and System Evaluation
Finally, a pilot operation of the final smart integrated system will be performed. The sensors will receive their final set up and they will be tested, evaluated and their final structure will be decided. A comparative performance evaluation will be performed, by comparing the results from the equipment's field testing with the experimental results.

4. Expected Outcomes
This study proposes clear research and operational goals, the fulfillment of which will lead to the following achievements: a. Database development for six types of construction equipment (For the Greek market, the analysis made by Petroutsatou and Giannoulis (2020), identifies that the most used and popular equipment are Crawler/Wheeled Excavator, Wheel Loader and Wheeled Backhoe Excavator), by combining the historical operation, maintenance and failure data with the data coming from the OEM's pre-installed sensors, b. The construction of specially designed sensors, used for the constant measurement of several values, i.e. temperatures, vibrations, pressures, fluid flow etc., which will be able to utilize the most recently developed communication networks, to feed the database with data, and c. The development of a smart system, based on machine learning techniques, able to manage

the processed data and propose solutions for the equipment's maintenance and its further use (Explainable Artificial Intelligence). Figure 1 depicts the above approach.

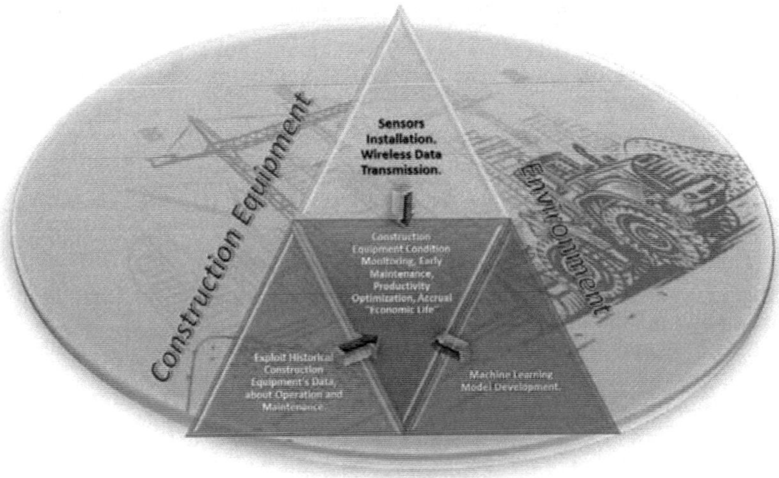

Figure 1. Research's Concept Development

5. Conclusion

This study contributes to the current body of knowledge in construction equipment and management by demonstrating a conceptual model for the development of an integrated smart system for monitoring the real-time status and productivity of six construction equipment types. It will contribute to the real-time collection of diagnostic data, regarding the type of maintenance needed, and it will decide if the equipment is still profitable or not, or it needs to be replaced. The achievement of the above goals will determine the final creation of the decision-making tool, which will allow, on the one hand timely prognosis and failures repairs, while on the other hand it will decide whether the equipment's operation and ownership is still financially efficient. Additionally, by utilizing Machine Learning methods, the proposed model will be able to demonstrate accurate predictions on several equipment's productivity indicators. Determining the feasibility of this methodology in a complex environment is the first step towards building a future tool that will apply to field practitioners, construction companies, repair companies, dealers and OEMs.

References

[1] Alshibani A and Moselhi O 2012 Fleet selection for earthmoving projects using optimization based simulation *Canadian Journal of Civil Engineering* 39 619-630
[2] Aziz R.F, Hafez S.M and Abuel-Magd Y.R 2014 Smart optimization for mega construction projects using artificial intelligence *Alexandria Engineering Journal* 53 591–606
[3] Bügler M, Borrmann A, Ogunmakin G, Vela P and Teizer J 2017 Fusion of Photogrammetry and Video Analysis for Productivity Assessment of Earthwork Processes *Computer-Aided Civil and Infrastructure Engineering* 32 107-123
[4] Fang Q, Li H, Luo X, Ding L, Luo H., Rose T.M and An W 2018 Detecting non-hardhat-use by a deep learning method from far-field surveillance videos *Automation in Construction* 85 1-9
[5] Gwak H.-S, Seo J and Lee D.-E 2018 Optimal cut-fill pairing and sequencing method in earthwork operation *Automation in Construction* 87 60-73
[6] Hamledari H, McCabe B and Davari S 2017 Automated computer vision-based detection of components of under-construction indoor partitions *Automation in Construction* 74 78-94
[7] Heidari B and Marr L.C 2015 Real time emissions from construction equipment compared with model predictions *Journal of the Air & Waste Management Association* 65 115-125
[8] Holt G and Edwards D 2015 Analysis of interrelationships among excavator productivity

modifying factors *International Journal of Productivity and Performance Management* 64 853-869

[9] Hongjo K, Youngjib H, Wontae K, Somin P and Hyoungkwan K 2019 Vision-based Nonintrusive Context Documentation for Earthmoving Productivity Simulation *Automation in Construction* 102 135-147

[10] Kim H, Ahn C.R, Engelhaupt D and Lee S 2018 Application of dynamic time warping to the recognition of mixed equipment activities in cycle time measurement *Automation in Construction* 87 225-234

[11] Kim H, Kim K and Kim H 2016 Vision-Based Object-Centric Safety Assessment Using Fuzzy Inference: Monitoring Struck-By Accidents with Moving Objects *Journal of Computing in Civil Engineering* 30

[12] Kisi K.P, Mani N, Rojas E.M and Foster E.T 2018 Estimation of Optimal Productivity in Labor Intensive Construction Operations: Advanced Study *Journal of Construction Engineering and Management* 144

[13] Lee S.S, Park S and Seo J 2018 Utilization analysis methodology for fleet telematics of heavy earthwork equipment *Automation in Construction* 92 59-67

[14] Marinelli M and Lambropoulos S 2012 Earth Loading and Hauling Optimal Trade-Off *Procedia – Social and Behavioral Sciences* 48 2325-2335

[15] Marinelli M and Lambropoulos S 2013 An algorithmic method for scraper load time optimization *Journal of Construction Engineering and Management* 139 459 - 465

[16] Memarzadeh M, Golparvar-Fard M and Niebles J.C 2013 Automated 2D detection of construction equipment and workers from site video streams using histograms of oriented gradients and colors *Automation in Construction* 32 24-37

[17] Park Y.J, Gwak H.S, Kim B.S, Lee D.-E 2016 Stochastic dozer productivity estimation method, *KSCE Journal of Civil Engineering* 21 1573–1580

[18] Petroutsatou K, Marinelli M and Lambropoulos S 2012 Earthmoving trucks condition level prediction using neural networks *Journal of Quality in Maintenance Engineering* 20 182-192

[19] Petroutsatou K and Naskoudakis I 2016 A thematic review of main researches on construction equipment over the recent years *Procedia Engineering* 164 206-213

[20] Petroutsatou K and Naskoudakis I 2016 A thematic review of main research on construction equipment over the recent years – Part II *Periodica Polytechnica Architecture* 47 110-118

[21] Petroutsatou K and Marinelli M 2018 *Construction equipment, operational analysis and cost of civil engineering works. 2th ed.* (Athens: Kritiki)

[22] Petroutsatou K and Giannoulis P 2020 Analysis of Construction Machinery Market and Construction Output: The case of Greece *International Journal of Construction Management,* https://doi.org/10.1080/15623599.2020.1741491

[23] Rezazadeh Azar E 2017 Semantic Annotation of Videos from Equipment-Intensive Construction Operations by Shot Recognition and Probabilistic Reasoning *Journal of Computing in Civil Engineering* 31

[24] Said H, Nicoletti T and Perez-Hernandez P 2014 Utilizing Telematics Data to Support Effective Equipment Fleet-Management Decisions: Utilization Rate and Hazard Functions *Journal of Computing in Civil Engineering* 30

[25] Vanraj, Goyal D, Saini A, Dhami S.S and Pabla B.S 2016 Intelligent Predictive Maintenance of Dynamic Systems using Condition Monitoring and Signal Processing Techniques - A Review, *International Conference on Advances in Computing, Communication, & Automation (ICACCA) (Spring)* 1-6

[26] Yip H, Fan H and Chiang Y 2014 Predicting the maintenance cost of construction equipment: Comparison between general regression neural network and Box–Jenkins time series models *Automation in Construction* 38 30-38,

IOP Conf. Series: Materials Science and Engineering 1218 (2022) 012007 doi:10.1088/1757-899X/1218/1/012007

IoT System for Maintenance Monitoring of Historic Buildings – Smart Monitoring

I Vidovszky[1] and F Pintér[2]
[1] Department of Construction Technology and Management, Budapest University of Technology and Economics, Budapest, Hungary
[2] Scientific Laboratory, Institute for Conservation, University of Applied Arts Vienna, Vienna, Austria

vidovszky.istvan@epk.bme.hu

Abstract. Maintenance monitoring is an essential task in the field of preservation of historic buildings. The continuously developing IoT-based systems for monitoring already exist in many areas, like the maintenance of complicated mechanical systems or large industrial facilities, but there are only a few examples of the application of such methods in the field of cultural heritage preservation. In this paper, the possible elements and system of an automatized maintenance monitoring network are tested by analyzing the possible options of the survey units and the system to be established in cases of altering circumstances of the investigated site, highlighting the factors which have to be considered in the case of developing a network on-site. Finally, typical application cases were distinguished, and configuration options were suggested for the automated maintenance monitoring systems.

1. Introduction

The monitoring of historic buildings is an essential task in the case of valuable but vulnerable historic building stock. Proper regular monitoring enables immediate reaction for the intervention needs, which results in less deterioration and loss of historic fabric (Forster-Kayan, 2009). The development of diagnostic methods for maintenance purposes enables more effective monitoring (Vidovszky, 2016).

There are already several examples for the application of IoTs in maintenance monitoring, as the monitoring of complex mechanical systems (Civerchia et al. 2017) or larger office buildings (Ko, 2009), but the time is seemingly arrived to use this technology in the field of the maintenance of historic buildings and sites too (Campos, 2009; Makadam, 2015; Perles 2018 et.al.; Bruno et al. 2018).

The usual monitoring approach needs a regular human presence, basically every single space of the historic sites for inspections. In contrast to that, in the case of applying IoT systems for maintenance purposes, several aspects can be monitored by IoT modules functioning as automated survey stations (Vidovszky-Pinér, 2020).

Automated IoT-based systems can support not just the maintenance activity, but even the control can be made more effective, resulting in a continuous survey and control on the most vulnerable or most problematic parts of the buildings. The advanced partially automatized monitoring has the advantage of the fastest possible reaction for the necessary interventions, which can result in even less need for change or loss of historic fabric.

In the case of automated maintenance monitoring, automatic survey stations and/or data loggers have to be established and organized to network in the investigated site. Depending on the location, the responsible organization, and the local circumstances, the system can be managed with local or remote control.

In this paper, the options of an automated maintenance monitoring system and the initial experiences with its possible configuration elements are analyzed.

2. Methods

Among the possible tools (survey stations), we can distinguish fixed and mobile units. Regarding the system's need, there are different ways of remote control of the survey stations. The system practically consists of IoT-based survey stations or data loggers, and in the case of larger systems, a central unit (i.e., data server). According to the circumstances, the server can be placed on the site or basically to any other location with a proper internet connection. The buildup of the system depends on many factors, like the site characteristics or the working method of the staff responsible for the monitoring.

2.1. Fixed units

Fixed survey stations are running surveys on a specific location and are equipped with sensors and/or cameras applied to measure or record environmental parameters. The sensors are applicable for measuring various factors, e.g., wind speed, climatic conditions, or the presence of different gases. The cameras can visually inspect the changes of various phenomena on the surface, like moisture, salt effloresces, crack movements, etc. There are many fields IoT based survey can be applied for on a heritage site:

- monitoring external weather conditions (temperature, rain, wind, air humidity, air pressure)
- monitoring of interior climatic conditions (temperature, air humidity, air quality)
- statical state measurements (measuring crack changes)
- monitoring maintenance needs (like gutter control)
- monitoring biological pests or fungus (mold control)
- monitoring deterioration state (moisture, salt, mold, stone detachment)
- seismic resonance measurement
- heat loss surveys
- etc.

For our examination, we applied two fixed survey stations. The first one was the unit prepared for our previous research (Vidovszky-Pinter 2020), a Raspberry Pi Zero W single-board computer, which was equipped with a Raspberry Pi V 2.1 camera module, a Grove Base Hat for the connection of the peripherals, a relay for the switching of the external lamp, and an SHT31 Grove I^2C temperature and humidity sensor (*Fig 1a*).

In the case of the second unit, we made some modifications. We chose a stronger Raspberry Pi 3A+ single-board computer equipped with a Raspberry Pi V 2.1 night camera module, a relay, and a DHT22 temperature and humidity sensor for measuring temperature and relative air humidity (*Fig 1b*).

2.2. Mobile units

Mobile units are similar to fixed ones but are also equipped with sensors and end-effectors, which making them able to change their location. Mobile units are applicable if the survey has to be done repeatedly in more than one location, not very far from each other in a site.

Mobile units can be programmed for certain autonomous works, but they can be applied by remote control too.

For our experiments, a Raspberry Pi 3B+ single-board computer based mobile unit was applied, which was equipped with a Raspberry Pi V 2.1 noir camera module and a Grove Pi to connect the

IOP Conf. Series: Materials Science and Engineering 1218 (2022) 012007 doi:10.1088/1757-899X/1218/1/012007

peripherals, like sensors, LCD, and motor driver (*Fig 1c*). The unit contains environmental sensors like DHT11 temperature and humidity sensor, moisture sensor, and light sensor. For safe movement, IR sensors were applied to avoid fell down in stairsteps, and an ultrasonic sensor enables obstacle avoidance.

2.3. Systems and server

In maintenance monitoring, the system records environmental and visual data that supports the monitoring of historical buildings for preservation purposes. The goal of this research was to test the options of the possible survey system network. We tested three units and two different scenarios.

The first unit was used for testing a scenario when only single survey stations are located on the site, with direct access for the remote connection.

This claim could be satisfied by establishing a VPN connection between the remote computer and the survey station.

The second unit simulates the situation when the units have an internet connection, but no full access bi-directional connection can be established between the computers on the site and remote computers.

For this situation, we established a server in our lab, with remote VPN access, that allows us to check it every now and then, where all the data is transferred by the on-site survey stations regularly joining to the server as clients.

Figure 1 demonstrates the built survey stations.

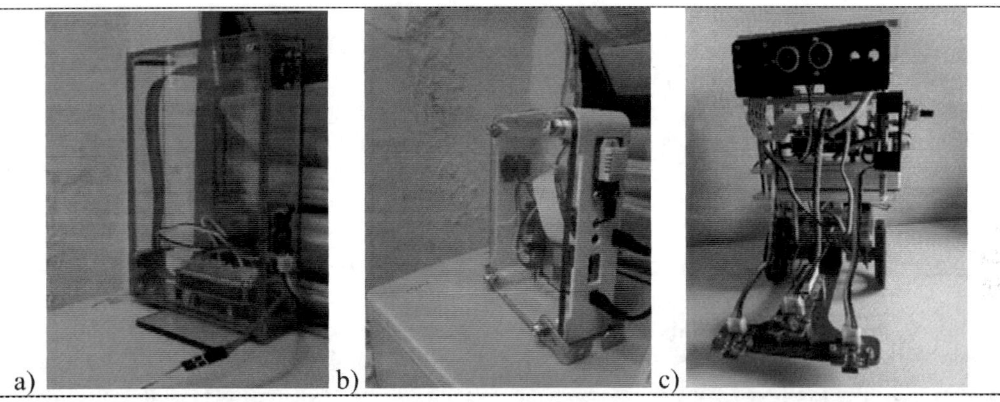

Figure 1. a) 1st state survey unit b) 2nd state survey unit c) mobile state survey unit

3. Discussion

Analyzing our experiments and the possible sites to investigate the possible claims, we defined categories, assuming some survey stations at each site that either worked alone or had to communicate with a server or other stations. We distinguished five cases within three main categories and suggested five configurations for the system (*Figure 2).*

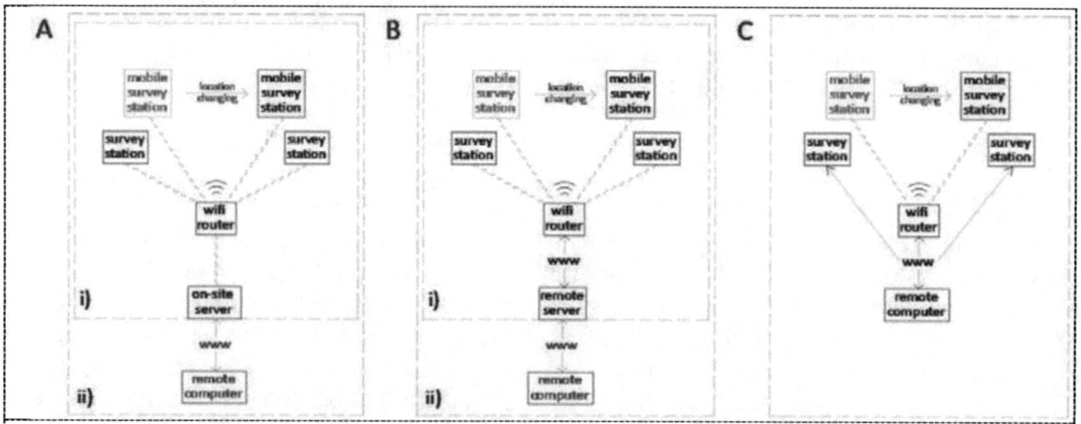

Figure 2. Network schematics a) on-site network b) remote server c) direct network control

3.1. *Experiences with the sample units*

In the case of the first survey station, two 30 days long tests were executed. The remote connection was provided through a wifi router, which needed fix IP address and had to be forwarded to the router. In this way, we established direct access to the unit with a VPN. The advantage of this method is that the unit can be programmed from a remote place too. It means that codes can be changed from any location with an internet connection (by mobile phone or computer), and we had continuous remote access to all the data on the survey station.

In the case of the second survey station, the system was equipped with a server. This system was tested for 30 days. The unit needs an internet connection, but it is insensitive to its type. The advantage of this configuration is that the local units do not need more than any stable wireless connection (local wifi). The disadvantage is that they cannot be programmed remotely. The serving application cannot be changed or updated, only in-situ. In this system, the survey stations have to be self-starting. Namely, the application has to start automatically at boot. There is an option for remote access to the server and the data stored on the server, but there is no remote access to the survey stations on site.

With the mobile unit, we had only preliminary experiments yet. Mobile survey stations might need remote control by an operator to change their places. For testing of such activity, we wrote a web-based application in python (Fig. 3.), which was used at a survey of an attic performed from the ground floor of the building.

In the case of autonomous work of this unit, the navigation has to be supported by various equipment. The application of lidar units or systems supporting artificial intelligence has excellent future potential but would make the survey stations expensive. In case of applying non-expensive elements, as we did in our research, the options are much more limited, like following marked out black lines by IR sensors or using ultrasonic-based obstacle avoidance methods. The options of autonomous work of the mobile unit are under testing at the moment.

3.2. *General considerations*

The system on a specific site has to be developed based on the given circumstances: the needs of the maintenance activities, the amount and character of the survey stations, and the technical factors, like the available internet connection.

The monitoring of a building depends on several factors, like the age, the material, the complexity, or the state of the historic fabric. The goal of the investigation also might give different tasks.

The collected data might be wanted to be used on the site, but in other cases, access might be needed to get the data from various places. For instance, if an international research group with diverse geographic locations is continuously working on the same collected database.

Figure 3. Navigation application for the mobile unit

The internet connection on the site is either a fixed network connection or a mobile network only. In some cases, not even the mobile web is stable enough to establish a proper connection to a remote computer.

According to the above mentioned and based on our investigations, five different situations were distinguished. The configurations applied for the five situations are demonstrated in *Picture 2* and in *Table 1.*

Table 1. The possible survey configurations on the site

configuration	description
A i)	An on-site server is established with a wifi router. The survey stations are communicating with the server and with each other via the local wi-fi-network (*Fig.2 a*)
A ii)	The same configuration as *A i)* but with an established VPN access to the on-site server (*Fig.2 a*)
B i)	A remote server is established. The survey stations are communicating with the server via VPN, and the nods are communicating with the server and with each other via the local wifi network (*Fig.2 b*)
B ii)	The same configuration as *B i)* but with an established external VPN access to the remote server too. (*Fig.2 b*)
C	No server is established, the survey station(s) are either forwarded by a local wifi router or connected directly to the internet, and VPN communications are established directly to survey stations (*Fig2 c*)

For configuration *A i)*, we do not need an established internet connection on the site. This configuration is applicable in an isolated heritage site, with no option for establishing a proper internet connection.

Configuration *A ii)* is recommended if the internet connection on the site enables a VPN connection to the server. It does not just allow the remote connection to the server, but via the wifi router, we can reach the survey stations too (assuming single-board computer-based survey stations with operation systems). The survey stations also can be mobile units.

Configuration *B i)* is a typical solution if there is an internet connection on the site, but no remote access can be established (either the survey stations work with microcontrollers instead of single-board computers, or it is not possible to get the right from the supervisor of the local network to reach the computers of the site from outside). In this case, our tools can connect to the internet via a local wifi network, but we cannot have direct access to them, and there is no option for remote work on our tools. We can still have regular feedback at the remote server, which is not on the site.

Configuration *B ii)* is applicable if we need access to the data collection from any place, then a VPN connection can be established to the server which is not located on the site.

Configuration *C* is applicable if we have a low number of survey stations. We can have access to each of our survey stations directly by static IP addresses. If the units have no LAN connectors or mobile survey units are applied (there is no option to be connected directly), we need the tools to be forwarded by a wifi router.

Table 2 summarises the options in the case of the various systems.

Table 2. The suggested configurations in the various cases

system	suggested number of the survey stations	need for reliable internet connection on the site	need for VPN access	remote access to the data	option for remote controlled mobile units in the system
A i)	2<	no	no	no	no
A ii)	2<	yes	yes	yes	yes
B i)	0<	no	no	no	no
B ii)	2<	yes	only at the remote server	yes	no
C	5>	yes	yes	yes	yes

4. Summary

Systems developed for smart monitoring of historic buildings enable a more accurate, more continuous maintenance monitoring on the heritage sites. This paper discussed the experience with some basic elements and some possible configurations of such a system. According to our observations, the optimal establishment of these systems predominantly depends on the characteristics of the historic site.

5. Possible extension of the research

The system demonstrated here can be tested adequately in a real environment, for which purpose our on-site tests will be continued to extend our experiences on the system and its elements, especially on the mobile survey units.

Acknowledgments
The research reported in this paper was supported by the BME Artificial Intelligence TKP2020 IE grant of NKFIH Hungary (BME IE-MI-SC TKP2020).

References

[1] Bruno S, De Fino M, Fatiguso F 2018 *Historic Building Information Modelling: performance assessment for diagnosis-aided information modeling and management (Automation in Construction Vol 86)* pp 256–276

[2] Campos J 2009 *Development in the application of ICT in condition monitoring and maintenance. (Computers in industry, Vol 60)* pp 1-20.

[3] Civerchia F, Bocchino S, Salvadori C, Rossi E, Maggiani L, Petracca M 2017 *Industrial Internet of Things monitoring solution for advanced predictive maintenance applications. (Journal of Industrial Information Integration Vol 7)* pp 4-12.

[4] Forster A M, Kayan B 2009 *Maintenance for historic buildings: a current perspective. (Structural Survey Vol 27)* pp 210-229

[5] Ko C H 2009 *RFID-based building maintenance system. (Automation in Construction Vol 18)* pp 275-284

[6] Madakam S, Ramaswamy R, Tripathi S 2015 *Internet of Things (IoT): A Literature Review. (Journal of Computer and Communications, Vol 3)* pp 164-173

[7] Perles A, Pérez-Marín E, Mercado R, Segrelles J D, Blanquer I, Zarzo M, Garcia-Diego F J 2018 *An energy-efficient internet of things (IoT) architecture for preventive conservation of cultural heritage (Future Generation Computer Systems Vol 81)* pp 566–581.

[8] Vidovszky I 2016 *Impact-based diagnostic approach for maintenance monitoring of historic buildings. (Procedia Engineering Vol 164)* pp 575-582.

[9] Vidovszky I, Pintér F 2020 *Automated State-Survey System for Monitoring Salt Damages on Plastered Wall Surfaces. Proc. Creative Construction e-Conference 2020 July (Online)* pp 8-13

Leveraging advances in automation to benefit the construction industry in the structural steel sector

S Elhouar and S Terreno

Bradley University, Peoria, IL USA

selhouar@bradley.edu

Abstract. The global structural steel market was reported to be valued at $96 Billion in 2018, and according to some projections, it will amount to about $140 Billion by 2025. This means that the structural steel industry is expected to have a Compound Annual Growth Rate (CAGR) of about 5.4% over this six-year period. Consequently, this sector is set to perform better than the construction industry as a whole, which is projected to have a CAGR of 4.2% over approximately the same period. This projection represents good news for the structural steel industry but it also magnifies some of the challenges that the industry has been facing or has faced in the past. For example, the US Bureau of Labor Statistics projects that overall employment of ironworkers will see an 11% growth from 2018 to 2028, which is much faster than the average growth for all occupations. On the other side of the equation, the industry is facing some challenges that may make it difficult for it to meet the projected demand. For example, there is a very well documented shortage in the construction labor force, and this shortage is already causing problems and driving construction costs upward. Moreover, there is an expectation of increased accountability towards the environment while dealing with generally larger and more complex projects. In this paper, the authors propose that increased automation may be the most effective way to solve the labor shortage problem in the steel construction sector. An overview of advancements in related automation technology is first presented, followed by an analysis of the areas of the industry that may benefit from an increased use or the introduction of automation. The authors conclude by presenting a set of recommendations for adding more automation in the production, servicing, fabrication, and erection of structural steel.

1. Introduction

Construction Automation (CA) has been defined and perceived in different ways, based on the context of use or interpretation [1]. Two main streams of thought exist – stemming from the divide between design and construction processes. Whilst automation may be viewed as automated design and planning processes, the context of construction leans more towards the use of robotic elements on site [2,3,4]. Still another context of automation proposed by Chen et al. [1], following a review of relevant literature and text-mining of the concept of construction automation; summarized the definition as the use of recent technologies to improve construction productivity. More recently, the concept of automation in construction has been expanded to integrate the concept of digitization. This stemmed from the Industry 4.0 concept developed by Oesterreich & Totenberg [5], where information modeling was linked with manufacturing concepts, and it is being applied in construction today. Terms such as "digital fabrication", "digitization", "3D printing", "simulation", "virtual/augmented reality" and "visualization" [1,6,7] are indicative of the fluid integration between computer based modeling and robotic technologies for construction manufacturing, fabrication, assembly and maintenance. Vaha et al. [7] take it further to introduce the concept of control systems and information technologies to apply industrial automation principles to construction.

Despite the potential of automation to improve construction industry site safety, productivity and production, Oesterreich & Totenberg [5] noted that the uptake of new technologies is still slow. Following a qualitative and quantitative content analysis of 268 articles from 10 countries on the subject, they identified a disconnect between research and practical implementation. Numerous technologies were identified, such as Building Information Modeling (BIM), Simulation and Modeling, The Internet of Things (IoT), Radio-Frequency Identification (RFID), Big Data, 3D-Printing/Additive Manufacturing, Smart Factory, Human-Computer-Interaction (HCI), Modularization, Cyber-Physical/Embedded Systems etc. A comparison of literature between conceptual scientific papers and case study applications showed a marked difference in volume, as only a small number of case studies were recorded. They identified the fact that the construction industry lags far behind other industries in digitization and automation.

However, following a renewed surge in digital and technological capabilities, numerous authors maintain optimism that the construction industry can still adopt and benefit greatly from automation. Castro-Lacouture's [8] survey of respondents from 24 countries identified benefits such as safety, enhanced working conditions, increased productivity, quality and reliability, savings in labor and lifecycle costs and standardization of components from the responses given. Bock [3] argued that the increasing potential of robots to work in unstructured environments and increasingly diverse fields, bodes well for construction. His study also noted that microsystem technologies, Single Task Construction Robots (STCR), IoT and smart home technologies have already become a part of construction and building technologies. His paper highlighted the concept of "technology disruption" and the potential of robotics in the construction process and in buildings. Whilst noting that CA is "capital intensive and machine-centered", the benefits appear to far outweigh its cost or initial limitations. Bock [3] analyzed the potentials and application of robotics in different contexts:

- Robot-Oriented Design (ROD) – design and management tools for the deployment of automation and robotics in construction
- Robotic Industrialization – automation and robotic technologies for customized prefabrication of "low-level" components such as steel, wood and concrete; differentiated from "high-level" building components such as larger modules e.g. prefabricated composite units.
- Construction robots – Single-Task Construction Robots (STCRs) and elementary technologies (simple systems that can perform single construction tasks in a repetitive manner).
- Site Automation – automated/robotic on-site factories (controlled, factory-like environments on the construction site).
- Ambient Robotics – technologies for maintenance, assistance and service such as merging robotic technologies with building components and furniture to produce "ambient intelligence" through the use of medical sensors for assisted living. Examples such as the Patient Transfer Assist and Care-O-Bot are in use.

A recent survey conducted by Cai et al. [9] analyzed the opinions of construction industry experts across numerous fields including academia, general contracting, earth and foundation works, steel structure works, mechanical, electrical, electronic, waterproof and thermal insulation, formwork and scaffold subcontracting etc. Their study combined the results of surveys with focus group sessions on two main themes comprising the developmental needs and influential factors of automation in high-rise construction. The top ten issues under each theme from the surveys were analyzed, with interesting results. The top ten developmental priorities resulted in higher proportions for the superstructure and safety issues, as opposed to other categories such as earthwork, finishes and material quality inspections. Their ranking of issues in order of priority are summarized in Table 1:

Table 1. Ranking of issues in order of priority for the automation of high-rise construction.

Rank	Developmental Priorities
1	Steel welding
2	Monitoring of the deformation and internal force of the steel structure
3	Steel coating
4	Safety control of work performed at high elevations
5	Vertical site logistics of materials and equipment
6	Prefabricated component assembly
7	Construction quality inspection
8	Safety control of construction equipment
9	Steel assembly
10	External wall painting

The results of the rankings on influencing factors are shown in Table 2, and these prioritized the level of technological development, uncertainty and support as the factors with the highest influence on deployment and success.

Table 2. Ranking of influencing factors on automation of high-rise construction.

Rank	Influencing Factors
1	Initial investment cost of automation and robotics technologies
2	Maturity and proven technology for robust performance and ease of use
3	Application of other information technologies (e.g. BIM, IoT etc.)
4	Uncertainty of the economic benefits of automation and robotics
4	The ability of on-site management
6	Governmental support for application
7	Governmental support for academic research
8	Social attention to occupational health and safety
9	Age structure and education of the workforce
9	Labor cost of construction

Results from the focus group sessions, which were attended by seven university professors, two senior engineers and two software company specialists from three countries echoed many of the prioritized areas from the survey. The top three influencing factors included the immaturity of technology, uncertainty of economic benefit and the incompatibility of existing construction patterns and robot applications. Thus, the highest development priorities were noted by this group to be prefabricated component assembly, safety control of work performed at high elevations of construction equipment, and steel works. In summary, steel works, vertical site logistics, and safety needs ranked highly amongst the development priorities of the study.

Though numerous authors have studied CA in the overall context of construction, there is a dearth of studies that focus on the potential of CA in the steel industry. This study aims to analyze developments and upcoming potential in the application of CA to steel production, fabrication, and assembly.

2. Automation in the steel production industry

The steel industry has come a long way in terms of automation since improved processes for mass producing steel were developed in the mid 1800's. The steel making industry was very dynamic and saw significant improvements in processes and production over short periods of time towards the end of the 19th century and beginning of the 20th century with production capacity at least doubling every few years [10] with better product quality and lower prices. As an example, Spoerl [11] compared the production of steel between 1867 and 1884. He stated that in 1867, 460,000 tons of wrought iron rails were made and sold for $83 per ton, while 2,550 tons of Bessemer steel rails were made and sold for $170 per ton. By 1884, iron rails were basically no longer available and rail steel was being produced at an annual rate of 1.5 million tons and selling for $32 per ton. But however impressive the progress of that era was, it is dwarfed by today's capabilities, which utilize fully automated computerized processes to produce more than 1.8 billion tons of steel (2018) [12] at a cost of about US$ 750 (4,830 CNY) per ton, or the equivalent of roughly $23.50 per ton in 1900 US dollars. Automation and other improvements in the making of structural steel did not only affect the rate and cost of production of the material, it also made it possible to produce many different grades of structural steel of a superior quality while improving the safety aspect of steel production and the effect on the environment.

3. Automation in the structural steel fabrication industry

The fabrication of structural steel components comprises a number of laborious tasks that include cutting, drilling, bending, coping, and welding of structural steel components to prepare them for erection. The success of the fabrication process is highly dependent on the detailing phase that precedes it. Steel detailers perform extensive geometric calculations and produce accurate drawings of steel components to allow fabricators to prepare these components for straightforward erection and assembly in the field. In the not so distant past, i.e. prior to the 1970's, all the tasks associated with the fabrication of structural steel components, such as layout, cutting, bending, drilling, and hole punching were performed manually. In contrast, nearly every structural steel fabrication process can be automated today. Morrall [13] provides a good summary of the evolution of structural steel fabrication technology, starting with the introduction of beam punch lines in the 1970's.

The 1980's and 1990's saw additional advancements in automation. First, there was automated drilling, coping, and weld preparation, and then it became possible to create details in 3D CAD systems and download the data directly into machines that were able to cut members to length and mark the locations of holes and attached parts. Before too long, automated systems, referred to as MSI for multisystem integration, were developed to handle and move steel pieces through different setups to perform numerous automated tasks including cutting to length, drilling holes, coping, and even welding connected parts, all with limited human intervention. Introducing automation to the fabrication of structural steel components benefited the industry on many fronts including workers safety, productivity, accuracy and quality of final products, economy, and effect on the environment. The structural steel fabrication shop of the future will see even more integration of emerging technology, such as AI, robotics, and additive manufacturing.

4. Automation in the erection of structural steel

On the construction site, steel erection is not only a big-ticket item in terms of price and logistics, it is also a heavy determinant of the schedule – stemming from its criticality to the superstructure. Liang and Kang [14] identified steel beam assembly as heavily reliant on manual labor, and a main part of the critical path. Steel beam erection has been especially identified as labor-intensive and challenging in terms of safety of workers [2, 14]. Several studies have been conducted with the aim of improving the automation levels of steel erection. Cai et al. [15] categorized superstructure erection systems into three groups: automated construction systems, automated lifting systems and robot-based steel assembly.

Automated Building Construction Systems (ABCS) generally stem from the concept of the Super Construction Factory (SCF), introduced by the Obayashi Corporation. This system comprises a steel structure with a roof and surrounding walls that is designed to simulate a factory on the construction site [16]. A climbing mechanism lifts the structure higher from floor to floor and can incorporate crane lifting mechanisms, automated welding and bolting etc. These reduce the dependence on human labor but require strict supervision. The Takenaka Corporation developed the Roof Push up Construction Method, where the top level of a structure is constructed on grade, and then gradually "pushed up" to accommodate each floor as these are assembled [17]. Other systems include the SMART system which is similar to the SCF and the Roof Push up method, as the automated system auto erects, assembles, lifts and joints structural components from floor to floor. Automated steel beam assembly was taken further by Chu et al. [18] and Jung et al. [2, 19, 20]; a team of Korean researchers that proposed the Robotic Beam Assembly System (RBA system). The RBA attached a robotic bolting device to a control system and robotic transport mechanism. The latter version of this allowed for the operator to manipulate the system remotely from outside of the cabin, via joysticks, multiple screens and other control equipment. Their main aim was to increase safety on the site for workers.

A clear limitation of these systems is the limited versatility, weight and costs, in addition to the complexity of set-up and tear down of the systems themselves [15]. Liang & Kang [14] developed an autonomous joint system for steel hauling and automatic beam assembly, intended to reduce the weight and complexity of the SCF system. Their system attached an aligning mechanism to each beam-column joint prior to hauling the beam, and comprised a magnetic clip to help with alignment prior to bolting with a robotic arm. Kang et al. [21] attempted to decrease the impacts of weight and cost by developing the Robotic Crane-based Automatic Construction System (RCA), where the crane was placed at the core of the building, independent of the CF. Kim et al. [22] also developed a cost/benefit analysis process, in addition to several CF alternatives to evaluate the economic viability of automated systems.

5. Structural steel construction-BIM integration

Building Information Modeling (BIM) has been gaining widespread use in building engineering and construction. The advantages of using BIM technology are numerous and have been vastly studied by researchers in the field. However, investigations related to the use of BIM to enhance means and methods of the construction of steel structures is still in its early stages. For example, while a general search for articles dealing with the 'use of BIM in structural steel' on Google Scholar produced more than 21,000 articles in the first week of June 2021, a search at about the same time requiring the title to have the two words 'BIM' and 'steel' only produced 113 hits, and a search requiring the title to have the two words 'BIM' and 'concrete' produced 123 hits. In contrast, a search requiring the title of the article to contain the words 'BIM' and 'design' produced close to 2,000 hits, and a search requiring the title of the article to contain the words 'BIM' and 'construction' produced close to 3,000 hits. This informal search analysis, provides some insights on where the focus of the use of BIM has been over the last decade. Note that a general search looks for articles that have any of the search terms in them.

Some of the investigations related to the design and construction of steel structures include the use of BIM data to achieve a sustainable steel design by estimating labor cost [23] and the use of BIM-based data mining to estimate job man-hour requirements in structural steel fabrication [24]. However, most investigations dealt with design and sustainability related topics, such as the development of a BIM extension for the sustainability appraisal of conceptual steel design [25], using BIM technology for the design of steel bridges [26], [27], BIM based connection design in steel structures [28], the design of steel box girders [29], and others.

It is clear that there are still many opportunities available to study the use of BIM for structural steel fabrication and erection and other phases of a project. As the needs for automation rise due to the continuous shortage in skilled labor, an integrated framework will need to be established to facilitate the management of structural steel construction projects from conceptual design through construction. This will best be done by expanding BIM technology to include information that will go beyond modeling for design and constructability and allow for the automatic production of construction details, automate material procurement activities, automate the process of transferring data from a BIM model to automatic fabrication machines, automate the task of generating construction sequences, incorporate the ability to control cranes and robots to perform various construction tasks from within a BIM platform, and enhance the safety aspects of a structural steel construction project.

6. Conclusions

The production of structural steel and the fabrication and erection of structural steel components and systems traditionally involved labor-intensive processes with considerable concerns for workers safety and continued well-being. The processes are often carried out in harsh environments and may necessitate the handling of heavy pieces of metal at elevated heights. Furthermore, repetitive actions by workers are often required for fabrication and erection of structural steel. This can lead to workers fatigue, which leads to elevated risks of injury and error and decreased quality and productivity.

Automation has been playing an incrementally important role in the manufacturing sectors for centuries but the accomplishments of the last few decades were especially phenomenal. The outstanding advances in computer and other technologies, such as laser and GPS, allowed for the development of automated processes that produce impressive results in terms of safety, quality, and speed of completion of tasks. As it relates to the steel construction industry, automation has been greatly taken advantage of in the steel production sector and incrementally taken advantage of in the structural steel fabrication sector. However, there are many advantageous technologies that are available but the use of which still needs to be leveraged for enhancing the fabrication and erection of structural steel. The authors propose that future progress in structural steel construction will be heavily dependent on advancements in the collection, dissemination, and use of information related to each project. This will need to be done by advancing the use of Building Information Modeling (BIM) to include information that can be used through the various project phases from design to detailing, procurement, fabrication, and then erection.

Acknowledgments

The publication of this article was made possible by the support of the Department of Civil Engineering and Construction at Bradley University in Peoria, Illinois, United States of America.

References

[1] Chen, Q., de Soto, B. G., & Adey, B. T., 2018. Construction automation: Research areas, industry concerns and suggestions for advancement. *Automation in Construction*, **94**, 22-38. https://doi.org/10.1016/j.autcon.2018.05.028.

[2] Jung, K., Chu, B., and Hong, D., 2013. Robot-based construction automation: An application to steel beam assembly (Part II). *Automation in Construction*, **32**, 62–79. https://doi.org/10.1016/j.autcon.2012.12.011.

[3] Bock, T., 2015. "The future of construction automation: Technological disruption and the upcoming ubiquity of robotics." Autom. Constr. **59** (Nov): 113–121. https://doi.org/10.1016/j.autcon.2015.07.022.

[4] Keating, S., & Oxman, N. 2013. Compound fabrication: A multi-functional robotic platform for digital design and fabrication. Robotics and Computer-Integrated Manufacturing, **29** (6), 439-448.

[5] Oesterreich, T. D., & Teuteberg, F., 2016. Understanding the implications of digitisation and automation in the context of Industry 4.0: A triangulation approach and elements of a research agenda for the construction industry. Computers in industry, **83**, 121-139.

[6] Willmann, J., Knauss, M., Bonwetsch, T., Apolinarska, A. A., Gramazio, F., & Kohler, M., 2016. Robotic timber construction—Expanding additive fabrication to new dimensions. *Automation in construction*, **61**, 16-23.

[7] Vaha, P., Heikkila, T., Kilpelainen, P., Jarviluoma, M., Gambao, E., 2013, Extending automation of building construction - Survey on potential sensor technologies and robotic applications. *Automation in Construction*, **36**: 168-178.

[8] Castro-Lacouture, D. 2009 Construction automation. In *Springer handbook of automation* (pp. 1063-1078). Springer, Berlin, Heidelberg.

[9] Cai, S., Ma, Z., Skibniewski, M. J., Bao, S., & Wang, H., 2020. Construction Automation and Robotics for High-Rise Buildings: Development Priorities and Key Challenges. *Journal of Construction Engineering and Management*, **146**(8), 04020096.

[10] Fitch, J. A., 1909, Labor in the steel industry-the human side of large outputs, The Annals of the American Academy of Political and Social Science, Vol. **33**, No 2, Labor and Wages, March 1909, pp. 83-91

[11] Spoerl J S, 2004 (accessed 5/30/2021), A brief history of iron and steel production, https://www.academia.edu/31060927/A_Brief_History_of_Iron_and_Steel_Production

[12] https://tradingeconomics.com/commodity/stee/

[13] Morrall, A. 2017. "A brief history of structural fabrication technology: The road to full automation." *The fabricator*, online article, accessed in June 2021 at https://www.thefabricator.com/thefabricator/article/automationrobotics/a-brief-history-of-structural-fabrication-technology-the-road-to-full-automation

[14] Liang, C. J., & Kang, S. C. 2014. Development of a steel beam hauling system for automatic steel beam assembly. In *Computing in Civil and Building Engineering*, pp. 1295-1302.

[15] Cai, S., Z. Ma, M. Skibniewski, J. Guo, and L. Yun. 2018. "Application of automation and robotics technology in high-rise building construction: An overview." In Proc., Int. Symposium on Automation and Robotics in Construction (ISARC), 309–316. Berlin: International Association for Automation and Robotics in Construction.

[16] Miyakawa, H., Ochiai, J., Oohata, K., & Shiokawa, T. 2000. Application of automated building construction system for high-rise office building. In International Symposium on Automation and Robotics in Construction (pp. 1-6).

[17] Prasad, K. N., & Agrawal, V. M. 2019. Automation and Robotics in the Construction Industry-a Review. *i-Manager's Journal on Future Engineering and Technology*, **14** (3), 49.

[18] Chu, B., Jung, K., Lim, M. T., and Hong, D. 2013. "Robot-based construction automation: An application to steel beam assembly (Part I)." Automation in Construction, **32**, 46–61.

[19] Jung, K., Kim, D., Bae, K., and Hong, D. 2007. "Development of the gripping control algorithm for wire-suspended object in steel construction." 24th International Symposium on Automation and Robotics in Construction, ISARC 2007, Madras, India, 151–155.

[20] Jung, K., Chu, B., Bae, K., Lee, Y., Hong, D., Park, S., and Lim, M. T. 2008. "Development of automation system for steel construction based on robotic crane." International Conference on Smart Manufacturing Application, Kintex, Gyeonggi-do, Korea, 486–489

[21] Kang, S. C., Miranda, E., and Chi, H. L. 2009. "Three-dimensional simulation and visualization of crane assisted construction erection processes." Journal of Computing in Civil Engineering, **23** (6), 363–371.

[22] Kim, T., Lee, U.-K., Yoo, W.S., An, S.-H., Cho, H., Lee, Y., Doh, N. 2010. Benefit/cost analysis of a robot based construction automation system. In International Conference on Control, Automation and Systems, Gyeonggi-do, Korea, pages 616-621.

[23] Mohsenijama, A., Lu M. 2016. Achieving Sustainable Structural Steel Design by Estimating Fabrication Labor Cost Based on BIM Data. *Procedia Engineering*, 145, 654-661, International Conference on Sustainable Design, Engineering, and Construction, Elsevier.

[24] Hu, X., Lu M., and AbouRizk, S. 2014 BIM-based data mining approach to estimating job man-hour requirements in structural steel fabrication, Proceedings of the Winter Simulation Conference 2014, pp. 3399-3410, doi: 10.1109/WSC.2014.7020173.

[25] Akponanabofa, H and Tizani, W. 2015. BIM extension for the sustainability appraisal of conceptual steel design. *Advanced Engineering Informatics*, **29**, 1, 28-46.Elsevier.

[26] Liu W., Guo, H., Li, H., Li, Y. 2014 Retracted: Using BIM to Improve the Design and Construction of Bridge Projects: A Case Study of a Long-Span Steel-Box Arch Bridge Project, *International Journal of Advanced Robotic Systems*, 11,125, DOI: 10.5772/58442

[27] Gu1, P., Tao1, F. and Wang, G. 2020 Application of BIM technology in fine design in steel structure bridges, *IOP Conference Series: Earth and Environmental Science*, **474**, 6. Environmental Engineering and Scientific Application

[28] Li, K., Gan, Y., Ke, G., and Chen, Z 2015. The Analysis and Application of BIM Technology in Design of Steel Structure Joints, *Proceedings of the 2015 4th International Conference on Sensors, Measurement and Intelligent Materials*, https://doi.org/10.2991/icsmim-15.2016.215

[29] Lee, J-K, Lee, H-M, You, J-M, Shin, H-M 2011, BIM based Design of Steel Box Girder, Journal of the Korean Institute of Building Information Modeling, Volume **1** Issue 2, pp. 6-11, https://doi.org/10.13161/kibim.2011.1.2.006

IOP Conf. Series: Materials Science and Engineering 1218 (2022) 012009 doi:10.1088/1757-899X/1218/1/012009

Slam and Beacon Data for Automation of Indoor Construction Progress Tracking

L Marcy[1] and I Iordanova[2]

[1] M.Eng., Department of Construction Engineering, Ecole de Technologie Superieure, Montreal, Quebec, leo.marcy.1@ens.etsmtl.ca
[2] PhD., Professor, Department of Construction Engineering, Ecole de Technologie Superieure, Montreal, Quebec, ivanka.iordanova@etsmtl.ca

Abstract. Construction progress tracking and monitoring is a complex process that is crucial for the successful execution of projects and the delivery of a high-quality product to the client. However, these tasks remain mostly manual - time-consuming and error prone, which often leads to suboptimal quality, cost and schedule overruns. The present research proposes the use of an autonomous rover for the data collection from construction site. The objective is to create a hybrid data processing system using point clouds from the Simultaneous Localization and Mapping (SLAM) algorithm used for robot navigation, and beacons' data integrated with BIM, to track construction progress and provide project managers with reliable information in quasi-real time. The paper is composed of four parts: first, a literature review of best practices regarding the technology used for progress tracking is performed. Second, we propose a framework for automated data collection and information processing for automated progress tracking and monitoring. Third, we present a real-world case study partially implementing the framework by using data acquired by an autonomous rover and BIM, and simulate a real-time reconstruction of the construction site status. Finally, the results are discussed, and future work is identified.

1. Introduction

Due to the nature of construction, with its uncontrollable environment, unique resources and fragmented operation, the construction industry suffers from low productivity, cost and schedule overruns, and recurring safety issues. Traditional methods of data acquisition and progress tracking remain manual, time-consuming and error prone. Yet these tasks account for a significant portion of construction workers' time. This explains why work progress never meets early estimates, resulting in schedule and costs overruns and dissatisfied clients. The recent development of real-time data acquisition technology tools could address these issues and significantly improve industry productivity [15]. A lot of research is being developed to automate the progress monitoring management. This involves providing the right, accurate, and reliable information to the right person in the right place on time. Having a quasi-real-time information about the site works and their context would allow the construction manager to make the right decisions and ensure that the project will be delivered within the defined parameters [10].

The study presented in this article is part of a larger research project on construction progress monitoring using autonomous rover and heterogenous data coming from mobile LiDARs and Beacons, all supported by BIM. The particular objective of this study is to develop an operational framework to recreate a model representing the as-built state of a construction site at the moment of the data collection by the rover, and compare it with the Dated 4D (D4D) model as planned. The paper is structured in four parts, namely: (1) literature review on new methods for progress tracking in construction (2) proposed theoretical framework for automated monitoring of construction progress through technology association and data fusion; (3) description of a case study based on BIM, mobile LiDARs and Beacons technologies, for partial validation of the proposed framework; (4) discussion and future work.

2. Literature review

This literature review explores the process of automated progress tracking by examining data acquisition technologies, and methods for automating progress tracking tasks.

There are two types of imagery data collection technologies, structured data provided by Terrestrial Laser Scanning (TLS) and unstructured data from mobile LiDARs and photogrammetry, both of which provide point clouds. For information transfer technologies, the most commonly used ones are Radio Frequency ID (RFID), QR Codes, Bluetooth Low Energy (BLE) and Ultra-Wide Band (UWB). To automate data acquisition, these two types of technologies can be used with supporting equipment such as Unmanned Aerial Vehicle (UAV) or Unmanned Ground Vehicle (UGV).

Structured point clouds data can be transformed into a mesh model which can be automatically reconstructed (using specialized software) to generate an as-built 3D model. in the case of unstructured data, the raw point clouds acquired by the system must be preprocessed, first to reduce the noise and outliers, and then to align the different sets of point clouds to obtain the complete dataset. Object recognition can be developed by recognizing elements on point clouds by their geometry [2], texture [5], color [1], or position. The segmentation can first be planar to recognize vertical and horizontal elements, and then the points clouds can be segmented using Machine Learning and Computer Vision techniques to detect all types of features directly on the point clouds [2]. Following the recognition of the objects, and to obtain surfaces and then a 3D model, the objects are reconstructed by creating mesh models. This process can be automated by using advanced algorithms.

As far as the Beacon data is concerned, it can be connected to BIM using the ID feature and linking a database with the BIM model to retrieve the information [11].

Various techniques can used to compare the as-built information provided by the scan with the as-planned provided by the 3D model. The comparison can be between the surfaces of two models [6], between two point clouds [4], or between a model and a point cloud [14]. Progress monitoring update is often referred to as an accomplishment of the research performed following object recognition and reconstruction. In the literature, three main techniques are used, firstly by calculating and comparing quantities, performing quantity extraction from both models [8][13]. Second, by recreating a schedule in relation to the detected object and comparing it with the original schedule [11]. Finally, the most widely used method is pattern labeling which provides an automatic visual comparison between as-built and as-planned surfaces [6].

Visualization of updated information between as-planned and as-built can be done by color-coded models [6], colored point clouds [3] to show the surfaces that do not correspond or with dashboards [13] to show a more detailed visualization using numerical information.

The diagram on Figure-2 provides a graphic summary of the literature and best practices review concerning the comparison and visualization methods for progress monitoring presented above.

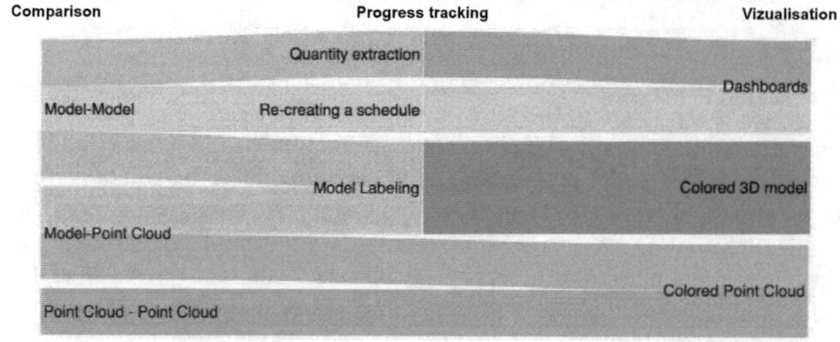

Figure 1. Associations of comparison and visualization methods for progress monitoring.

3. Proposition of a framework

Based on the previous section and prior work of our team [7], a general framework of the research is presented in this section. The team has developed an autonomous rover that acquires data on construction site based on semantic information from the BIM model. The objective is to update the BIM information from the collected data. The point clouds are captured by a mobile LiDAR, which can generate a large unstructured point cloud of the building from the SLAM (Simultaneous Localization and Mapping) algorithm used for robot navigation. We are also able to generate an RGB-mapped point cloud using cameras installed on the rover. These point clouds will be used to detect building elements on site.

To identify some objects which cannot be 'seen' by LiDAR or cameras, a beacon-based system is developed and used for the detection of MEP equipment or secondary elements, the objective being to provide the maximum information to the BIM model, to determine the progress of the work on site.

In the following part, we present the whole framework (presented on Figure-2), from data acquisition to dashboards, in three major parts: the LiDAR-based system, the beacon-based system and the methods to update the Dated 4D model (D4D).

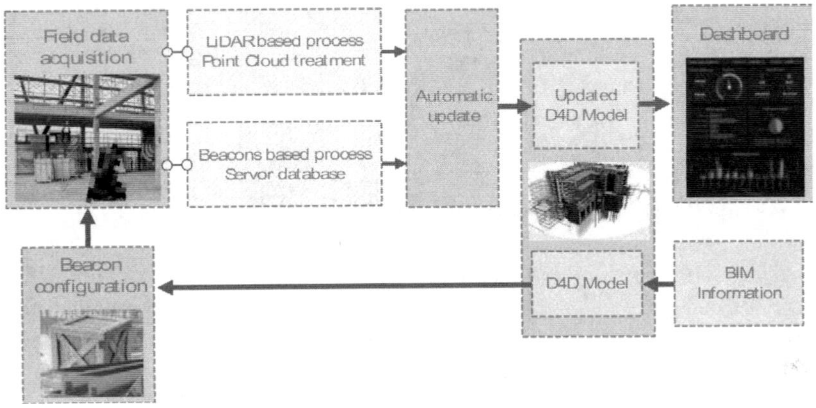

Figure 2. Proposed framework.

3.1. *LiDAR based – process*

From the on-site acquisition with the autonomous rover, two types of point clouds are collected, SLAM-mapped point clouds from LiDARs and RGB-mapped point clouds from RGB cameras.

These point clouds are less accurate and less dense than the typical point clouds used in construction from Terrestrial Laser Scanning or even from photogrammetry. To overcome this problem, the point clouds are pre-processed to improve their quality. The element detection system needs a certain quality to reconstruct the element to form an as-built model.

The point clouds need to be processed individually, creating a study area, a bounding box which can be a floor or more precisely a part of the building that needs to be explored. Noise reduction is applied with the (Statistical Outlier Removal) SOR Filter to remove points with outliers in each point cloud. Point clouds from different sources are aligned with the Iterative Closest Point (ICP) algorithm, which creates a single point cloud with all the information retrieved from the scans. The RANSAC algorithm is tested to segment the point cloud into different planes, which can help the element detection system, but the low density does not allow for proper detection of the planes and thus of the element. However, after processing the point clouds, it could be determined that with this type of point cloud, only structural elements can be detected.

To enable element detection, after pre-processing the point cloud from SLAM, a patching technique from the BIM model is used to enhance the point cloud without losing the scanning information. The structural elements are isolated on the Revit Model, and with a Dynamo script, we export the studied area into a mesh model and then into a dense point cloud, thus creating a point cloud of the 'as planned'.

After aligning the as-planned and as-built point clouds with the ICP algorithm, these point clouds are compared using a cloud-to-cloud distance method. Then, with the scalar field corresponding to the distance between the two models, we can create an as-built point cloud with good density and better quality than the raw point cloud data.

By applying the RANSAC algorithm after this process, we can analyse the quality differences between the raw laser data and the data acquire with the proposed method for processing the point cloud.

3.2. *Beacon-based process*

This step of the system consists of assigning information to the physical resources, in order to detect the position of the elements that cannot be detected with the LiDAR-based process. This association is supported by attaching tags to the concerned items in order to provide a unique identification (ID) for each resource and to link the information to these IDs, and thus, to the resource. All kinds of information can also be added, such as product name, manufacturer, dimensions, installation instructions, delivery date, etc. The whole system works on the basis of the rewritable nature of the chosen technology. For the purposes of the project and the system, two technologies that can support this type of use are BLE tags and UWB tags, as they are both composed of rewritable tags and have readers with long enough range to remotely and wirelessly collect field data.

For the purposes of the data collection, a tag reader is placed on the autonomous rover to detect the position of elements. The coordinates (x,y,z), and their ID can be exported to a database, and compared to the same properties exported with a Dynamo script from the BIM model.

3.3. *Updating the D4D model*

The first step in updating the BIM information using the scans and the tags, is to create an as-built model. Once the point clouds are pre-processed and the tag locations are detected and stored, it is possible to update the BIM model with real-time information.

An as-built model can be created manually or automatically. Manual reconstruction consists of rebuilding a model from the geometric information of the point cloud aligned to the BIM model. For automatic reconstruction, machine learning or computer vision techniques for element recognition will have to be used and can be developed in future work. The structural information (walls, columns, ceilings, floors) is reconstructed from the information from the LiDAR-based process.

In parallel, the beacon-based process compares the position of the detected elements (x,y,z) with those of the BIM model, by exporting the coordinates of the BIM elements to a database and comparing them with those from the beacon system on the rover. This information will complement the as-built model and provide information about all element categories.

To compare with the as-built model, a D4D model is created. It is generated during the 4D simulation, on the day of the scan, to have 'live' information during the construction. This D4D model is exported, and it is this one that is compared with the as-built model created from the rover-collected data. The 4D simulation is created with the Work Breakdown Structure (WBS) of the selected elements, the tasks of the planning are associated with the elements in the model.

For the visualization of the results, Autodesk Navisworks and BIM360 software can be used. With the "compare" tool, it is possible to have a colored visualization of the model to see which elements are well built or not. With this, it is possible to export a list with the name and the parameters. Different properties are used to compare the models: for structural columns - the height; for walls - the length; and for floors and ceilings - the area. The exported lists are then compared and thanks to the parameters, it is possible to calculate a percentage of progress at the chosen date and to see which elements are not as expected. Furthermore, with the WBS, it is possible to associate the elements that are not built as planned and to update the tasks as "done" or "not done", to automatically update the status of the tasks on the progress schedule.

With all this information available after the comparison, dashboards are created to facilitate the understanding of the construction team. The objective is to have a dashboard with the percentage of progress, the tasks that are not well executed and the elements that need special attention.

4. Case study

A case study was developed for the partial validation of the proposed framework. Tests have been performed in a university building in Montreal. The data acquisition was supported by a Clearpath Jackal rover. The following equipment was mounted on it: two Ouster LiDARs (one to capture data around the rover, the other – above); one Velodyne LiDAR; and three Intel RealSense RGB cameras. This system allows the acquisition of different point clouds on site.

Difficulties arose with the RGB-mapped point cloud from the cameras as the acquired data had a lot of noise and the scan was not usable even with preprocessing. For this reason, the case study was conducted exclusively with the point cloud of the mobile LiDARs namely the Velodyne LiDAR and the Ouster. We carried out experiments to see what kind of elements can be detected with this type of point cloud. As explained in the previous section, the quality of the point cloud was improved by using the BIM model and comparing it with the scan.

With this system, we could more easily detect the elements and create an as-built BIM model. The structural elements were well detected, so we continued the experiment to the other structural parts of the building. An as-built 3D model was generated manually, from the pre-processed point cloud, with the structural elements that can be detected, namely: walls, ceilings, floors and structural columns.

Several tests were performed with BLE beacons (Confidex Viking model) to try the detection of the elements. The remote reader for the beacons was an Android smartphone equipped with a BLE scanner application. The beacons were placed on elements in the university and manually configured with their corresponding ID. The result of the beacons revealed a high percentage of errors between the position predicted by the Revit model and the position obtained by the scanner. The beacon information, such as the ID, was well detected but the distance was completely wrong, due to the presence of many metallic obstacles that interfered with the signal. In the next iteration of our system, a different and more advanced system with UWB beacons (Pozyx System) is developed to detect smaller equipment on the construction site.

The comparison system is currently used with the as-built model from the LiDAR-based process, while the information from the beacon-based system will be added in our future work.

The D4D model was generated with the schedule of the structural portion of the construction. After creating this model, the as-built model was imported, aligned and then compared with Autodesk Navisworks and BIM360 "Compare" tool. The results are very encouraging, the differences between the D4D model and the as-built model are well detected, and a list of "non-conforming" elements can be exported to create dashboards for the project manager.

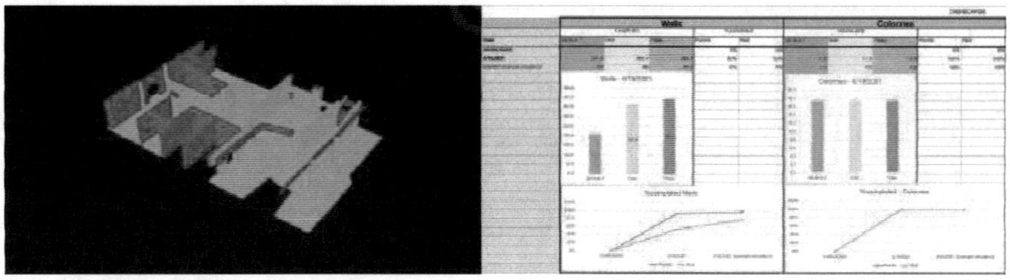

Figure 3. Results of the comparison, colored model and dashboards

5. Discussion and future works

In this research, we explore the possibilities for developing an automatic reconstruction and progress tracking system based exclusively on unstructured data, without using Terrestrial Laser Scanning, or photogrammetry which are manual and time-consuming processes. The objective is to obtain enough information to recreate an as-built model and compare it with the as planned D4D model.

The main problems encountered are the insufficient quality of the data acquired on site and, for the LiDAR and Beacon data, a significant amount of pre-processing. However, this pre-processing allows the system to detect many categories of elements on construction site.

In our future research, we will focus on generating an RGB-mapped point cloud that can be used to complete the automatic reconstruction of the as-built model from the preprocessed point cloud. More investigation on planar and semantic segmentation, and more resources in terms of machine learning will help automate the process as much as possible. For mechanical equipment or secondary elements, we found that simple BLE detection with smartphone is not enough to detect the elements on the construction site, so a new UWB-based system under development will be integrated into the proposed framework.

References

[1] Adán A, Quintana B, Prieto SA, Bosché F 2018 *Advanced Engineering Informatics* **37** 119-38.
[2] Bassier M, Vergauwen M 2020 *Automation in Construction* **120** 103338.
[3] Bonduel M, Bassier M, Vergauwen M, Pauwels P, Klein R 2017 *International Archives of the Photogrammetry, Remote Sensing and Spatial Information Sciences* XLII-2/W8 45-52.
[4] Chen J, Cho Y 2018 *International Conference on Computing in Civil and Building Engineering*
[5] Dimitrov A, Golparvar-Fard M 2014 *Advanced Engineering Informatics* **28**(1) 37-49.
[6] Golparvar-Fard M, Peña-Mora F, Savarese S. 2015 *Journal of Computing in Civil Engineering* **29**(1) 04014025.
[7] Karimi S, Iordanova I. 2021 *Arch Computat Methods Eng* [Internet].
[8] Kielhauser C, Renteria Manzano R, Hoffman JJ, Adey BT 2020 *Infrastructures* **5**(11):98
[9] Kim S, Kim S, Lee D-E 2020 *Remote Sensing* **12**(9) 1457.
[10] Matthews J, Love PED, Heinemann S, Chandler R, Rumsey C, Olatunj O 2015 *Automation in Construction* **58** 38-47.
[11] Pučko Z, Šuman N, Rebolj D 2018 *Advanced Engineering Informatics* **38**:27-40.
[12] Teizer J, Wolf M, Golovina O, Perschewski M, Propach M, Neges M 2017 *ISARC Proceedings of the International Symposium on Automation and Robotics in Construction* Waterloo, Canada: IAARC Publications 1-7.
[13] Turkan Y, Bosché F., Haas CT, Haas R, 2013 *Journal of Construction Engineering and Management* **139**, 423–433.
[14] Tuttas S, Braun A, Borrmann A, Stilla U 2017 *PFG – Journal of Photogrammetry, Remote Sensing and Geoinformation Science* **85**(1):3-15.
[15] Valero E, Adán A, Cerrada C 2015 *Sensors* **15**(7) 15988-6008.

IOP Conf. Series: Materials Science and Engineering 1218 (2022) 012010 doi:10.1088/1757-899X/1218/1/012010

Tightly-Packed Repetitive Schedules: A Tetris Challenge

T Hegazy, and K Mostafa

Civil and Environmental Engineering Department, University of Waterloo, Waterloo, ON, N2J 3C1, Canada.

Corresponding Author: tarek@uwaterloo.ca

Abstract. A large portion of new and rehabilitation projects for the civil infrastructure involves repetitive tasks that mandate high degree of synchronization among the resources to meet project constraints. Despite their large need for support, few tools exist commercially. This is due, in part, to the computational and visualization challenges of existing repetitive scheduling techniques, which complicate the process and make it less understood, particularly when the repetitive units are not identical and when resources are limited. This presentation introduces enhancements by dealing with repetitive scheduling as a game of Tetris, where tasks can change geometry to fit tightly together to save project duration. This is demonstrated through novel mathematical formulations, as well as a new heuristic scheduling algorithm (First-Come-First-Serve). It will be shown that the proposed geometrical adjustments are more advantageous in improving repetitive schedules than using expensive accelerations, and the new visuals make the schedule more legible. This research is a step towards making repetitive scheduling a mainstream tool, to benefit the construction industry.

1. Introduction

Repetitive construction projects, by definition, require the construction crews to repat the same work across a number of units. As such, the key for efficient scheduling of repetitive projects is allowing the crews to move from one unit to the other with minimal interruptions. This has numerous benefits such as establishing a good work momentum which in turn leads to time and cost savings [1, 2].

To consider the minimal interruptions objective and work continuity constraints, many repetitive scheduling models have been developed since the 1960s. Examples include Vertical Production method [3], Line of Balance, LOB [4], Flowlines [5], Linear Scheduling Model [6], Repetitive Scheduling Method [7], Productivity Scheduling Method [8], Location-Based Management and Scheduling (LBMS) [9], and others [e.g., 10, 11]. Alas, most of the presented scheduling models focus only on the case of identical units. Although there are new methods developed to address the more general case of non-repetitive units [12-14], repetitive scheduling still has challenges that are yet to be addressed and require careful planning to determine the necessary number of crews to achieve the project deadline while maintaining work continuity.

1.1. Repetitive Scheduling as a game of Tetris

Repetitive scheduling can be seen as a visual process where the activities can take on different geometries based on the number of crews used and their arrangement (Figure 1a). It can be seen in Figure 1b-d that having time gaps in the schedule leads to extended project durations. As such, repetitive scheduling can be viewed as a game of Tetris where the successive activities represent the blocks that need to be arranged next to each other with minimal gaps to achieve minimal project duration. This is not an easy endeavour in the case of complex construction projects that involve a large number of tasks and a variety of complex constraints. Such complexity often forces project managers to use more costly time-cost trade-off solutions to meet the necessary deadlines. As such, this paper proposes a combination of mathematical computations and scheduling algorithms to systematically produce tightly-packed schedules that meet the project necessary deadlines while maintaining an efficient use of the available crews and project resources.

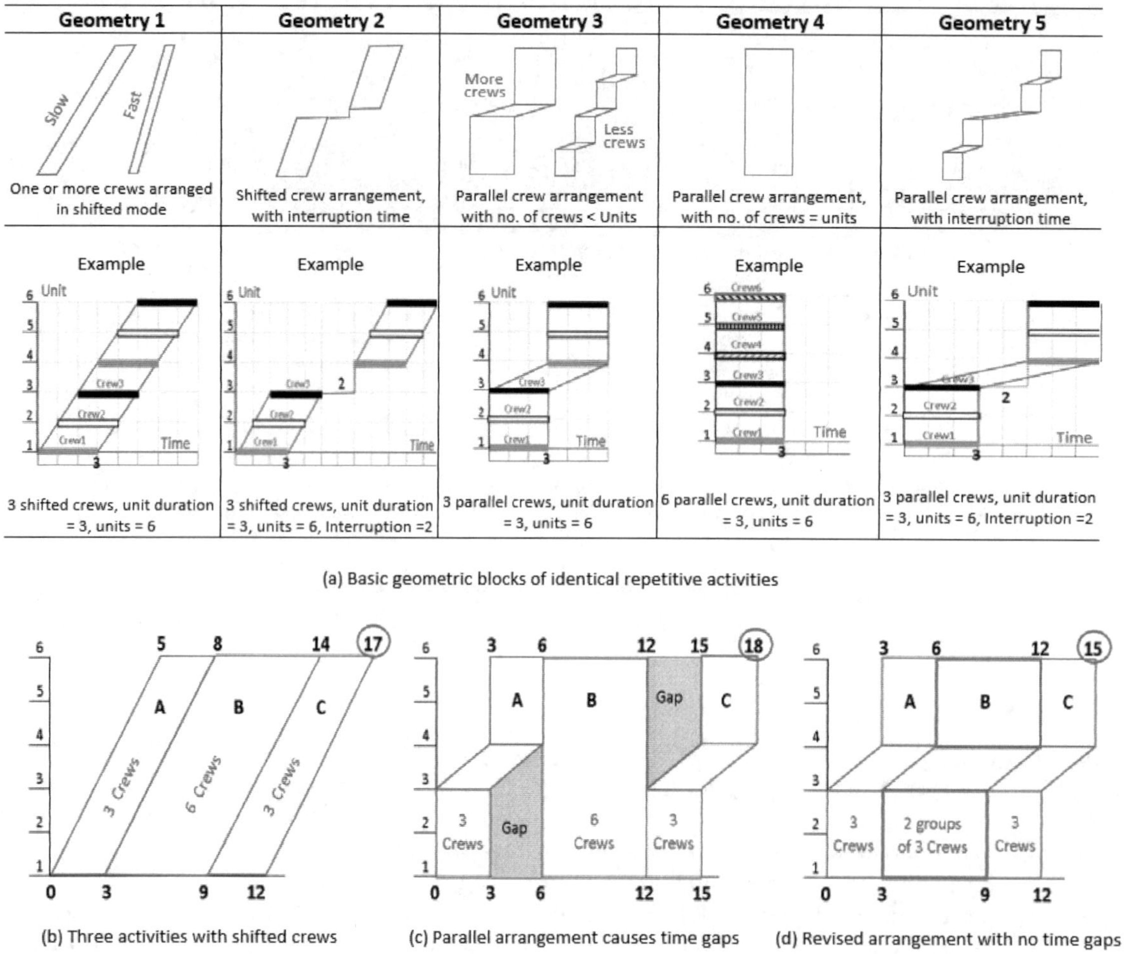

Figure 1: Effect of geometric arrangement on schedule duration

2. Current CPM/LOB Calculations and Persistent Challenges

The first step in existing CPM/LOB calculations is use conventional CPM formulae to calculate the time needed to finish one unit (T1). Then, since the all the units (N) need to be completed before the deadline (DL), the rate of delivery of the units (R) is calculated using Equation 1. Equation 1 incorporates the total float of the activity (TF) as an adjustment factor that reduces the delivery rate of non-critical activities and thus requires less crews. Accordingly, the number of required crews is calculated based on required rate (R) and the duration of the activity (D) using Equation 2, and then the rate of delivery is adjusted based on the new number of crews after rounding (because having fractions of a crew is impractical) through Equation 3. An example can be seen in Figure 2.

[1] $$\textit{Task i desired rate } (R_i) = \frac{N-1}{(DL - T_1 + TF_i)}$$
[2] $$\textit{Task necessary crews } (C_i) = \textit{Roundup } (D_i \times R_i)$$
[3] $$\textit{Task actual rate } (R_i) = C_i / D_i$$

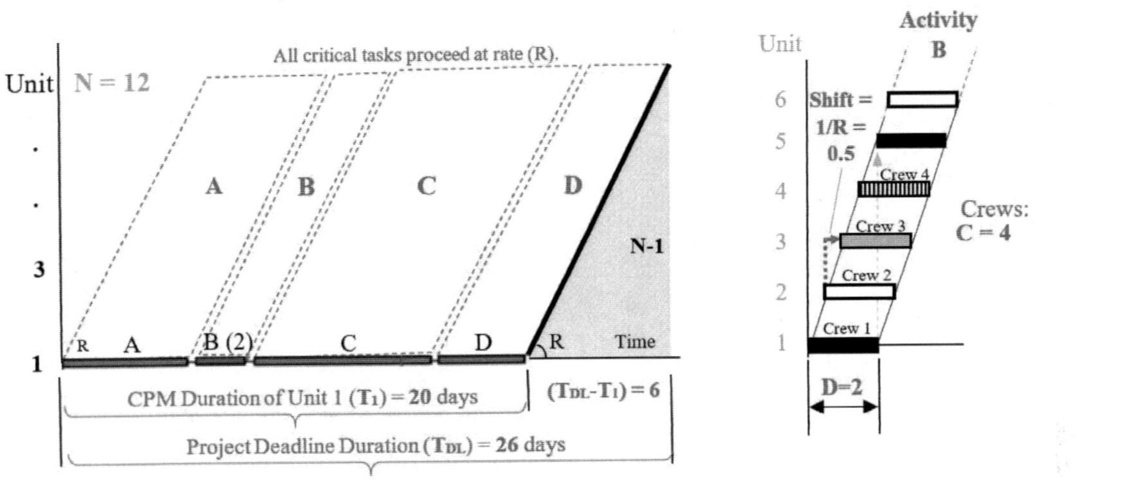

(a) Calculation of desired rates based on project deadline (b) Enlarged schedule of task B

Figure 2: CPM/LOB Analysis of the tasks' required shifted crews to meet deadline

Repetitive schedules, however, often exhibit deadline violations even when the necessary computations are applied. This is because of the multiple reasons such as rounding of crews, rounding of start times, crew availability limitations, or other reasons that change the geometry of the activity/crew assignment and thus introduces schedule gaps that lead to project duration extensions. As such, this paper presents novel computations and algorithms to produce tightly packed schedules that meet the necessary deadlines. The contributions of this paper are threefold: (1) propose a formulation for designed interruptions that adjust the task geometry and delivery rate to reduce the schedule gaps by synchronizing the delivery rates of all the tasks; (2) introduce a validation loop to avoid deadline violations in relaxed schedules; and (3) propose a novel crew assignment algorithm (First-Come-First-Serve) that reduces the schedule gaps by treating the schedule as a game of Tetris on the activity level.

3. Formulation for Designed Interruption

As seen in Figures 1b and 2a, a schedule that has no time gaps is a schedule whose tasks operate with the same rate of delivery. However, this is not always practical because of multiple reasons such as crew availability constraints and/or rounding of crews (see Equation 2). A more vivid example can be seen in Figure 3a, where activity B has a higher delivery rate than the other activities which creates huge schedule gaps. It can then be seen in Figure 3b that deliberately interrupting activity B to bring its overall rate of delivery closer to its counterparts reduces the schedule gaps which in turn reduces the overall project duration.

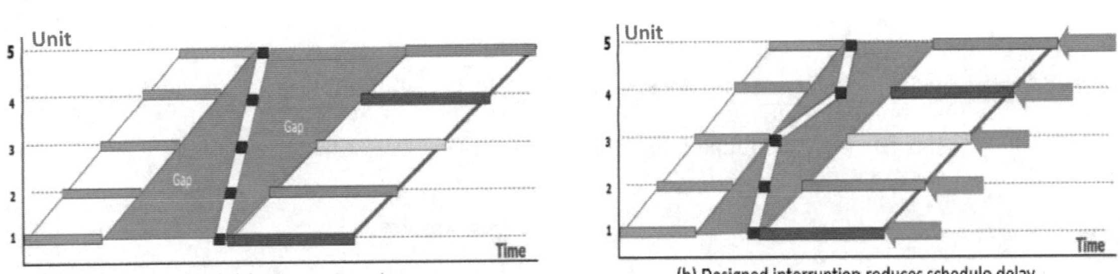

(a) Unsynchronized task delivery rates introduce gaps (b) Designed interruption reduces schedule delay

Figure 3: Effect of Unsynchronized Delivery Rates

To present a formulation for achieving better synchronization among the activities' delivery rates through designed interruption, the case in Figure 4 is used, where a faster activity (B) follows a slow one (A). As schematically shown in Figure 4a, the start-time for activity B in case of no interruptions (STO_B) is shown. Assuming one interruption at N/2 is sufficient, the interruption time that allows the bottom half of activity B to start earlier is shown on the figure. As shown, STO_B = Y + Duration of B + Interruption time, therefore:

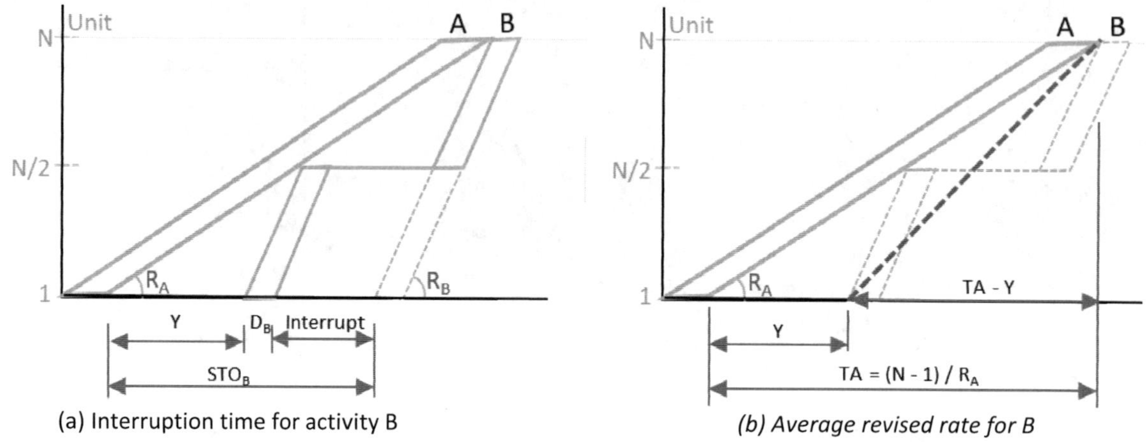

Figure 4: Formulating Task Interruption Time

[4] *Interruption Time* = *STO of B* – *Y (STO of B for N/2 units)* – *Duration of B* =

$$= (N-1)\left(\frac{1}{R_A}-\frac{1}{R_B}\right) - ((N-1)/2)\left(\frac{1}{R_A}-\frac{1}{R_B}\right) - D_B$$

Thus, applying the interruption, the bottom part activity B (with N/2 units) starts with an offset of Y:

[5] *Start-Time Offset of B with interruption* = $Y = ((N-1)/2)\left(\frac{1}{R_A}-\frac{1}{R_B}\right)$

Using this formulation, it can be seen that the lower part of activity B started early, thus, changing the average delivery rate of B (dashed line in Figure 4b) to a lower rate than the original rate of B, and becoming more closer in its rate to its predecessor (A). This average modified rate of B can also be formulated, based on the presentation in Figure 4b, as follows:

[6] *Average rate of B with interruption* = $\left(\frac{N-1}{TA-Y}\right) = \left(\frac{N-1}{\frac{N-1}{R_A}-Y}\right)$

4. Deadline Checking Loop

Consider the example in Figure 5, with a deadline of 20 days and 5 repetitive units. Figure 5 includes the standard CPM calculations using Equations 1-3 as well as the produced schedule. It can be seen from the schedule at the bottom of Figure 4 that the standardized CPM/LOB formulations produced a schedule that is 24 days long. This is unacceptable as it violates the original 20-day deadline. While introducing interruptions using Equations 4-6 does help reduce the project duration, it does not reduce it enough to meet the deadline (unless an interruption is placed after every unit, which defeats the work-continuity purpose of using repetitive scheduling techniques). It can also be seen that the actual duration of the first unit is 12 days, not 8 as calculated earlier.

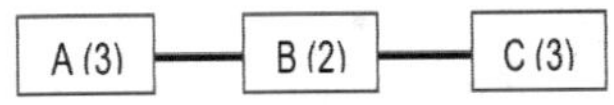

Deadline (DL) = **20** days, CPM (T_1) = 8 days, Units (N) = 5

	Duration **D**	Desired Rate, Eq.1, **R**	Necessary crews Eq. 2, **C = D x R**	Actual Rate R = C / D	Shift 1/R
A	3	4/(20-8) = 0.33	Roundup (3 x 0.33) = 1	1 / 3 = 0.33	3
B	2	4/(20-8) = 0.33	Roundup (2 x 0.33) = 1	1 / 2 = 0.50	2
C	3	4/(20-8) = 0.33	Roundup (3 x 0.33) = 1	1 / 3 = 0.33	3

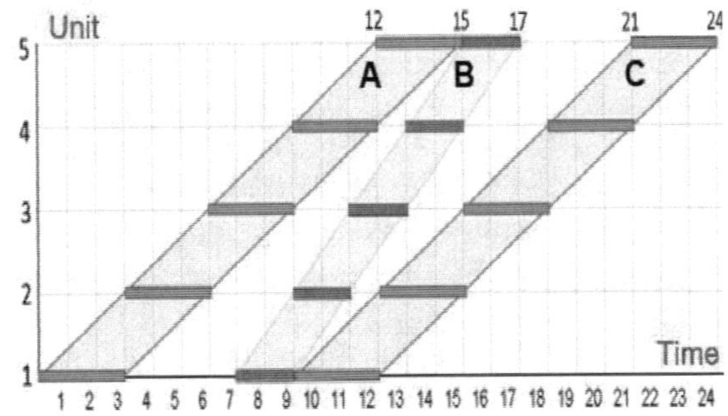

Figure 5: Original CPM/LOB Calculations Exceed the 20-day Deadline

Computationally, the calculation and assignment of the different crews in repetitive schedules is controlled by one main parameter which is the rate of delivery (R), which is a function of the deadline and CPM duration as seen in Equation 1. Based on the example shown in Figure 5, the duration of one unit (T_1) was higher than what was originally calculated. Based on these observations, it can be concluded that using a higher value for T_1 would lead to improved CPM/LOB scheduling procedures and prevent deadline violations. However, having higher delivery rate corresponds to the need for more crews. Hence, there is a need to strike a balance between avoiding deadline violations as well as avoiding the overuse of available resources.

As such, this paper proposes a simple loop that iterates through different values of T_1 to see if the deadline is met. This loop is introduced in Figure 6, where the difference between the deadline (DL) and T_1 is divided into five intervals of length G. Hence, if the schedule produced using the original value of T_1 is violating the deadline, T_1 is increased by G days and the LOB calculations are reperformed. The loop terminates once a schedule is reached that satisfies the project deadline.

Applying the deadline checking loop on the example introduced in Figure 6, G is calculated to be (20-8)/5 = 2.4 days. Since the schedule created using the original T_1 (8 days) did not satisfy the deadline, T_1 is increased by G days and a new schedule is created using the new T_1 (T_1=8+2.4 = 10.4 days). Based on the new values, a new schedule is created that happened to satisfy the original 20-day deadline as seen in Figure 6. Hence, the loop terminates. Otherwise, the LOB scheduling calculations would have been performed with T_1 = 10.4 + 2.4 = 12.8 days.

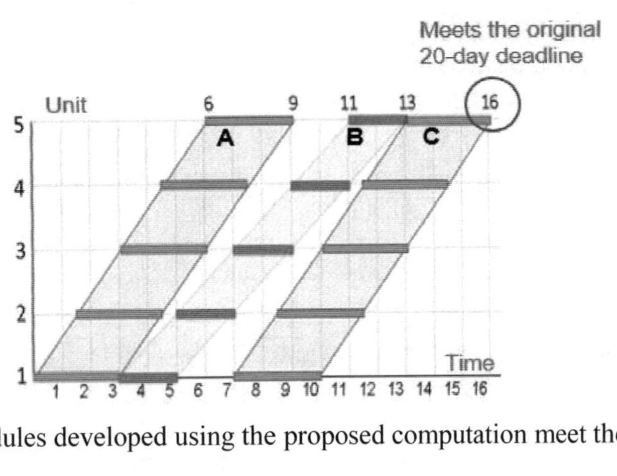

Figure 6: Schedules developed using the proposed computation meet the original deadline

5. First-Come-First-Serve (FCFS) Crew Assignment

It can be noticed from the figures and equations proposed earlier is that they are most effective when the repetitive units are identical. As such, non-identical units make the formulations less effective, and the overall scheduling process more difficult. For example, Figure 7a shows a repetitive task where unit2 has a smaller duration than the typical unit, while unit4 has a larger duration. A scheduling method has been developed to handle such a case [14] where the successor activity is scheduled in isolation, and then "shifted" back to be directly after its predecessor as seen in Figure 7. The shift is calculated based on the smallest difference (delta) between the start of the successor task (in that random point in the future where it was scheduled) and the finish of its predecessor. As such, this approach is commonly referred to as the "Delta-Shift" approach. While this approach offers an effective way to deal with non-identical tasks by treating the schedule as a Tetris game on the activity level, the produced schedule still exhibits significant gaps. As such there is a need for a new method that treats the schedule as a Tetris game on the unit level to reduce the gaps and provide a more efficient schedule.

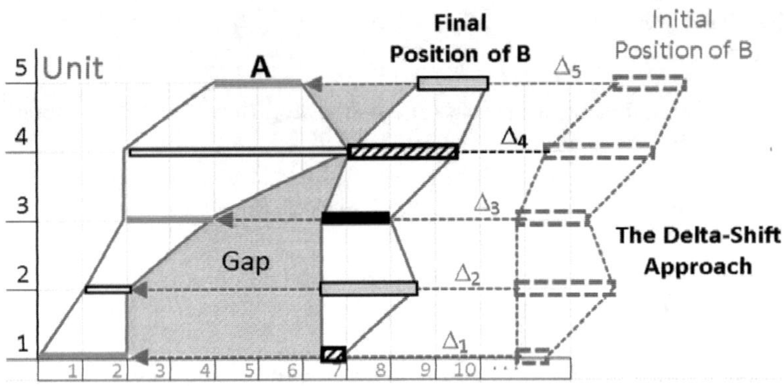

Figure 7: Schedule Gaps Produced by the Delt-Shift Approach for Non-Identical Units

As opposed to the Delta-Shift approach dealing with each task as a large block, the proposed FCFS process deals with each crew at a time, not the task as a whole, as can be seen in Fig. 8 on an example of scheduling a successor activity (B) after a predecessor (shaded task A). The process is as follows:

1. In the first step, all crews are available. Hence, crew 1 is assigned to the first unit and its finish time is updated to reflect the finish time of activity B in the first unit.
2. In step 2, the finish time of the predecessor to activity B in the second unit is greater than the updated finish time of the crew 1. Hence, crew 1 is assigned to the second unit. To maintain work continuity, the start of crew 1 at the first unit is delayed by the difference between the finish time of the second unit and the scheduled finish of the first unit. Finally, the scheduled finish of crew 1 is updated to reflect the finish of activity B in unit 2.
3. In step 3, the finish time of the third unit is smaller than the scheduled finish of crew 1. Crew 2 is assigned to that unit and its finish time is updated accordingly. The predecessor finish time at the fourth unit is less than the finish time of crew 1 but equals to that of the crew 2. Therefore, crew 2 was assigned to unit 4.
4. Similarly, in step 4, crew 1 is assigned to unit 5, the start dates of its previous units (units 1 and 2) are delayed to remove work interruption, and its finish time is updated to reflect the finish of the activity in unit 5.
5. The procedure is continued until all units of activity B have been scheduled, and the process is repeated for other successors activities until all activities have been scheduled.

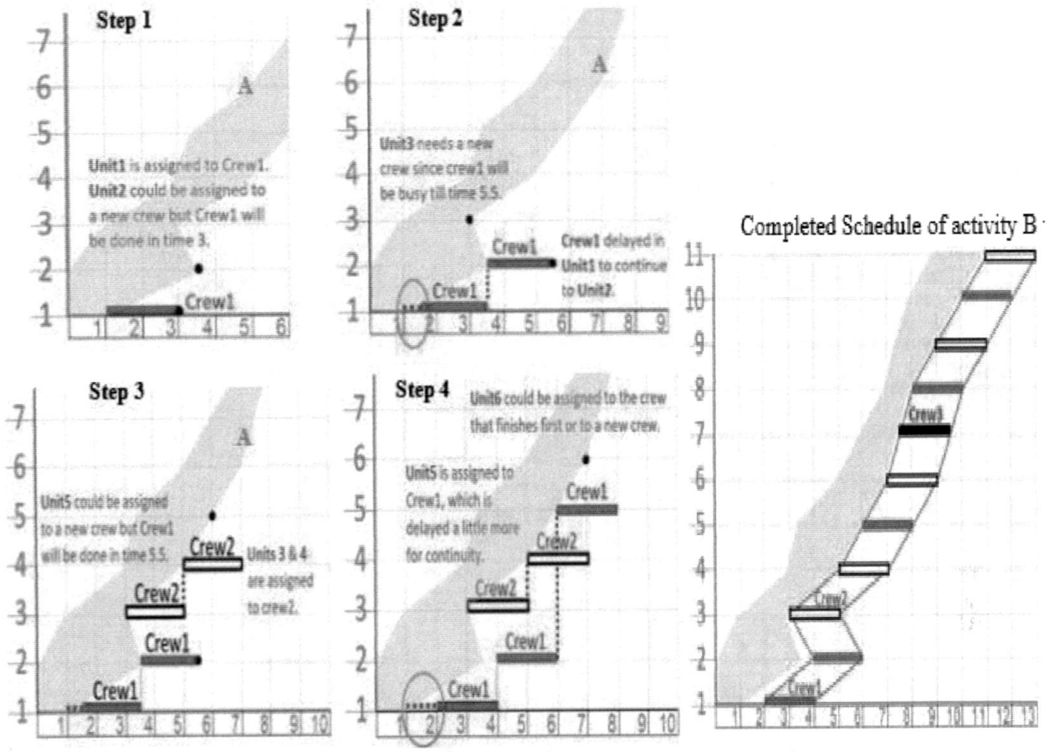

Figure 8: Steps of the Proposed FCFS Crew Assignment Process

6. Example Application

To demonstrate the resulting schedule improvements of the proposed computations and algorithms, as well as the Tetris analogy in repetitive scheduling, an example project is considered, as shown in Figure 9. The example consists of four activities with CPM duration being 12 days. In this example, 10 road sections need to be completed within a total project duration of 24 days. All the units have the same duration with the exception of units 5 and 6, whose durations are 0.5 and 1.5 times the typical duration, respectively. Furthermore, Activity A must be performed from both sides of the road. In other words, units 1 and 10 need to start simultaneously, and their crews move from one unit to the next accordingly (e.g., one crew moves through sections 10 to 6 while the other moves through sections 1 to 5).

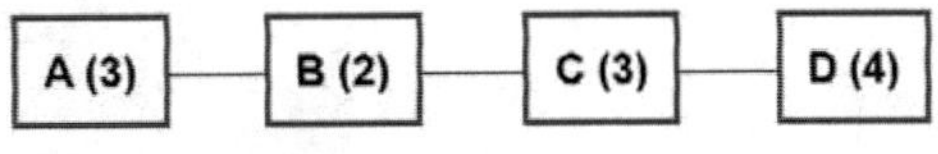

Deadline (*DL*) = **24** days, CPM (T_1) = **12** days, Units (N) = **10**.

Sections 5 and 6, task durations are multiplied by 0.5 and 1.5, respectively.

	Duration D	Desired Rate R	Necessary crews C = D x R
A	3	9/(24-12) = 0.75	Roundup (3 x 0.75) = **4**
B	2	9/(24-12) = 0.75	Roundup (2 x 0.75) = **2**
C	3	9/(24-12) = 0.75	Roundup (3 x 0.75) = **3**
D	4	9/(24-12) = 0.75	Roundup (4 x 0.75) = **3**

. To save time, **Activity A** must use 2 crews from each side of the road, thus,
two crews go from section 10 to 9 to 8 to 7 to 6, while the other **two crews go from 1 to 2 to 3 to 4 to 5**.

Figure 9: Data and CPM/LOB Calculation Table of the Example

The table in Figure 9 shows the CPM/LOB calculations of necessary crews, and the steps in producing the schedule, task-by-task, are shown in Figure 10. Initially, the schedule developed based on a T_l value of 12-days (CPM duration), and the schedule (Figure 10) becomes 27 days, violating the deadline.

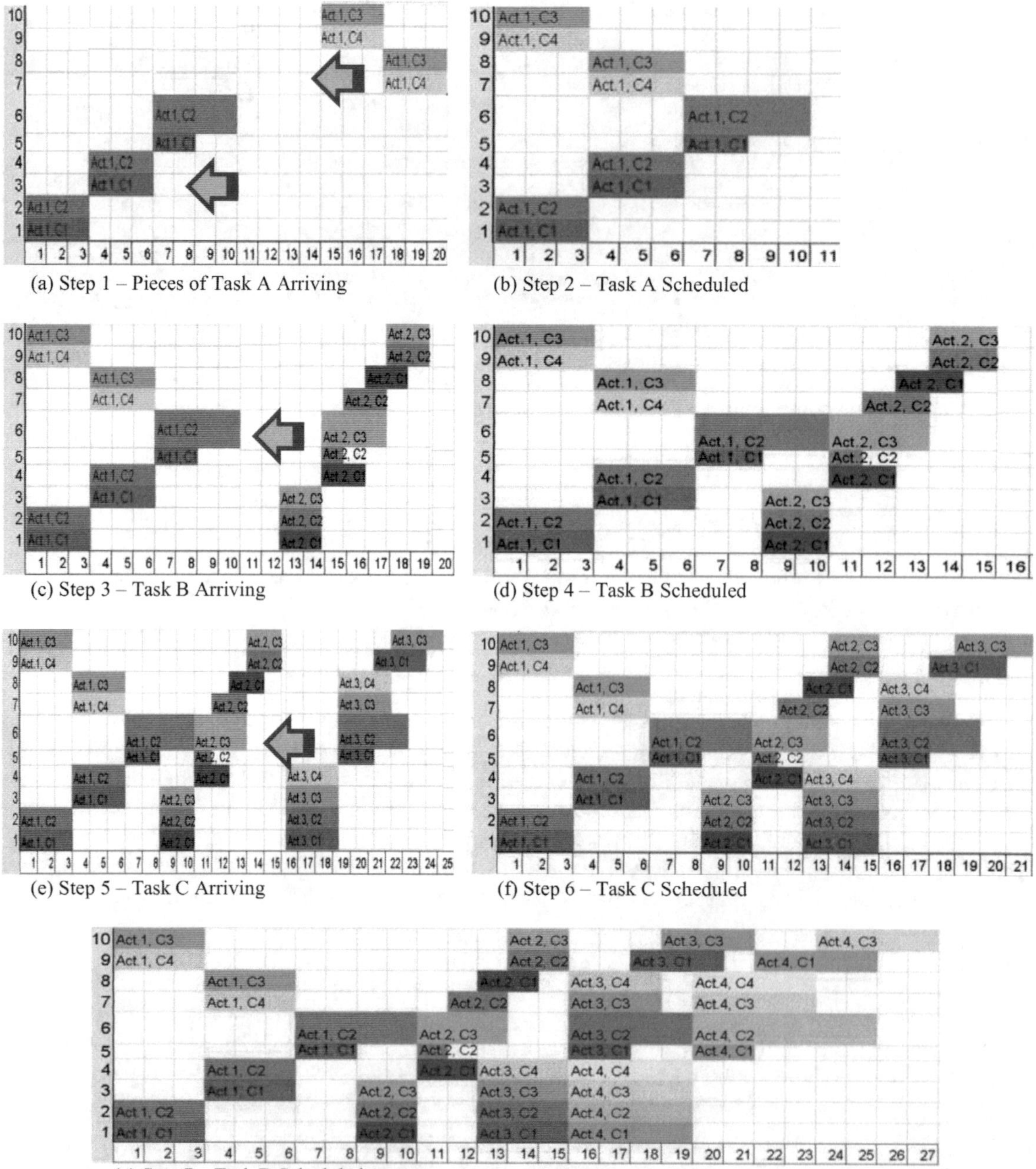

(a) Step 1 – Pieces of Task A Arriving (b) Step 2 – Task A Scheduled

(c) Step 3 – Task B Arriving (d) Step 4 – Task B Scheduled

(e) Step 5 – Task C Arriving (f) Step 6 – Task C Scheduled

(g) Step 7 – Task D Scheduled

Figure 10: Tetris-like Steps of Generating the Schedule

It can be noticed that the schedule in Figure 10 has been created using the Delta-Shift approach. Using the same crew information, a schedule is developed using the proposed FCFS algorithm in Figure 11. The developed schedule has a duration of 26 days, which is smaller than the Delta-shift based schedule in Figure 10. This proves the efficiency of the proposed approach and its potential time and cost savings due to creating more compact schedules.

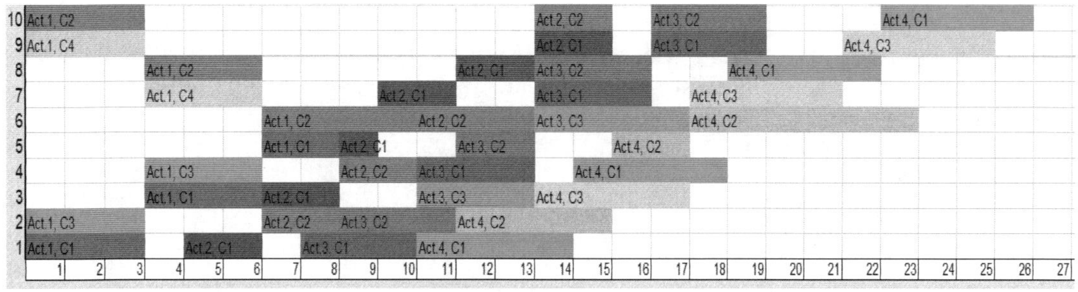

Figure 11: FCFS Schedule using Original Crew Calculations

Since the FCFS schedule alone still did not meet the 24-day deadline, the calculations loop was then introduced and the parameter G is calculated to be (24-12)/5 = 2.4 days, thus, the new value for T_1 becomes 14.4 days. Accordingly, the revised number of crews were assigned in a FCFS arrangement, and a new schedule was generated in Figure 12, where the crews for activity 4 increased to 4 instead of 3. The schedule satisfies the 24-day deadline constraint. The schedule of Figure 12 thus proves that using the FCFS crew assignment with the validation loop produces schedules with minimal gaps, to meet the project deadline, thus validating the efficiency of the proposed approach.

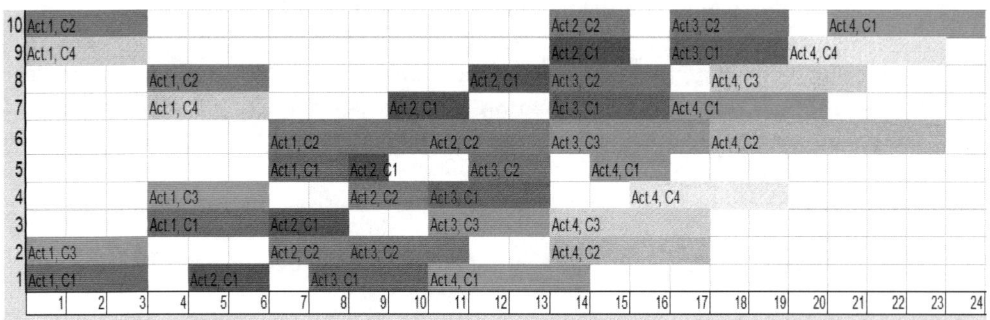

Figure 12: Final FCFS Schedule using Validation Loop Calculations

7. Conclusion

Repetitive scheduling is like a Tetris puzzle whose best solution is one that tightly packs all the activities next to each other, thus achieving minimum project time and maximum work continuity. To that end, this paper presented a variety of novel computations and algorithms to produce more efficient schedules by removing the schedule gaps produced by conventional CPM/LOB calculations. First, a formulation to calculate designed interruption was introduced to help reduce project durations by synchronizing the task delivery speeds and adjusting the task geometry. Second, a deadline validation loop was implemented to avoid deadline violations. Third, a novel scheduling algorithm (First-Come-First-Serve) was developed where the schedule is treated as a Tetris puzzle, thus producing minimal schedule gaps. The proposed contributions were implemented on a prototype example to demonstrate their effectiveness and versatility in terms of dealing with non-identical units and other practical project requirements and constraints. The deadline checking loop allows for adjusting the project schedule by manipulating one parameter only (T_1) as opposed to tweaking every activity manually. The implementation shows that is possible to produce repetitive schedules with shorter durations by modifying the task geometry and arrangement without the need for costly crashing alternatives.

Overall, repetitive scheduling is a key project management skill that is suggested to be an integral part of project management education so that these powerful techniques can become mainstream in managing large-scale repetitive projects, which are the majority of infrastructure projects. To support this educational objective, interested readers can download useful education tools and a spreadsheet program to implement repetitive scheduling from the first author's dropbox at: https://www.dropbox.com/s/dxtzo9bjznfwego/Repetitive%20Scheduling.zip?dl=0. Many areas of research could also improve existing techniques and provide better decision support for these projects. These including integration with BIM models, simulation studies, large-scale optimization, and better visualization of the resulting schedules.

References

[1] Altuwaim, A., and El-Rayes, T. (2018). "Optimizing the Scheduling of Repetitive Construction to Minimize Interruption Cost." *Journal of Construction Engineering and Management*, 144(7): 04018051.

[2] Bakry, I., O. Moselhi, and T. Zayed. (2014). Optimized Acceleration of Repetitive Construction Projects. *Automation in Construction*. 39: 145–151

[3] O'Brien, J. J. (1975). ''VPM scheduling for high-rise buildings.'' *Journal of the Construction Division*, ASCE, 101(4), 895–905.

[4] Arditi, D., and Albulak, M. Z. (1986). Line-of-balance Scheduling in Pavement Construction. *Journal of Construction Engineering and Management*, 112(3), 411–424.

[5] Russell, A. D., and Wong, W. C. M. (1993). ''New generation of planning structures.'' *Journal of Construction Engineering and Management*, ASCE, 119(2), 196–214.

[6] Harmelink, D. J., and Rowings, J. E. (1998). Linear Scheduling Model: Development of Controlling Activity Path. Journal of Construction Engineering and Management, 124(4), 263–268.

[7] Harris, R. B., and Ioannou, P. G. (1998). Scheduling Projects with Repeating Activities. Journal of Construction Engineering and Management, 124(4), 269–278.

[8] Lucko, G. (2008) Productivity scheduling method compared to linear and repetitive project scheduling methods, *Journal of Construction Engineering and Management*, 134(9) 711–720,

[9] Kenley, R., and Seppänen, O. (2010). Location-based Management System for Construction: Planning, Scheduling and Control. Spon Press, London and New York

[10] Su, Y. & Lucko, G. (2016). Linear Scheduling with Multiple Crews Based on Line-of-Balance and Productivity Scheduling Method with Singularity Functions. *Automation in Construction*, 70: 38-50.

[11] Shah, R. K. (2014). A New Approach for Automation of Location-Based Earthwork Scheduling in Road Construction Projects. Automation in Construction, 43: 156-169.

[12] Elbeltagi, E., and Dawood, M. (2011), "Integrated visualized time control system for repetitive construction projects". Automation in Construction (20), pp. 940–953.

[13] Hegazy, T., and E. Kamarah. (2008). "Efficient Repetitive Scheduling for High-Rise Construction." Journal of Construction Engineering and Management, ASCE, 134(4), 253–264.

[14] Long, L. and Ohsato, A. (2009). "A genetic algorithm-based method for scheduling repetitive construction projects" Automation in Construction, (18), pp. 499–511.

IOP Conf. Series: Materials Science and Engineering 1218 (2022) 012011 doi:10.1088/1757-899X/1218/1/012011

Blueprint for Construction 4.0 Technologies: A Bibliometric Analysis

A Ammar[1]and H Nassereddine[2]

[1]University of Kentucky, Lexington, USA, ashtarout.ammar@uky.edu
[2]University of Kentucky, Lexington, USA, hala.nassereddine@uky.edu

ashtarout.ammar@uky.edu

Abstract. To survive in an increasingly complex environment, the Architectural, Engineering and Construction (AEC) industry needs a game-changing innovation. As, the industrial sector has provided compelling evidence that Industry 4.0 gives the opportunity to gain competitive advantages, the construction industry started embracing what is now acknowledged as Construction 4.0. While Building Information modeling (BIM) has been extensively discussed and is perceived as a core enabler of Construction 4.0, this next revolution in the AEC industry goes beyond BIM. As Construction 4.0 is gaining momentum in the construction body of knowledge, there is still no consensus on what Construction 4.0 really means and what technologies it encompasses. Building on insights collected from the extant research corpus through a systematic literature review of the published scientific research in the last ten years, this paper proposes a blueprint for Construction 4.0. Bibliometric analysis was employed to map the currently discussed Construction 4.0 technologies, group them into clusters based on their occurrences, and determine the least researched ones. This blueprint sheds light on how well-defined the concept of Construction 4.0 is, detects the least researched technologies, and provides AEC stakeholders with a reference to better understand this transformation and where it stands today.

1. Introduction

As one of the oldest and most traditional industries, the Architectural, Engineering and Construction (AEC) industry has been recognized for its resistance to change and obstinacy to technology, leading to lagging productivity growth, limited digitization, frequent insolvencies, and low customer satisfaction [1]. However, influenced by the industrial sector and the gains that resulted from the fourth industrial revolution or what is known as Industry 4.0, the AEC industry began to follow suit and embraced this new transformation to improve its performance [2]. As a result, the term Construction 4.0 emerged in the construction research corpus [3].

Although Industry 4.0 is commonly and frequently used among practitioners and academics, there exists no clear definition of its concept [4]. The literature reveals more than 100 different definitions of Industry 4.0 [5]. Like its counterpart, there is also no consensus on the definition of Construction 4.0 and its elements [6]. Some authors refer to Construction 4.0 as the digitization of the AEC industry [7], others consider it a representation of a transformative framework towards a globally connected, smart, and active construction process integration [8].

Based on a recent report published by McKinsey & Company, 80% of the surveyed respondents believe that the AEC industry will change drastically in the next 20 years and that the COVID-19 pandemic will likely accelerate the transformation of the industry. Also, it is expected that the AEC industry will harness new digital technologies to accelerate the transformation process [1]. In this context, it is critical to understand recent efforts undertaken toward achieving this transformation and investigate the current trends of adopting emerging technologies in the AEC industry.

Some implemented technologies are well investigated such as BIM [9] [10], modular and off-site construction [11] [12], virtual and augmented reality [13] [14]; however, they are not representative of all the recent efforts and trends toward digitizing the AEC industry. Moreover, most research efforts focus on the relatively mature technologies and ignore the least adopted ones. Thus, for a holistic understanding of Construction 4.0, it is vital to direct research toward investigating the less mature technologies to allow for a better forecast of future priorities. The work presented in this paper provides a review of the recent literature that investigated the technologies that are imbedded under the umbrella of Construction 4.0. A bibliometric analysis of the scientific literature since 2010 is presented to map out the currently discussed Construction 4.0 technologies and group them into clusters, forming the blueprint of Construction 4.0. Consequently, the findings of this paper shed light on the least addressed Construction 4.0 technologies, providing opportunities to further investigate this aspect of Construction 4.0. This paper assists both researchers and practitioners in directing their research efforts to accelerate the transformation of the AEC industry.

2. Recent Work on Construction 4.0

After the emergence of the concept of Construction 4.0, several researchers conducted a systematic literature review to investigate the efforts undertaken toward transforming the AEC industry by questing technologies that compose Construction 4.0. These research efforts and the identified Construction 4.0 technologies are summarized in Table 1.

Table 1. Overview of the existing Construction 4.0 technologies literature reviews

Reference	Enabled Construction 4.0 technologies
Oesterreich and Teuteberg (2016)[15]	Additive Manufacturing; Automation; Augmented/Virtual/Mixed Reality (AR/VR/MR); Big Data; BIM; Cyber-Physical Systems (CPS)/Embedded systems; Cloud Computing; Digitization; Human-Computer Interaction (HCI); Internet of Things (IoT)/Internet of Services (IoS); Mobile Computing; Modularization/Prefabrication; Product-Lifecycle Management (PLM); Radio Frequency Identification (RFID); Robotics; Simulation tools/ Simulation models; Social Media
Dallasega et al. (2018)[16]	BIM; Cloud Computing; Digitization; E-supply Chain Management; Tracking and Localization; 3D Printing
Maskuriy et al. (2019)[17]	Additive Manufacturing; AR/VR; Big Data; BIM; Cloud Computing; CPS; Geographic Information System (GIS); IoT; Mobile Computing; RFID
Boton et al. (2020)[18]	Additive manufacturing; AR/VR; Big Data; BIM; Cloud computing; CPS; IoT; RFID; 3D Printing
Forcael et al. (2020)[19]	*Technologies that are essential to understand Construction 4.0 at present time:* Big Data; IoT; VR; 3D Printing
Perrier et al. (2020)[20]	Additive Manufacturing; AR/VR/MR; BIM; Blockchain; Cloud Computing; Drones; GIS; Laser Scanners/Point Cloud/Photogrammetry; Sensors (RFID-Infrared- data transfer sensors); Machine Learning /Neural Networks; Mobile Robotic Units
Sawhney et al. (2020)[8]	Actuators; Additive manufacturing; Artificial Intelligence (AI); AR/VR; Big Data and Analytics; BIM; Blockchain; Cloud-based Common Data Environment (CDE); Cloud-based Project Management; Cybersecurity; IoT/equipment

	with IoT; Laser Scanners; Off-site Construction; Robotics and Automation; Sensors/ workers with wearable sensors; Unmanned Aerial Systems (UAVs)
You and Feng (2020)[21]	AR/ VR; Big Data; BIM; Cloud Computing; Intelligence computing (a branch of AI); IoT; Machine learning; Reasoning Technology; 3D Reconstruction
Turner et al. (2021)[22]	AI; Big Data Analytics; Digital Twin; Industrial connectivity achieved by IoT; Robotics and Automation; Sensors and wearables

3. Research approach

This research paper is part of an ongoing effort to investigate the least discussed Construction 4.0, their requirements, and potential integration with other Construction 4.0 technologies. This overarching objective is achieved through a series of intermediate objectives: (1) identify the existing research work that categorized Construction 4.0 technologies; (2) find the most common technologies from the work identified in task 1; (3) conduct a systematic literature review of the detected technologies and provide a bibliometric analysis to cluster the technologies based on their occurrences; (4) identify the least researched technologies, extract the requirements of their implementations divided into three tiers : people, process, and technology i.e. who are the people who will use this technology and what are their needs, the process of implementing the technology, and the required technologies; (5) identify the challenges that hinders the implementation of the least researched technology; and (6) investigate if the technology can be implemented to both vertical and horizontal construction. This paper only covers the first three tasks.

3.1. Data collection

As outlined in Table 1, different research efforts resulted in different sets of Construction 4.0 technologies. Based on the results in Table 1, the most frequently mentioned technologies, sorted in alphabetical order, are listed as follows: AI, Additive Manufacturing, AR/VR, Automation and Robotics, Big Data, BIM, Cloud Computing, CPS, Digital Twin, GIS, IoT, Laser Scanners, Modular Construction, Offsite construction, Prefabrication, Sensors, and UAVs/Drones. This list of Construction 4.0 technologies was then used to conduct a systematic literature review to develop the blueprint of Construction 4.0.

Web of science was considered as the data collection source. The unit of analysis used was "term extracted from title or abstract" which is considered one of the most used bibliometric techniques [23]. Based on the Web of Science settings, all research written in English from 2010 to 2021 were selected. To ensure a good representative selection of the collected data, different queries were used. Moreover, to better focus the scope of the search and limit unnecessary burden, the focus was on articles that contained the searched query term in their title. Table 2 summarizes the queries used for the data collection.

Table 2. Summary of the queries used for the data collection

"Construction" AND	"Additive Manufacturing"; "Artificial Intelligence"; "AI"; "Augmented Reality"; "AR"; "Automation"; "Robotics"; "Building Information Modeling"; "BIM"; "Big Data"; "Cyber-Physical System"; "CPS ";"Cloud Computing"; "Digital twin"; "Geographic information System"; "GIS"; "Internet of Things"; "IoT"; "Laser scanners"; "Laser Scanning"; "Offsite Construction"; "Sensors"; "Drones"; "UAV"; "Unmanned Aerial System"; "Unmanned Aerial Vehicle"; "Virtual Reality"; "VR"

3.2. Quality assessment and normalization

Each query search was conducted separately to ensure better quality assessment and quality control. After each query search, the abstracts were investigated, and, if the topic of the research was found to be relevant to the current study, the research was selected. This selective process enabled the selection of the research work that is only specific to the construction industry and one of the selected Construction 4.0 technologies. As a result, a total of 2204 articles were collected.

3.3. Data analysis

The collected data was exported to both text files and CSV files. The selected research data was imported to VOSviewer, an open-source network visualization software application, which allows for the visualization of research work based on their occurrences and separates them into clusters. The VOSviewer setting of "creating a map based on text data" was employed where terms are extracted from the title and the abstract. Moreover, to filter the extracted terms, a VOSviewer thesaurus file was applied. For instance, the terms "BIM", "Building Information Modeling", "Building Information Modelling", and "BIM Models" were replaced by Building Information Modeling. Also, binary counting (i.e., only the presence or the absence of the term is counted and the number of occurrences of a term in the document is not considered) was selected. This selected option allowed for the reliable analysis of the data occurrences. Finally, only terms that represent Construction 4.0 technologies were selected to allow a better visualization of the results. The resulting map is presented in Figure 1.

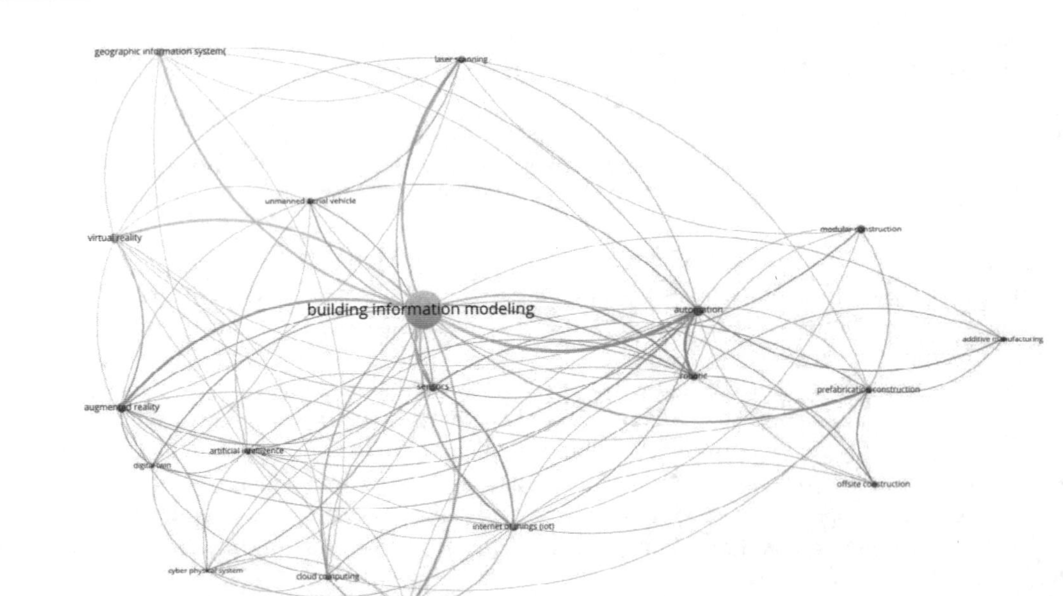

Figure 1. Network of normalized Construction 4.0 technologies' occurrences

Each cluster is identified by a color and relative technologies/points are grouped together based on highly connected terms that share similar fields. A total of five clusters were identified. The total number of occurrences and the resulting clusters are summarized in Table 3.

The first cluster includes a wide range of technologies available in Construction 4.0 which represent automation of the construction production, the second cluster represents storing and processing of construction data, the third cluster refers to construction data analysis and visualization, while the fourth cluster defines construction information modeling and design, and finally the fifth cluster depicts construction surveying and photogrammetry. According to the results, the most researched technology in the construction industry is BIM with 1289 occurrences. On the other hand, the least researched Construction 4.0 technologies are Digital Twin with 13 occurrences and Cyber-Physical Systems with 15 occurrences.

Table 3. Construction 4.0 technologies clusters and occurrences (n=number of occurrences)

Cluster 1	Cluster 2	Cluster 3	Cluster 4	Cluster 5
*Additive Manufacturing (n=28) *Automation (n=98) *Offsite Construction (n=32) *Modular Construction (n=59) *Prefabrication (n=41) *Robotics (n=51)	*Big Data (n=74) *Cloud Computing (n=33) *IoT (n=55) *Sensors (n=73)	*AI (n=28) *AR (n=69) *CPS (n=15) *Digital Twin (n=13)	*BIM(n=1289) *GIS (n=74) *VR (n=79)	*Laser Scanning (n=46) *UAV (n=47)

Based on the data analysis the least occurring technologies are Digital Twin and CPS and they are both clustered together (Cluster 3) with AI and AR. This observation is supported by findings of [23] who stated that the concept of Digital Twin emphasizes a holistic characterization of the operating facility by contributing with CPS to optimize bidirectional data flow and data interoperability. Moreover, the concept of Digital Twin in the AEC industry goes beyond BIM [24]. It integrates with several technologies, positioning it as the hub of all other technologies that are encompassed within the concept of Construction 4.0, including autonomous construction vehicles, aerial drones, sensors, BIM models, IoT, semantic web technologies, and AI [23] [25].

4. Conclusion, limitations, and further studies

This paper investigated the research corpus to categorize the technologies that encompass Construction 4.0. Bibliometric analysis and clustering of technologies based on their occurrences were conducted. Results showed that Digital Twin and CPS are the least investigated technologies and they were clustered with AI and AR. It should be noted that the results presented in this paper are limited to the web science database. Additional databases could be added to the conducted analysis.

Based on the findings of this paper, and part of the ongoing research effort, future work will investigate the requirements of Digital Twin (people, process, and technology) with a focus on the technologies that were clustered with Digital Twin. The use of Digital Twin in the AEC industry will be demonstrated through case studies in the vertical and horizontal construction sector. Providing an understanding of a promising technology that is key to connectivity in Construction 4.0 will help direct research toward investigating the interrelations between the technologies and better forecast of the digital future of the AEC industry.

5. References

[1] Ribeirinho M J, Mischke J, Strube G, Sjödin E, Blanco J L, Palter R, Biörck J, Rockhill D and Andersson T 2020 *The next normal in construction How disruption is reshaping the world's largest ecosystem* (McKinsey & Company)

[2] El Jazzar M, Urban H, Schranz C and Nassereddine H 2020 Construction 4.0: A Roadmap to Shaping the Future of Construction 37th International Symposium on Automation and Robotics in Construction (Kitakyushu, Japan)

[3] Rastogi D S 2017 CONSTRUCTION 4.0: THE 4th GENERATION REVOLUTION 12

[4] Buer S-V, Strandhagen J O and Chan F T S 2018 The link between Industry 4.0 and lean manufacturing: mapping current research and establishing a research agenda *Int. J. Prod. Res.* **56** 2924–40

[5] Moeuf A, Pellerin R, Lamouri S, Tamayo-Giraldo S and Barbaray R 2018 The industrial management of SMEs in the era of Industry 4.0. *Int. J. Prod. Res.* **56** 1118–36
[6] Craveiro F, Duarte J P, Bartolo H and Bartolo P J 2019 Additive manufacturing as an enabling technology for digital construction: A perspective on Construction 4.0 *Autom. Constr.* **103** 251–67
[7] García de Soto B, Agustí-Juan I, Joss S and Hunhevicz J 2019 Implications of Construction 4.0 to the workforce and organizational structures *Int. J. Constr. Manag.* 1–13
[8] Sawhney A, Riley M and Irizarry J 2020 *Construction 4.0: an innovation platform for the built environment* (Abingdon, Oxon; New York, NY: Routledge)
[9] Li X, Wu P, Shen G Q, Wang X and Teng Y 2017 Mapping the knowledge domains of Building Information Modeling (BIM): A bibliometric approach *Autom. Constr.* **84** 195–206
[10] Yin X, Liu H, Chen Y and Al-Hussein M 2019 Building information modelling for off-site construction: Review and future directions *Autom. Constr.* **101** 72–91
[11] Razkenari M, Fenner A, Shojaei A, Hakim H and Kibert C 2020 Perceptions of offsite construction in the United States: An investigation of current practices *J. Build. Eng.* **29** 101138
[12] Wang M, Wang C C, Sepasgozar S and Zlatanova S 2020 A Systematic Review of Digital Technology Adoption in Off-Site Construction: Current Status and Future Direction towards Industry 4.0 *Buildings* **10** 204
[13] Nassereddine H, Veeramani D and Hanna A 2019 Augmented Reality-Enabled Production Strategy Process 36th International Symposium on Automation and Robotics in Construction (Banff, AB, Canada)
[14] Nassereddine H, Hanna A S and Veeramani D 2020 Augmented Reality in the Construction Industry: An Industry's Perspective of Its Users, Phases, and Future Trends *Construction Research Congress 2020* Construction Research Congress 2020 (Tempe, Arizona: American Society of Civil Engineers) pp 743–52
[15] Oesterreich T D and Teuteberg F 2016 Understanding the implications of digitisation and automation in the context of Industry 4.0: A triangulation approach and elements of a research agenda for the construction industry *Comput. Ind.* **83** 121–39
[16] Dallasega P, Rauch E and Linder C 2018 Industry 4.0 as an enabler of proximity for construction supply chains: A systematic literature review *Comput. Ind.* **99** 205–25
[17] Maskuriy R, Selamat A, Ali K N, Maresova P and Krejcar O 2019 Industry 4.0 for the Construction Industry—How Ready Is the Industry? *Appl. Sci.* **9** 2819
[18] Boton C, Rivest L, Ghnaya O and Chouchen M 2020 What is at the Root of Construction 4.0: A Systematic Review of the Recent Research Effort *Arch. Comput. Methods Eng.*
[19] Forcael E, Ferrari I, Opazo-Vega A and Pulido-Arcas J A 2020 Construction 4.0: A Literature Review *Sustainability* **12** 9755
[20] Perrier N, Bled A, Bourgault M, Cousin N, Danjou C, Pellerin R and Roland T 2020 Construction 4.0: a survey of research trends *J. Inf. Technol. Constr.* **25** 416–37
[21] You Z and Feng L 2020 Integration of Industry 4.0 Related Technologies in Construction Industry: A Framework of Cyber-Physical System *IEEE Access* **8** 122908–22
[22] Turner C J, Oyekan J, Stergioulas L and Griffin D 2021 Utilizing Industry 4.0 on the Construction Site: Challenges and Opportunities *IEEE Trans. Ind. Inform.* **17** 746–56
[23] Boje C, Guerriero A, Kubicki S and Rezgui Y 2020 Towards a semantic Construction Digital Twin: Directions for future research *Autom. Constr.* **114** 103179
[24] El Jazzar M, Piskernik M and Nassereddine H 2020 Digital Twin in construction: An Empirical Analysis *EG-ICE* Workshop on Intelligent Computing in Engineering pp 501–10
[25] Tao F, Cheng J, Qi Q, Zhang M, Zhang H and Sui F 2018 Digital twin-driven product design, manufacturing and service with big data *Int. J. Adv. Manuf. Technol.* **94** 3563–76

IOP Conf. Series: Materials Science and Engineering 1218 (2022) 012012 doi:10.1088/1757-899X/1218/1/012012

Cost of supply and ventilation concepts in shield-driven tunnelling in due consideration of the engine type and drive technology

M Hoffeller

Research Assistant, Technical University Munich, Germany

E-Mail: m.hoffeller@tum.de

Abstract. Supply logistics as a supporting sub-process of tunnel driving is significantly responsible for a smooth construction process and, thus, for the desired project success. The main task of supply logistics is to provide the required logistical goods for the tunnel-boring machine. Available are track-bound and tyre-bound systems, which are driven by conventional diesel engines or alternatively by hybrid or electric motors. A multitude of regulations set exhaust emission limits, which can only be met by sufficiently dimensioned ventilation. In particular, the use of diesel engines can result in high costs for ventilation, which must not be neglected for an economic optimization of the supply concept. In order to compare the economic efficiency of different systems, a calculation concept is presented, which shows possible economic, but also ecological and logistical advantages of the individual systems by quantifying relevant factors. The costs for the different types of vehicles are examined separately for chassis and drive train, as well as for ventilation. The information and results obtained with the calculation provide a basis for the decision-making process when choosing an optimal logistics concept. Due to the quantities of logistical goods involved, the focus is on supply logistics in shield tunnelling.

1. Introduction

In order to ensure an optimal construction process under consideration of existing constraints - in terms of costs, deadlines and quality - and complying with existing restrictions, construction logistics concepts focus primarily on the optimization of material and information flows to be handled [1]. The possible savings and optimization potential of construction logistics over the entire construction process are the main reasons for a more intense analysis of the subject. Above all, processes with a particularly high influence on the adherence to schedules and costs of a project take an important position in the context of project execution. Nevertheless, logistic plays a seemingly undervalued role in the construction industry - despite it's evident importance for project success – in particular compared to other branches of industry [1].

The aim of construction logistic concepts is optimizing the material and information flows [2]. Construction logistics can be broken down into the segments supply logistics, disposal logistics, information logistics and construction site logistics [3].

The focus of this paper is the development of a calculation concept for determining cost of supply logistics in shield tunneling. The concept is developed in a way that it can be applied to any project. The application is exemplarily shown for a tunnelling project.

2. Logistics as a sub-process in shield-driven tunnelling

The decision for a suitable supply concept depends on the type of tunnelling, the existing construction and the project-specific circumstances [4]. In shield tunneling, the logistical goods comprise in particular the tubbings, the annular gap mortar, pipes for extending the supply lines and personnel. The objective of all efforts is the undisturbed operation of these core processes. The supply of a shield machine with the required materials can be carried out either on trackless or track-bound systems. So-called multi-service vehicles (MSVs) are used as trackless systems. With regard to the drive train systems can be divided into diesel, electric or hybrid-driven systems. As a result of increasingly strict regulations in the context of occupational health and safety guidelines, the use of alternative drives has also gained importance for tunnel construction. In addition to the regulations for exhaust gas limit values, so-called workplace limit values were specified. In order to comply with these prescribed limits, active ventilation is required in the tunnel during the construction period. Alternative drive concepts with hybrid or purely electric drives can significantly reduce exhaust gases and pollutants in the tunnel, which has an impact on the dimensioning of the ventilation system. This can save cost for the operation of the ventilation system, especially with increasing tunnel length.

3. Calculation methodology

In the following, a methodology based on physical principles is developed which allows different concepts for the supply of shield machines to be evaluated in terms of economic efficiency by quantifying all relevant factors. The total cost for supply logistics consist of various components. For each of those components a methodology for calculating the cost is presented below. On the first level, the cost can be divided into cost of supply vehicles and cost for ventilation. The second level breaks down the supply vehicle cost into explicit vehicle cost and fuel cost. The ventilation cost can be further divided into infrastructure cost, which consist of the cost for ducts, clamps and fans, and energy cost. All components are developed separately for the variants shown in Table 2:

Table 1: Variations of concepts.

	Trackbound	**Trackless**
Diesel- driven	Diesel-trackbound	Diesel-trackless
Hybrid- driven	Hybrid- trackbound	Hybrid- trackless
Electric drive	Elektro-trackbound	Elektro trackless

Since the underlying calculation steps are based on physical principles - and thus independent of a specific project - this model can be used to simulate the supply logistics for any tunnel construction project and to predict the cost taking into account all relevant influencing project-specific conditions.

3.1. Supply vehicle cost

To calculate the vehicle cost the required number of vehicles must be known, which depends on the speed of excavation and the cycle time. The definition of the cycle time is as follows:

The cycle time describes the time required by the supply unit to be loaded at the tubbing storage facility, to travel to the tunnel machine (TM), to be unloaded there and to travel the distance back to the tubbing storage facility.

Hence, the cycle time can be determined by the tunnel length and the vehicle speed. The speed is given by a speed profile for trackless and track-bound vehicles depending on the gradient and weight. Therefore, the tunnel must be divided into sections of equal gradient. Here, a separation must be made between outward and return journeys due to the gradient and the mass of the vehicle. The time t_n required for the individual sections can be calculated as follows:

$$t_n = \frac{s_n}{v_n} \tag{1}$$

t_n = time of section n

s_n = length of section *n*
v_n = speed of section *n*

To calculate the time for the total distance to be covered, all sections n and the time t_{VE} for the individual supply sub-processes are added. After loading the tubbings, the mortar box is filled (t_{load}). The time for driving the vehicle into the TM must also be taken into account (t_{TM}). Unloading of the segments and the mortar box are done simultaneously (t_{unload}). After this, the driver of the MSV has to change the driver's cab (t_{change}); this process does not exist for track-bound concepts.

$$t_{VE} = t_{load} + t_{TM} + t_{unload} + t_{change} \tag{2}$$

The cycle time t_c can be calculated by adding up the individual sections and the individual subprocesses:

$$t_c = \sum_{i=1}^{n} t_n + t_{VE} \tag{3}$$

The number of vehicles required is determined by the cycle time t_c and the time t_V to set a tubbing and advance a drilling stroke. This results from the advance rate and the width of the tubbing:

$$t_V = t_{tubbing} + v_{advance\ rate}/b_{tubbing} \tag{4}$$

Finally, the number of required vehicles *N* is

$$N = t_c/t_V \tag{5}$$

As the tunnel length continues to increase over the entire project duration, more vehicles will be needed as the project progresses. One advantage of track-bound vehicles is already obvious: by attaching wagons, no extra vehicles need to be provided for the transport of staff and pipes. In case of using trackless systems, an additional vehicle must be used for pipes and lines, as well as two vehicles for transporting staff members. Regardless of the system, another vehicle is required as a rescue vehicle. It should be noted that cost for power supply are not affected, since the transport power consumption is not significant compared to TM.

3.2. Supply vehicle – fuel costs

To calculate fuel costs, the energy required to move a vehicle needs to be known. This is a purely physical quantity required to compensate for the driving resistance which can be calculated as follows:

$$F_{FW} = F_{Luft} + F_{Roll} + F_{Steig} + F_B \tag{6}$$

$F_{FW} = vehicle\ resistance\ [N]$
$F_{Luft} = air\ resistance\ [N]$
$F_{Roll} = rolling\ resistance\ [N]$
$F_{Steig} = climbing\ resistance\ [N]$
$F_B = acceleration\ resistance\ [N]$

The required energy is independent of the drive technology. The various drive technologies differ only in their efficiency and in the ability of recuperating the energy released. The air resistance increases with the square of the vehicle speed. Due to the low vehicle speed in the scenario considered here, the proportion of air resistance is negligibly small and can be neglected. Since the vehicles assumedly will usually not start off in the steepest (decisive) section, the acceleration resistance is also neglected. The remaining driving resistances is:

$$F_{roll} = c_{Roll} \cdot m \cdot g \cdot \cos\alpha \tag{7}$$

$F_{roll} = rolling\ resistance\ [N]$
$c_{Roll} = rolling\ resistance\ coefficient\ [-]$
$= 0{,}014 - 0{,}02$ for trackless systems and $0{,}001 - 0{,}002$ for trackbound systems
$m = vehicle\ mass\ (incl.\ load)\ [kg]$
$g = gravity = 9{,}81\ [m/s^2]$
$\alpha = pitch\ angle\ [rad]$

$$F_{steig} = m \cdot g \cdot \sin\alpha \tag{8}$$

$F_{steig} = climbing\ resistance\ [N]$
$m = vehicle\ mass\ (incl.\ load)\ [kg]$
$g = gravity = 9{,}81\ [m/s^2]$
$\alpha = pitch\ angle\ [rad]$

Due to the influence of the load and the gradient, the tunnel also needs to be divided into individual sections for the calculation of fuel costs. If the speed and vehicle load are known, the time required for covering the total length of the individual section can be calculated, based on these considerations: The daily advance of the machine consists of several drilling strokes. For each drilling stroke, the TM is to be supplied with a tubbing and other materials. Consequently, one trip of the supply train is necessary for each drilling stroke. The distance to cover is not constant, however, increases with each drilling stroke by the length of the drilling stroke or by the daily advance each day.

If the corresponding driving resistance is known (equation (6)), the power and the resulting energy required for the individual sections can be calculated. Since the process is a translational motion, the ZUGHAKENLEISTUNG("pulling power") must be calculated. This is the maximum power available after subtracting all losses at the towing device and is a frequently used characteristic value for towing vehicles in rail and road traffic. It is calculated as:

$$P_n = F_{W,n} \cdot v_n \tag{9}$$

For starting and braking as well as speed changes, the acceleration energy also needs to be taken into account:

$$E_{acc} = {}^1/_2 \cdot m \cdot v_n^2 \tag{10}$$

As a first step, the model calculates the physical energy (E_{ideal}) required to move a vehicle with a known weight and payload for a given distance using the equations shown above:

$$E_{ideal} = E_{acc} + P_n \cdot t_n \tag{11}$$

Then, the required total energy is weighted by the corresponding efficiency values and cumulated over all sections. Finally calculating the fuel quantity for the diesel drives, the specific fuel consumption of the installed unit is used, leading to cost on the basis of an exemplary diesel price of1.00 €/l at the considered construction site.

3.3. Ventilation – energy cost

A physical model representing the ventilation system is being developed, which can be used to calculate the maximum pressure and the maximum amount of air that needs to be delivered (figure 1).

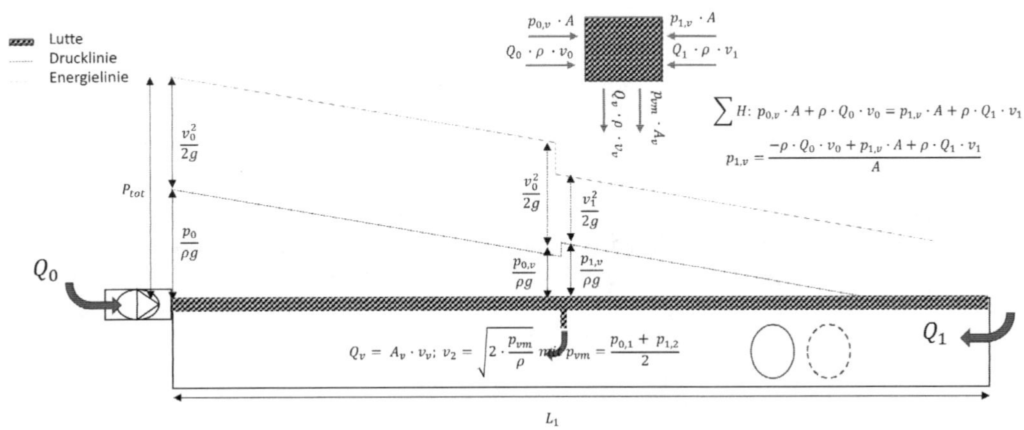

Figure 1: model of the ventilation system

The tunnel ventilation is based on pipe hydraulics. The duct is regarded as a pipe and the fans as a pump. The decisive factor for the required fan power is the total pressure present. This can be calculated for ideal, frictionless flows using the Bernoulli principle. In real flows, internal friction is given by the viscosity of the fluid, which leads to thermal losses which need to be taken into account via a corresponding loss height h_v. The total pressure at the fan is calculated by the Bernoulli equation extended by the loss height:

$$P_{tot} = \frac{1}{2} v_1^2 \cdot \rho + p_1 + \frac{1}{2} v_1^2 \cdot \rho \cdot (\lambda \cdot \frac{L_1}{d} + \Sigma \zeta_{i,1}) \tag{12}$$

$\lambda = coefficient\ of\ friction\ = 0{,}015[-]$
$d = diameter\ of\ the\ duct\ [m]$
$L = lenght\ of\ the\ duct\ [m]$
$\zeta = loss\ coefficient\ = 0{,}7[-]$

The speed v can be calculated by the volume flow Q and the cross-sectional area A:

$$v = \frac{Q}{A} \tag{13}$$

The principle of linear momentum can be used to determine the pressure as shown in figure 1:

$$P_1 = \frac{p_2 \cdot A + \rho \cdot Q \cdot v_2 - \rho \cdot Q \cdot v_1}{A} \tag{14}$$

In this model, as a simplification atmospheric pressure conditions are assumed to prevail at the exit point in the tunnel. In fact, however, a certain overpressure is generated by the ventilation at the tunnel face, which is reduced along the tunnel axis to the portal. In relation to the pressure within the ducts this is negligible due to the low flow velocities along the tunnel axis.

The required power of a fan can be calculated as follows:

$$N_V = \frac{p_{tot} \cdot Q}{\eta_M \cdot \eta_V\ 10^3} \tag{15}$$

$N_V = required\ power\ of\ the\ fan\ [kW]$

$Q = required\ air\ to\ be\ transported\ \left[\frac{m}{s}\right]$

$p_{vent} = total\ pressure\ at\ the\ fan\ [Pa]$
$\eta_M = motor\ efficiency$
$\eta_V = fan\ efficiency$

As can be seen from the formulae, the required capacity is strongly dependent on the volume of air to be transported. This is precisely determined by various regulations of the individual countries, which needs to be taken into account when dimensioning the ventilation system. According to figure 1 loss due to leakage in the ducts must also be added to the required fresh air volumes. The loss quantity is calculated using the Toricelli's law and the mean hydrostatic pressure. Since this in turn depends on the volume of air to be transported, it is necessary to iterate. Figure 2 shows a summary of the used calculations to determine the required power for a fan:

Calculation of the power of the fan				
		Diesel	Hybrid	Electic
Section 2	Q2 [m^3/s]	87,00	51,00	35,00
	v_2 [m/s]	17,72	10,39	7,13
	λ*L/D [-]	42,00	42,00	42,00
	$v_2^2 *\rho/2$ [kg/s^2m]	188,47	64,77	30,50
$=v_2^2 *\rho/2*(\lambda*L/D+\Sigma\zeta)$	p2 [Pa]	7.981,85	2.742,87	1.291,82
Loss	Pvm	7.490,56	2.574,05	1.212,31
=square root of (2*pv/ρ)	v_A	111,73	65,50	44,95
	Q_A	65,82	38,58	26,48
Section 1				
	Q_1 [m^3/s]	152,82	89,58	61,48
	v1 [m/s]	31,13	18,25	12,52
	$v_1^2 *\rho/2$ [kg/s^2m]	581,50	199,83	94,11
$=(P_v*A+\rho*Q_2*v_2-\rho*Q_1*v_1)/A$	P_1 [Pa]	7.195,79	2.472,75	1.164,60
$=P_1+v_2^2 *\rho/2*(\lambda*L/D+\Sigma\zeta)$	P_{0S} [Pa]	31.822,34	10.935,38	5.150,27
$=v_1^2 *\rho/2$	P_{0D} [Pa]	581,50	199,83	94,11
	P_{0tot} [Pa]	32.403,84	11.135,21	5.244,38
	L [kW]	**6.189,79**	**1.246,89**	**403,02**

Figure 2: used equations to determine the power of a fan

The necessary energy [kWh] can be calculated by multiplying the excavation time [h] required for the section under consideration by the power [kW] required along the section.

3.4. Ventilation – infrastructure costs
The number and size of the fans are based on the required power. The total length of the ducts required is derived from the tunnel length. Accordingly, the number of required clamps can also be calculated.

4. Case Study of an Exemplary Project
The example used here is a twin-tube tunnel of length 8,200 m per tube, where tire-bound vehicles were used. Official market prices were obtained to quantify the vehicle costs in order to obtain significant results. For not used variants, offers from various companies were available. For vehicle cost, a buy-back option after the end of the project of 20% for track-bound and 35% for tire-bound vehicles was taken into account according to the offers. The prices of the MSVs and the track-bound vehicles also include a fixed flat rate for wear and tear and maintenance.

For the determination of the energy costs 150 m long sections of the ducts are selected. With a tunneling speed of 20 m/d, this results in a duration of 180 h. With a known electrical power price, which is 0.19 €/kWh for the project under consideration, the aeration costs can finally be determined. For all variants, a diameter of the ducts of 2.5 m was assumed.

For the calculation of the infrastructure costs, the prices for the fans were estimated at €100,000 for the diesel variant, €75,000 for the hybrid variant, and €50,000 for the electric variant, according to information from a company. The price for the ducts was estimated at 17,23 €/m and for the clamps at 32,54 € per piece.

Considering the total cost of the supply concept (see figure 3), it can be noted that the fuel cost and the infrastructure cost of tunnel ventilation are playing a minor role. The vehicle cost are by far making up the largest part.

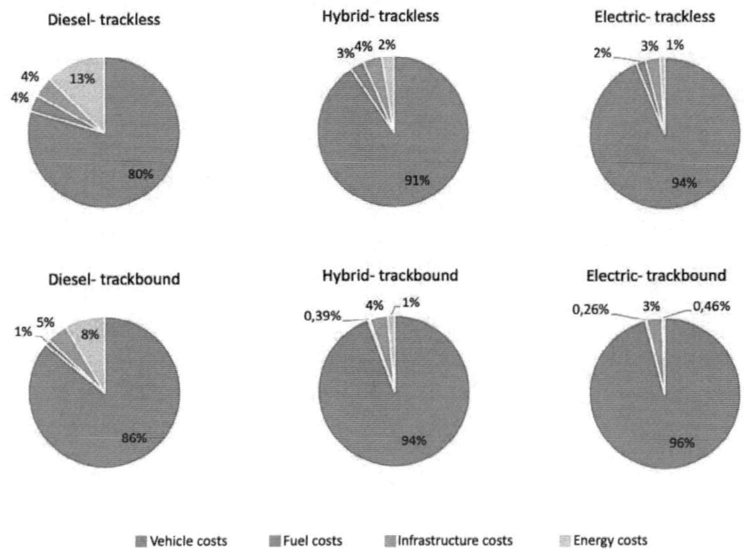

Figure 3: Results for the different systems

Due to the lower speeds and transport capacities, more vehicles are required for tire-bound systems as the tunnel length increases (from approx. 4,200 m, six vehicles for tire-bound and four vehicles for track-bound systems). That is why the vehicle cost for tire-bound systems increase faster than for track-bound systems as the tunnel length increases. This also results in an increasing exhaust gas production and thus an increased amount of fresh air required. Using alternative drives, energy cost increase not that much with rising tunnel length.

Alternative drives produce far fewer exhaust gases, which is why the dimensioning of the ventilation concept is approx. 25 % - 50 % lower, depending on the drive type. E.g., using diesel-powered MSVs, leads to energy cost for the reference project of approx. 13 % of the total costs, while using MSVs with an electric drive, they only account for a share of approx. 1 %. However, the higher initial investment cost of alternative drives are not compensated just by their lower energy and fuel cost. As conventional drives are significantly less expensive to purchase, they are the best choice from a cost-oriented point of view - despite higher operating costs - for the reference project in the case of track-bound concepts up to a tunnel length of approx. 7,600 meters. For track-bound concepts, the use of alternative drive concepts would only be profitable for a tunnel length of more than approx. 7,300 m. In terms of cost, tire-bound concepts are to be preferred to track-bound concepts with conventional drive up to a tunnel length of approx. 4,600 m (figure 4).

The differences between tire- and track-based concepts in regard to the energy cost turn out to be significant, amounting to 28.9 % for the diesel variant, 13.9 % for the hybrid variant and only 12.1 % for the electric variant of the energy cost of the tire-bound counterparts. The reason for a higher monetary efficiency of track-bound systems is their higher transport performance as well as the rolling resistance of rail wheels and rubber tires, which is about a magnitude lower.

It should be noted that the vehicle cost are not including wages for drivers. Due to the higher number of vehicles required for trackless systems, it can be assumed that track-bound systems are better solution from an economic point of view. In addition, the maintenance costs for trackless vehicles are higher than those for track-bound concepts.

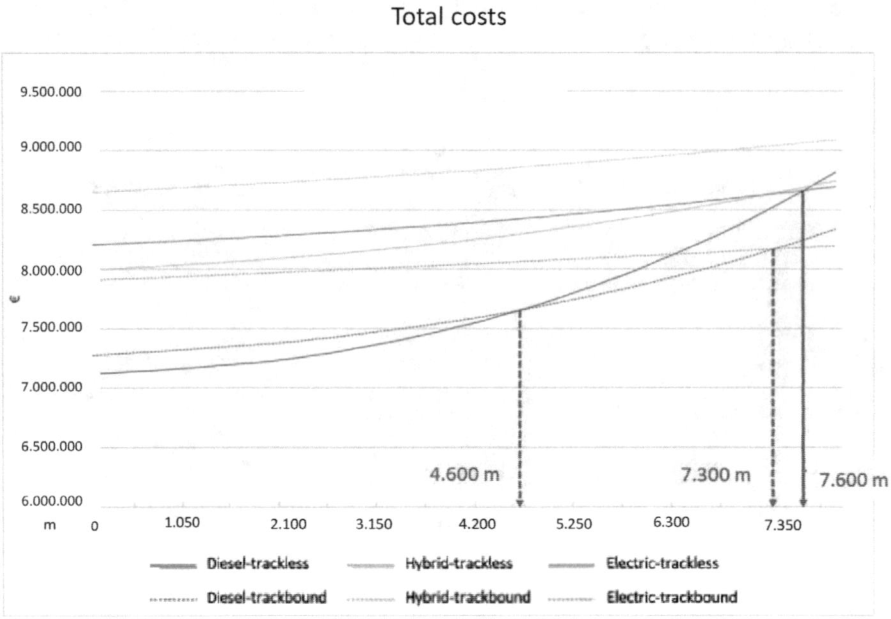

Figure 4: Total costs of the different concepts

5. Conclusion

The evaluation of the logistics costs showed that the initial investment costs of the vehicles account for the largest share of the total costs. As the tunnel length increases, the remaining cost items become increasingly important, but the vehicle cost remain decisive.

Whereas the initial investment costs for the vehicles and for the infrastructure of the ventilation are linear with increasing tunnel length, the operating costs have a strongly exponential character. As the number of vehicles increases - and so the difference in the exhaust gas production increases - the operation costs of diesel and hybrid-driven systems rise much faster. The increase in costs over the length of the tunnel is much more exponential for diesel drives.

Generally speaking, the costs of tire-bound concepts rise more sharply over the length of the tunnel than those of track-bound concepts. The reason for this is the strong increase in ventilation costs (which account for the majority of aeration costs) and fuel costs for tire-bound concepts due to the lower transport capacity and the lower driving resistance.

The energy costs linked to the tunnel ventilation can be drastically lowered by using alternative drives. Here, alternative drive concepts offer great saving potentials. For example, the use of hybrid-driven systems in the example project presented here results in a cost saving of approx. 81% in the energy cost of tunnel ventilation.

References

[1] Günthner W and Zimmermann J 2008 *Logistik in der Bauwirtschaft - Status quo, Handlungsfelder, Trends und Strategien*. (Nürnberg) S. 11

[2] Zimmermann J 2009 (Hrsg.): *Baulogistik- Innovationspotenziale für die Bauwirtschaft*. Lehrstuhl für Bauprozessmanagement und Immobilienentwicklung. (München: Technische Universität München) S. 13

[3] Zimmermann J and HAAS B 2011 *Unerschlossene Potenziale in der Logistik von Bauunternehmen in Straßen- und Tiefbau*, vol 65 Nr. 10. S. 6-9

[4] Kanther C 2006 *Untersuchung der Prozesse im Schutterungsbetrieb bei Tunnelbaustellen und deren wesentliche Parameter, wie z.B. Leistungswerte und Kosten*. (München: Technische Universität München) S. 68

Creating a tiny home building code to positively affect long-term sustainability

K Boyer[1] and A Simons[2]

[1]Program Graduate, BSCI, Auburn University, US
[2]Assistant Professor, BSCI, Auburn University, US

Email: ellisap@auburn.edu

Abstract. There is an ongoing effort in the tiny home construction industry in the United States to create a tiny home specific building code. To date, there has been little data collected to reflect how the creation of a building code could affect the tiny home industry. Mixed methods research was conducted, utilizing both quantitative and qualitative data. Interviews were conducted with tiny home experts, builders, owners, advocates, insurers, financers, and building officials. The results of the research indicated that creating a tiny home building code would positively affect the sustainability of the tiny home movement. Of the different types of tiny homes identified in the research, those associated with tiny homes on wheels (THOW) would likely benefit the most from a creation of a tiny home specific building code. The majority of those surveyed (76%) were in support of creating a THOW specific building code. Five main categories were identified as necessary for inclusion in a THOW specific building code including trailer foundations, trailer to structure connections, framing, electrical, and sanitation. Eleven subcategories, or specific items, were identified as essential for inclusion in a THOW specific building code. Future research should validate the data collected and ensure all pertinent code items are addressed.

1. Introduction and background

The tiny home movement has gained notoriety in the United States over the last several years due to increased publicity of the lifestyle through television programs, internet blogs, and other forms of social media. Tiny homes generally range in size from 60 to 400 square feet with most near the 200 square foot mark [1]. Tiny homes offer the potential for increased financial freedom. The cost of living in the United States doubled between 1992 and 2017 [2]. This sharp increase (145%) in cost of living makes tiny homes an attractive housing option for many people. One study found that an unaffordable housing market in the U.S. ignited new housing choices, potentially altering housing trajectories in the future [3]. Additionally, the U.S. Department of Housing and Urban Development (HUD) reported that 580,466 people experienced homelessness in the United States on a single night in 2020, an increase of 12,751 people, or 2.2 percent, from 2019 [4]. With the rising cost of living and rising rate of homelessness in the U.S., tiny homes offer a potential solution worth exploring.

Many subcategories of tiny homes exist. Though the following list is not all-inclusive, most tiny homes fall into one of the following five categories: tiny house on foundations (THOF), accessory dwelling units (ADU), tiny house on wheels (THOW), park model recreational vehicles (PMRV), and recreational vehicles (RV). The research presented herein focused on tiny homes on wheels (THOW). Advocacy groups supporting these various types of tiny homes have been formed in recent years to help advance and promote the tiny home movement. The goal of the American Tiny House Association

(ATHA) is to is to support tiny house enthusiasts who are seeking creative and affordable housing as part of a more sustainable and self-reliant lifestyle [5]. The Tiny Home Industry Association (THIA) is a non-profit organization on a mission to advocate for regulation changes, develop standards, and promote best practices in the construction, placement, and widespread use of tiny homes as permissible and permanent housing [6]. Other formal groups and individuals have advocated for the advancement of the tiny home movement in recent years. These advocacy groups offer a pool of individuals who are well aware of the challenges that exist within the tiny home movement, even those that may not be well known among the general population.

Several known barriers to the tiny home movement exist, making it difficult for some to build or purchase a tiny home. State and local governments have resisted the tiny home movement due to lack of regulations. One study identified lifestyle, land, financing, and classification as early barriers to the tiny home movement [7]. This study recommended that future research should include interviewing building officials, builders, and others to close the gap between building code and uncovering solutions for tiny home owners acquiring land, solutions to standardize building codes for tiny homes and even serve as education on the tiny home movement [7]. Another study concluded the two greatest barriers to tiny homes are building codes and zoning ordinances that treat them as illegal or illegitimate types of housing [8]. As noted in another article, tiny homes fall between cracks in the current building code [9]. The aforementioned barriers has also made it difficult for some to procure insurance for a tiny home [10]. A standardized building code could help support legalization of tiny homes as a form of long term housing. To date, there has been little data collected to reflect how the creation of a building code could affect those involved with tiny homes including builders, owners, insurance agents, financial partners, and others. Additionally, little research has been done to identify specific areas of building code which need to be written or specified for tiny homes.

Tiny homes offer one potential solution to rising home prices and the increased rate of homelessness in the U.S. today. The tiny home movement includes several subtypes of tiny homes that have garnered support from various advocacy groups. The movement has also seen adversity, with various barriers to tiny home ownership. Previous research has identified several known barriers to the tiny home movement, one of which is the lack of related building code [7], [8]. The research presented herein sought to determine whether a tiny home specific building code would be positive for the movement and identify specific areas of building code to be included.

2. Methodology

Mixed methods research was utilized to complete this study. Qualitative research was conducted by interviewing industry professionals, groups, and associated trade professionals including insurance and financing companies. These interviews were conducted via phone and engaged tiny home builders, developers, social media advocates, designers, and owners across the country. Each interview took approximately one hour to complete. Interviewees were asked a series of questions that required open-ended discussion. The questions included background about the individual, their position on current and future building regulations, their thoughts on current inspection services for tiny homes, and how financing and insurance for tiny homes had affected their business. Notes were taken during all phone interviews and answers to the interview questions were compared and analysed to find trends or distinctions among answers.

The initial interviews described above helped refine a standard set of eleven survey questions which were distributed to builders associated with the tiny home movement. This survey is herein referred to as Survey 1. Qualtrics survey software was used to collect and process the survey data. The tiny home builder survey focused primarily on the five categories of code issues frequently mentioned during the initial interviews. Builders of varying experience and involvement were invited to participate in this survey. This survey was disseminated as widely as possible through email and social media.

After analysing the data collected from Survey 1, and comparing it to the qualitative data collected during the interviews, the survey was then updated and expanded for use as a quantitative means of data collection. A new survey was then created in Qualtrics that consisted of 24 questions including multiple

IOP Conf. Series: Materials Science and Engineering 1218 (2022) 012013 doi:10.1088/1757-899X/1218/1/012013

choice, ranking, multiple answer, and fill-in-the blank. This survey is herein referred to as Survey 2 and was open to all members of the tiny home movement. Links to Survey 2 were distributed via email and social media and respondents were encouraged to share the link with others. Upon closure of the survey the data was analysed using Qualtrics reporting software.

3. Results and discussion

As previously mentioned, the research was conducted in 3 parts. First, qualitative data was gathered through phone interviews with tiny home designers, builders, owners, and advocates to develop a set of survey questions for Survey 1. Survey 1 was sent to 11 tiny home builders to further refine and validate the survey questions. The refined survey, Survey 2, was open to anyone associated with the tiny home movement and was disseminated as widely as possible through email and social media.

3.1. Interviews

A total of 14 interviews were conducted including 8 professional tiny home builders, 2 do-it-yourself builders, 1 tiny home insurance agent, and 3 tiny home advocates. After completing all 14 interviews, the notes taken during the process were reviewed and a common trend was discovered among tiny home builders. Of the 10 builders interviewed, all of them shared the same opinion that a building code or standard needed to be written for tiny homes to further the advancement of the tiny home movement. Furthermore, when asked if there were specific code concerns, interviewees focused on five categories of code compliance issues: trailer foundations, trailer to structure connections, framing, electrical, and sanitation. These five categories of code compliance formed the basis of the Survey 1 which was distributed to tiny home builders. Additionally, the initial interviewees commonly stated that tiny homes on wheels (THOWs) were the most unique brand of tiny homes and currently most in need of a common code. Therefore, the remainder of the research focused solely on THOWs.

3.2. Survey 1

As previously mentioned, a builder specific survey was created based on results from the initial interviews. The survey questions themselves are excluded from this paper for brevity. The survey included both pre-determined and open-ended questions which focused primarily on the 5 categories of code issues identified during the initial interviews. These categories included trailer foundations, trailer to structure connections, framing, electrical and sanitation. The results of Survey 1 questions revealed subcategories, representing more specific areas of deficiencies in current building code as it relates to THOWs, within each of the five main categories. As shown in Figure 1, eleven total sub-categories representing deficiencies within current building code were identified by the survey respondents.

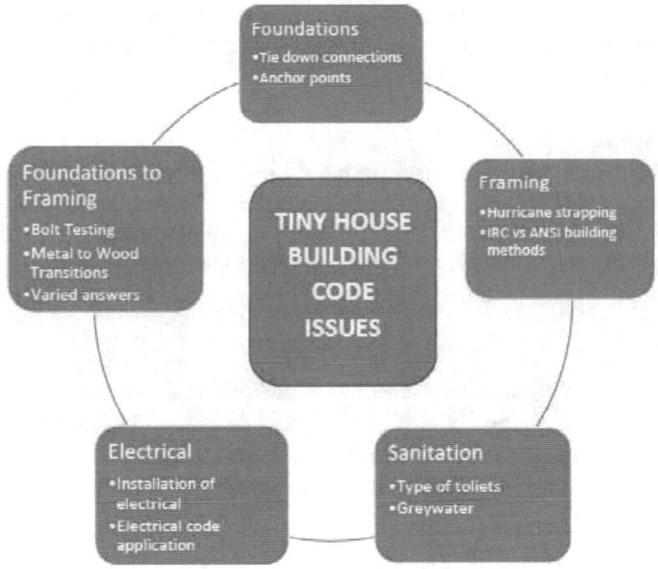

Figure 1: Tiny home building code categories and builders survey sub-categories

Based on the extent of the perceived lack of clarity in current building codes, which do not specifically address some of the unique features of tiny homes, the creation of a tiny-home specific building code could prove beneficial by improving safety, quality, and consistency of tiny home construction while providing much needed guidance to builders. This type of code could benefit both professional and DIY tiny home builders.

3.3. Survey 2

Survey 2 was available for all members and supporters of the tiny home movement. The purpose of Survey 2 was to substantiate the findings from Survey 1 while also comparing perceptions between builders, developers, potential and current owners. When considering the development of a tiny home specific building code, the concerns of all stakeholders should be taken into account, including consumers, suppliers, financers, and potentially others.

Questions 1-4 were intended to classify the respondent based on experience and location. Questions 5-6 were used to further qualify the experience of builders while questions 7-8 were meant for tiny home owners. Questions 9-12 were used to assess the availability of and/or the need for both insurance and financing for tiny homes. Questions 13-18 related to specific building inspection services and codes, while questions 19-23 focused on the top five general areas of code deficiency found during the interviews and substantiated by Survey 1.

Survey 2 was conducted over the course of two months and amassed 72 responses. Question 1 responses were recorded as follows: 16 owners, 26 potential owners, 6 advocates, 16 builders, 2 business entities, and 6 who identified as “other.” Question 2 responses were as follows: 18 respondents had been involved with the tiny home movement for 1 year or less, 28 respondents for 2-3 years, 12 respondents for 4-5 years, 9 respondents for 6-9 years, and 5 respondents for 10 or more years. Question 3 results indicated that 21 survey respondents chose to “go tiny” for financial freedom, 16 for both lifestyle change and business opportunity, 10 to downsize, and 9 for “other” reasons. Upon further analysis, the owners and potential owners accounted for the majority of the responses for both financial freedom and lifestyle changes. Builders accounted for the majority of the business opportunity responses. In response to question 4, 22 respondents were located in the Southwest, 20 in the Southeast, 15 in the Northeast, 7 in the Northwest, 7 in the Midwest, and 1 on the West Coast. This indicated a good geographical amongst those who completed the survey. In response to question 5, 10 builders completed at least 1 tiny home,

8 builders completed 21 or more tiny homes, 4 completed 2-3 tiny homes, 3 completed 4-5 tiny homes, 2 completed 16-20 tiny homes, and 1 completed 6-10 tiny homes.

In response to question 9, 16 of 29 respondents stated that they did have insurance on their tiny home. The 16 responses claiming to have insurance were then asked to identify what type of insurance they had. Of the 11 responses recorded listing the type of insurance, 6 of the answers identified RV insurance as the policy type. Additional answers were area specific or different types of tiny home specific insurance such as homeowner's insurance on a THOF. In response to question 10, 5 out of 21 answered "yes" to having financing for their tiny home. Of the financing type recorded, RV financing was the most common answer. This finding supports the idea that financing is an underlying issue within the tiny home movement. Although insurance has become readily available for tiny home owners, financing has yet to offer options to the movement beyond small credit unions or personal equity companies.

To further identify how insurance and financing have impacted the tiny home movement, questions 11 and 12 posed a scenario about the impact on building or buying a tiny home if there were more readily-available options for insurance and financing, respectively. A resounding 50 of 72 "yes" responses were recorded for question 11 regarding insurance, while 13 responded "maybe," and 9 responded "no." Even more overwhelmingly, 52 of 72 responded "yes" to question 12 regarding financing, while there were 10 responses each for "maybe" and "no" for the same question.

Questions 15 asked if those surveyed are in favour of creating a uniform code for tiny homes on wheels (THOWs). The response to question 15 shows that 76% (55 of 72) of those surveyed are in favour of creating a THOW building code. With less than 7% of survey takers answering "No" to the question there was overwhelming support that a building code would be widely supported in the tiny home community. Similarly, question 16 asked if the respondent would be more likely to buy or build a THOW if a tiny home specific building code were created. Nearly 70% responded "yes" to this question while 22% said "maybe" and 8% "no."

Question 17 provided interesting results. As shown in Figure 2, there were great inconsistencies in which building code those surveyed believed to be most applicable to THOWs. The results provided in this figure give credit to the postulation that the formation of a consistent standard could be beneficial for many reasons.

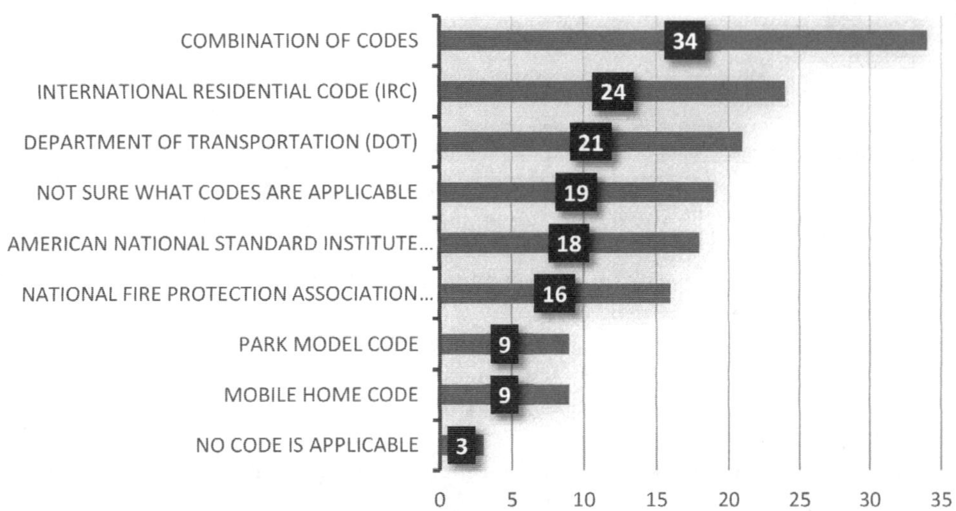

Figure 2: Current building codes cited as most important for THOWs

Questions 19-23 utilized a modified format of the builder survey (Survey 1) to pinpoint specific code issues perceived to be most important to respondents. These questions correspond to the perceived biggest issues/omissions in current code combinations, or lack thereof, for foundations, trailer-to-structure connections, framing, sanitation, and electrical items and results are summarized in Table 1, below.

Table 1: Specific items in need of code inclusion

Major Category	Minor Category	Responses
Foundations	Tie-Downs/Securing	32
Trailer-to-Structure	Metal to Wood Connections	18
	Spacing/Frequency of Connector	20
	Type of Connector Used	16
Framing	Use of Hurricane Straps	23
	Type of Members Used	15
	Energy Code Requirements	11
Sanitation	Proper Disposal of Waste Removal	33
	Composting Systems	10
Electrical	Installation Qualifications	32
	Wire Protection	15

As shown in Table 1, there were several consistent responses noted within each of the 5 major categories of building code. For foundations, potential owners and builders accounted for the majority of the tie down/securing responses. There were three significant responses in the trailer to structure category including metal to wood connections, connector type, and spacing/frequency of connectors. These results indicate a significant need for specific code guidance in this category. When framing a THOW, the underlying assumption is that these homes must withstand large wind forces during tow. Therefore hurricane strapping and type of framing members used make sense in the context of a THOW. Energy code requirements was also a significant response in the framing category. Within the sanitation category, the common responses of proper disposal of waste removal and composting systems which makes sense in the context of a THOW. Respondents indicated that installer qualifications should be standardized and also listed wire protection as a deficiency in current code combinations.

4. Conclusions

Among those interviewed and surveyed for this research, there was positive support for the creation of a standard building code specific to tiny homes on wheels (THOWs) with over 76% (55 of 72) of those surveyed in favour of creating a THOW specific building code. Initial interviews led to selection of 5 overall categories of emphasis for a THOW building code, including trailer foundations, trailer to structure connections, framing, electrical, and sanitation. Two follow-on surveys identified 11 subcategories within each of the five main categories including the following: tie-downs/securing; metal to wood connections; spacing frequency of connector; type of connector used; use of hurricane straps; type of members used; energy code requirements; proper waste removal; composing systems; electrical installation qualifications; and wire protection. A THOW specific building code should focus, at a minimum, on these 11 items.

The creation of a building code of this type would remove much of the ambiguity from the THOW building process, thus leading to more standard design and construction procedures for this type of home. A THOW code also has the potential to facilitate legalization of THOWs as building officials would have a standard for inspection. This could ease the current difficulties some individuals have had obtaining financing and/or insurance.

IOP Conf. Series: Materials Science and Engineering 1218 (2022) 012013 doi:10.1088/1757-899X/1218/1/012013

Future research should seek to expand and validate the current data set and ensure all relevant items are included. Additionally, future research should document the creating and implementation of a THOW specific building code.

5. References

[1] Mitchell R 2020 How big can a tiny house be without breaking the law *The Tiny Life* https://thetinylife.com/tiny-house-dimensions-what-size-can-a-tiny-house-be-without-breaking-the-law/

[2] Consumer Pricing Index (CPI) 2018 https://www.statista.com/statistics/190974/unadjusted-consumer-price-index-of-all-urban-consumers-in-the-us-since-1992/

[3] Carras M 2019 Tiny house big impact?: an investigation into the rise of the tiny home lifestyle (THL) in the United States University of St Andrews thesis https://eds-a-ebscohost-com.spot.lib.auburn.edu/eds/detail/detail?vid=6&sid=ad2fc231-d75c-42da-a55b-547d1bd46ebd%40sdc-v-sessmgr01&bdata=JnNpdGU9ZWRzLWxpdmUmc2NvcGU9c2l0ZQ%3d%3d#AN=edsble.777301&db=edsble

[4] U.S. Department of Housing and Urban Development (HUD) 2021 Press release HUD No. 21-041 https://www.hud.gov/press/press_releases_media_advisories/hud_no_21_041

[5] ATHA American Tiny House Association https://americantinyhouseassociation.org/our-mission-2/

[6] THIA Tiny Home Industry Association https://tinyhomeindustryassociation.org/

[7] Simons A, Moore E, and Kramer S 2018 Barriers to the tiny house movement RICS Cobra London UK

[8] Brown E 2016 Overcoming the barriers to micro-housing: tiny houses, big potential University of Oregon graduate capstone http://hdl.handle.net/1794/19948

[9] Stanton T 2015 Timber trails Tiny house building codes http://www.timbertrails.tv/Tiny-House-Building-Codes

[10] Wells A 2017 Insurance Journal Tiny homes: how agents, insurers are helping to lay a foundation for a growing insurance market https://www.insurancejournal.com/magazines/mag-features/2017/03/06/442981.htm

IOP Conf. Series: Materials Science and Engineering 1218 (2022) 012014 doi:10.1088/1757-899X/1218/1/012014

Evaluation of defect in hospital buildings

A Olanrewaju[1], W X Tan[1] and Z Gou[2]

[1]Universiti Tunku Abdul Rahman, Jalan Universiti, Bandar Barat, 31900 Kampar, Perak D.R., Malaysia

[2]Wuhan University, Wuhan, Hubei Province, P.R.China. 430072

abdullateef.olanrewaju@ymail.com

Abstract. There are many complaints and criticisms in media and research literature on defects in hospital buildings despite the consistent increase in maintenance expenditures. This research investigated the nature of defects in private hospital buildings. Primary data for the study were collected through a survey questionnaire administered to hospital building users. Based on the findings, the defects that often require frequent maintenance are leaky faucets, defective doors, defective windows, broken lights, and faulty ventilation. The major conclusion drawn from the survey was that although there are many different defects in the buildings, the urgency at which they require maintenance interventions varies. Therefore, in order to avoid value mismatch and resource misallocations, resources and attention should be directed to the defects that are more important to the users and the other should be included in the next maintenance rolling programme. Maintenance management should be implemented to achieve good maintenance, capable to meet the expected requirements of the end-user and increase building performance and reduce the maintenance backlog.

Keywords: building users, Satisfaction, Building Performance, Maintenance organisation

1. Introduction

A hospital building is one of the most complex buildings. The corporate objectives of the hospital organization, placed the building in a strategic position Thus, any inadequacy with the building facilities, will seriously affect the achievement of the prime objectives of the hospital. The condition of the hospital building is correlated with the satisfaction or dissatisfaction of the users. Based on the media report and examinations of government reports, there are a lot of complaints from hospital users on the performance of the hospital buildings on account of the defects. Defects in the building could cause psychological and illness problems to the users and will increase the users' dissatisfaction. The occurrence of defects in hospital buildings is due to lack of maintenance, poor design, and poor workmanship that can lead to psychological and illness issues to the users. About 75% of the building cost in use is attributed to defects in the building works and about 90% of the lifetimes of building projects require maintenance works. Maintenance is required when the building performance is not meeting the required quality and standards [1]. The degree of maintenance demand differs mainly on whether private or public hospitals [2]. It also differs based on procurement strategies and maintenance expenditure. In Malaysia, the ratio of maintenance expenditure is 1: 5 for a public and private hospital. In other words, the private hospital invested 5 times that of public hospitals for maintenance annually. High maintenance costs, poor satisfaction, and loss of productivity could be avoided or reduced with systemic maintenance process. Therefore, the research investigated the defects in private hospital buildings in an effort to frame the maintenance management process for the hospital maintenance organization.

2. Background and theoretical framework

Hospital buildings are procured to create a suitable, conducive, and adequate environment that can support and encourage patients' recovery and allow medical and on-medical staff to perform their duties productively. A failure in the supply of these essential services is a loss in value to hospitals and customers. There is an increasing demand for hospitals in Malaysia. To meet the increasing demand, new hospitals are established and the existing hospital is extended and refurbished to meet the growing demand and increased customer requirements. There are 144 public hospitals/ special medical institutions and 210 private licensed hospitals in Malaysia [3]. However, despite the increase in investment in the construction of new hospitals and upgrading and expansions of the existing hospitals, hospitals have remained inadequate. This inadequacy is affecting the condition and performance of hospital buildings [4]. However, academic literature, media reports, and government-sponsored research revealed that there are recorded cases of defects in the hospital buildings [5,6,7,8]. Results from a fire safety audit conducted on 46 hospitals built more than 50 years ago, revealed the existence of many defects in the buildings including faulty electrical networks and outdated systems [9]. Due to the increasing rate of defects in the hospital buildings, the hospital buildings are now described as "Rumah Sakit' (or sick house). Although research on hospital buildings is receiving attention recently, there is a dearth of research on the evaluation of building defects in hospitals.

2.1. Defect in hospital buildings

Words including deficiency, damage, default, deterioration, and decay are often used to describe the defect. Here, a building defect is defined as an undesirable or inadequate condition in the buildings that affect the serviceability, performance, structural condition, or appearance of the buildings [10]. Building defects can manifest within the structure, fabric, services, and other facilities of the defective building. Examples of defects include cracked, damaged, or deteriorated brick/block walls, clogged water closets and faulty lifts faulty windows and doors, and leaking showers. Defects are contagious because if it is not addressed on time, it would not only get worse, but will also destroy the adjourning components, elements or building parts. Building defects can be identified by the users themselves or thorough inspection by those concerns with maintenance management. Buildings, as Olanrewaju et al [10] explained, are not procured for their own sake rather for the services (i.e., comfort, protection, and esteem) that the buildings provide to the users. Hence, users should be the objectives of all defect's management philosophy and science. Building users are the entity or group of individuals or the organization who are interested in the condition and performance of the building. Users are affected by the performance of the building and the building is also affected by the activities of users. The degree of defects in the building is a measure to evaluate the functional performance of the buildings. Although building defects can be generally attributed to a lack of good construction practices, but are often cause and in fact aggravated by a lack of good maintenance practices Defect set up inner psychological tension in the users if the defect is not rectified in terms of quality and time. Therefore, information on nature, degree, and kind of defects will dictate when to undertake the repair work and allow future works to be programmed and financed as part of the maintenance rolling programme [11]. A significant aspect of building maintenance management \ is defect management.

3. Method of data collection and analytics

The questionnaires were administered to the building's users through an online survey (14/07/2020 to 28/08/202020). The building users were asked based on evidence to tick the degree to how often the specific defect appears /exist in their buildings. The defects were measured on a 6-point continuum scale of 1 to 6, where 1 denotes Not at all, 2 denotes Least often, 3 denotes Less often, 4 denotes Often, 5 denotes Very Often and 6 denotes Extremely Often and standard deviation. The severity of the defects is determined by the Average Frequency Index (AFI) (Equation 1). The average relative is based on the cumulative weighting of the initial frequency score of each of the constructs. This method is preferred because it takes into account the effect of the weight the building users attached to their decision

$$AFI = \frac{\sum_{i=0}^{6} a_i x_i}{6\sum_{i=0}^{5} x_i} \times 100 \qquad \text{(Equation 1)}$$

Where constant expressing the weight given to the group; x_i is the frequency of the responses, i = 1, 2, 3, 4, and 5 and 6 described as below: $x_1, x_2, x_3, x_4, x_5, x_6$ are the frequencies of the responses corresponding to $a_1 = 1, a_2 = 2, a_3 = 3, a_4 = 4, a_5 = 5$, $a_6 = 6$ respectively. For interpretation, an AFI score of 1.00 – 16.67 denoted not at all, 16.68-33.33 denoted least often, 33.34 - 50.00 denoted less often, 50.01– 66.67 denoted often, and 66.67-83.33 denoted very often and 83.34 – 100 denoted extremely often. There is a pooled difference of 1.0 between each of the scales. The defects with the highest AFI score are considered as the defects requiring more maintenance. All the constructs were positively worded. The questionnaire was piloted before the final version was finalised. One-way t-test, Omega, convergent validity, mode, a standard deviation was performed. Bartlett's test was conducted to enhance the correctness of the instruments. For the one-way t-test, the null hypothesis was that each of the defects does not exist in the hospital building (H0: U=U0) and the research hypothesis was that each of the defects exist in the hospital building (Hr: U≥U0). U_0 is the population mean. The critical level was set as 5. The small standard errors were also computed. The standard error measures the accuracy of the extent to which the sample mean is close to the population mean. A small standard error is an indication that the sample mean is a more accurate replication of the actual population mean. 17 defects that affect buildings and engineering services were identified and addressed to the building users. The defects were important sources of complaints that hospital maintenance organizations received from building users in their respective hospitals. The building users are encouraged to add any other defect they considered critical to the hospital building. This also includes how the defects specifically affect their service delivery and other functions that hospital buildings are expected to perform:

- Not at all often; if the user agreed that the particular defect has not occurred before and will not require any maintenance.
- Least often; if the user agreed that the particular defect occurred at the lowest time since using the building and maintenance is slightly required.
- Less often; if the user agreed that the particular defect occurred a few times since using the building and maintenance is less required
- Often; if the user agreed that a particular defect occurred frequently and lead to regular maintenance demand.
- Very urgent; if the user agreed that the particular defect has a very significant impact on the operation of the building and the maintenance demand is very urgent.
- Extremely urgent; if the user agreed that the particular defect has an extreme impact on the operation of the building and the kind of maintenance requires is greatly urgent.

4. Analysing the results of the survey

The survey pooled 538 forms and 207 completed responses were received during the survey period spanning over one month. The results of which are presented and discussed in the following sections.

4.1. Analysing the respondents' profile

The users held key positions and have diverse experiences with the buildings (Table 1). More than 60% of the building users are medical staff and the remaining are non-medical and executive staff. More than 50% of the building users have been using the hospitals for more than 4 years (Figure 1).

Table 1. Positions of the Respondents

Positions	Frequency	Percentage (%)
Doctor	36	17.4
Nurse	77	37.2
Therapist	14	6.8
Pharmacist	6	2.9
Cashier	6	2.9
Cleaner	10	4.8
Food Service Worker	8	3.9
Security Guard	6	2.9
Hospital Engineer	10	4.8
Information Technology Staff	6	2.9
Receptionist	4	4.8
Technician	5	2.4
Executive	13	6.3
Admin	6	2.9
Total	207	100.00

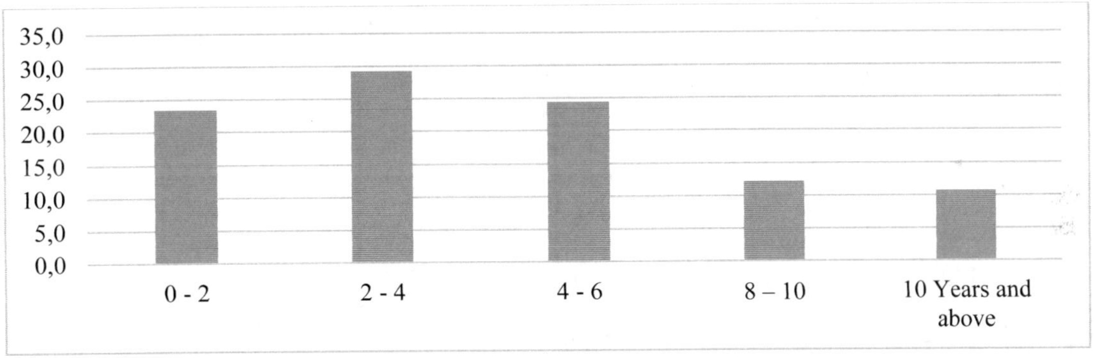

Figure 1. Respondent's experience with the buildings

4.2. Results of the descriptive statistics

The Omega value of the reliability for the defects is 0.851 and the validity ranges from 0.6 to 0.89. The Kaiser's Measure of Sampling Adequacy was significant $\chi 2$ (91) = 857.560, $p<0.001$), N=.830) indicating the data were drawn from the same population and that the defects were related. These data also mean there are no multicollinearity problems in the data. The descriptive statistics on how often the defects occurred in the buildings are contained in Table 3. The survey found that approximately 37% of the users measured that the defects were rare. But according to some 6% of the users, the defects least often occurred in the building even as 17% of the building users measured that defects less often occurred. Meanwhile, about 23% of the hospital building users have experienced the defects in the buildings often. Approximately, 17% of the building users, however, measured that the defects occurred in the buildings very or extremely often. The cumulative AFI score for all the defects is 46.18%, while the cumulative standard deviation (SDV) was 17.02 %. To interpret, while the defect may be less in some of the hospital buildings, they are very rampant and critical in some of the hospital buildings.

Table 2. Descriptive of the types of performance services.

Defect	1	2	3	4	5	6	SD	AFI	Occurrence
Leaky Faucet	9	20	59	78	36	5	20.50	60.18	1
Defective Doors	30	0	79	79	53	0	19.14	60.01	2
Defective Windows	42	11	22	66	61	5	19.25	58.68	3
Broken Light	12	16	73	79	23	4	19.42	57.84	4
Plugged toilet	31	8	40	92	36	0	20.50	57.51	5
Faulty Ventilation	41	7	50	84	22	3	23.85	53.84	6
Wall Cracks	42	25	47	55	38	0	18.32	51.84	7
Peeling Paint	63	21	40	49	34	0	20.72	48.18	8
Defective Lifts	87	0	19	62	39	0	20.84	47.34	9
Cracked Tiles	84	10	45	18	49	0	18.85	44.84	10
Electrical Problems	101	11	25	47	23	0	18.97	40.34	11
Leaky Roof	102	17	32	22	34	0	16.35	39.51	12
Ceiling Collapse	129	15	27	11	25	0	11.87	33.01	13
Water Seepage	148	8	19	5	28	0	8.72	30.51	14
Roof Corrosion	150	10	6	16	25	0	4.85	30.34	15
Gas Leakage	166	17	0	11	13	0	10.17	24.84	16

5. Discussion of defects in the hospital building

Because of space constraints, only the first 7 frequent defects will be briefly discussed. The findings show that the most frequently defects in hospitals are related to faucets. Faucet is common in the hospital because water is required for various types of functions. Therefore, it is not surprising to find that faucet leakage is a perennial problem in hospitals. Most of the leakage is due to use or misuse which is compounded by poor maintenance. The impact of faucet leakage is huge. For instance, it may lead to water seepage, tile failure, and water wastage. The data show that 78 of the users agreed that faucet often needs repair work. In fact, more than 45 of the building users believe that faucet leakage occurred very often or extremely often in the buildings. Defective Doors are the second common defect in hospital buildings. A total of 53 of the building users indicated that doors need repair work very often. While 45 of the building users agreed that doors required less repair work. The remaining 30 building users agreed that the door does not need repair work at all. The most common problems with the door are that the doors do not open properly. It was found that most of the hospitals face defective doors. The causes of defective doors were classified into design, workmanship, materials, and lack of protection (Isa et al., 2011).

The results of the survey show that the majority of building users agree that windows often need repair work. In fact, 66 building users measured it to require maintenance and 61 of the building users believed that door defects occur very often in the buildings. However, this seems to be counter-intuitive because widows at the hospitals are not often open. The causes of defective windows included poor design, use of unsuitable materials, poor workmanship during installation and fabrication (Isa et al., 2011). Seventy-nine of the building users agreed that lighting often needs repair work. Seventy-three of the users believe that the defect in lighting less often occurred. Light fittings and appliances need repair work very often. Lightning in the hospital is very important because doors and windows are often closed. Proper lighting in the hospital would help to improve the user's productivity. Poor lighting will limit concentration levels significantly which could lead to reduced performance levels (Bermudez, 2014). Faulty ventilation systems turned out to be a common defect in hospitals. In fact, about 60% of the building users measured faulty ventilation as often, very often, and extremely often. Hospital buildings need adequate ventilation systems to refresh all the air in order to reduce airborne infections (Villafruela et al., 2019). Therefore, it is not surprising that the defect in the ventilation was measured by users as critical. The results of the survey show that the majority of users resolved that cracks in walls often occurred. About 50% measured that defect walls often and very often occurred. Cracks on internal walls may cause by poor initial design, use and poor maintenance.

6. Conclusion

This study has been able to identify, quantity, and categorized defects in the university buildings. The identification and analysis of the degrees of the defects contribute to directing university maintenance organizations on maintenance management. This is crucial if universities desire to succeed in modelling a systemic maintenance management system for their buildings and engineering services. Defect classification is a very strategic function of maintenance organizations. In order to maximize the maintenance budget, the resource should direct to the extremely urgent defects then the very urgent, and finally to the ones that are not very urgent. Probably it is the only way, that the ever-inadequate fund will be spent judiciously and users' satisfaction can be achieved and maximized accordingly. The defects addressed in this paper do not include defects that relate to substructure works (foundations cracking, foundation bowing, foundation settlement).

7. References

[1] Desmyter, J. and Huovila, P., 2010. Performance Indicators for Health, Comfort and Safety of the Indoor Environment. In W084-Special Track 18th CIB World Building Congress May 2010 Salford, United Kingdom (p. 39).

[2] Olanrewaju, A.A., Fang, W.W. and Tan, Y.S., 2018. Hospital building maintenance management model. International Journal of Engineering and Technology, 2(29), pp.747-753.

[3] Ministry of Health Malaysia, 2019. Malaysia National Health Accounts, Health Expenditure Report 1997-2017

[4] Olanrewaju, A, Wong, F.W., Yahya, N.N and Im, P. L (2019) Proposed research methodology for establishing the critical success factors for maintenance management of hospital buildings. AIP Conference Proceedings 2157, 020036 (2019); https://doi.org/10.1063/1.5126571.

[5] The Star Online (2007), "Authorities aware of problems at Ampang Hospital", available at: http://www.thestar.com.my/news/nation/2007/03/14/authorities/aware/of/problems/at/ampang/hospital/ (accessed 12 March 2017).

[6] The Star Online (2007), "Hospital is one sick building", available at: http://www.thestar.com.my/news/nation/2007/03/13/hospital-is-one-sick-building/ (accessed 3 December 2017).

[7] The Star Online (2012), "Hospital elevators need better maintenance", available at: http://www.thestar.com.my/news/community/2012/06/07/hospital-elevators-need-better-maintenance/ (accessed 7 June 2017).

[8] The Star Online (2016), "Broken lifts force patients to use stairs at Sabah Hospital", available at: http://www.thestar.com.my/news/nation/2016/1130/broken-lifts-force-patients-to-use-stairs-at-sabah-hospital/ (accessed 12 March 2017).

[9] MOH, 2017b

[10] Olanrewaju, A., Khamidi and M. F and Arazi, I. (2010), "Building Maintenance Management in a Malaysian University Campus: A Case Study", Australasian Journal of Construction Economics and Building, Vol 10, No 1/2 (2010)

[11] Olanrewaju, A.L. and Abdul-Aziz, A.R., (2015). Building maintenance processes and practices: The case of a fast-developing country. Springer.

Acknowledgments

The research has been carried out under Fundamental Research Grant Scheme project FRGS/1/2018/TK06/UTAR/02/ provided by Ministry of Higher Education of Malaysia.

Tracking and Monitoring Technologies to Support Airport Construction Safety

B J Hubbard[1] and S M Hubbard[2]

[1] School of Construction Management Technology, Purdue University, West Lafayette, IN 47907, USA

[2] School of Aviation and Transportation Technology, Purdue University, West Lafayette, IN 47907, USA

E-mail: bhubbard@purdue.edu

Abstract. This paper discusses how advanced tracking and monitoring technologies can be used to increase safety for construction activities at airports. Airport construction is a significant infrastructure sector and in 2020 the Airport Improvement Program funded $300 million for airport projects in the United States. This includes funding for pavement reconstruction, airfield upgrades to meet current standards (e.g., wider runways), the construction of new terminal facilities, and additional capacity (e.g., new runways). The airfield environment presents special hazards for constructors, including proximity to active aircraft operations, and the requirement to comply with many additional rules such as those required by Federal Aviation Administration (FAA) and the Transportation Security Agency (TSA); many of these rules are required to ensure safety and security. Airport construction activities require close oversight, additional inspections by airport operations personnel, and stringent rules regarding access and operation in different areas of the airfield. Access and operating requirements affect both construction workers and construction equipment, and there are often different rules for different areas at the airport. One of the most restricted areas on an airport is the runway. Access to the runway and the safety area around a runway requires explicit permission from Air Traffic Control (ATC). If a vehicle or pedestrian (or aircraft) crosses the boundary without permission, it is considered a runway incursion. A runway incursion could cause a major incident at the airport and could result in injuries and fatalities. Runway incursion events are recorded by the FAA and represent a serious safety violation for the airport. Runway incursions and the technology to prevent incursions are discussed as an example that illustrates the need for and benefits of accurate monitoring of construction equipment and personnel. Technologies that may be used to enhance tracking and support situational awareness are present to illustrate the possible ways that enhanced monitoring can be used at an airport. Solutions may include the use of advanced technologies such as real-time vehicle location sensors, which may be combined with robust electronic maps to enhance safety for construction equipment and personnel location in the airfield environment. A discussion of opportunities and challenges of these systems is detailed.

1. Introduction

Airport construction is large sector and growth in this sector will continue since airport construction supports the movement of both passenger and cargo. U.S. airports expected an average of $22 billion dollars of construction projects each year from 2019 through 2023 [1]. Future funding is expected to remain strong, and in the coming years from 2021 to 2025, the U.S. National Plan for Integrated Airport Systems (NPIAS) identifies $34.3 billion dollars for airport projects related to reconstruction, standards, safety and capacity (2020) [2]. Airfield construction is critical to our aviation system, however, it can create operational challenges since airport construction is typically undertaken while aeronautical activities continue. This can create challenges for both constructors and airport users.

Different areas of the airport have different restrictions for use, and the runway is an area with the most restrictions. At airports with an air traffic control, all airport users must have permission to enter the runway or the safety area around the runway. If any vehicle, pedestrian or aircraft enters this area without permission, it is considered a runway incursion. Runway incursions are a major concern for airports due to the safety threat, and because the occurrence of a runway incursion implies a "near miss". The official definition of a runway incursion is provided by FAA as "Any occurrence at an aerodrome involving the incorrect presence of an aircraft, vehicle or person on the protected area of a surface designated for the landing and take off of aircraft" [3]. Although there are many other operating constraints at an airport, examination of runway incursions provides a good case study to illustrate the operational constraints at an airport, which presents unique constraints, rules and regulations for constructors. This paper presents data on runway incursions when there is airport construction activities underway and then explores new technology that constructors and airports can employ to help reduce runway incursions during construction activities.

2. Background

Airfield construction is an important activity but it also introduces a number of operational and safety considerations. Airport construction affects many airport users, including pilots for airlines and general aviation aircraft (these are smaller airplanes used by individuals and businesses), air traffic controllers (ATC), personnel who ensure the safety of the runway and airports (often called airport operations personnel), airport tenants such as aircraft maintenance facilities and fixed base operators (FBOs), and emergency response personnel, called aircraft rescue and fire fighting (ARFF) personnel at airports. Not all airport construction activities have the same impact, and characteristics such as the number of construction workers, the kind of project, the project duration, the construction material and equipment, and the location of the construction project on the airfield may all affect the impact of the construction project on airport activities. In many cases, the construction project has defined stages and phasing throughout the duration of the project. Some characteristics such as material storage location, access points, and construction site boundaries may change throughout the project, which will have an impact on the airport operations, as well as safety, security [4] and cost. There is relatively little published research regarding airport construction on an active airfield, although some of the issues that have been mentioned in the literature include security escort requirements, night work, short closures, segmenting of work, provision of barricades and fencing, maintaining operational surfaces free of FOD (Foreign Object Debris), maintaining operational surface zoning requirements, protecting workers from jet blast, the need for flexibility to adapt to changing circumstances, unusual weather and labor disputes [5]. The limited research regarding the safety of airport construction is in sharp contrast to research in other transportation infrastructure sectors, such as the roadway sector, where there have been extensive studies related to work zones, including the associated costs, risks, and accident characteristics (e.g.[6 - 9]).

3. Data Pertaining to Runway Incursions During Construction

One metric for the impact of airside construction activities is the incidence and cause of runway incursions during airport construction [10]. FAA publishes runway incursion data in the Runway Incursion Database. Examination of data from the last two decades indicates there were 612 runway incursions that had "construction" in the narrative (for the years from 2001 to 2020). Details about these runway incursions are shown in Figure 1. There are three primary causes for runway incursions: either a pilot makes an error, called a pilot deviation (PD), air traffic control makes an error, which is called an operational error (OE) or a pedestrian or the driver of a vehicle makes an error (VPD). For construction events, 46% are vehicle or pedestrian deviations (VPD), 33% are operation error (OE, caused by ATC) and 20% are pilot deviation (PD). VPD and OE are much more likely for events with "construction" in the narrative than for all runway incursion events, which are dominated by pilot deviations (60%). This suggests that there may be a need to provide better tracking of people and vehicles, which includes construction workers and construction equipment, when there is an active construction project at the airport. There may also be a need to provide additional support for air traffic controllers. One way to meet both these goals may be to provide better tracking of construction personnel and equipment, with the potential to make this information available to ATC.

Figure 2. provides information regarding the severity of runway incursions and Figure 3 provides definitions of these severities. FAA ranks runway incursions from severity A (most severe) to severity D (least severe). Fortunately, severe runway incursions (A and B) are very unusual event and represent less than 2% of all construction runway incursions. Most runway incursions pose no risk of collision, with 15% Type C and 32% Type D incursions. Approximately half of the construction runway incursions did not have a designated severity. A further breakdown of VPD, OE, and PD next to the definitions are shown in Figure 3. Vehicle or pedestrian deviation for type C and D are the most prevalent. Of the 612 construction runway incursions, 266 indicated a vehicle and 32 indicated a pedestrian (in the aircraft flight code columns); this suggests that vehicles may be a greater concern than pedestrians during airfield construction activities.

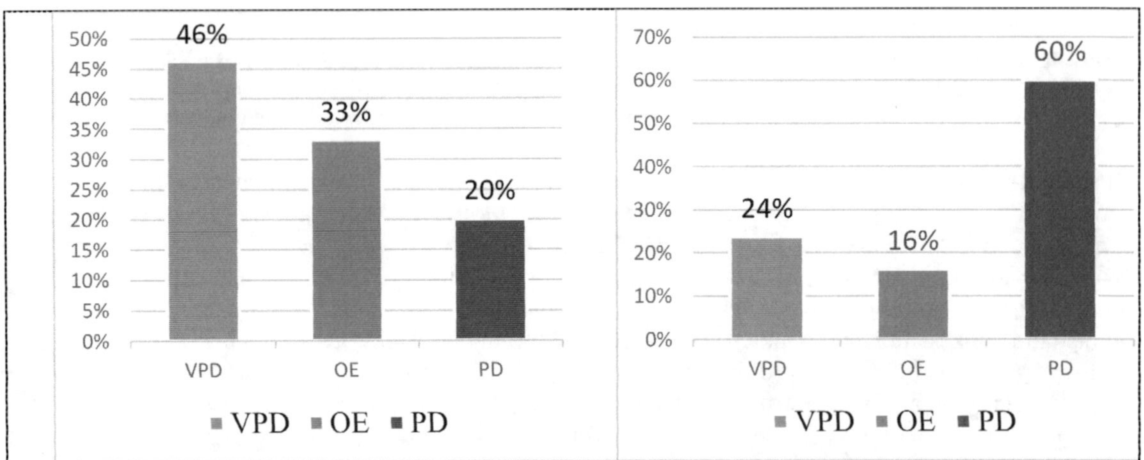

"Construction" in Narrative (n=612) All events (n = 25,584)

Figure 1. Distribution of Incident Type for Runway Incursions from 2001 to 2020.

IOP Conf. Series: Materials Science and Engineering 1218 (2022) 012015 doi:10.1088/1757-899X/1218/1/012015

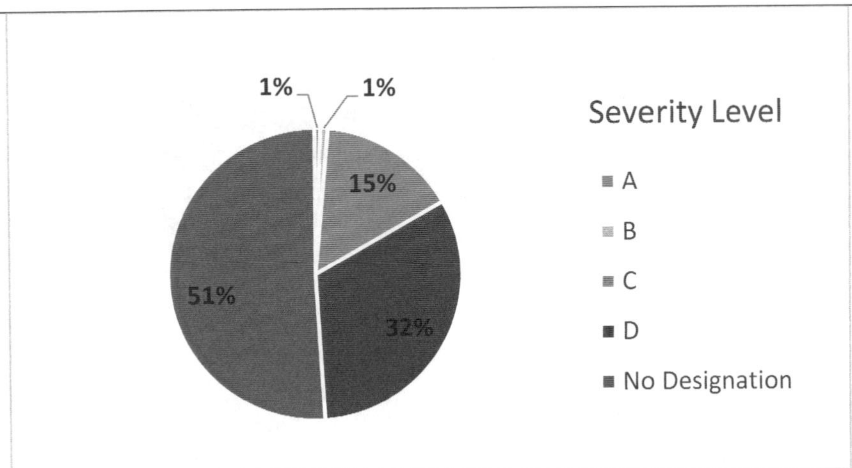

Figure 2. Percent of Runway Incursion Events with "Construction" in Narrative Based on Severity.

4. Advanced technology to prevent runway incursions

To prevent runway incursions there are a number of technologies that can track and monitor people and vehicles in the airport construction environment. These technologies can be programmed to provide an alert when the vehicle and/or person is approaching and/or in proximity the proximity of an area that is restricted and/or has special operating rules. Technologies in both the construction and aviation sectors are being used for these types of purposes and offer solutions to airport construction issues such as runway incursions. While many of these technologies in both construction and aviation share common technology platforms the use, accuracy, robustness and applicability to the issue of runway incursions varies. In addition to the capability and applicability of the systems, the systems have to operate under FAA and airport operators' rules and regulations and not interfere with plane and airport location and communication systems. In this section some potential technologies and systems used in both the construction sector and aviation sector are presented that could be applied to prevent runway incursions.

4.1. Construction Technologies to Prevent Runway Incursions

Construction companies have utilized many off the shelf commercially available systems to track people and equipment and with recent technology advances in tracking has become more feasible and economical. Tracking in construction is often used to support increased productivity and monitor safety issues. Initial productivity studies were done with data generated from the job jobsite and analysed later. As technologies have improved these systems have become real-time and are able to provide alerts to construction supervisors and safety personnel for safety issues and real-time production information.

4.1.1. GPS Positioning. Constructors have utilized low cost commercially available Global Positioning System (GPS) equipment to identify and track equipment activity [11]. GPS locations coupled with accurate mapping have the ability to provide real-time location and detect the proximity to airport protected surfaces to reduce the risk of a runway incursion. Location tracking could also be used on construction personnel similar to equipment. The real time data could be used to alert equipment operators, construction personal, and/or airport officials of a potential runway incursions and take appropriate action. GPS has the advantage over many other system because it is receiving signals from satellites and cell phone towers (Assisted GPS) and not sending or generating a signal. The generation of additional signals could interfere with plane and airport operations and would require additional FAA and airport approval.

Severity	
A A serious incident in which a collision was narrowly avoided.	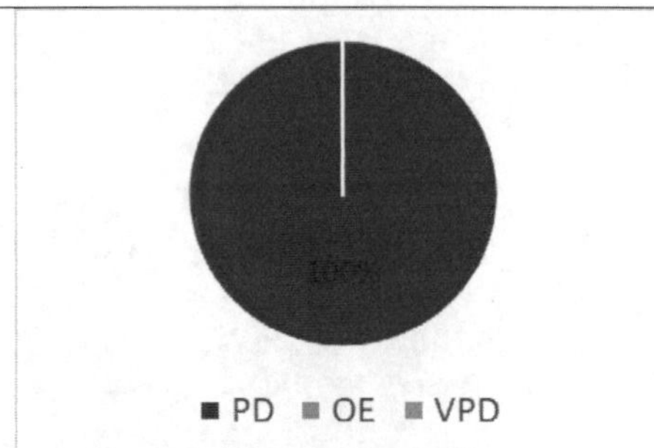
	n = 4
B An incident in which separation decreases and there is a significant potential for collision, which may result in a time critical corrective/ evasive response to avoid a collision.	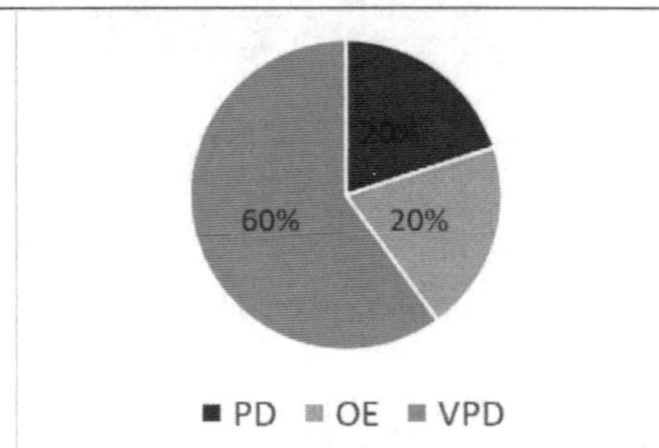
	n = 5
C An incident characterized by ample time and/or distance to avoid a collision.	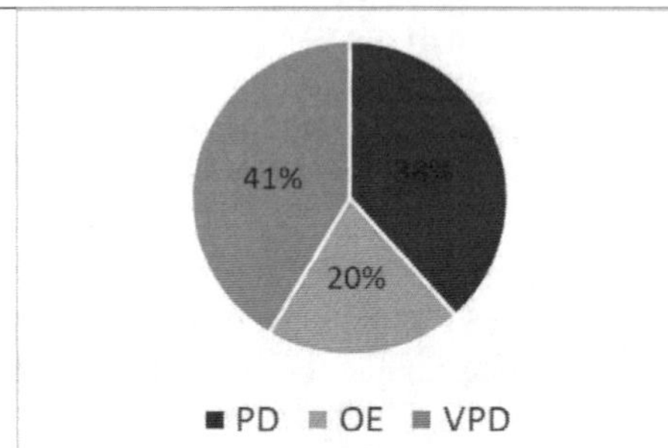
	n = 94
D Incident that meets the definition of runway incursion such as incorrect presence of a single vehicle/person/aircraft on the protected area of a surface designated for the landing and take-off of aircraft but with no immediate safety consequences.	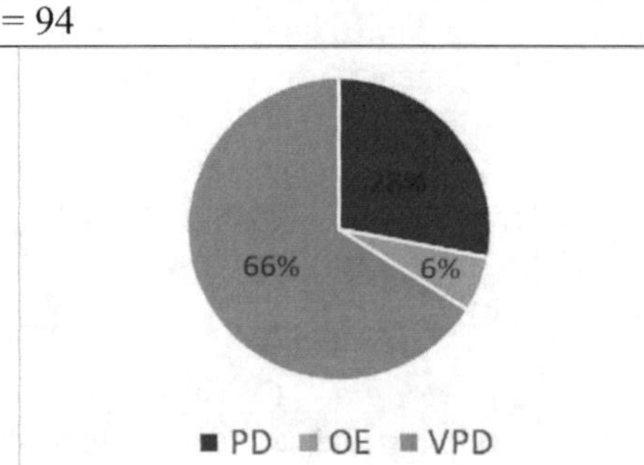
	n = 197
Figure 3. Severity Definitions [3] and Breakdown by Severity Type	

4.1.2. UWB Positioning. Ultra Wideband (UWB) systems have also been utilized to provide real-time location information for construction workers and equipment. UWB systems use multiple receivers to triangulate a position [12]. The system consists of a central processing unit, receivers, UWB tags for the equipment and workers and a reference location. UWB could be used to detect a worker or equipment leaving the construction site and getting close to or entering a protected surface in real-time. While UWB has been used for indoor tracking at airports in Beijing and Estonia [13-14], setting up a UWB system in the runway area may be an issue because of signal interference with airport and airplane cockpit instrumentation. UWB is advancing quickly such as the development of the 5G cellular network however electromagnetic interference with aircraft systems is still a concern [15].

4.1.3. Multiple systems positioning (Wi-Fi, RFID, GPS and Bluetooth). Commercial systems have been developed that use Wi-FI along with other signals such as RFID (Radio Frequency Identification), GPS, and Bluetooth to accurately determine positions. Commercially available tracking systems such as Aeroscout [16] have developed systems based on Wi-Fi positioning that also use RFID and Bluetooth signals to triangulate a position. These systems could be used on their own but increased accuracy is provided with multiple systems. These systems have been used in airport environments such as in Davis-Monthan Air Force base [17] and Munich Airport [18].

4.2. Aviation Technologies to Prevent Runway Incursions
The airport environment is focused on safety and have for years been developing systems to keep airplanes safe and to keep the air and ground operation running smoothly including the prevention of runway incursions. To ensure the safety of the airport environment and prevent runway incursions, air traffic controllers track and monitor the location of the planes. In the airport environment, planes are monitored by a transponder that is known as ADS-B. ADS-B provides information to air traffic control with information on the identity of the plane, location, speed, altitude and additional data [19]. At some airports, an ADS-B system is also used for tracking ground vehicles; this and other systems to prevent runway incursions are currently being used in the airport environment and may have applicability to construction activities.

4.2.1. ADS-B out squitter. ADS-B out squitters are also known as ADS-B squitters, squitters, or Vehicle Movement Area Transmitters (V-MAT or VMAT)]. ADS-B out squitters can be used in the airport environment on ground vehicles, and this is the same kind of ADS-B transponder that is used on aircraft [20]. The ADS-B out squitter identifies the vehicle, and provides speed, similar to the information provided by ADS-B equipped aircraft. FAA limits the number of ground vehicles that use ADS-B out squitters, and the ADS-B system is used to support airport ground surveillance by ATC and is connected to the Airport Surface Detection System. The 35 largest airports have ASDE-X, which is the Airport Surface Detection Equipment - Model X; a newer version of this is ASSC, Airport Surface Surveillance Capability (ASSC) which has been installed at seven airports and provides enhanced capabilities in all kinds of weather to improve situational awareness and surface surveillance. Both systems use radar, multilateration, and satellite technologies to track the surface movements of aircraft and ground vehicles on the airfield and translate this information to a visual overlay of the airfield map that provides the real-time locations [21].

A primary goal of this technology is to prevent runway collisions and reduce critical Category A and B runway incursions by alerting ATC to any hazardous movements [22]. These systems also may reduce surface delays and improve situational awareness. However, the ADS-B systems for ground vehicles are expensive, require FAA approval, are geared for use by airport vehicles, and allow only a limited number of vehicles to use ADS-B squitters at the airport. Even if airports obtained permission to use ADS-B squitters for constructor equipment, it is very expensive and there are other more cost-effective solutions. Furthermore, the system is not set-up to track personnel in the airfield environment, only equipment. Another consideration is that since the system is intended for aircraft and a limited number of airport vehicles, the additional signals from construction system may overwhelm the system capacity cause confusion for air traffic controllers and pilots using cockpit displays. To use ADS-B squitters, both the airport authority and the FAA would need to approve and authorize the use of this equipment for monitoring construction activities.

4.2.2. Fleet Online Airside. Fleet Online Airside is used for tracking and location of ramp equipment and was developed by AdaptaliftGSE [23]. The system uses sensors to identify the current position and track each piece of airport ground vehicles equipped with a transponder. The software provides driver verification, vehicle control and monitoring, geofencing, safety overrides, and supports mobile communications. The geofencing aspects of the system would be useful for the construction process to ensure vehicles do not stray outside of their designated areas. Also, since the system is currently being used in the airport environment it has gone through the scrutiny of airport operations and would require less approval process to be used in the airfield environment. One drawback is it is set up for vehicles and not for workers.

4.2.3. Runway Incursion Warning System (RIWS). Similar to the GPS systems noted in the construction section, there are commercially available systems that combine the GPS signal with up to date airport GIS mapping information such as arriving and departing planes along with NOTAMs (Notice to Airmen, a safety alert system for the airport). These specialised systems, such as [24], provide on-screen airport information and can notify drivers of vehicles when there is a potential for a runway incursion.

5. Discussion

Technology in the construction industry for tracking equipment and people has become feasible and economical. Many commercial systems could be used to help avoid runway incursions by alerting equipment operators or personnel that they are going into an unauthorized location. Systems that use available signals such as GPS and cell phone towers signals to provide location information would be immediately applicable in the airport environment because they are only receiving signals and do not require additional systems to send out a signal.

The use of a Building Information Modelling (BIM) in construction that provides construction locations with specific coordinate details provides supporting information for GPS location systems and other location devices. Information from a BIM Model can be used to plan areas for construction along with acceptable and restricted movement areas and pathways.

Other proposed systems such as a local UWB system for highly accurate location detection would be difficult to use in airport construction environment because of concerns with signal interference and the process to get a system approved. However, as cellular companies begin to develop 5G UWB networks, these systems could be used in the airport environment as long as the signal is available. Accuracy is reported to be 10 meters in urban environments [25] which would make it feasible for both equipment and workers to have the technology available in their smart phone or other device.

Tracking and monitoring technology in the aviation industry is very advanced and there are many systems to support tracking. Aviation has been more focused on equipment (planes and ground equipment) and most of the systems are equipment based and do not support the tracking and location of individuals. While the software and hardware are available to allow construction equipment to be incorporated into the system, the addition of construction equipment would be a burden to the ATC and may not be allowed. The systems also tend to be expensive and may require extensive periods working in the airport environment or multiple jobs to make the equipment cost effective.

One requirement that is creating change in the ADS-B manufacturing industry is that all small planes [26] as of 2020 are required to have ADS-B system. This has created a bigger market for ADS-B systems and the cost has reduced. Also, there is discussion that drones (aka Unmanned Aircraft Systems (UAS)) also should be equipped with an ADS-B type transponders [27]. Because of the increase in number of signals the airport may receive there are also changes being made to the transponders that may only provide a signal when they are near an aircraft or close to a runway/airport (site UAS). These recent movements may make it feasible in the future for construction equipment to be equipped with a ADS-B transponder to provide better monitoring of the airport environment.

There are also a significant number of systems in aviation that work similar to the construction industry location and tracking systems. Since these systems are already used in the airport environment they could be easily employed for construction activities. Constructors may also want to consider these types of devices since airport operation and the FAA are more familiar with their operation.

6. Conclusion

Airport construction during active airport operations requires an additional level of safety and monitoring of equipment and personnel to ensure the safety of both the planes and construction personnel. Runway incursions due to construction are significant concern and advanced technologies are available to help prevent runway incursions. Both the construction industry and airport industry have technologies available to help reduce the number of construction runway incursion events. Many of these systems are immediately applicable to help reduce the potential for injuries and fatalities of an airport incident.

References

[1] GAO 2020 *Airport Infrastructure: Information on Funding and Financing for Planned Projects* (United States Government Accountability Office Report to Congressional Requesters) GAO-20-298 Available from: https://www.gao.gov/assets/gao-20-298.pdf
[2] FAA 2020 *National Plan of Integrated Airport Systems 2021-2025* Available from: https://www.faa.gov/airports/planning_capacity/npias/current/media/NPIAS-2021-2025-Narrative.pdf
[3] FAA 2020 *Runway Incursions* Available from: https://www.faa.gov/airports/runway_safety/resources/runway_incursions/
[4] Khalafallah A and El-Rayes K 2008 Minimizing construction-related security risks during airport expansion projects *Journal of Construction Engineering and Management* 134(1) 40-48
[5] Stewart C 2001 Airfield Safety: Developing Operational Procedures and Schedules for Airside Construction Projects *In Advancing Airfield Pavements* 309-321
[6] Saha P 2020 An optimization model to determine an appropriate budget for improving work zone safety *IATSS Research*
[7] Schrock S D, Fitzsimmons E J, Lindheimer T, Wang M H, and Bai Y 2014 Work zone positive protection policy guidance: Synthesis of devices and state of practice *In Transportation Research Board 93rd Annual Meeting* No. 14-5574
[8] Chen E, and Tarko A P 2012 Analysis of crash frequency in work zones with focus on police enforcement *Transportation Research Record* 2280(1), 127-134
[9] Li Y, and Bai Y 2009 Highway work zone risk factors and their impact on crash severity *Journal of Transportation Engineering* 135(10), 694-701
[10] Hubbard S, Hubbard B, and Sobieralski J 2021 Minimizing The Negative Impacts Of Airport Construction *21st International Symposium on Aviation Psychology* pp 362-356 Available from: https://doi.org/10.5399/osu/1148
[11] Pradhananga N, Teizer J 2013 Automatic spatio-temporal analysis of construction site equipment operations using GPS data. *Automation in Construction*. Jan 1;29:107-22.
[12] Cheng T, Venugopal M, Teizer J, Vela P 2011 Performance evaluation of ultra wideband technology for construction resource location tracking in harsh environments *Automation in Construction* Dec 1;20(8):1173-84
[13] Nanjing Woxu Wireless Co. 2019 *Breakthrough Application Of UWB RTLS In Beijing New Airport* July 12 Availble from: https://www.uwbleader.com/news/breakthrough-application-of-uwb-rtls-in-beijin-25173540.html
[14] Undusk S 2019 *Eliko's indoor positioning system at Tallinn Airport* Oct 15 Availble from: https://www.eliko.ee/indoor-positioning-system-for-airports/

[15] Ely J 2005 Electromagnetic interference to flight navigation and communication systems: new strategies in the age of wireless *AIAA Guidance, Navigation, and Control Conference and Exhibit* Aug 15 pp 6361
[16] Aeroscout Industrial (n.d.) *The Future of Industrial Real-Time Location* Availble from: https://www.aeroscout.com/index.html
[17] Swedberg C 2008 Air Force Base Deploys Wi-Fi/GPS RFID System Across 2,500 Acres *RFID Journal* Nov 11 Availble from: https://www.rfidjournal.com/air-force-base-deploys-wi-fi-gps-rfid-system-across-2500-acres
[18] Wessell R 2009 Munich Airport Says RFID Improves Dolly Management *RFID Journal* Oct 21 Availble from: https://www.rfidjournal.com/munich-airport-says-rfid-improves-dolly-management
[19] FAA 2020 *Ins and Outs* Jan 2 Availble from: https://www.faa.gov/nextgen/equipadsb/capabilities/ins_outs/#:~:text=ADS%2DB%20Out%20works%20by,can%20immediately%20receive%20this%20information.
[20] Hubbard S, Voyles R, Souders D, Yang H, Hart J, Brammell S 2020 *Advanced Ground Vehicle Technologies for Airside Operations* ACRP Research Report 219
[21] FAA Sept 9 ADS-B Airport Surface Surveillance Capability (ASSC) Availble from https://www.faa.gov/nextgen/programs/adsb/atc/assc/
[22] FAA 2014 *Airport Surface Detection Equipment, Model X (ASDE-X)* Aug 11 Availble from: https://www.faa.gov/air_traffic/technology/asde-x/
[23] AdaptaliftGSE. (n.d.). Fleet Online & Telemetry Availble from: https://www.aalgse.com.au/hello-world-2/
[24] Leonardo (n.d.) *Aeroboss Runway Incursion Warning System* Availble from: https://www.leonardocompany-us.com/atc/products/aeroboss-riws
[25] Ericsson 2020 *5G Positioning: What You Need to Know* Dec 18 Availble from: https://www.ericsson.com/en/blog/2020/12/5g-positioning--what-you-need-to-know#:~:text=Consumers%20with%20a%20cellular%20phone,positioning%20accuracy%20can%20be%20achieved.
[26] FAA 2021 *Equip ADS-B* Mar 15 Availble from: https://www.faa.gov/nextgen/equipadsb/
[27] Moore J 2021 *ADS-B for Drones Broadcasts Conditionally* AOPA Apr 29 Availble from: https://www.aopa.org/news-and-media/all-news/2021/april/29/ads-b-for-drones-broadcasts-conditionally

A blockchain-based framework for supporting BIM-based building code compliance checking workflow

H Gao[1] and B Zhong[1]

1 School of Civil and Hydraulic Engineering, Huazhong University of Science and Technology, Wuhan 430074, Hubei, China.

E-mail: dadizhong@hust.edu.cn

Abstract. Code compliance checking of building designs is an essential task during the design and construction process, this process is time-consuming and laborious, and there is a risk of human manipulation, collusion and tampering with the results during the process. Existing BIM automated compliance checking technologies can reduce human errors and automatically output drawing review reports in some domains. But there are still few provisions that can be automatically reviewed, and most of the provisions still require human intervention. This research proposes a blockchain-based framework for managing both automated and manual compliance review processes in a BIM environment. The framework is developed to help improve the traceability and transparency of the compliance review process in the BIM environment. The framework includes on-chain and off-chain solutions for addressing the management issues in the code checking process. The prospects of the solution, challenges in implementation and future research directions were discussed.

1. Introduction

Design review is to identify defects and shortcomings of the design before construction takes place[1]. The aim of the design review is to ensure the design quality and detect design issues before construction. With the development of BIM technology, the data is presented in a machine-readable format and supports model checking applications in different disciplines[2–6]. In current design review practice, there are limited provisions that can be automatically reviewed, and many tasks still require human intervention. It leads to the risk of human manipulation, collusion and tampering with the results during the design review process.

Blockchain is an emerging technology that can improve data security, transparency, and traceability, it has attracted attention from different industries such as healthcare, logistics, finance, and agriculture[7–11]. The concept of blockchain was invented by Satoshi Nakamoto in 2008 to solve the security and privacy issues of digital currency transactions (Bitcoin)[12]. Blockchain is a distributed ledger that chronologically and securely records all data shared among all participants in a Peer-to-Peer network[13]. Data is uploaded to the distributed ledger through the consensus mechanism of different nodes in the network and updated in real time[14]. The peer-to-peer propagation mechanism makes records that cannot be tampered with. At present, various types of blockchains have been developed, these blockchains are divided into three types, public blockchain, private blockchain and consortium blockchain. They are different in the degree of openness, transaction speed, consensus mechanism, and scalability. Due to the requirements of limited participants, high transaction speed, privacy protection and scalability, consortium blockchain is widely studied in the research of the construction industry[8,15,16].

The study is to explore the potential of consortium blockchain in design review by developing a blockchain-based framework for managing compliance review processes in a BIM environment. It aims to provide a new solution for overcoming management issues in the design review practice, such as tracing the source of design errors and reducing the risk of collusion and human manipulation.

2. Statement of the problem in the design review process

Through the analysis of the workflow of existing government design review agencies in China, there are some common management issues in the current practice of design review: 1) The design review process is not transparent, and there is a risk of collusion between the owner and the reviewer. 2) The review comments and revision process of the design drawings were not recorded, and quality problems occurred during delivery, and it was difficult to determine the legal liability. 3) The data is stored on a centralized database, and the data is at risk of tampering, manipulation, unauthorized access. In order to solve these problems, this study proposes a design review management framework based on blockchain, and discusses the role of blockchain in design review process management.

3. Creation of a blockchain-based BIM design review framework

The main goals of developing the framework are to i) integrate the blockchain into the BIM design review workflow, ii) strengthen the liability control and data security in the BIM review process through the blockchain. This framework is based on the following technologies.

3.1. Related technologies

3.1.1. Automatic Code Checking(ACC). In some advanced countries, the governments require the delivery of the BIM files in the design stage. Manually performing code reviews for BIM data is not only time-consuming and labor-intensive but also prone to errors. Some of the countries are promoting automatic code checking applications. At present, BIM-based automatic code checking technology is being extensively studied[17–19], which can improve the accuracy and speed of compliance review in some areas. Most studies use the Industry Foundation Classes (IFC) to extract building object information as the source information to perform design compliance checking. The limitation of automatic code checking is that many parameters required by design rules are not defined in IFC format[19]. Hence, this study proposes a blockchain-based design review environment which both considering the process of automatic code checking and design review by human expert.

3.1.2. Hyperledger Fabric. Hyperledger Fabric is one of the leading open-source consortium blockchain systems, it was associated with Linux Foundation[20]. Different to public blockchain such as Ethereum, Bitcoin, Fabric focuses on developing private blockchain solutions in B2B applications. In Fabric, all participants on the blockchain network should get valid identities through MSP (Membership Service Provider), MSP is a mechanism that enables participants to convert their identities into digital certificates that can be recognized by a permissioned blockchain network. In this way, it can guarantee that only the permitted participants can access the design review process record. Fabric provides channel mechanism to put different participant nodes (such as designer, owner, reviewer, and government regulator) on the same blockchain, each node on the chain will hold the same copy of the design review process record which is known as a ledger in blockchain, the ledger records the transaction details (such as model modifications, the digital signature of nodes, time stamps, etc.) of each operation in the activities, it has the characteristic that once the ledger is established, it cannot be changed. The mentioned features of Fabric make that the data in the design review process is secure and can be traced.

3.1.3. IPFS. In a BIM-based design review environment, large files submitted by designers such as BIM files, design simulation and calculation reports need to be stored off-chain because it will increase the transaction speed and storage cost in a P2P network. Many blockchain solutions use cloud storage or local databases to record the transaction information off-chain, but the lack of a decentralized storage structure will also bring side effects on the management of design files. Interplanetary File System (IPFS) is a decentralized P2P file system to share unchangeable information in the blockchain network[21]. IPFS eliminates the role of centralized storage server and replaces it with a P2P file transfer protocol, files are distributed in different storage nodes which avoids a single point of failure. In the BIM-based design review environment, IPFS chooses to provide secure off-chain storage for large design files or documents.

3.2. System architecture

This paper proposes a system architecture based on the consortium blockchain. In this architecture, Hyperledger Fabric is selected to implement the blockchain network due to the requirement of data privacy and permission mechanism. As is shown in **figure 1**, the architecture is composed of four key components which are the application layer, consensus layer, network layer, storage layer.

Application layer: The application layer involves the key applications in the design review activities, such as design review process management, record storage, version management of BIM models, personnel identity management, etc.

Consensus layer: The consensus layer enables highly dispersed nodes to reach a consensus on the block data. It is a multi-party collaboration way to coordinate the multi-participants to reach an agreement about the process data in the design review activities and to ensure the record is difficult to be deceived.

Network layer: The blockchain network layer is to realize the information exchange between different participants and to make the data of the design review process transparent and traceable.

Storage layer: The layer is about the storage of the process data, it can be divided into on-chain data storage and off-chain data storage. Considering that storing large files on-chain will cause delays and increase storage costs. This study uses a combination of on-chain and off-chain storage. In the design review process, the BIM models and unstructured large design documents are stored off-chain with the help of IPFS, the hash value of these files is stored on-chain.

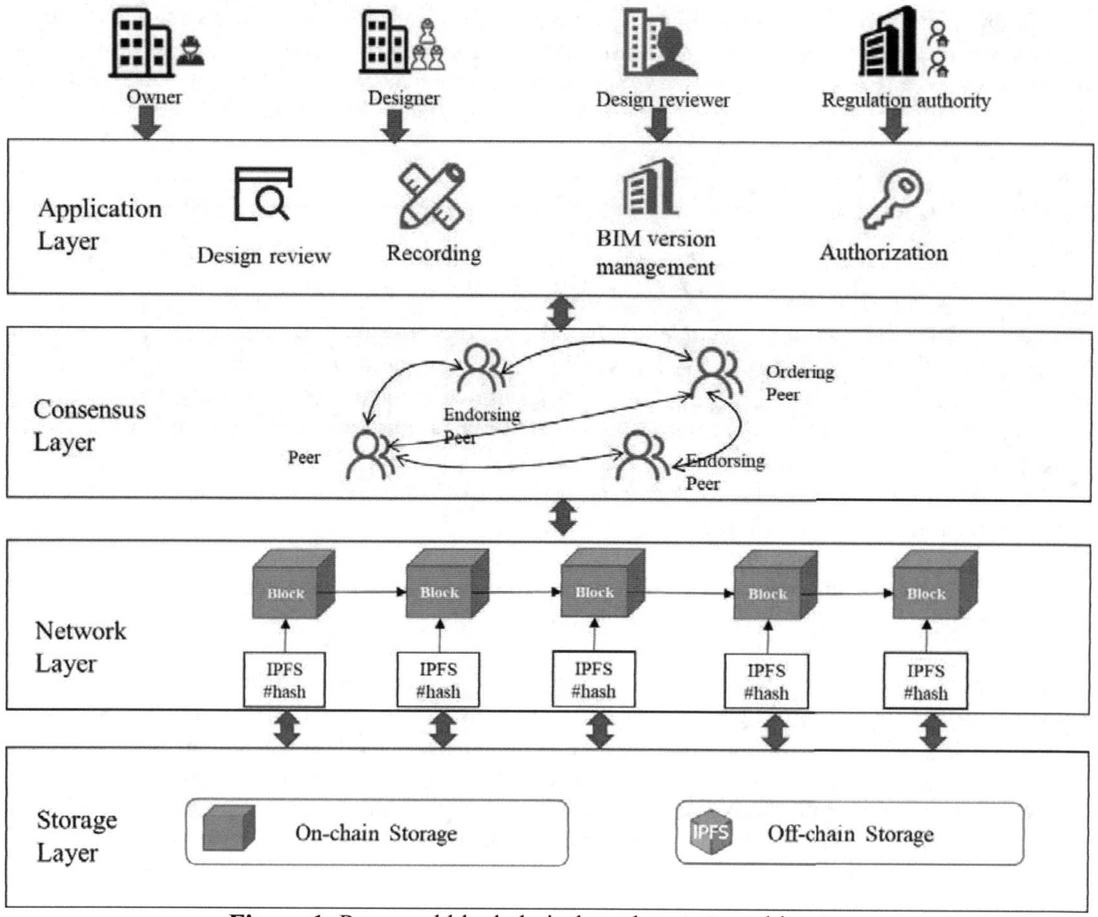

Figure 1. Proposed blockchain-based system architecture

3.3. Information flow during the design review process

The design review system based on the blockchain can be developed into a DApp(Decentralized Application), as is shown in **figure 2**, the designed workflow of the DApp is as follows:

The owner obtains a digital identity through Fabric CA(Certification Authority), then logs in and uploads the design files to the DApp. DApp uploads the design files to the IPFS database and generates the hash value corresponding to the files, and uploads the hash value and transaction information（timestamp, file type, digital signature, etc.）to the Fabric blockchain network. When each node on the blockchain reaches a consensus, a block is formed and stored on the chain.

The reviewer accesses the DApp through a digital entity, then starts the design review process. Design review is divided into review by human expert and automatic code checking by computer program. Some provisions can be automatically reviewed by computer programs, others can be reviewed manually. First, the BIM-based code checking algorithm will automatically check and output unqualified building elements and the corresponding violations. The records will be directly upload to the blockchain network through the peer node directly, it can prevent the possibility of tampering. After the reviewer has completed the design review, the reviewer needs to upload the design review report and add comments with digital signatures and time stamps. Regardless of whether the design review is passed or not, the review report and original design files will be uploaded to the blockchain network as immutable proof.

The designer logs in to the client through the digital identity granted by the Fabric CA, and downloads the review report of the unqualified design file, and makes changes. Then the designer uploads the modified files into the DApp, the transaction information will be broadcast after each node reaches a consensus, then the reviewer can download the modified design file in the IFPS database and repeat the design review process. In this process, each modification of the design file will be stored as transaction information to the blockchain network as an unchangeable record.

During the review process, the regulation authority does not participate in the actual activities. As a node on the blockchain, the node's ledger records all the review processes.

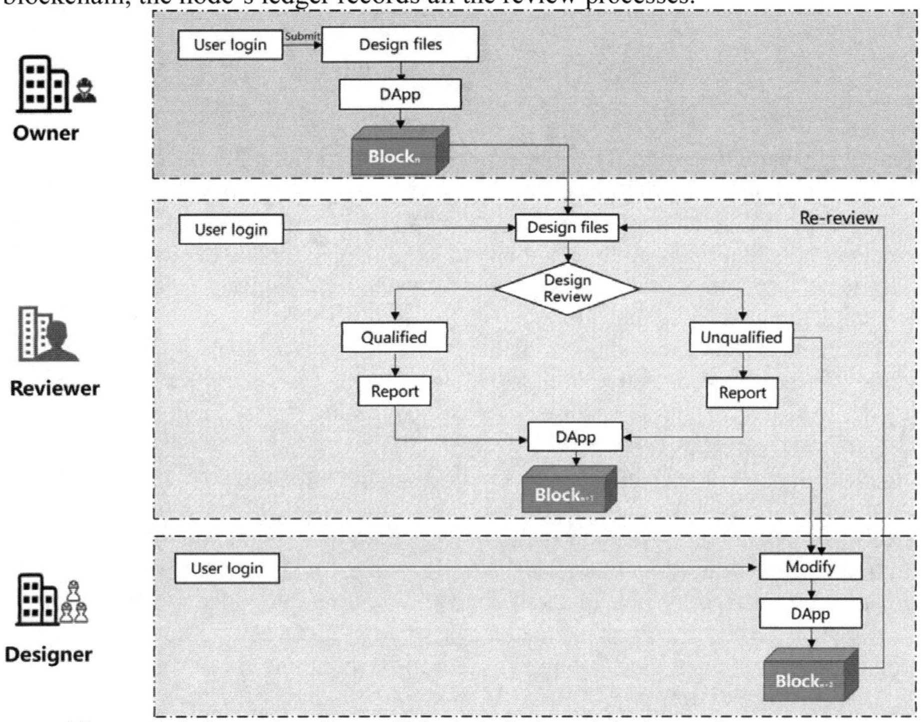

Figure 2. Information flow of blockchain-based design review process

3.4. Off-chain data management

This study uses IPFS as off-chain data storage, the participant in the blockchain network can share files through IPFS, the files are stored distributed in the P2P network, the peer-to-peer protocol of IPFS makes that the files cannot be modified. Combine asymmetric encryption with IPFS, design files can be shared among specific participants. For example, in **figure 3**, the designer is sharing BIM model with the reviewer, the following steps describe the process:

1. The designer desires to share the BIM file to IPFS and solely provide access to the design reviewer.
2. The designer encrypts the BIM file with the public key of the reviewer.
3. The designer uploads the encrypted BIM file to IPFS, and IPFS generates a hash value for the file.
4. The designer uploads the hash value and transaction information (time-stamp, file name, etc.) to the blockchain network.
5. The reviewer downloads the encrypted BIM file on IFPS through the hash value.
6. The review decrypts the BIM file with his or her private key.

IPFS a peer-to-peer network that allows data to be served based on its content, not its location. The address of the file is represented by a hash, once the content of the BIM file is changed, the hash value of the file will also change, and the historical version of the BIM file will still exist and cannot be modified. At the same time, during the file transfer process, the malicious third party does not have the private key of the reviewer, and cannot obtain the BIM file.

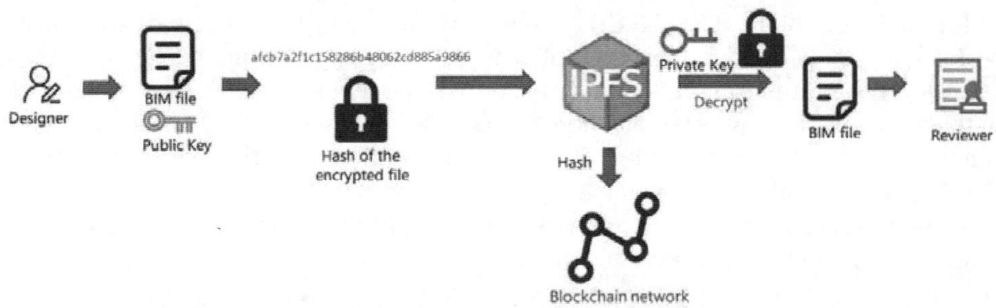

Figure 3. Design files sharing on IPFS through asymmetric encryption

4. Discussion

The design review panel is required to be independent of the local authority and the developer. But due to the funding relationship, sometimes, the design review process may not be completely objective. There exists a risk of collusion between the reviewer and other related parties. This section discusses the potential benefits that the proposed framework can bring to the design review process.

- Provide transparency in the design review process. In the traditional design review system, the system usually records the final qualified design documents without the process information. The content-based addressing mechanism of IPFS provides version management function for design files. It can provide a transparent environment to record the design review reports and model modifications in the process. Thereby reducing the possibility of collusion between the designer and government supervisor.
- Improve the design liability control. In the blockchain-based process, immutable proofs contribute to supporting accountability and traceability of design problems. If there occur quality issues, the designer and the reviewer's electronic signature and time stamp can be used to trace the cause of the problem in the design stage and help improve design liability control.
- Ensure system security. The blockchain-based solution integrated with IPFS makes that the data is stored on a P2P network, it is independent of any third-party storage. The sharing of the design files is securely transmitted through asymmetric encryption. It can ensure that the original design files and design review process records will not be tampered with.

At the same time, some challenges may hinder the deployment of the proposed blockchain-based system architecture in the practice. From a technical perspective, the distributed storage mechanism of the blockchain will make each node to store a copy of the transaction information. As the number of projects increases, it may cause transaction delays and reduce the efficiency of data queries on the blockchain. From a management perspective, the construction industry is considered a traditional industry that is unwilling to accept new technological changes. It is a challenge to persuade government agencies to transform a centralized database into a distributed storage database.

5. Conclusion and future work

This study explores how blockchain technology can improve traceability and transparency in BIM-based design review workflow. The study analyzes the existing problem in the design review process and provides a framework integrated consortium blockchain with IPFS to strengthen the liability control and data security in the design review process. In the presented framework, the blockchain technology is used for tracking data exchange in the design review workflows and different versions of BIM model are unchangeably stored on the IPFS network. Blockchain technology will help create an environment of trust and improve design responsibility control and reduce collusion among stakeholders and ultimately contribute to achieving better design quality. The DApp related to the proposed framework is under development, future work will integrate the IFC-based automatic BIM version difference detection algorithm into the blockchain solution to improve the process tracking ability.

References

[1] Nawari N O 2020 Generalized Adaptive framework for computerizing the building design review process *J. Archit. Eng.* **26** https://doi.org/10.1061/(ASCE)AE.1943-5568.0000382.

[2] Tan X, Hammad A and Fazio P 2010 Automated code compliance checking for building envelope design *J. Comput. Civ. Eng.* **24** 203–11 https://doi.org/10.1061/(ASCE)0887-3801(2010)24:2(203).

[3] Zhang J and El-Gohary N M 2016 Integrating semantic NLP and logic reasoning into a unified system for fully-automated code checking *Autom. Constr.* **73** 45–57 https://doi.org/10.1016/j.autcon.2016.08.027

[4] Malsane S, Matthews J, Lockley S, Love P E D and Greenwood D 2015 Development of an object model for automated compliance checking *Autom. Constr.* **49** 51–8 https://doi.org/10.1016/j.autcon.2014.10.004.

[5] Martins J P and Monteiro A 2013 LicA: A BIM based automated code-checking application for water distribution systems *Autom. Constr.* **29** 12–23 https://doi.org/10.1016/j.autcon.2012.08.008.

[6] Qi J, Issa R R A, Asce F, Olbina S, Asce A M, Hinze J and Asce M 2014 Use of building information modeling in design to prevent construction worker falls *J. Comput. Civ. Eng.* **28** 1–10 https://doi.org/10.1061/(ASCE)CP.1943-5487.0000365.

[7] Wang Z, Wang T, Hu H, Gong J, Ren X and Xiao Q 2020 Blockchain-based framework for improving supply chain traceability and information sharing in precast construction *Autom. Constr.* **111** 103063 https://doi.org/10.1016/j.autcon.2019.103063.

[8] Elghaish F, Abrishami S and Hosseini M R 2020 Integrated project delivery with blockchain: An automated financial system *Autom. Constr.* **114** 103182 https://doi.org/10.1016/j.autcon.2020.103182.

[9] Shi S, He D, Li L, Kumar N, Khan M K and Choo K K R 2020 Applications of blockchain in ensuring the security and privacy of electronic health record systems: A survey *Comput. Secur.* **97** https://doi.org/10.1016/j.cose.2020.101966.

[10] Patil A S, Tama B A, Park Y and Rhee K H 2018 A framework for blockchain based secure smart green house farming *Lect. Notes Electr. Eng.* **474** 1162–7 https://doi.org/10.1007/978-981-10-7605-3_185.

[11] Helo P and Hao Y 2019 Blockchains in operations and supply chains: A model and reference implementation *Comput. Ind. Eng.* **136** 242–51 https://doi.org/10.1016/j.cie.2019.07.023.

[12] Nakamoto S and Bitcoin A 2008 A peer-to-peer electronic cash system *Bitcoin.–URL https//bitcoin. org/bitcoin. pdf* **4**

[13] Miglani A, Kumar N, Chamola V and Zeadally S 2020 Blockchain for Internet of Energy management: Review, solutions, and challenges *Comput. Commun.* **151** 395–418 https://doi.org/10.1016/j.comcom.2020.01.014.

[14] Yu C, Jiang X, Yu S and Yang C 2020 Blockchain-based shared manufacturing in support of cyber physical systems: concept, framework, and operation *Robot. Comput. Integr. Manuf.* **64** 101931 https://doi.org/10.1016/j.rcim.2019.101931.

[15] Nawari N O and Ravindran S 2019 Blockchain technology and BIM process: Review and potential applications *J. Inf. Technol. Constr.* **24** 209–38 https://www.itcon.org/2019/12

[16] Zhong B, Wu H, Ding L, Luo H, Luo Y and Pan X 2020 Hyperledger fabric-based consortium blockchain for construction quality information management *Front. Eng. Manag.* **7** 512–27 https://doi.org/10.1007/s42524-020-0128-y.

[17] Choi J, Choi J and Kim I 2014 Development of BIM-based evacuation regulation checking system for high-rise and complex buildings *Autom. Constr.* **46** 38–49 https://doi.org/10.1016/j.autcon.2013.12.005.

[18] Patlakas P, Livingstone A, Hairstans R and Neighbour G 2018 Automatic code compliance with multi-dimensional data fitting in a BIM context *Adv. Eng. Informatics* **38** 216–31 https://doi.org/10.1016/j.aei.2018.07.002.

[19] Eastman C, Lee J min, Jeong Y suk and Lee J kook 2009 Automatic rule-based checking of building designs *Autom. Constr.* **18** 1011–33 https://doi.org/10.1016/j.autcon.2009.07.002.
[20] Hyperledger Fabric n.d. https://www.hyperledger.org/.
[21] IPFS Powers the Distributed Web n.d. https://ipfs.io/.

A preliminary SWOT evaluation for the applications of ML to Cyber Risk Analysis in the Construction Industry

Dongchi Yao[1], Borja Garcia de Soto[1]

[1] S.M.A.R.T. Construction Research Group, Division of Engineering, New York University Abu Dhabi (NYUAD), Experimental Research Building, Saadiyat Island, P.O. Box 129188, Abu Dhabi, United Arab Emirates

dy2037@nyu.edu, garcia.de.soto@nyu.edu

Abstract. Construction 4.0 is driving construction towards a data-centered industry. Construction firms manage significant amounts of valuable digital information, making them the target of cyberattacks, which not only compromise stored information but could cause severe harm to cyber-physical systems, personnel, and products. Therefore, it is critical to conduct cyber risk analyses to manage construction information assets to ensure their confidentiality, integrity, and availability. Traditional risk analysis methodologies like Fault Tree Analysis have limitations in dealing with the rapidly evolving cyber risks. As an alternative, Machine Learning (ML) methods are finding their way into the risk analysis field. ML models developed for cybersecurity purposes can learn from past results to make reliable predictions while removing the laboriousness of the traditional risk analysis. This article reviews ML techniques used for cyber risk analysis in different industries in recent years. Based on that, we investigate how ML techniques could be used for cyber risk analysis. Afterward, a SWOT analysis is conducted to identify the Strengths, Weaknesses, Opportunities, and Threats regarding the applications of ML in cyber risk analysis in the construction industry, and recommendations to address the weaknesses and threats are presented. Finally, future research areas using ML to prevent cyberattacks in the construction industry are proposed.

Keywords: *Artificial Intelligence; Construction 4.0; Cybersecurity; Machine learning*

1. Introduction

With the advent of Construction 4.0, big data, BIM, Machine Learning (ML), and cloud computing, to name a few, are finding their way into the different phases of the lifecycle of construction projects [1]. While digitalization undoubtedly facilitates the efficiency of project management, emerging challenges on data protection have to be addressed [2]. Apart from confidential data loss due to cyberattacks for profitability, harms to cyber-physical systems, personnel, and assets might also be incurred. A breach of blueprints in the planning phase [3], malicious attacks to BIM models during the design phase [4], or attacks on automatic control systems in the construction phase [5], data leakage due to information exchange among various parties [6] are all examples of cyber incidents. According to [7], more than 75% of respondents in the construction industry had experienced a cyber incident since 2019, and it is projected that cybercrime will cost businesses approximately $6 trillion per year on average through 2021[7]. However, the awareness and readiness on cybersecurity in the construction industry are still quite low, and in general, the construction sector treats cybersecurity as a lesser business priority compared to other industry sectors [8]. Therefore, considering cyber risk analysis is necessary to give stakeholders insights about preventing cyberattacks and minimizing potential losses. Traditional cyber risk analysis methods, such as Evidence Theory, and Analytic Hierarchy Process, rely excessively on experts' subjective judgment, and the analysis processes have to be started over when placed in different risk scenarios, which is time-consuming [9]. Fortunately, with artificial intelligence (AI) advancements, ML techniques can address some of these problems and generate results more swiftly and objectively. The rest of this paper is organized as follows. Section 2 reviews recent publications on the application of ML in cyber risk analysis in other industries, and shows its applicability in the construction industry. Based on the findings from existing publications, Section 3 includes a summary of a SWOT (Strengths, Weaknesses, Opportunities, Threats) analysis to get a preliminary evaluation of the applications of ML for cyber risk assessment in the construction industry; Section 4 lists corresponding recommendations regarding the weaknesses and threats identified; Section 5 concludes this study and suggests areas of future work.

2. Literature review

Cybersecurity risk analysis refers to the process of evaluating the security attributes of a cyber system, such as confidentiality, integrity, and availability as per the relevant information security technology and management standards. A typical framework of cyber risk analysis is shown in Figure 1. It can be seen that there are three main elements: (1) Risk identification: Consisting of asset, threat, and vulnerability identification, (2) Risk estimate: Including risk level determination and risk score calculation, and (3) Risk management: Consisting of several tasks. For purposes of this study, we focus on Threat and Vulnerability Identification, Risk Estimate, and Risk Monitoring (the shared boxes in Figure 1). The selection of the first three is evident. The last one was chosen because it can provide real-time feedback on a cyber system's risk posturing, which should be prioritized [10].

ML learns from existing data to forecast future behaviors, outcomes, and trends rather than through programming. In the cyber risk analysis context, the past data are the results of previous threat and vulnerability identifications, cyber risk estimates, and risk management plans. In general, cybersecurity problems can be classified into three categories. (1) For classification category, supervised learning algorithms along with DL (Deep Learning, a branch of ML) algorithms are suitable, (2) for Clustering and (3) for Dimension Reduction categories, unsupervised learning algorithms are mainly adopted.

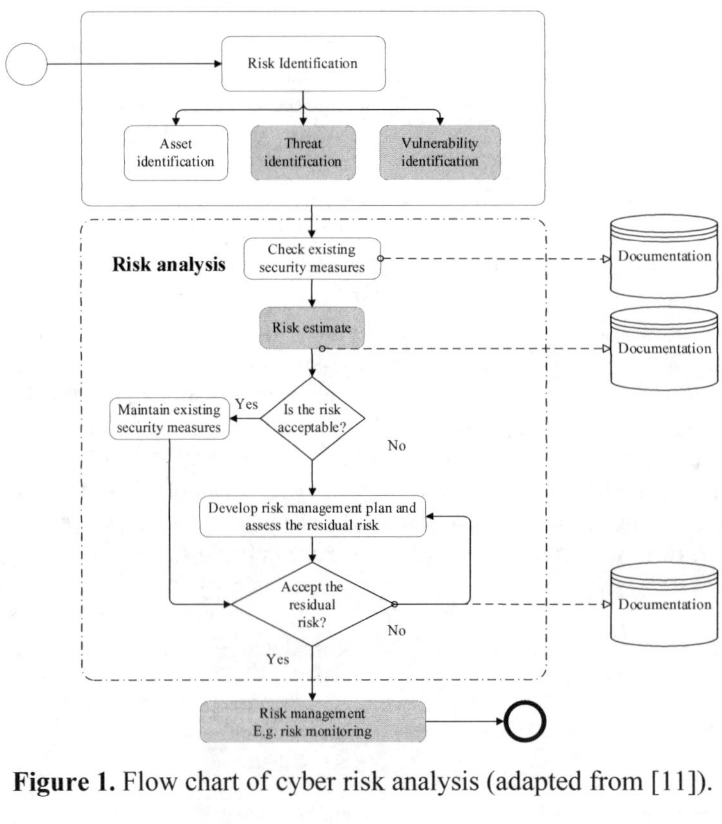

Figure 1. Flow chart of cyber risk analysis (adapted from [11]).

2.1. Machine learning in risk identification

Threats can exploit vulnerabilities existent in a cyber system to attack the corresponding asset. For example, in the construction industry, threats can be data exfiltration, contagious malware spread, denial of service attack, etc. In addition, vulnerabilities can come from collaborative working processes without strict rules, coding mistakes, software design defects, and unregulated data aggregation in BIM [12]. However, threats are becoming increasingly complex, and vulnerabilities are becoming more covert, making the manual identification of threats implausible. In this case, ML can fit in very well due to its automation ability. Examples of relevant recent (from 2016 to 2021) publications that investigate the use of ML for risk identification (in particular threat and vulnerability identification) in various fields include the following. Pang et al. [13] proposed a prediction technique targeting the early identification of potentially vulnerable software components by combining the Support Vector Machine (SVM) algorithm and ensemble learning strategy. Varma et al. [14] combined Bat Optimization Algorithm for Wrapper-based Feature Selection (BOAWFS) and the Random Forest (RF) algorithm to identify Android malware. Yuan et al. [15] presented a novel insider threat detection method with Deep Neural Network (DNN)-specifically, the Long Short-Term Memory-Convolutional Neural Network (LSTM-CNN) framework-to find employee's anomalous behavior. Russell et al. [16] employed a Deep Representation Learning algorithm and leveraged the wealth of C and C++ open-source code available to develop a large-scale function-level vulnerability detection system. Ebrahimi et al. [17] proposed a Transductive Support Vector Machines (TSVM) algorithm and a deep bidirectional Long Short-Term Memory (LSTM) algorithm for semi-supervised threat identification and labeling. Medina et al. [18] employed Inception-ResNet-V2 architecture and VGG16 architecture on transfer learning and fine-tuning to detect vulnerabilities in industrial control systems. Of course, these are not an exhaustive list but give insights into how ML techniques could be used in the construction industry in risk identification, such as identifying an attack by detecting abnormal data flow in computers, identifying a phishing email received by contractors by analyzing the text, identifying an intentional data-stealing of building plans by detecting the excessive attempts of log-in, etc.

2.2. Machine learning in risk estimate

Following the identification processes for threats and vulnerabilities, an asset's risk level or score should be determined (i.e., a risk estimate can be made). For example, in the construction industry, when a kind of threat to the BIM software appears and is detected, relevant personnel should calculate the score of risk this threat incurs or should classify the risk to a specific risk level, based on which a risk management measure is to be decided to deal with this threat. Since the risk level classification and risk scoring process align well with the functions of ML and with the advancement of Neural Network-based DL algorithms, new models based on the ML are being developed for achieving the goal of automatic risk estimate calculations. From a survey of existing literature from various fields focused on ML for risk estimates, we summarize some of the relevant work found. Matsika et al. [19] developed a risk estimate tool based on Multi-layer Artificial Neural Networks to calculate the risk of Metro and Light Rail Systems imposed by a potential terrorist attack. Bilge et al. [20] adopted fully and semi-supervised learning algorithms to create the RiskTeller system that can classify the risk level of a computer according to each computer's usage patterns. Jiao [21] adopted the Genetic Algorithm based Back Propagation Neural Network (GABP-NN) model to calculate the cyber risk score of a computer where 12 risk factor indexes are established as the input and a risk level from 1 to 5 is set to be the output. Zhang et al. [22] proposed a DL-based model for industrial cyber risk level probabilistic classification (5 levels from "very low" to "very high") using the real-world industrial dataset. Sun et al. [23] proposed Siamese Network Classification Framework (SNCF) algorithm that can classify the network machine to a specific risk level given the imbalanced original data. Kalinin et al. [24] proposed a 3-layer-perceptron NN algorithm to classify the risk level of a given infrastructure within the context of Smart City Construction, using the information of the network nodes as the input. Using ML methods, repetitive risk estimate processes are no longer needed, saving much time and labor. These frameworks can be applied to the construction field to manage cyber risk, such as predicting the risk level/score incurred by a malicious threat, not patching a vulnerability of a construction software, IT personnel's maloperation, the incompleteness of certain contract text, etc.

2.3. Machine learning in risk management

Within the scope of risk management, risk monitoring is an important topic. Continuous cybersecurity monitoring provides real-time visibility into the network ecosystem, allowing incident response teams to stay ahead of impending cyber threats and mitigate information security risks. For example, in the construction industry, a complete risk monitoring system can, at any time, analyze the data flow from the computers in the site office, from the sensors in the field, from the equipment operating, etc. Any abnormal pattern of the data flow can be analyzed and judged whether it signifies an attack occurs, and then the monitoring system can automatically choose which action to take or remind personnel to make decisions to mitigate the risk. Recent publications of ML applications on risk monitoring include the following. Chung et al. [25] integrated a Naïve Q-Learning algorithm into the cybersecurity monitoring system of a cyber infrastructure for modeling the decision-making process. He et al. [26] proposed a DL-based scheme by employing Conditional Deep Belief Network (CDBN) to monitor the high-dimensional temporal behavior features of the unobservable False Data Injection (FDI) attacks. Subroto et al. [27] compared various ML algorithms (Naïve Bayes, K-Nearest Neighbors, SVM, Decision Tree, ANN) to monitor cyber risks of social media. Li et al. [28] combined distributed streaming processing mechanism and ensemble prediction algorithm to monitor the abnormal network traffic of a cyber-physical power system. Sakthivel et al. [29] employed a Recursive Neural Network (RNN) algorithm to monitor and prevent the cyber system from cyberattacks in a manufacturing company. Garrido et al. [30] adopted a graph learning algorithm to detect intrusion, score anomalous activities and monitor security in industrial automation systems. These applications give the construction industry useful insights into how to create a monitoring system to identify any threat or vulnerability in real-time and simultaneously calculate the risk, based on which an action can be automatically elicited. For example, a risk monitoring system can be installed in the computer controlling the 3D printing process. This system synchronously monitors the network data flow of the computer, parameters of the 3D model in STL form, and parameters of the physical printing components. Once these parameters are tampered with and recognized, the system itself is able to decide whether to stop printing based on the risk value it calculates.

From this literature review, it can be seen that the automation of risk analysis and the use of ML techniques, especially DL, is gaining much attention. It is then reasonable to assume that AI applied to cyber risk analysis in construction has a bright future.

3. SWOT analysis

To get a better understanding of the potential of AI applied to cyber risk analysis in construction, we conducted a preliminary SWOT analysis to facilitate a realistic approach to evaluate the construction industry's position regarding the applications of ML in cyber risk analysis by identifying strengths, weaknesses, opportunities (i.e., positive external factors) and threats (i.e., negative external factors) of such applications. In addition, we considered internal (financial and human resources, tangible and intangible assets, operational efficiencies, etc.) and external (policies and regulations, market and technology changes, suppliers, etc.) factors and current and future strengths and weaknesses of the construction industry. The results are summarized in Fig 2.

3.1. Strengths

Strengths here are regarded as the internal factors rendering ML techniques competitive in cyber risk analysis. Since the construction industry is marching towards digitalization and automation, we consider ML techniques as the backbone of cyber risk analysis in the future due to these strengths. The main benefits of ML are as follows: (1) ML can handle the volume. ML can analyze a very large volume of activities across a construction company's network and the massive volume of digital information generated, such as emails, files, etc., quickly. (2) ML can learn over time. ML can identify malicious attacks based on the behaviors of applications and the network as a whole. Over time, ML cybersecurity solutions learn about a network's regular traffic and behaviors, providing a reactive approach to attacks. (3) ML can identify unknown threats. By using clustering algorithms, ML can successfully identify previously unknown attacks and update itself with the new information of these attacks.

3.2. Weaknesses

Weaknesses, like strengths, are inherent features of ML techniques and indicate the shortages that hinder its applicability and advancement. Only by identifying the weaknesses and solving them could ML techniques develop up to a higher level. Identified weaknesses include: (1) The hardship of high-quality data acquisition. The quality of input data is the cornerstone of ML applications. However, obtaining high-quality data is difficult, and the process is complex [31]. (2) The cost of error. False positives require analysts to spend expensive time inspecting, and the damage caused by false negatives is also non-negligible [32]. (3) The difficulty of deploying and maintaining ML modules. Some tools used for implementing ML algorithms were newly launched (e.g., TensorFlow and PyTorch packages were launched in 2015 and 2017, respectively), which might not be compatible with the older versions of computer configuration; thus, the deployment of ML might necessitate the redesign of the relevant business processes, which is expensive for construction companies [33]. (4) The lack of skilled personnel in construction. Since ML is a relatively new technology, the construction industry might lack the skilled personnel familiar with the implementation and utilization of ML, which requires the organization to add new positions to existing staff or hire new staff, thus increasing the cost.

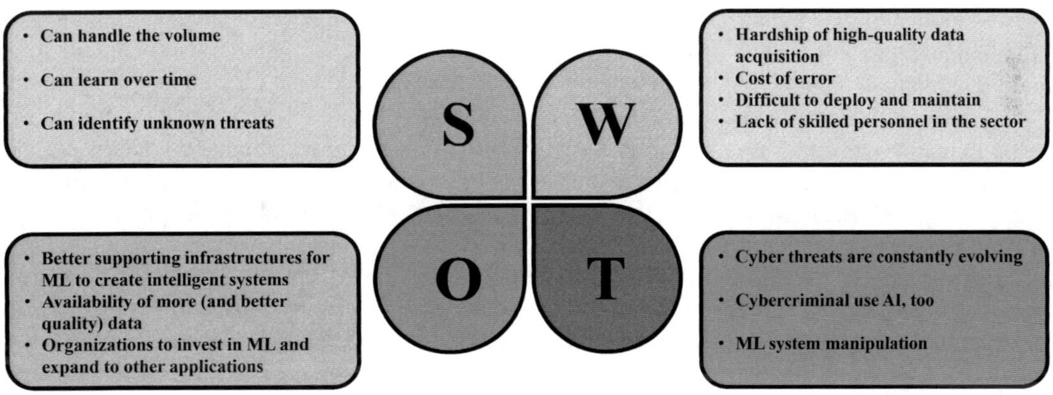

Figure 2. SWOT analysis of ML application in cyber risk analysis in the construction industry.

3.3. Opportunities

Opportunities here are positive external factors that help realize the set goals of ML techniques or advance them to achieve higher performance in cyber risk analysis in construction. Identifying these opportunities and seizing them at the right moment is critical for realizing the automation of cyber risk analysis. In this study, opportunities are identified as follows: (1) Better supporting infrastructures for ML to create intelligent systems. With the upgrading of software and other configuration that support ML implementation, such as better CPUs and GPUs, utilizing ML techniques to create a fully intelligent system to perform part or the whole cycle of cyber risk analysis is prospective. For example, ML-IDS (Intrusion Detection System) integration can be realized and applied to digital devices like sensors in the construction field and data-rich computers to detect intrusions [34]. (2) The availability of more data. ML is a data-driven technique, thus to achieve higher performance, more data is needed. Like in other industries, digitalization makes it possible for the construction industry to store data of virtually any software, equipment, and processes, such as CDE (Common Data Environment), BIM software, and SCADA systems, which could be sufficiently utilized to feed the ML algorithms. (3) The willingness of organizations to invest in ML. To facilitate the growth of organizations in the digitalized era, the decision levels are more willing (forced) to take action against cyberattacks swiftly. Knowing the prospect of ML, they tend to invest in better preparing existing staff, hiring new staff, purchasing advanced ML software, etc., thus iteratively propel the development of ML in the cyber risk analysis area [35].

3.4. Threats

Threats here refer to external factors that can negatively affect the implementation or the performance of ML techniques from the outside. It is vital to anticipate these threats and take action against them before ML becomes a victim of them and its growth stalls. The threats identified in this study include: (1) Cyber threats are constantly evolving. As a significant number of new threats emerge, security solutions that use ML have to be regenerated to keep up, thus increasing labor and cost. (2) Cyber criminals use AI, too. Cybercriminals can use ML techniques for illegal information gathering of people and IT assets, malicious user impersonation through phishing emails, unauthorized access to social accounts or machines, etc. (3) ML system manipulation. A true ML system never ceases to learn for improvement by taking constant feedback from the environment. The attacker can misuse this property by steering the system in the wrong direction by providing falsified data as feedback to the system [36].

4. Recommendations

Focusing on the weaknesses and threats from the SWOT analysis (Section 3), we propose some recommendations for construction project managers that could be considered as a starting point to overcoming the weaknesses and threats when considering ML techniques for cyber risk analysis in their projects

4.1. Recommendations to address weaknesses

4.1.1. Recommendations to address the hardship of high-quality data acquisition. All too often, the data collected for cybersecurity risk analysis is unstructured. To get high-quality data, data processing is necessary. The data processing phase consists of all the operations, from collecting the raw data to forming the final data set fed to the modeling tool [37]. These operation tasks may be performed multiple times in no fixed order, which includes: (1) Standardize the data collection process in the construction industry; (2) Clean and normalize the data to make the range of data value consistent; (3) Discard the redundant information by dimensionality reduction analysis; (4) Perform feature transformation to create structured data; (5) Conduct sample segmentation and sample balancing operations; (6) Augment the data to increase the diversity of data sources.

4.1.2. Recommendations to address the cost of error. To increase the precision of the ML model and prevent false positives and false negatives, several methods can be considered: (1) For a model whose accuracy rate is below 85% (or any threshold based on the organization's risk tolerance), personnel should take part in the decision-making stage, combined with the decision AI suggests. (2) Consider results from different models and take the final decision with the most votes. (3) Construction companies can install an extra firewall to achieve the goal of multi-protection purposes. (4) Improve the performance of the model [38], which includes performing sufficient data processing as mentioned in section 4.1.1, adopting the K-fold cross-validation method to utilize the samples as many times as possible, designing different loss functions to find the best loss function, and adjusting the parameters consistently, including hyper-parameters, ways of optimization.

4.1.3. Recommendations to address the difficulty of deploying and maintaining ML modules. There are three recommendations regarding this weakness: (1) For middle-sized and larger construction companies, the board members should enhance their awareness of cybersecurity issues [39] and proactively take actions to improve their existing systems to accommodate ML modules; (2) For companies that are prone to cyberattacks, directives should hire information security professionals to manage cyber risks to prevent losses ahead of time. (3) When ML modules are being designed, the structure of the model should align with the goal of the task. ML modules should not be excessively complex if simpler structures meet the demands as well.

4.1.4. Recommendations to address the lack of skilled personnel. There are two ways to address this issue, and construction organization should base their choice on available budget and internal organizational structure. The first recommendation is that the training sessions can equip the existing employees with ML knowledge on its implementation and maintenance procedures. The second is that the company can hire (or subcontract) new ML professionals to deal with relevant cyber risk analysis tasks.

4.2. Recommendations to address threats

4.2.1. Recommendations to address the constantly evolving threats. Facing the rapidly changing threats, besides improving the ML models' performance, construction companies must strengthen IT security mechanisms [40]. These mechanisms are listed as follows: Make sure systems, applications, and users are patched; Make specific regulations regarding the information sharing among owners, contractors, suppliers, and other parties; Implement a consistent security model regardless of user location or device type; Practice the rule of least access; Embrace the advanced terminal approach; Require secure applications; Sufficiently use threat intelligence; Train the employees to fully understand threat environment; Proactively use new security technologies.

4.2.2. Recommendations to respond to cybercriminals' usage of ML. Facing different usages of ML by cybercriminals, different preventive actions should be taken. (1) To prevent illegal information gathering of people and IT assets, personal information and properties of confidential IT assets must not be available in open sources. (2) To avoid being the victim of malicious impersonation of users, email/text receivers should double-check with the users who send suspicious messages through other channels. (3) To prevent unauthorized access, construction personnel can choose a captcha like MathCaptcha or its alternatives for websites and machines and simultaneously use complicated passwords and any form of multi-factor authentication (MFA) [41].

4.2.3. Recommendations to address ML system manipulation. There are several ideas to prevent ML system manipulation: (1) ML experts should prevent the damage by minimizing the amount of training data cybercriminals can control. (2) ML engineers can address this issue by maintaining a record of data ownership and streamlining and securing system operations. In addition, (3) ML engineers can design new functions to prevent data extraction and model function extraction by hackers to avoid system poisoning [42].

5. Conclusions and Outlook

This study reviews recent publications on the applications of ML techniques in risk identification, risk estimate, and risk monitoring and shows its potential in the construction industry for cyber risk analysis. Based on the literature review, a preliminary SWOT analysis was performed, addressing the strengths, weaknesses, opportunities, and threats that might be faced by the construction industry when considering ML techniques. Regarding the weaknesses and threats, we propose some recommendations for addressing them. From the literature review, it is apparent that NN-based DL techniques are obtaining an increasing amount of attention, mainly due to their capability of dealing with Natural Language Processing (NLP) and high accuracy for task classification. To expand this preliminary study, the authors' work focuses more on image classification for threat identification and text analysis for vulnerability identification by integrating current DL algorithms like RNN-based LSTM algorithm, Probabilistic Latent Semantic Analysis (PLSA), and simultaneously developing new architectures to create an intelligent integrated system combining all processes of risk identification, estimate, and management in the long run.

Acknowledgments
The authors want to thank the support provided by the Center for Cyber Security at New York University Abu Dhabi (CCS-AD).

References

[1] Kling R and Turk Z 2021 Construction 4.0 – digital transformation of one of the oldest industries *Econ Bus Rev* **21**(3) pp 393–496

[2] Mantha BRK and de Soto BG 2019 Cyber security challenges and vulnerability assessment in the construction industry *Proc. of Creative Construction Conf. 2019* pp 29–37

[3] Mott G. 2018 A critical reflection on the construction of the cyberterrorist threat in the United Kingdom of Great Britain and Northern Ireland *Nottingham Trent University*

[4] Zheng R, Jiang J, Hao X, Ren W, Xiong F and Zhu T 2019 CaACBIM: A context-aware access control model for BIM J. *Inf.* **10**(2) pp 1–15

[5] Lin CT, Wu SL and Lee ML 2017 Cyber attack and defense on industry control systems *2017 IEEE Conf. on Dependable and Secure Computing* pp 524–6

[6] Stapleton KAJ, Gledson BJ and Alwan Z 2014 Understanding technological interoperability through observations of data leakage in Building Information Modelling (BIM) based transactions *Proc. of the 32nd eCAADe Conf.* pp 515–24

[7] The rise of cyber risk in the construction industry 2020 *Gallagher* [cited 2020 Mar 15] Available from: https://www.ajg.com/us/news-and-insights/2020/jun/cyber-risk-within-construction-industry/

[8] Cyber security breaches survey 2021 *GOV.UK* [cited 2021 Jun 2] Available from: https://www.gov.uk/government/statistics/cyber-security-breaches-survey-2021/cyber-security-breaches-survey-2021#chapter-3-awareness-and-attitudes

[9] Cherdantseva Y, Burnap P, Blyth A, Eden P, Jones K, Soulsby H, et al 2016 A review of cyber security risk assessment methods for SCADA systems *J. Comput. Secur.* **56** pp 1–27

[10] Williams C. Risk monitoring: 6 considerations for understanding this make or break moment for ERM [Internet]. *ERM Insights*. 2019 [cited 2021 Jun 2]. Available from: https://www.erminsightsbycarol.com/risk-monitoring/

[11] Zhao G 2020 *Information Security Management and Risk Assessment 2nd ed.* ed. Liu X (Beijing: Tsinghua University Press) p 48

[12] Kurtz S 2019 Cybersecurity vulnerabilities in the construction industry [Internet]. *Total IT Information Technology* [cited 2021 Mar 19] Available from: https://totalit.com/cybersecurity-vulnerabilities-in-the-construction-industry/

[13] Pang Y, Xue X and Namin AS 2016 Early identification of vulnerable software components via ensemble learning *2016 15th IEEE Int. Conf. on Machine Learning and Applications (ICMLA)* pp 476–81

[14] P RKV, S KRM, K JS and D PL 2017 Bat optimization algorithm for wrapper-based feature selection and performance improvement of android malware detection *Proc. of the 33rd Annual Computer Security Applications Conf.* pp 362–72

[15] Yuan F, Cao Y, Shang Y, Liu Y, Tan J, Fang B 2018 Insider threat detection with deep neural network *International Conference on Computational Science* pp 43–54

[16] Russell R, Kim L, Hamilton L, Lazovich T, Harer J, Ozdemir O, et al 2019 Automated vulnerability detection in source code using deep representation learning *17th IEEE Int. Conf. on Machine Learning and Applications (ICMLA)* pp 757–62

[17] Ebrahimi M, Nunamaker JF and Chen H 2020 Semi-supervised cyber threat identification in dark net markets: a transductive and deep learning approach *J. Manag. Inf. Syst.* **37**(3) pp 694–722

[18] Blanco-Medina P, Fidalgo E, Alegre E, Vasco-Carofilis RA, Jañez-Martino F and

Villar VF 2021 Detecting vulnerabilities in critical infrastructures by classifying exposed industrial control systems using deep learning *J. Appl Sci.* **11**(1) pp 1–14

[19] Matsika E, Robinson M and Neill CO 2016 Risk assessment tool for analysing terrorist attack *European Transport Conf. 2016 Association for European Transport (AET)* pp 1–22

[20] Bilge L, Han Y and Dell'Amico M 2017 RiskTeller: Predicting the risk of cyber incidents *Proc.of the 2017 ACM SIGSAC Conf. on Computer and Communications Security* pp 1299–311

[21] Jiao J 2018 Discussion on the neural network model of comprehensive evaluation of computer network security *J.Inf. Commun.* **10** pp 14–5

[22] Zhang X, Lashkari AH, Maleki N, Mudge J and Ghorbani AA 2019 Probabilistic classification for industrial cyber risk using deep neural network *the 2019 Int.Conf. on Security and Management* (Las Vegas) pp 208–15

[23] Sun D, Wu Z, Wang Y, Lv Q and Hu B 2019 Risk prediction for imbalanced data in cyber security: a Siamese network-based deep learning classification framework *Int. Joint Conf. on Neural Networks (IJCNN)* pp 1–8

[24] Kalinin M, Krundyshev V and Zegzhda P 2021 Cybersecurity risk assessment in smart city infrastructures *J.MDPI* **9** (4) p 78

[25] Chung K, Kamhoua CA, Kwiat KA, Kalbarczyk ZT and Iyer RK 2016 Game theory with learning for cyber security monitoring *2016 IEEE 17th Int. Symp. on High Assurance Systems Engineering (HASE)* pp 1–8

[26] He Y, Mendis GJ and Wei J 2017 Real-time detection of false data injection attacks in smart grid: a deep learning-based intelligent mechanism. *J. IEEE Trans. Smart Grid* **8**(5) pp 2505–16

[27] Subroto A and Apriyana A 2019 Cyber risk prediction through social media big data analytics and statistical machine learning *J. Big Data* **6**(1) pp 1–19

[28] Li Q, Meng S, Zhang S, Wu M, Zhang J, Taleby Ahvanooey M, et al 2019 Safety risk monitoring of cyber-physical power systems based on ensemble learning algorithm. *IEEE Access* **7** pp 24788–805

[29] Sakthivel RK, Nagasubramanian G, Al-Turjman F and Sankayya M 2020 Core-level cybersecurity assurance using cloud-based adaptive machine learning techniques for manufacturing industry *J. Trans. Emerg. Telecommun. Technol.* p e3947.

[30] Garrido JS, Dold D and Frank J 2021 Machine learning on knowledge graphs for context-aware security monitoring. *arXiv Prepr arXiv2105.08741* Available from: http://arxiv.org/abs/2105.08741

[31] Goh GD, Sing SL and Yeong WY 2021 A review on machine learning in 3D printing: applications, potential, and challenges *J. Artif. Intell. Rev.* **54**(1) pp 63–94

[32] Yellenki RB 2020 Top 8 challenges for machine learning practitioners *Towards Data Science* [cited 2021 Apr 1] Available from: https://towardsdatascience.com/top-8-challenges-for-machine-learning-practitioners-c4c0130701a1

[33] Paleyes A, Urma RG and Lawrence ND 2020 Challenges in deploying machine learning: a survey of case studies *arXiv Prepr arXiv* pp 1–21. Available from: https://arxiv.org/abs/2011.09926

[34] Gassais R, Ezzati-Jivan N, Fernandez JM, Aloise D and Dagenais MR 2020 Multi-level host-based intrusion detection system for Internet of things *J. Cloud Comput.* **9**(1) pp 1–16

[35] Das S, Dey A, Pal A and Roy N 2015 Applications of artificial intelligence in machine learning: review and prospect *Int. J. Comput. Appl.* **115**(9) pp 31–41.

[36] Tatwadarshipn 2021 Security threats to machine learning systems *Analytics Vidhya* 2021 [cited 2021 Jun 4] Available from: https://www.analyticsvidhya.com/blog/2021/01/security-threats-to-machine-learning-systems/
[37] Rassam MA, Maarof MA and Zainal A 2017 Big data analytics adoption for cyber-security: a review of current solutions, requirements, challenges and trends *J. Inf. Assur. Secur.* **11** pp 124–45
[38] Schelter S, Biessmann F, Januschowski T, Salinas D, Seufert S and Szarvas G 2018 On challenges in machine learning model management *Bull. IEEE Comput. Soc. Tech. Comm. Data Eng.* pp 5–13
[39] Armenia S, Angelini M, Nonino F, Palombi G and Schlitzer MF 2020 A dynamic simulation approach to support the evaluation of cyber risks and security investments in SMEs. *J. Decis. Support. Syst.* **June 2020** p 113580
[40] Goldstein M 2020 4 Ways small businesses can protect themselves from cyber attacks [Internet]. *The Intersection of Business and Technology* [cited 2021 May 27]. Available from: https://www.coxblue.com/4-ways-small-businesses-can-protect-themselves-from-cyber-attacks/
[41] Polyakov A 2021 Machine learning for cybercriminals 101 *Towards Data Science* [cited 2021 Jun 4] Available from: https://towardsdatascience.com/machine-learning-for-cybercriminals-a46798a8c268
[42] Jagielski M, Oprea A, Biggio B, Liu C, Nita-rotaru C and Li B 2018 Manipulating machine learning: poisoning attacks and countermeasures for regression learning *2018 IEEE Symp. on Security and Privacy (SP)* pp 19–35

A Theoretical Assessment of the Impacts of Poor Maintenance of Public Infrastructure

Ncamiso Matse[1], Nokulunga Xolile Mashwama[1], Didibhuku Thwala[2] and Clinton Aigbavboa[1]

[1]Dept of Construction Management and Quantity Surveying, University of Johannesburg, Johannesburg, South Africa
[2]SARChI in Sustainable Construction Management and Leadership in the Built Environment, Faculty of Engineering and the Built Environment, University of Johannesburg, Johannesburg, South Africa

Nomiso01investment@gmail.com, nokulungam@uj.ac.za, Didibhukut@uj.ac.za & clinton@aigbavboa.com

Abstract. The study examines previous literature on the impacts of poor maintenance of public infrastructure in the Kingdom of Eswatini. In the past two decades the kingdom of Eswatini has experienced a huge growth of public building development. Construction of new facilities alone is not enough but the most critical activities in public buildings is to provide adequate and proper maintenance for the structures to accomplish their intended purpose. It is worth noting that building maintenance prevents accumulated decay and deterioration, which is contributing to health concerns and reducing the profitability of occupants. Therefore, absence or poor performance of facility management in an institution is accompanied by negative impacts. Moreover, to correct these negative consequences a massive financial cost and worst severe financial loss would be generated to those institutions concerned. The study is conducted with reference to existing theoretical literature, published and unpublished research. The study is mainly a literature review on the impacts of poor maintenance of public infrastructure. One of the primary findings regarding the impact of poor building maintenance are: that it reduces the value of existing public buildings and it also increases budgets for untreated maintenance faults, increased costs for emergency repairs, insufficient provision and delivery of public services. The study explores the effects of poor maintenance of public infrastructure and presents a robust background on the theories of poor infrastructure maintenance. This study will enable the public facility management and the government at large to know the importance of maintaining a building.
Key words: Impacts, Maintenance, Public Infrastructure

1. INTRODUCTION

Abdullah et al. (2015:363) observed that in the past few years there has been an increase in building construction in most developing countries as part of development. Increase in the requirement of public services triggered the demand for the construction of public buildings in order for the public institution to render the various services needed by the public such as health, education, welfare and many others. Growth in infrastructure development will lead to an increasing demand for maintenance works in order to enable the buildings to perform their intended purpose. Hence, such development created demand for building maintenance (BM) and to become a major activity in the construction industry. Moreover, Olufunke (2011:290) emphasized that, considering construction of new facilities alone is not enough but the most critical activities in PB is to provide adequate and proper maintenance for the structures to accomplish their intended purpose.

2. LITERATURE REVIEW

2.1 Early decaying of public building

According to Olufunke (2011:290) public buildings became valueless due to lack of strong maintenance culture in the country. Therefore, public buildings (PB) which were meant to benefit a large number of people are decaying prematurely, rotting without any good use as the public is not willing to make such facilities sustainable. Furthermore, PB collapses before their expected life span due to lack of proper maintenance. Tijani et al., (2016:27) emphasis that the attitudes within people, that nobody owns government properties brings a negative impact on the buildings.

2.2 Non-operational use of buildings

Wuni et al. (2018:11) argued that most PB are found abandoned due to the fact that most public institutions are unable to undertake major rehabilitation. Furthermore, it is noted that financing renovations or major rehabilitation are a great threat to the upcoming financial sustainability of the institutions and the involved buildings are abandoned. Kamarazaly (2014:111) emphasized that most government institutions are not profit making and they are rendering many services to the public. Therefore, PB renovations budget will have to compete with many services to be offered by the government which may be valued high, and result in facilities being non-operational.

2.3 Reduce the economy of the country

Most public institutions are facing negative financial impacts due to poor facility management (Kamarazaly et al., 2013:140). Hamid et al. (2010:80) argued that poor PB management creates pressure on the future government budget because excessive budget allocation will be needed to correct the wrongs of the facility management

2.4 Insufficient provision and delivery of public services

According to Hopland and Kvamsdal (2019:141) insufficient maintenance level and deprived facility management have a negative impact in delivering quality services to the public. Tijani et al. (2016:24) argued that equipment on PB are poorly maintained which resulted in regular failure, utilization rate becoming low and production schedule being lowered.

2.5 Low production due to hindered productivity

The situation of PB not maintained is not different with the education department. In most rural areas it is common to find dilapidated public school buildings being occupied by pupils. Lessons are being disturbed by the collapsing classes with leaking roofs, without ceiling which makes the environment unsuitable for learning (Olufunke, 2011:292). Breakdown of machines and industrial equipment as a result of poor maintenance had obstructed production in many industries. Negative effects of production in terms of capacity decrease, poor quality services, growth in production cost and public dissatisfaction among others are the results of poor production resources maintenance (Tijani et al., 2016:28).

2.6 Public building loses its physical appearance

The physical appearance of PB is highly affected by the decision and action taken in the history of the building regarding its maintenance (Ofori et al., 2015:185). Ackah, (2016) argued that the public pride or indifference is normally seen by the appearance, quality and physical condition of the buildings. These judge the prosperity level in the area, social values and conduct. Improper maintained PB in certain areas and streets degrade the beauty of the surroundings.

2.7 Reduces the convenience and safety use of building

According to Oluwatoyin et al. (2017:1) buildings are designed to provide safe and conducive accommodation for humans to undertake their various activities. Therefore, it is necessary to take measures to deteriorate the building component to ensure that buildings preserve the initial designed functions which are to provide safety and convenience to the building users. Hamid et al., (2010:83) argued that deferred maintenance of PB increases the safety hazard, reduces productivity of employees and affects business operations.

2.8 Increases cost for emergency repairs

Richards (2018) argued that the cost for planned repairs are low compared to emergency maintenance and the emergency maintenance usually occurred during peak production hours. Therefore, reactive repairs need to be performed as soon as possible for production to resume quickly. This kind of maintenance is usually outsourced with escalated costs instead of being done by in-house maintenance team.

2.9 Increased budgets for untreated maintenance fault

Ihsan and Alshibani (2018:303) argued that unattended small maintenance problems accumulate to become worse. Minor building faults like leaking roofing, if not treated would cause damages on the ceiling, stain wall, affect light fitting, etc. Hopland and Kvamsdal (2019:141) agreed that deferred maintenance on public buildings creates expenditures backlog that would be covered in future. Therefore, to keep government buildings on fully operational condition more resources will be required.

2.10 Reduces the value of the building

Deferred maintenance results in the building to physical depreciation, losing its initial value due to the postponement of maintenance. Normally, physical depreciation is cured by applying the required maintenance and improvements. Such building depreciation is called curable obsolescence or curable physical depreciation (Fakhrudin et al., 2011:173).

3. RESEARCH METHODOLOGY

The research was carried out with reference to the published and unpublished theoretical literature from online journals, Conferences, Proceedings dissertations. Again, articles from Google scholar, Emerald, Science direct which were all related to the impacts of poor maintenance of public infrastructure were also used. Moreover, the study is mainly a theoretical literature review. The current methodology falls within the qualitative research methodology. Qualitative approaches to data collection, analysis, interpretation, and report writing differ from the traditional, quantitative approaches. Qualitative data usually not numerical therefore it cannot be statistically analyzed. (Creswell, 2014:23 and Muijs, 2004:1)

4. LESSON LEARNT

This research work assisted to reveal the impacts of poor maintenance of public infrastructure. Poor maintenance results in physical deterioration of the building, losing its initial value due to maintenance postponement. Usually, physical depreciation is cured by applying the required maintenance and improvements. Moreover, poor maintenance increases budgets for untreated maintenance faults. Unattended small maintenance problems, they accumulate to become worse. Furthermore, poor maintenance on PB creates expenditures backlog that would be covered in future. Therefore, to keep government buildings on fully operational condition more resources will be required.
The cost for planned repairs are low compared to emergency maintenance and the emergency maintenance usually occurred during peak production hours. Hence, reactive repairs need to be performed as soon as possible for production to resume quickly. This kind of maintenance is usually out sourced with escalated cost instead of being done by in-house maintenance team.

5. CONCLUSION

The literature reviewed revealed that construction of new facilities alone is not enough but the most critical activities in public buildings is to provide adequate and proper maintenance for the structures to accomplish their intended purpose. Furthermore, the literature disclosed that absence or poor performance of the maintenance management in an institution is accompanied by negative impacts regarding their buildings. Moreover, to correct these negative consequences a massive financial cost and worst severe financial loss would be generated to those institutions concerned.

References

[1] Abdullah S., Razak A. A. and Rashid Z. Z. A. (2015). Towards the Development of Maintenance Culture: An Evaluation on Users Understanding towards the Building Maintenance Aspects. *Australian Journal Basic & Applied Science*, vol. 9, Issue 7: pp 363-368.

[2] Ackah D. (2016) The Poor Mindset of Citizens on Maintenance aspect of Infrastructure Projects in most Developing Countries; as Compare to the Advanced Countries Available at https://www.linkedin.com/pulse/poor-mindset-people-maintenance-aspect-infrastructure-ackah-ph-d accessed on 15 July 2020

[3] Creswell J. W. (2014) Research design: qualitative, quantitative, and mixed methods approaches, 4th edition. Available from http://fe.unj.ac.id/wp-content/uploads/2019/08/Research-Design_Qualitative-Quantitative-and-Mixed-Methods-Approaches.pdf accessed on 06 September 2020: pp 1-342

[4] Fakhrudin I. H., Suleiman M. Z. and Talib R. (2011) The need to implement Malaysia's Building and Common Property Act 2007 (Act 663) in building maintenance management. *Journal of Facilities Management.* vol. 9, issue 3: pp 170-180

[5] Hamid Y., Alexander K and Baldry D. (2010) The cause and effects of deferred maintenance on higher education buildings: pp 78-88
[6] Hopland A. O and Kvamsdal S. (2019) "Building conditions in Norwegian local governments: trends and determinants", *Facilities*, vol. 37, issue 3/4: pp 141-156
[7] Ihsan B and Alshibani A. (2018) Factors affecting operation and maintenance cost of hotels. Property Management. vol. 36, issue 3: pp 296-313
[8] Kamarazaly, M. A. (2014) Challenges in Strategic Facilities Management: Analysis of Problems Faced by Universities. *PhD Thesis Submitted to the School of Engineering and Advanced Technology, Massey University, Newzealand*: pp 518
[9] Kamarazaly, M. A., Mbachu, J., Phipps, R., Kamarazaly, M. A., Mbachu, J., & Phipps, R. (2013) Challenges faced by facilities managers in the Australasian universities. *Emerald Journal of Facilities Management*, vol.11, issue 2: pp 136–151.
[10] Muijs D. (2004) Doing Quantitative Research in Education with SPSS, Available from http://modares.ac.ir/uploads/Agr.Oth.Lib.23.pdf accessed on 07 September 2020: pp 228
[11] Ofori I, Duodu P. M and Bonney S. O (2015) Establishing Factors Influencing Building Maintenance Practices: Ghanaian Perspective. *Journal of Economics and Sustainable Development*, vol. 6, issue 24: pp 184 -193.
[12] Olufunke A. M. (2011) Education for maintenance culture in Nigeria: Implications for community development. *International Journal of Sociology and Anthropology*, vol. 3, issue 8: pp 290-294
[13] Oluwatoyin O. I. M., Olotuah A.O., Adeyemi E. and Ayo-Vaughan E. K. (2017) Maintenance of Public Secondary School Buildings: Users' Practices in Nigeria: pp 9
[14] Richards T. (2018) Costly Dangers of Deferred Maintenance in Facilities. Available at https://www.chthealthcare.com/blog/deferred-maintenance accessed on 17 July 2020
[15] Tijani S. A., Adeyemi A. O. and Omotehinshe O. J., (2016) Lack of Maintenance Culture in Nigeria: The Bane of National Development. *Civil and Environmental Research*. Vol. 8, issue 8: pp 23-30
[16] Wuni I. Y., Agyeman-Yeboah S. and Boafo H. K. (2018) Poor Facility Management in the Public Schools of Ghana; Recent Empirical Discoveries. *Journal of Sustainable Development Studies*, vol.11, issue 1: pp 30

An artificial neural network analysis of the satisfaction of hospital building maintenance services

AL Olanrewaju and W X Tan

Universiti Tunku Abdul Rahman, Jalan Universiti, Bandar Barat, 31900 Kampar, Perak, Malaysia

abdullateef.olanrewaju@ymail.com

Abstract. Modern hospital buildings are very large, complex, sophisticated, and costly structures. The productivity of the hospital building users is influenced by the building performance. Maintenance services are conducted on the buildings to ensure optimum building performance and users' satisfaction. However, many instances, have been recorded dissatisfactions among the hospital building users with respect to the performance of the maintenance services. This research investigated the satisfaction of private hospital building users with respect to maintenance services through a survey questionnaire that included ten satisfaction of maintenance services. The data revealed that the users are most satisfied with the standard of workmanship, cleanliness of the maintained area, communication with users during maintenance works, and degree of politeness and kindness of the maintenance workers. Using the nine/ten as the independent variables, an artificial neural network model to predict the satisfaction of users with maintenance services was presented. The model revealed that cleanliness and tidiness of the maintained area and the quality of maintenance works are the main predictors of the total maintenance service delivery of the maintenance organisations.

Keywords: private hospitals, building performance, benchmarking, maintenance organisation, productivity.

1. Introduction

Hospitals are building-based organisations because the expenditure on the building operation is more than 50% of their asset base. Apart from the salaries of the staff, the operational cost of hospital buildings is the most significant item on a hospital's balance sheet. Hospitals are large enterprises and are one of the most challenging buildings to construct and operate. The users of the hospital buildings, especially the patients depend on the building performance to facilitate their recuperation process and procedures. The productivities of the medical and non-medical staff depend on the condition and performance of the building. For the continued operation of the hospital to happen, maintenance functions are required. Therefore, the standard of maintenance depends on the services that the building offers. As a result, hospital building maintenance organisations are under severe pressure. A maintenance organization is a service provider whose purpose is to provide elegant maintenance services with minimum disturbance and disruptions at less cost. In Malaysia, despite the increasing expenditure on investment in hospital buildings, the performance of the hospital buildings is low. While the backlog of maintenance demand is increasing, the levels of criticisms and complaints in the media and research literature on the poor performance of the hospital buildings in Malaysia due to maintenance-related practices and processes are plenty and increasing [1, 2,3,4] and expectations of users have not been met. Therefore, there is a need for empirical research to examine the building maintenance services of hospitals. Because the hospital maintenance organisation needs to know whether or not their users are satisfied or not with the services provided. Although there is increasing awareness among hospitals about the importance of maintenance, the hospitals lack a logical, holistic and consistent approach to conduct maintenance that focuses on user value systems as the key reference point for maintenance management decisions and actions.

2. Background and theoretical framework

Hospitals are generally divided into public and private hospitals. Government is the largest provider of healthcare in most countries through its extensive network of hospital and primary care clinics in the public sector. There are 144 public hospitals and 210 private hospitals in Malaysia [5]. A private hospital is profit-oriented and is funded through payment for medical services. A well-maintained building is critical for the delivery of a hospital's corporate business objectives. Hospital buildings are procured to create a suitable, conducive, and adequate environment that can support and encourage patients' recovery and allow staff to perform their duties productively. When maintenance of a building is not well managed, it often leads to buildings that are very expensive to own or operate in the long run. Hospital organisation that does not own their buildings, the rents are usually the highest payments in the hospital's annual balance sheets. However, there are a lot of defects in hospital buildings [1,7]. Defect in buildings compromises the performance of the buildings. Hence, it is a necessity to conduct maintenance in hospital buildings to meet their intended design and construction function as well as maintain the user's satisfaction levels [1]. Satisfaction is the pleasure that felt after a need is fulfilled [8]. Satisfaction has been described to include an outcome and a process. This research is more concerned with the process in the satisfaction of the maintenance service delivery.

2.1. Users' satisfaction in building maintenance

The hospital building is any person, group, or organization that uses the building as an occupier, owner, tenant, visitor or other stakeholders. To continue to justify the investment or uptake in the building maintenance, the performance of the buildings and the service delivery requires evaluation by the building users. Because the building is a capital good [9]. Therefore, the evaluation of maintenance services is a strategic tool to evaluate building performance and maintenance services. The user's evaluation can be carried out using different methods, including surveys, interviews, and case studies. Largely, it involves inciting a response from the building users on the different aspects of the service delivery process. What to include on the list of evaluation criteria depends on the purpose of the evaluation. User satisfaction is a technique to increase user experience and provide feedback to stakeholders for future development. It measures the difference between experience and expectation. Building user satisfaction is evaluated with reference to expectations. Based on an extensive review of the literature on the maintenance services [9,10,11] and discussion with hospital building users, the following services may determine the level of satisfaction of hospital building user regarding the service provided by the hospital maintenance department condition of buildings, service quality, speed of service delivered, failure to meet the needs and expectations, the ineffective scope of works, building appearance, well-functioning of building equipment, attitude and behaviour of building management staff, execution time, operating hours, convenient to the user, interaction with maintenance personnel, cleanliness after repaired, the staff appearance. Therefore, the maintenance service criteria for the buildings are expected to cover all the above features, and a failure in the delivery of the performance requirements will have severe consequences on the satisfaction of the users with the maintenance service and building performance.

3. Method of data collection and analytics

The primary data were collected through online survey questionnaires based on snowball sampling. Snowball sampling is a non-probability sampling technique that is used if there is less information on the potential respondents. However, one of the limitations of this technique is that the number of respondents will not be known to the researcher. However, the technique is an inductive technique.

Snowballing is a data collection method where the survey is administered to available, accessible, and willing respondents [12]. The respondents were requested to help forward the form to friends and associate that are competent to provide an answer to the questions. The method is appropriate where sufficient information on population size and sample frame is not available. However, its findings may not be generalizable, but with large respondents, the finding can be a representative [12]. The questionnaires were administered to the hospital building's users through an online survey from 14/07/2020 to 28/08/202020. The building users were asked based on pieces of evidence to tick the degree on how they are satisfied with the services that the maintenance organisation provides to the hospital building users. In order to evaluate the user's level of satisfaction, Likert Scales of 5 points were used. Where 5 = extremely satisfied, 4 = very satisfied, 3 = satisfied, 2 = less satisfied and 1 = least satisfied. Two and 3 are located in between.

Artificial neural networks (ANN) technique was used to create a model for maintenance services. ANN is a type of non-parametric artificial intelligence model. ANN is suitable for systems that demonstrate a definite relationship between the dependent and independent variables, but one that is too complex to analyze using traditional techniques of correlations, regression analysis, and group differences. ANN method has a higher capability to model nonlinear and functions accurately. It can accommodate multiple parameters without any computation problems. However, unlike fuzzy system models where the internal workings are transparent, ANNs do not readily reveal how a particular answer was obtained [13]. In addition, ANNs do not indicate the direction of the relationship between dependent and independent variables. In other words, it is not clear from the results of the relationship between the dependent variable and the independent variable is negative and positive. Artificial neural networks have not been applied to model building maintenance services. One hidden layer seems adequate in our ANN model estimation. First, Multilayer Perceptron (MLP) was selected as the network to be trained. Automatic architecture selection was then selected, which automatically identifies the best possible network architecture with only one hidden layer. The minimum and maximum numbers of hidden layer units were selected to be 3 and 50 respectively. The dataset is divided into three samples. Firstly, the training sample used to optimize the network weights (supervised learning process). Secondly, the testing sample used to check the error during the learning process. Finally, the holdout sample used for validation; after the training of the network done, the holdout sample assesses the predictive accuracy of the network in a ratio of 70% 20% and 10% for the divisions.

4. Analysing the results of the survey

The survey pooled 538 forms and 207 completed responses were received during the survey period spanning over one month. The data revealed that 60% are medical staff (i.e., doctors, nurses, therapists, and pharmacists). The remaining 40% include (i.e., executive, cashier, cleaner). More than 50% of users have been using the hospitals for more than 4 years.

4.1. Results of the descriptive statistics

The Omega value of the reliability of the factors is 0.851 and the validity ranges from 0.6 to 0.89. The Kaiser's Measure of Sampling Adequacy was significant $\chi2$ (91) = 857.560, p<0.001), N=.830) indicating that the data were drawn from the same population and that the services were related. The results of the one-sample t-test revealed that all the services are statistically significant (Hr: U≥U0). The small standard errors are small which means that the measurements are representative. The survey found that more than 50% of the users measured that they were either very satisfied or extremely satisfied with the maintenance services. 31% measured that they are satisfied, but 19% acknowledged that they were least satisfied or less satisfied with the maintenance services. The cumulative ASI score for all the services is 73.11% while the cumulative standard deviation (SD) was 16.45%.

4.2. Results of the ANN Model

All the 10 services were used to create an ANN model. In this model, 74.4.0%of the total 207 datasets were used for training, 16.9% for testing, and 8.7% percent are stored as a holdout sample. The Relative Error during training (0.181%), testing (0.198%), and holdout (0.199%) confirmed the correctness of the model for predicting the maintenance service delivery. The sum of squares error for the training is 18.81 and 4.45 for testing.

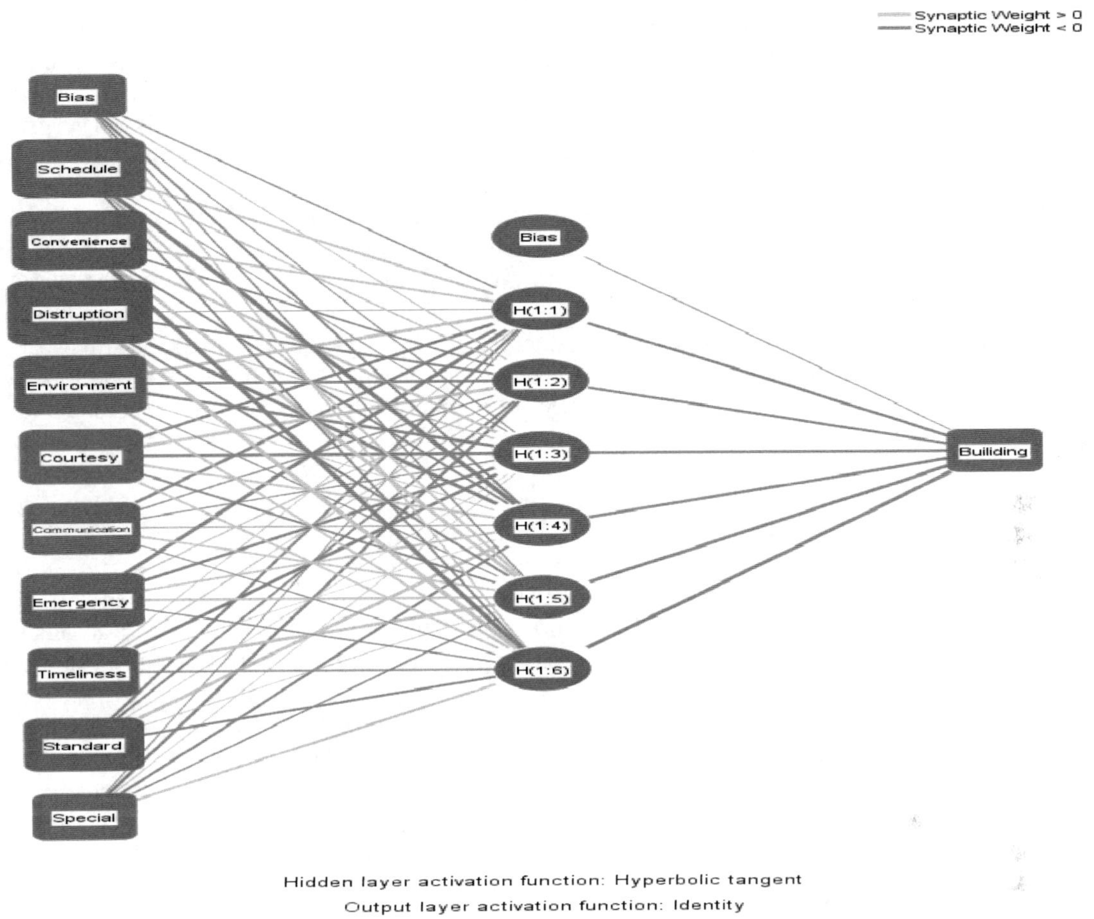

Figure 1 The architecture of the neural network

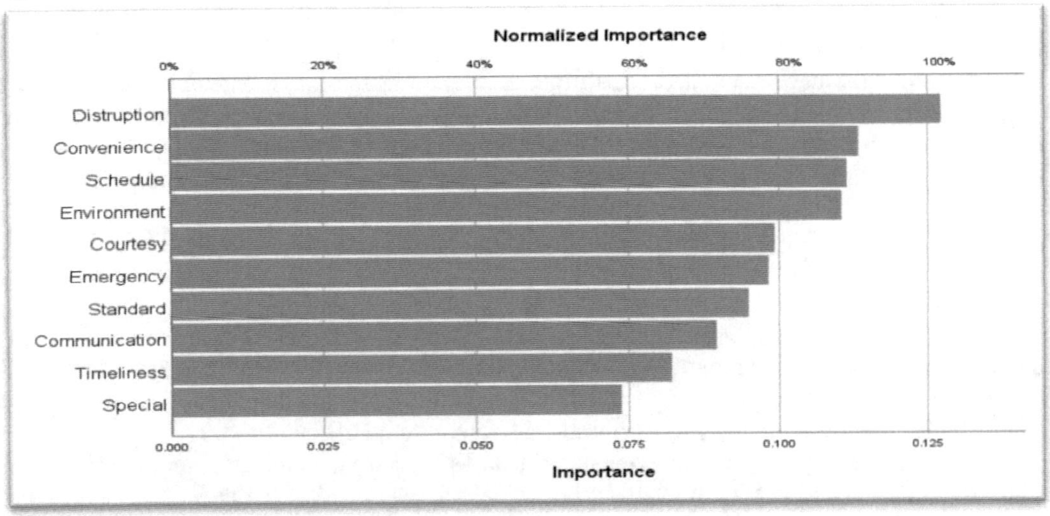

Figure 2 The architecture of the neural network

5. Discussion of the ANN model

Because of space constraints, a brief discussion is provided. The average relative errors were 18%, 20% and 19% for the training, testing, and holdout, which show that the precision of the model is relatively high. The relative errors are relatively constant across the training, testing, and the holdout. This is a confirmation that the model is not overtrained and that the error in future cases will be near to the error revealed in this research. The model suggests that the services can predict the overall service that the maintenance organisation provides the lowest minimum importance of about 60%. The primary drivers of satisfaction are the level of disruption due to maintenance work that is allowable in the hospital is very minimum. One of the differential factors between new construction and maintenance is that maintenance is conducted when users/occupants are still using the building. Because of this, maintenance is often disruptive to the user's/occupant's activities. To reduce this, the maintenance organisation must plan and execute the repair with minimum disruption and distraction to the users. Hence, the convenience of users is very critical. The ease to make a service request is a major predictor. It is imperative for the building users to know who to call when maintenance demand arises, especially in the emergency maintenance inwards, operation rooms, and theatre. Lack of information on personal that are responsible for repair and their availability could complicate the already tense situation at the hospital.

The data also revealed that the duration that the maintenance organisations take to complete the repair is critical to the users. Maintenance demand can happen at any time because the building materials and components have a finite life span and the materials and components deteriorate at different times. While some defects may be anticipated, some may not be anticipated. When maintenance demand arises, the users desire prompt response [10]. This is not surprising because of the tense condition in the hospitals, especially in critical areas and times. There has been a lot of complaining that, once repaired conducted, used /damaged materials and components are sometimes left behind. Leaving the users with the odious responsibilities to do the cleaning of the repaired materials/components and area. However, the data revealed many of the users are not satisfied with this particular service offer by the maintenance organisation. Additionally, many of the users are not satisfied with the time taken to complete the repair works. The maintenance organisation must make certain that the cleanliness of the maintained area remains a priority. The maintenance organisation needs to ensure that the standard of cleanliness is increased. The courtesy of the maintenance organisations to the building users is very important to the users. How users are treated by the maintenance organisation/ employees has been identified as a critical factor to improve building user satisfaction by the maintenance organisations [9]. Therefore, it is interesting to find that the hospital building users measured courtesy as a major predictor.

6. Conclusion

There is literature on the development of performance measurement for maintenance services. The theory behind performance measurement of maintenance is maintenance service should be measured based on outcomes with respect to the maintenance service delivery. Hospital organisations whether public or private seek to provide quality services with minimum resources. The relationship between building defect and building performance and ultimately users' productivity and satisfaction is well posited. In other words, building maintenance is progressively recognized as a factor of production like other resources. The building and its management need to be considered as an investment rather than as liabilities. On the one hand, university buildings are factors of production. On the other hand, the building cannot remain new through its entire life span. The value of the building decreases unless maintenance is administered to the building. Maintenance should be positively planned, strategically organize, proactively lead, holistically controlled, and dynamically implemented if the best value is critical. In order for the building to serve its purpose effectively and efficiently, there is the need for a change in mindset in the way they are the view and the ways they are managed. Hospital buildings are procured to provide an enabling environment for health care delivery. Thus, the need to have a capable maintenance organization is on the increase. In that regard, the essence of the maintenance department should be to enhance users' satisfaction and to improve productivity. This research uses 9 satisfaction factors to create an ANNs model to predict the performance of the building and satisfaction with the service that the building maintenance organisation provides in the hospitals. Although there has been substantial research concerning patient satisfaction with medical care, there have been no attempts to empirically investigate hospital building users with maintenance services. It is hard to determine how much satisfaction or how happy the building's user with the maintenance services

7. References

[1] Olanrewaju A A, Fang WW and Tan, Y S 2018 Hospital building maintenance management model. *International Journal of Engineering and Technology*, 2(29), pp.747-753.

[2] Rani N AA, Baharum M R, Akbar A R N and Nawawi A H 2015. Perception of maintenance management strategy on healthcare facilities. *Procedia-Social and Behavioral Sciences*, 170 pp.272-281.

[3] Wong WF., Olanrewaju A. and Lim PI., 2021. Value-Based Building Maintenance Practices for Public Hospitals in Malaysia. Sustainability, 13(11), p.6200.

[4] Othman N L Mustapha M J Harun W M W and Ibrahim,F 2015 A case study on workmanship issues and public hospital building defects. *Advances in Environmental Biology*, 9(5), pp.147-149.

[5] Ministry of Health Malaysia. 2019. Healthcare Facilities (Government), available at: http://www.moh.gov.my/ (accessed 30 December 2019).

[6] Ministry of Health Malaysia, 2019. Malaysia National Health Accounts, Health Expenditure Report 1997-2017.

[7] Olanrewaju, A, Wong, F.W., Yahya, N.N and Im, P. L (2019) Proposed research methodology for establishing the critical success factors for maintenance management of hospital buildings. *AIP Conference Proceedings* 2157, 020036 (2019); https://doi.org/10.1063/1.5126571.

[8] Ali W 2016. Understanding the concept of job satisfaction, measurements, theories and its significance in the recent organizational environment: A theoretical framework. *Archives of Business Research*, 4(1).

[9] Olanrewaju A L and Abdul-Aziz AR 2015 *Building maintenance processes and practices: The case of a fast developing country*. Springer

[10] Chanter B and Swallow P 2007 *Building Maintenance Management*, Oxford: Blackwell Publishing,

[11] Wordsworth P and Lee R 2001 *Lee's building maintenance management.* London: Blackwell Science

[12] Sekaran U and Bougie R 2016 *Research methods for business: A skill building approach*. John Wiley & Sons.

[13] Shephard R B 2006 *Quantifying Environmental Impact and Assessment Using Fuzzy Logic*, Springer, USA

Acknowledgments
The research has been carried out under Fundamental Research Grant Scheme project FRGS/1/2018/TK06/UTAR/02/ provided by Ministry of Higher Education of Malaysia.

IOP Conf. Series: Materials Science and Engineering 1218 (2022) 012020 doi:10.1088/1757-899X/1218/1/012020

Analysis of Challenges to BIM Adoption in Mega Construction Projects

E C Akcay

Department of Civil Engineering, Atılım University, Ankara, Turkey

*caner.akcay@atilim.edu.tr

Abstract. Megaprojects are large-scale and complex investments, including a long construction period, as a result need efficient management skills during their life cycle. The success rate of the megaprojects is lower compared to the traditional projects due to their complex characteristics. By adopting new technologies, such as Building Information Modeling (BIM), more efficient project management can be performed for mega construction projects. Hence, the success of projects could be increased in terms of time, cost and quality. In this study, the challenges to BIM implementation in mega construction projects are investigated. To achieve this aim, the challenges for BIM adoption extracted through a literature review. Then, a questionnaire survey was designed to collect the data from practitioners. Furthermore, the challenges were analyzed using descriptive statistics.

Keywords: BIM, construction, megaproject, project management

1. Introduction

Mega construction projects can be described as "large-scale, complex ventures that typically cost US$1 billion or more, take many years to develop and build, involve multiple public and private stakeholders, are transformational, and impact millions of people" [1]. Some examples of the megaprojects include railways, airports, wastewater projects, power plants tunnels. They have an important role in meeting the diverse needs of people [2]. Rapid global urbanization has led to an increase in demand for megaprojects, especially in developing countries [3]. As a result, the number of megaprojects in the world has rapidly increased [4-5]. Although their numbers have increased so much, their success rates have remained low compared to traditional projects. The reason for having such low success rates is their uncertain and complex characteristics [6]. To reduce the impact of the complex and uncertain nature on the success of the project, more efficient project management needs to be performed. This can be achieved by adopting new technologies.

Building Information Modeling (BIM) is one of the most common information technology based approach that enables users to create virtual model of construction for the planning, design, construction and maintenance phases [7-8]. It allows engineers, constructors, and architects to recognize any problems associated with design, construction and operation stages [9]. In addition to these, using BIM helps projects achieve the desired targets in terms of time, quality and cost. It is clear that using BIM solves most of the problems encountered in projects, especially in megaprojects. It also increases the success rates of projects. Hence, its use in the AEC industry is also growing rapidly. Although it has many benefits and is widely used in many countries, it has not reached the desired level of use in some countries. Turkey is one of the country that the implementation of BIM has not reached the desired levels especially in megaprojects. Therefore, the main objective of this research is to explore the barriers that hinder the implementation of BIM in mega construction projects.

2. Literature Review

BIM technology has become one of the most important topics in construction management literature for the last two decades. There are large number of BIM studies that particularly focus on the challenges to the implementation of BIM technology in different countries. Al-Hammadi and Tian [10] analyzed the critical challenges to BIM technology adoption in the Saudi Arabian construction industry. "Lack of demand", "lack of experts" and "poor awareness of BIM benefits" were listed as the major barriers. Farooq *et al.* [11] identified 12 barriers that hinder the BIM implementation in Pakistan. They also used an Interpretive Structural Modeling (ISM) to investigate the interrelationships between the barriers. Doan *et al.* [12] explored 14 potential barriers to the adoption of BIM technology in the New Zealand construction industry. According to this study, "a lack of BIM understanding", "a lack of expertise", "high economic investment", "a lack of collaboration and coordination" and "legal issues" were listed as the major barriers. Roy and Firdaus [13] investigated the barriers for BIM implementation in Indonesia. They summarized the top five barriers to BIM adoption are "lack of BIM training", "lack of BIM experience and capability", "no client demand", "high cost in software and hardware acquisition", and "inadequate information technology (IT) facilities". Chan *et al.* [14] examined the barriers to BIM implementation in Hong Kong. According to them, the major barriers impeding the adoption of BIM technology are "resistance to change", "inadequate organizational support and structure to execute BIM" and "lack of BIM industry standards". Charef *et al.* [15] highlighted the major barriers hindering the adoption of BIM technology in EU countries. Cao *et al.* [16] listed the barriers to BIM implementation in the Canadian construction industry. Hatem *et al.* [17] investigated the potential barriers impeding the BIM adoption in Iraq. "Weakness of the government's efforts", "poor knowledge about the benefits of BIM" and "resistance to change" were listed as the top three barriers hindering the BIM adoption. Sreelakshmi *et al.* [18] identified the major challenges faced by Indian construction companies in adopting BIM technology. Wu *et al.* [19] concluded from their study that the major barriers hindering the adoption of BIM technology for industrialized building in China are "capital-related factors" and "lack of support from owners". In the light of the background research, this research aims to explore and analyze the barriers hindering the BIM technology adoption for megaprojects in Turkey.

3. Research Methodology

This study aims to investigate the barriers for megaprojects in Turkey in three steps such as survey design, data collection, and data analysis. In the first step, a questionnaire was designed to evaluate the impact of the barriers. To design the questionnaire, a set of barriers was identified through a literature survey (Table 1). It should be noted the published papers (both in journals and conferences) related to BIM technology barriers were collected using Scopus and Web of Science databases for this identification stage.

Table 1. Barriers to BIM implementation

Barriers	References
B1.Lack of knowledge about BIM	10, 20, 21, 22
B2.High cost of BIM software	13, 15, 22, 23, 24
B3.Lack of experience in using BIM	16, 17, 19,25
B4.Lack of standards and government regulations	11,15,16,25,30
B5.Lack of skilled personnel	11, 13, 21, 26
B6.BIM Licensing problems	13, 22, 27
B7.Lack of demand	10, 11, 15, 25, 28
B8.Interoperability problems	14, 20, 24, 28
B9.Resistance to change	21, 22, 26, 28

B10.Inadequate studies on BIM	11, 17, 20
B11.High cost of training	15, 26, 27, 29
B12.Lack of tangible BIM benefits	14, 16, 22, 23, 28

The questionnaire includes two parts. First part includes questions asking the demographic information about the participants, whereas the second part covers the barriers to BIM technology adaption. In this part, a five-point Likert scale was adopted to assess the impact of each barrier. After the questionnaire was designed, it was distributed to the participants via e-mail. As it is important to determine the proper respondents, the population of this research includes consultants, contractors and clients who have been involved at least one mega construction project in Turkey.
In the last step, the collected data were statistically analyzed to elicit the ranking of each barrier. For this purpose, the mean score method was adopted. In this method, the ranking of each barrier was found using the following formula:

$$MS = \frac{\Sigma(f\ x\ s)}{N} \qquad (1)$$

Where MS is the mean score for each factor; f is the frequency of response to each rating for each factor; s is the score of each factor; N is the total number of responses.

4. Results

4.1. Profiles of Respondents

A total number of 73 questionnaires have been distributed. 52 responses were received, which shows a response rate of 71.2%.

- 48% of the respondents were contractors having a majority, 29% were consultants, and 23% were clients.
- 77% of the respondents had bachelor's degree , 19% had master's degree, and 4% had doctorate degree.
- 67% of the respondents were civil engineers , 13% of the respondents were architect, 12% of the respondents were mechanical engineers, and 8% of the respondents were electrical engineers.
- 21% of the respondents have 0-10 years of working experience , 62% of the respondents have 10-20 years of working experience, and 17% of the respondents have more than 20 years working experience.
- 61% of the respondents have 0-5 year experience in BIM technology, 33% of the respondents have 5-10 years experience in BIM technology, and 6% of the respondents have more than 10 years experience in BIM technology.

4.2. Ranking of Barriers

Table 2 shows the respondents' perceptions for 12 barriers that hindering the implementation of BIM. The results showed that respondents ranked "B5. Lack of skilled personnel" as the most important barrier with a mean value of 4.38. It is one of the biggest barrier to BIM technology implementation, especially for developing countries. Although the number of BIM professionals in Turkey has rapidly increased in the last decade, it is still limited. As the need for skilled personnel in megaprojects is higher than traditional projects, this barrier takes the priority among all respondents. The second most highly rated barrier was "B1. Lack of knowledge about BIM" with a mean value of 4.25. This is another major barrier that hinder the BIM implementation. Most people working in the construction industry do not have enough knowledge about BIM. This barrier has also been reported as one of the most important barriers in many studies [12,13,17]. "B4. Lack of standards and government regulations" was ranked third with a mean value of 4.19. Another barrier to overcome to reach the desired level of BIM techology use is the preparation of the necessary standards and legal regulations. "B3. Lack of experience in using BIM" is another important barrier that was ranked fourth with a mean value of 4.08. The fifth most important barrier hindering the BIM technology adoption is "B9. Resistance to change" with a mean value of 3.96. "B6. BIM Licensing problems" is the least import barrier with a mean value of 2.29. Contrary to similar studies, financial barriers (B11.High cost of training and B2.High cost of BIM software) remained at the bottom of the ranking. This can be explained by the fact that the megaprojects have a larger budget compared to the traditional projects. Clients group and contractors group ranked "B9. Resistance to change" as the most important barrier, whereas contractors group ranked it eighth. Contractors group ranked "B5. Lack of skilled personnel" as the first priority whereas clients group ranked it third an consultants group ranked it second. According to the clients, the top three barriers are as follows: B9. Resistance to change (MS = 4.58, Rank = 1), B1. Lack of knowledge about BIM (MS = 4.42, Rank = 2), and B5. Lack of skilled personnel (MS = 4.33, Rank = 3). According to the contractors, the priority of barriers are as follows: B5. Lack of skilled personnel (MS = 4.44, Rank = 1), B4.Lack of standards and government regulations (MS = 4.28, Rank = 2), and B1. Lack of knowledge about BIM (MS = 4.33, Rank = 3). According to the consultants, the priority of barriers are as follows: B9.Resistance to change (MS = 4.40, Rank = 1), B5. Lack of skilled personnel (MS = 4.33, Rank = 2), and B1. Lack of knowledge about BIM (MS = 4.27, Rank = 3).

Table 2. Ranking of Barriers

	Overall		Client		Contractor		Consultant	
	MS	Rank	MS	Rank	MS	Rank	MS	Rank
B1.Lack of knowledge about BIM	4.25	2	4.42	2	4.16	3	4.27	3
B2.High cost of BIM software	2.65	11	3.83	6	2.24	12	2.40	11
B3.Lack of experience in using BIM	4.08	4	3.92	5	4.12	4	4.13	4
B4.Lack of standards and government regulations	4.19	3	4.08	4	4.28	2	4.13	4
B5.Lack of skilled personnel	4.38	1	4.33	3	4.44	1	4.33	2
B6.BIM Licensing problems	2.29	12	1.67	12	2.56	11	2.33	12
B7.Lack of demand	3.69	7	3.42	7	3.80	7	3.73	7
B8.Interoperability problems	2.96	10	2.75	11	2.92	10	3.20	10
B9.Resistance to change	3.96	5	4.58	1	3.40	8	4.40	1
B10.Inadequate studies on BIM	3.23	9	3.17	10	3.24	9	3.27	9
B11.High cost of training	3.69	7	3.25	9	4.04	5	3.47	8
B12.Lack of tangible BIM benefits	3.87	6	3.33	8	4.04	5	4.00	6

5. Conclusion

This study reports the results of a survey questionnaire investigating barriers that hinder the BIM technology implementation in megaprojects. Potential barriers to BIM implementation were initially identified through a literature survey. Using these barriers, a questionnaire was prepared and applied to the clients, contractors and consultants. The results were analyzed by adopting a mean score method. The findings of this research show that the top five most important barriers are lack of skilled personnel, lack of knowledge about BIM, lack of standards and government regulations, lack of experience in using

BIM, and resistance to change. Based on the findings of this study, future work can develop some strategies to overcome the identified barriers that hinder the BIM technology implementation. The sample size of the respondents is the limitation of this study. As the number of megaprojects is limited in Turkey, it is difficult to find suitable respondents, have experience in megaprojects, for the questionnaire survey. Future studies can be performed by increasing the sample size.

References

[1] Flyvbjerg B 2014 What you Should Know about Megaprojects and Why: An Overview *Project Management Journal* **45** 2 pp 6–19 doi: 10.1002/pmj.21409
[2] Hwang B-G, Shan M, Zhu L and Lim W-C 2020 Cost control in megaprojects: efficacy, tools and techniques, key knowledge areas and project comparisons *International Journal of Construction Management* **20** 5 pp 437-449 doi:10.1080/15623599.2018.1484976
[3] Hu Y, Chan A P C, Le Y and Jin R-z 2015 From Construction Megaproject Management to Complex Project Management: Bibliographic *Analysis J Manage Eng* **31** 4
[4] He Q, Luo L, Hu Y and Chan A P C 2015 Measuring the complexity of mega construction projects in China—A fuzzy analytic network process analysis *International Journal of Project Management* **33** 3 pp 549-563 https://doi.org/10.1016/j.ijproman.2014.07.009
[5] Mok K Y, Shen G Q and Yang J 2015 Stakeholder management studies in mega construction projects: A review and future directions *International Journal of Project Management* **33** 2 pp 446-457 ISSN 0263-7863 https://doi.org/10.1016/j.ijproman.2014.08.007
[6] Othman A 2013 Challenges of mega construction projects in developing countries *Organisation, Technology and Management in Construction* **5** pp 730-746
[7] Bryde D, Broquetas M and Volm J M 2013 The project benefits of Building Information Modelling (BIM) *International Journal of Project Management* **31** 7 pp 971-980 https://doi.org/10.1016/j.ijproman.2012.12.001
[8] Gu N and London K 2010 Understanding and facilitating BIM adoption in the AEC industry *Automation in Construction* **19** 8 pp 988-999
[9] Azhar S 2011 Building Information Modeling (BIM): Trends, Benefits, Risks, and Challenges for the AEC Industry *Leadership and Management in Engineering* **11** 3
[10] Al-Hammadi M A and Tian W 2020 Challenges and barriers of building information modeling adoption in the Saudi Arabian construction industry *Open Construction and Building Technology Journal* **14** 1 pp 98-110
[11] Farooq U, Ur Rehman S K, Javed M F, Jameel M, Aslam F and Alyousef R 2020 Investigating bim implementation barriers and issues in Pakistan using ism approach *Applied Sciences (Switzerland)* **10** 20 1-21.
[12] Doan D T, GhaffarianHoseini A, Naismith N, Ghaffarianhoseini A, Zhang T and Tookey J 2020 Examining critical perspectives on Building Information Modelling (BIM) adoption in New Zealand *Smart and Sustainable Built Environment*
[13] Roy A F V and Firdaus A 2020 Building Information Modelling in Indonesia: Knowledge, Implementation and Barriers *Journal of Construction in Developing Countries* **25** 2 pp 199-217
[14] Chan D W M, Olawumi T O and Ho A M L 2019 Perceived benefits of and barriers to Building Information Modelling (BIM) implementation in construction: The case of Hong Kong *Journal of Building Engineering* **25**
[15] Charef R, Emmitt S, Alaka H and Fouchal F 2019 Building Information Modelling adoption in the European Union: An overview *Journal of Building Engineering* **25**
[16] Cao Y, Zhang L H, McCabe B and Shahi A 2019 The benefits of and barriers to BIM adoption in Canada *Proceedings of the 36th International Symposium on Automation and Robotics in Construction ISARC* pp 152-158
[17] Hatem W A, Abd A M and Abbas N N 2018 Barriers of adoption building information modeling (BIM) in construction projects of Iraq *Engineering Journal* **22** 2 pp 59-81
[18] Sreelakshmi S, Kantilal B S, Roshan M and Gopinath S 2017 A study on the barriers to the implementation of building information modeling *International Journal of Civil Engineering and*

Technology **8** 5 pp 42-50

[19] Wu P, Jin R, Xu Y, Lin F, Dong Y and Pan Z 2021 The analysis of barriers to bim implementation for industrialized building construction: a China study *Journal of Civil Engineering and Management* **27** 1 pp 1-13 https://doi.org/10.3846/jcem.2021.14105

[20] Olanrewaju O I, Chileshe N, Babarinde S A and Sandanayake M 2020 Investigating the barriers to building information modeling (BIM) implementation within the Nigerian construction industry *Engineering, Construction and Architectural Management* **27** 10 pp 2931-2958

[21] Marefat A, Toosi H and Hasankhanlo R M 2019 A BIM approach for construction safety: applications, barriers and solutions *Engineering, Construction and Architectural Management* **26** 9 pp 1855-1877

[22] Ahmed S 2018 Barriers to implementation of building information modeling (BIM) to the construction industry: a review *Journal of Civil Engineering and Construction* **7** 2 pp 107–113 https://doi.org/10.32732/jcec.2018.7.2.107

[23] Aibinu A and Venkatesh S 2014 Status of BIM adoption and the BIM experience of cost consultants in Australia *J. Prof. Issues Eng. Educ. Pract.* **140** pp 1–10, https://doi.org/10.1061/(ASCE)EI.1943-5541.0000193

[24] Yaakob M, James J, Nawi M N M and Radzuan K 2018 A Study on Benefits and Barriers of Implementing Building Information Modelling (BIM) in Malaysian Construction Industry *Proceedings of the International Conference on Industrial Engineering and Operations Management* Paris France July 26-27

[25] Gamil Y and Rahman I A R 2019 Awareness and challenges of building information modelling(BIM) implementation in the Yemen construction industry *Journal of Engineering, Design and Technology* **17** 5 pp 1077-1084

[26] Abubakar M, Ibrahim Y M, Kado D and Bala K 2014 Contractors perception of the factors affecting building information modelling (BIM) adoption in the Nigerian construction industry *Computing in Civil and Building Engineering* pp 167-178

[27] Liu S, Xie B, Tivendal L and Liu C 2015 Critical barriers to BIM in the AEC industry International *Journal of Marketing Studies* **7** 6 pp 162–171 https://doi.org/10.5539/ijms.v7n6p162

[28] Saka A and Chan D 2020 Profound barriers to building information modelling (BIM) adoption in construction small and medium-sized enterprises (SMEs): an interpretive structural modelling approach *Construction Innovation* **20** 2 doi:10.1108/CI-09-2019-0087

[29] Li P, Zheng S, Si H and Xu K 2019 Critical challenges for BIM adoption in small and medium-sized enterprises: evidence from China *Advances in Civil Engineering* pp 1-14 doi:10.1155/2019/9482350.

[30] Mehran D 2016 Exploring the adoption of BIM in the UAE construction industry for AEC firms *Procedia Eng* **145** pp 1110–1118

Approach for Data-driven Controlling of Construction Projects using the example of Claim Management

L T Lenz, J-L Schmidt, A Riechmann and M Gralla

Chair of Construction Management, Department of Architecture and Civil, Engineering TU Dortmund University

Ingenieurbüro für Bauberatung Univ.-Prof. Dr.-Ing. Mike Gralla GmbH

lisa.lenz@tu-dortmund.de

Keywords: Business Intelligence, Claim Management, Data-driven Controlling

Abstract. Current methods for controlling of construction projects are characterized by manual processes, which are failure-prone, tedious and for that time-consuming and cost-intensive. Furthermore, an additional challenge is caused by the absent dynamic in conventional controlling processes, which is of high importance to avoid damages and ensure a proactive management of a construction project. As construction projects are highly complex due to various influences which imply an enormous amount of information and are designated by dynamic processes, conventional controlling methods are insufficient and not able to meet these requirements. Therefore, current data-driven developments which can deal with an enormous amount of information are indispensable for a sufficient controlling of complex and dynamic projects. Accordingly, within this paper an approach for a data-driven controlling solution is presented, in which common data environments, databases and business intelligence applications are combined to ensure an efficient and proactive controlling. To investigate the advantages of business intelligence solutions for data-driven controlling the exemplary scenario of claim management controlling is used. It is shown that the user is able to speed up the processes and secure a proactive controlling by using a consistent database throughout the project. To achieve these targets an automatic data orchestration and visualization of the claim offers is exploited. Consequently, the claim manager is able to predict developments, identify risk potential, investigate KPIs or take actions to minimize negative effects e.g., time effort and costs. The requirements for these purposes are considered in a recommendation for action differentiated regarding the individual elements of a socio-technical system (human, organization, technology). The outlook includes an overview of potential technologies to ensure real-time analytics to shorten reaction and damage times related to the claims by immediately initiating countermeasures.

1. Introduction

The foundation for controlling is represented by financial accounting, as it generates the relevant information which management needs to fulfil its functions. It incorporates procedures for the systematic acquisition and evaluation of all measurable relationships and processes for the purposes of planning, steering and evaluating operations of a company. For this purpose, internal and external information in the form of data has to be gathered and evaluated to provide a decision basis for the management of the company within the framework of controlling [1]; [2].

A so-called control loop model for controlling is used (see fig. 1) to be able to react to dynamic changes and includes the detection of changes based on a continuous matching of target and actual data, as well as a steering action mechanism. Planning represents the target definition of all relevant requirements for the execution of the construction project, such as the qualities, construction methods, scheduling, costs, etc. Steering is used to define the management of the construction processes on the construction site. Evaluating summarizes the process steps recognize, check, analyze and forecast. The recognition corresponds to the actual recording of all events on the construction site using the standard construction site reporting procedures, such as daily construction reports, hourly rate records, etc. The processes checking, analyzing and forecasting are defined as the verifications of various target data to actual data. The results of these checks initially identify any existing target-actual deviations, which must be analyzed in terms of the area of responsibility from which they arise and what effects, for example on time schedules or costs, can be forecasted as a result [3].

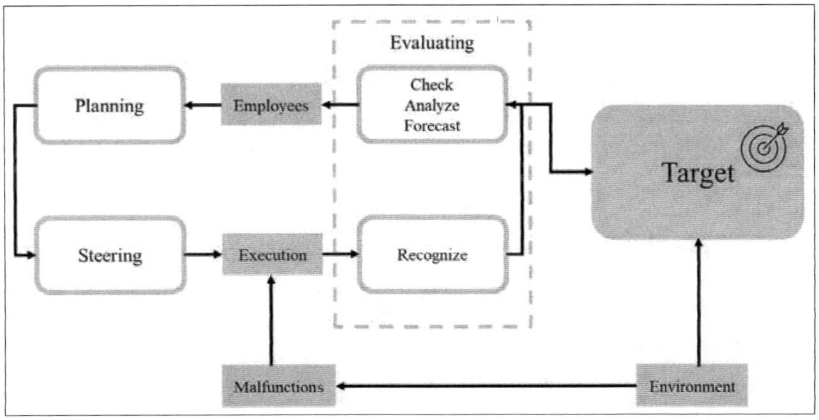

Figure 1. Control loop model for controlling [3]; [4]

Within this context, the employees configure the planning processes and integrate any necessary changes out of the control loop. The execution represents the realization of the steering specifications on the construction site. Whereas disruptions can have a direct effect on them or can be caused by environmental influences; these disruptions can be e.g., changes, additions, modifications of the contents of concluded contracts or environmental influences such as administrative instructions that cause an addendum.

The term of controlling refers to coordination in corporate management with the aim of maximizing profits. Controlling processes regarding construction management are orientated towards a specific project goal, primarily in compliance with specified targets e.g., costs, time and qualities. This distinguishes the process in construction management significantly from industrial production with its recurring processes [14]. The application of the controlling loop mentioned above, is therefore to be carried out in defined time cycles. The interval length of each time cycle depends on various

specifications like the project status (planning or construction phase), the project complexity and associated project risks as well as the project duration [5].
To implement a successful controlling within construction management it is firstly important that transparent and comprehensive project goals (target values) are given. Thus, it is crucial to have a cost target based on the project calculation with budget items, deadlines based on a time schedule including milestones as well as quality targets based on e.g.,, construction descriptions from lease contracts, specifications from the design or requirements for sustainability goals associated with certifications. Each time the controlling loop is running, project controlling must request the current key figures (actual values) from project management and evaluate them accordingly. If major deviations from target values are identified, the necessary control measures must be initiated immediately in coordination with the project management. The prerequisite for a functional and effective controlling is that all information is transparently available to those involved in the controlling processes [6]. Various approaches and technologies are available for specific domains in controlling for construction projects, which will be introduced in more detail in the following chapter.

2. State of Research

Nowadays the required information acquisition, sensing and analysis for controlling in construction management in the field of claim management in construction industry are characterized by manual processes, which present a number of disadvantages (e.g.,, cost and time consuming). Compared to the industry, some approaches and trends have already been developed in research in automated project performance in the past, monitoring and control with the focus on gathering real-time data. Navon developed a concept of automated project performance control (APPC) to ensure real-time feedback for managers [7]; [8]. Further developments dealt with the idea of closing the control loop by reporting leading performance indicators, such as equipment productivity by monitoring the movements in real-time [9]; [10]; [8]. Nevertheless, from a technical point of view, after various developments in data acquisition methods, there was a lack of platforms for the symbiosis of data with appropriate data management systems [11]; [12]; [8]. From a conceptual point of view, the first approaches were replaced by methodologies based on pull production planning and control [13]; [14]; [8]. Besides that, a BIM-based claims management system as a centralized information repository for extension of time claims was developed [15].

However, a concept for a framework combining a platform with business intelligence solutions to optimize the control loop model by using the example of claim management would be helpful to visualize the project status and support the claim manager. Especially in major projects, which are often characterized by numerous changes that lead to claims, a platform in combination with business intelligence solutions is necessary, to manage the project sufficiently and to predict forecast scenarios for future projects. This research paper, therefore, focuses on using Business Intelligence Solutions for the controlling processes of claim management in construction projects. In the following section application fields for controlling in construction management are presented to identify Key Performance Indicators (KPIs) that are important for the development of a database for controlling purposes in claim management.

3. Application Fields for Controlling in Construction Management

Controlling in construction management can be characterized as one area of an overall 'construction management concept' which serves the purposes of construction measures for individual projects as well as long-term goals of a company and its employees. Accordingly, all steps of planning, controlling and managing the economic success regarding construction projects are involved in this process. The application fields for controlling in construction management however can be subdivided further and therefore identified as activities and results of cost planning, budgeting, contract and claim management etc. (see fig. 2) [5].

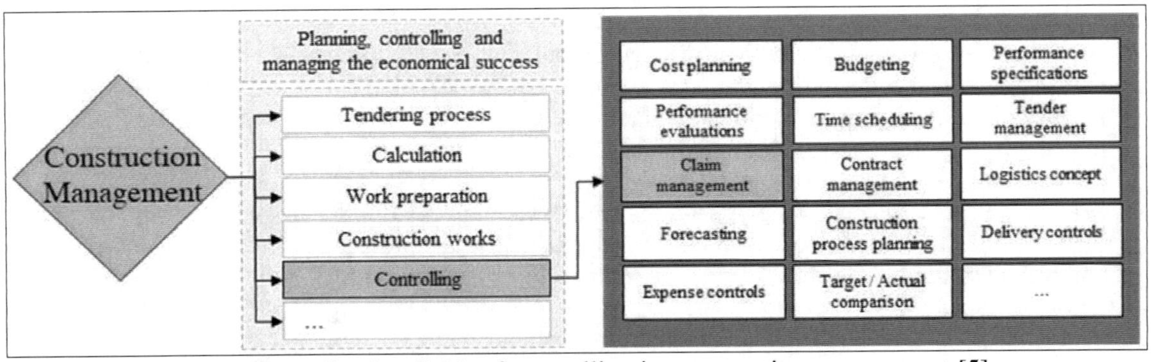

Figure 2. Application fields of controlling in construction management [5]

This research paper focuses on the controlling process using the example of claim management. Claim management is defined as the enforcement of, as well as the defense against addendum offers (claims). Usually a 'claim' covers all constellations in which the contractor demands higher remuneration due to changed and/or additional services and/or longer construction periods. A general differentiation is made between: random quantity changes, performance modifications like removed, changed or additional services as well as claims due to delayed construction periods [16]. In order to ensure an efficient and proactive controlling it is necessary to define a mutual goal, which typically is the economically success, as well as to identify the required Key Performance Indicators (KPIs) for controlling purposes in this regard. KPIs in operational controlling are supposed to represent the given complex facts in a quantitatively compressed way [6]. They can be summarized in Key Result Areas (KRAs), which represent the superordinate categories that are decisive for the successful completion of a project. Typically, those include e.g.,, quality, occupational safety, environmental management etc. However, neither the categories of KRAs, nor the KPIs that are to be derived from those are prescribed or predetermined generally and therefore need to be identified individually for each project, business model and industry and thus for the purpose of controlling in claim management as well [17]; [18]. KPIs are designed to elicit a specific desired behavior from the performers or a visible project outcome. For this purpose, they need to be 'smart', i.e. specific, measurable, executable, relevant and available on time [18].

In the context of controlling in claim management a general distinction is to be made between two processes. The first part of the process is the audit of an individual claim which is an autonomous process in itself and can be divided into four process steps [19]. Various results can be derived from each audit that serve as KPIs. The resulting KPIs are to be defined for each project individually (see fig. 3). The second part of the process reflects the controlling of the audited claims. The KRAs in claim management can be differentiated by disciplines (structural work, façade, etc.) or cost groups (costs for the building structure, costs for technical installations, etc.).

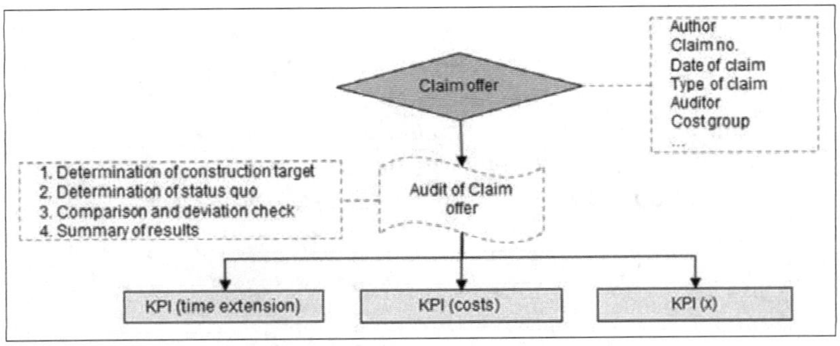

Figure 3. Audit of individual claims

A constant overview for the project manager, to guarantee a comprehensive cost tracking and cost management should always be provided and maintained [20]. Therefore, it is crucial to define the key performance indicators that must be monitored in order to ensure a comprehensive controlling in claim management.

Within conventional controlling procedures the relevant information for those processes is available in form of numerous documents, which are created by different project participants. For a data-driven controlling solution the central storage of required information is crucial and must be ensured.

4. Controlling Databasis

Data is a valuable asset for managing and controlling purposes. The usage of information can e.g., secure advantages in competitions and generate explicit market benefits [21]. Data therefore requires a defined concept, in which digital representations of analogue information in the form of digital symbols are arranged in a certain sequence [22]. It represents the solid base of an information system and serves as an information carrier for the transport and processing of facts and information on a digital level [23]; [24]. At this stage no semantic references can initially be allocated.

An essential part and the basis of using data in a targeted manner in relation to the controlling of claims is the 'data quality', which is defined as the totality of all dimensions of a data resource regarding its suitability, in order to fulfil certain requirements. Those dimensions are to be used as evaluation criteria and represented in diverse theories. According to Wang and Strong 15 quality dimensions are defined, which are assigned to four superordinate categories (see fig. 4). Each of the individual 'information quality many IQ dimension' (IQ) includes factors that are responsible for the success of an individual information system. To ensure the success and the quality of an information system, it is therefore crucial to fulfil as many dimensions as possible [21].

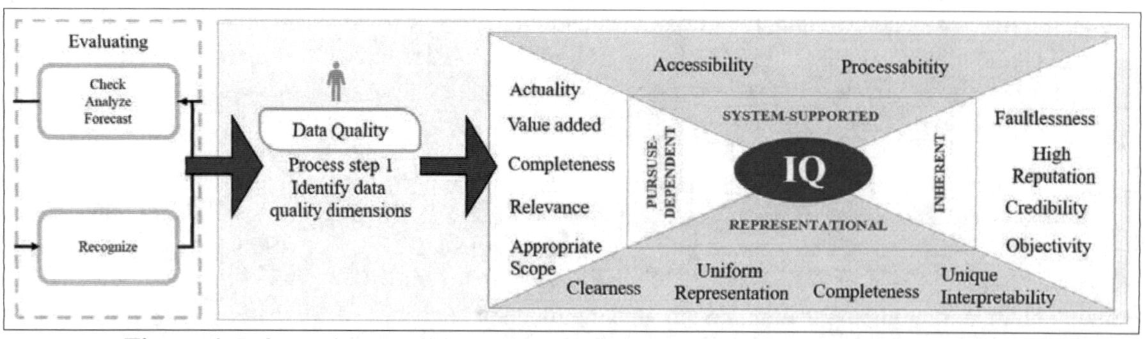

Figure 4: Information Qualities Dimensions according to Wang and Strong [21]

To transform information into data, it first has to pass through a 'data processing procedure' as presented in figure 5. Therefore, identification of data quality dimensions represents the start of that procedure. Data gathering in this regard describes the collection of data, taking into account the interdependencies of data, the dependency of technical aspects and the influence of the 'data gatherer' [21]. The data gatherer is both, the source and responsible for the input of data. Therefore, data gathering pictures the second step of the data processing procedure and defines the form of data input, which is elementary for the goal of using data for controlling purposes.

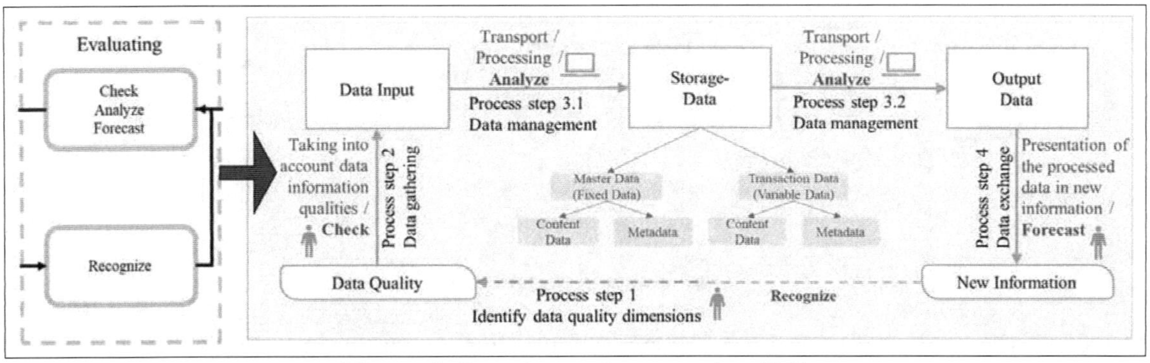

Figure 5: Data processing procedure [21]

The third step in the data processing procedure defined as 'data management' describes the organizational function that is responsible for the controlling processes. Therefore, time control for a simultaneous provision of information and data as well as their fast and flexible processing have become critical factors for the success of a company [23]. In this context, it is elementary that entered data can be used immediately as well as stored for usage for an indefinite period of time. Database systems are used to store that data systematically and without contradictions [25]. The stored data is to be divided into 'master data', which is data that rarely changes, as well as 'transaction data', which is data that changes frequently [21].

Within the data analyses, data is repeatedly checked to identify errors, even in already existing and checked data collections. This for example is done by using suitable query models in the database [21]. With regard to controlling purposes as one field of action, 'Business Intelligence Solutions' represent a convenient instrument to replace individual analyses of data, based on defined rules with programmed automated compilations. In relation to figure 5, this can be referred to as step 4. The following chapter identifies types of software systems and explains the approach of their usage for controlling in claim management referring to individual examples.

5. Approach for data-driven Controlling

To ensure a comprehensive controlling in claim management three areas have to be monitored:

1. The total amount of claims overall,
2. The total amount of claims for each contractor,
3. The total amount of claims for each status and contractor.

By controlling those areas, the project manager has the possibility to get a global overview and the possibility to quickly take countermeasures if necessary. In this paper three different types of software solutions are presented as steps for the controlling process in claim management. Each software respectively each step leads through an independent data processing procedure as described above. These are then transferred one after the other into a superordinate overall processing, which is explained schematically in the following (see fig. 6).

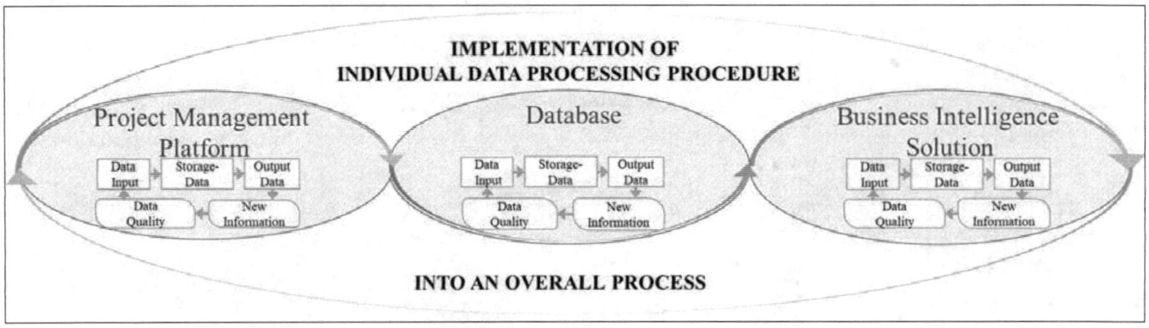

Figure 6: Three software solutions for controlling process in claim management

The first step contains the process of data extraction from a project management platform. Within this platform all information like e.g., correspondence, contractual documents, defect management as well as claims etc. are uploaded. Accordingly, all relevant information is stored centrally. Important for the controlling process in this regard are the reviewed claims, the following process applies: Each contractor uploads its claims onto the project management platform. Due to a given deadline this triggers the review of the responsible auditor, with the aim to either approve or decline the claim, respectively provide comments fur further negotiations. After the deadline is due and the audit is done, each claim is to be considered within the controlling process. In the presented use case three status are to be assigned as result for a reviewed claim, which are defined as:

Declined: Claims that are rejected and declined based on audit results.
Approved: Claims that are ordered and approved based on audit results and the requirements on the construction site.
In Negotiation: Claims that are approved on the merits but still in negotiation e.g., due to the amount or in need of further evidence and information.

With the three-status assigned here, it is possible for the project manager to check in which direction the project is developing and to intervene immediately in case of deviations from the declared project goals. By defining the according KRAs in which the reviewed claims are differentiated, it is also possible to control very precisely in which area problems might occur. Within the second step the second software applies. All reviewed claims are imported automatically from the project management platform into a database. Required adjustments, for the import from the project management platform into the database are solved and done automatically by corresponding queries that ensure the quality and right form of information to be implemented. In conventional projects the information must be transferred manually into software by the data gatherer, which is associated with a high expenditure of time and error rates, due to the manual data input. The automatic data acquisition allows a controlling in real time. As third step a business intelligence solution (BI solution) is used to present the results of the evaluated data visually and comprehensibly. The BI solution is used for the enrichment of previously defined and evaluated data (see fig. 7). All previously performed analyses and subsequent definitions are combined and prepared individually for reporting purposes. Data can be categorised and linked via filter functions and thus presented in the form of an individual dashboard configured for the purpose of presentations.

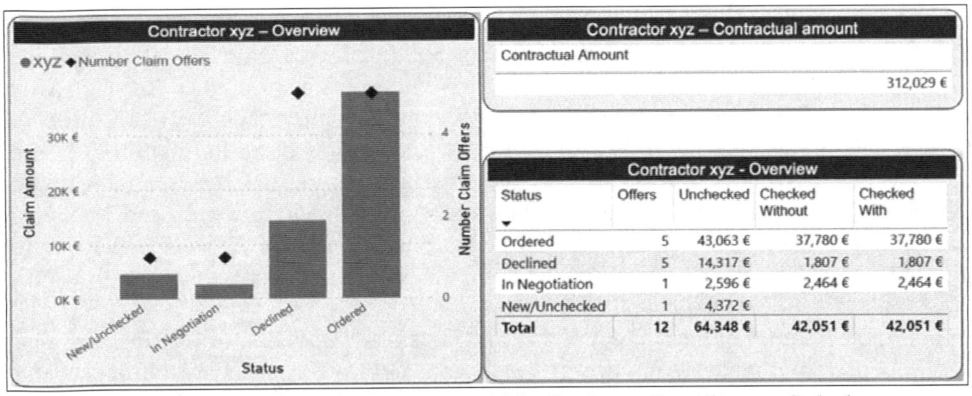

Figure 7: Example of contractor within Business Intelligence Solution

6. Findings and Conclusion

The overall process of data-driven controlling in claim management shows an interaction of individual data processing procedures into a superordinate process by automating traditionally manual process steps through the connection of software solutions. By avoiding repetitive data input of the same information by the user the susceptibility to errors is significantly reduced. As a result of higher-quality database generated in this way, the automatic evaluation of this data on basis of KPIs can represent a more precise evaluation of the actual state of claims. Furthermore, the process is accelerated at the same time and the solutions are available within a short time, whenever they are needed. Besides the technical perspective, the corresponding requirements out of the human and organizational subsystems should also be considered.

A socio-technical system is an approach to investigate the interdependencies of the subsystems technology, organization and human of a production process. In this context a detailed analysis of the subsystems to derive the requirements of the functional relations and interfaces must be conducted, because they are used as main design parameter [27]. The target is to create a system which is coordinated at the interfaces of the three subsystems [28]. In this paper the production process is figuratively the production of KPIs for controlling in construction management using the example of claim management (see fig. 8).

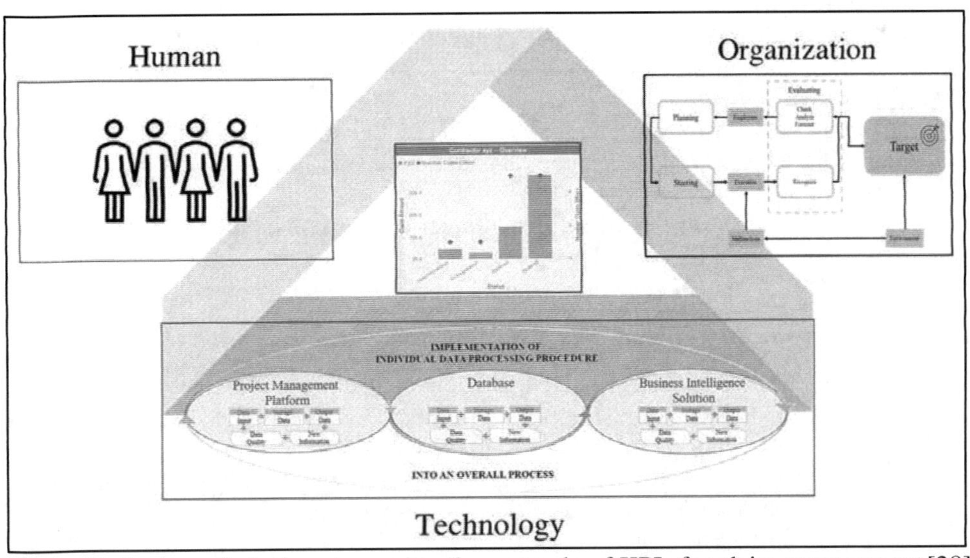

Figure 8: Socio-technical system using the example of KPIs for claim management [28]

The basis for the organizational subsystem is introduced is adapted from the control loop model for controlling in chapter 1. The consolidation with the technical requirements is described in chapter 3 and in chapter 5 the overall process for a data-driven approach for controlling with all necessary software components is finally presented. The subsystem human, respectively the ability of humans to interpret data is currently not fully reflected in the technology used. In relation to the application case of controlling in claim management in the construction industry, the creation of KPIs and the classification of the final output of analysed data in a context should be mentioned at this point. Thus, although controlling of claims cannot entirely be automated with the use of data-driven processes at present, but the execution components of the human being can be increasingly minimised.

In summary, the findings of the paper are, that the processes within controlling in claim management with an data-driven approach can become faster and more efficient, repetitive data gathering and mistakes can be avoided, the data quality can be increased and the availability of data respective KPIs for controlling of claim management improved.

7. Outlook

Soon, controlling processes in construction industry could evolve even further towards automated solutions. Artificial Intelligence (AI) based on machine learning could be used to observe human decision making and thus adapt it later or generate forecast scenarios, that the user can proactively interact to avoid further claims and not just react when they arise. However, this would require large amounts of observable and well documented controlling process data.

Especially by considering a higher level of automation the human needs should be enhanced. There have to be surveys what regarding controlling KPIs in construction management is important for humans, what kind of configurable interface they need and overall it has to be an user friendly solution, in order to ensure the user trusting the KPIs out of the system.

References

[1] Cf. Springer Gabler Verlag (ed.) 2013 *Gabler Wirtschaftslexikon, Stichwort: Rechnungswesen und Controlling* online: http://wirtschaftslexikon.gabler.de/Archiv/72018/rechnungswesen-und-controlling-sachgebietstext-v6.html
[2] Cf. Meinen H 2020 *Baubetriebstabellen (Baubetrieb, Bauwirtschaft, Bauprozessmanagement)* (Köln: Reguvis Verlag) Gralla M (ed.) chapter 1 Bauwirtschaft 3.6 Rechnungswesen und Controlling pp 1.30-1.33 Ab
[3] Cf. Gralla M 2020 *Baubetriebstabellen (Baubetrieb, Bauwirtschaft, Bauprozessmanagement)* (Köln: Reguvis Verlag) Gralla M (ed.) chapter 5 Kalkulation 4 Prozesse in der Auftragsabwicklung pp 5.16-5.25
[4] Cf. Martinsen U 2017 *Kostenrechnung in der Bauwirtschaft* (Berlin: Ernst & Sohn) p 227 ff.
[5] Cf. Seyfferth G 2003 *Praktisches Baustellen-Controlling: Handbuch für Bau- und Generalunternehmer* p 14 ff.
[6] Cf. Becker W Feichtinger C 2018 *Controlling in Jungen Unternehmen Controller Magazin Mai/Juni* Vol 3 p. 85 ff.
[7] Cf. Navon 2005 *Automated project performance control of construction projects* (Autom. ConStruct.) 14 (4), pp 467-476
[8] Cf. Sacks R Girolami M Brilakis I 2020 *Building Information Modelling*, Artificial Intelligence and Construction Tech (Elsevier) Developments in Built Enivoronment
[9] Cf. Navon R Sacks R 2007 *Assessing research issues in automated project performance control* (APPC) (Autom. ConStruct.) 16, pp 474-484
[10] Cf. Sacks R Navon R Brodetskaia I 2006 *Interpretation of automatically monitored equipment data in construction control* (J. Comput. Civ. Eng.) 20 pp 111–120 https://doi.org/10.1061/(ASCE)0887-3801(2006)20:2(111).
[11] Cf. Al Quady M Kandil A 2010 *Concept relation extraction from construction*

documents using naturla language processing (J. Construct. Eng. Manag.) 136 (3) pp 294-302
[12] Cf. Brilakis I Soibelman I Shinagawa Y 2005 *Material-based construction site image retrieval* (J. Comput. Civ Eng.) 19 (4) pp 341-355
[13] Cf. Ballard G 2008 *The lean project delivery system: an update* (Lean Constr. J.) pp 1-19
[14] Cf. Kenley R Seppänen O 2010 *Location-Based Management for Construction: Planning, Scheduling and Control* (Spon Press)
[15] Cf. Babar A Hafiz Z Abdur Rehman N Ahsen M Rai Waqas Azfar K Khwaja Mateen M 2020 *BIM-based claims management system: A centralized information repository for extension of time claims* (Elsevier: Automation in Construction) Vol 110
[16] Cf. Würfele F Gralla M Sundermeiser M 2012 Nachtragsmanagement: Leistungsbeschreibung, Leistungsabweichung, Bauzeitverzögerung p 673 Vol 2
[17] Cf. Schlabach C 2018 *Key Perrformance Indicators zur Steuerung des Tagesgeschäfs in der Baubranche in Fiedler M Lean Construction – Das Managementhandbuch: Agile Methoden und Lean Management im Bauwesen* p. 363 ff.
[18] Cf. Marosszekey M 2010 *Public sector experience with the use of KRA and KPI frameworks on alliances – to incentivise performance in non-cost areas of delivery.*
[19] Cf. Würfele F Bielefeld B Gralla M 2017 *Bauprojektüberwachung: Kosten – Qualitäten – Termine – Organisation – Leistungsinhalt – Rechtsgrundlagen – Haftung – Vergütung* Chapter 8 Nachtragsmanagement p 185
[20] Cf. Girmscheid G Motzko C *Kalkulation, Preisbildung und Controlling in der Bauwirtschaft* 2013 Vol. 2 p 407 et seq.
[21] Cf. Hildebrand K Gebauer M Hinrichs H Mielke M Editorial in: Hildebrand K Gebauer M Hinrichs H und Mielke M (eds.) 2018 *Daten- und Informationsqualität. Auf dem Weg zur Information Excellence* Wiesbaden: Springer Vieweg
[22] Cf. Daten available: www.itwissen.info/Daten-data.html, 2019, 14.02.2020
[23] Cf. Dippold R Meier A Schnider W Schwinn K 2005 *Unternehmensweites Datenmanagement. Von der Datenbankadministration bis zum Informationsmanagement. (Zielorientiertes Business Computing)*, Wiesbaden: Springer Vieweg
[24] Cf. Helmus M Meins-Becker A Kelm A Kaufhold M Khorrami N 2013 *Entwicklung einer idealtypischen Soll-Prozesskette zur Anwendung der BIM-Methode im Lebenszyklus von Bauwerken, ZukunftBAU (Hrsg.), F 3094, Bundesinstitut für Bau-, Stadt- und Raumforschung (BBSR) im Bundesamt für Bauwesen und Raumentwicklung (BBR)*, p 30
[25] Cf. Bühler P Schlaich P Sinner D 2019 *Datenmanagement. Daten. Datenbanken. Datensicherheit.* Berlin/ Heidelberg: Springer Vieweg
[26] Cf. Lenz L T 2020 *Bewertungssystem zur Entscheidungsunterstützung von Fabrikgebäudeanpassungen auf Basis von Building Information Modeling*, unv. Diss. 2019, Mike Gralla (Hrsg.), Schriftenreihe des Lehrstuhls Baubetrieb und Bauprozessmanagement der Technischen Universität Dortmund, Band 6, Köln: Reguvis Verlag p 54
[27] Cf. Hirsch-Kreinsen H ten Hompel M. 2016 *Social Manufacturing and Logistics, Arbeit in der digitalisierten Produktion* Bundesministerium für Wirtschaft und Energie (BMWi) (Eds.), Berlin available: https://www.bmwi.de/Redaktion/DE/Publikationen/Digitale-Welt/arbeiten-in-der-digitalen-welt.pdf?__blob=publicationFile&v=6, 02.02.2019 p. 6
[28] Cf. Zink K J 1997 *Soziotechnische Ansätze* Luczak H Volpert W (Eds.) Handbuch Arbeitswissenschaft Stuttgart Schäffer-Poeschel p. 74-77

IOP Conf. Series: Materials Science and Engineering 1218 (2022) 012022 doi:10.1088/1757-899X/1218/1/012022

Are we ready for a rational discussion? The existence of biases in construction dispute negotiation

K Li[1] and S O Cheung[2]

[1] Research fellow, Future of Work Institute, Faculty of Business & Law, Curtin University, Perth, Australia, 6000; Email: Keyao.li@curtin.edu.au.

[2] Corresponding author. Professor, Construction Dispute Resolution Research Unit, Department of Architecture and Civil Engineering, City University of Hong Kong, Hong Kong. Email: saion.cheung@cityu.edu.hk.

Abstract. Embedded in a network constellation of multi-stakeholders with various professional backgrounds and differing goals, the construction industry is featured with adversarial relationships and inevitable disputes. Therefore, high quality communication and negotiation among the parties are essential in achieving speedy and less costly dispute resolution. As rational evaluation underpins quality negotiation decisions, are construction disputing parties rational as they assumed? Are there any biases that could possibly prohibit them from making prudent judgement? This study seeks to understand the existence and impact of bias in construction dispute negotiation (CDN). The existence of four types of bias in CDN were discussed: preconception, self-affirmation, optimism and interest-oriented. Three categories of de-biasing strategy were also suggested to support the efficient settlement of construction dispute. Vast resources would be saved and amicable relationship among the collaborating parties could be developed when the impact of bias was curbed. Stepping into the digital era, construction dispute resolution professionals should develop the capabilities needed to harness the benefits of technologies for innovative ways of designing a bias-free dispute resolution mechanism.

1. Rational conversation is the key to successful dispute negotiation

The construction industry, encompassing real estate, infrastructure, and industrial structures is the largest ecosystem in the global economy, accounting for 13% of the world's GDP [1]. With the ever-growing in scale, construction projects have also become more and more complex and sophisticated, requiring close collaboration from different organizations and even from different countries. The differences in organizational and personal interests set the multiple stakeholders up for inevitable conflicts and disputes that might jeopardize successful project delivery [2, 3]. Construction professionals are not unfamiliar with dispute resolution, as the process is renowned for its lengthy, protracted and costly attributes [4, 5], especially formal legal procedures such as arbitration and litigation could take longer time and more resources [6]. Moreover, processes such as arbitration and litigation have the power to damage business relationship among the collaborating parties and harm long-term cooperative opportunities [7]. Thus, it is well recognized that negotiations and neutral-assisted resolutions are the collective beneficial resolutions for relationship maintenance and overall project success [8, 9].

Due to the nature of construction project, it is common that disputing parties are involved in negotiation situations where large amount of money, long-term opportunities and earning potentials are put at stake [10]. High-stake negotiation could easily heat up the situation and put the parties under great

IOP Conf. Series: Materials Science and Engineering 1218 (2022) 012022 doi:10.1088/1757-899X/1218/1/012022

desperation to reach a quick deal [11]. Being steered by emotions, anger or disappointment, disputing parties might overlook notable chances of settlement, in extreme situations even win-win options are forsaken [12]. Therefore, as rational evaluation is considered a prerequisite for quality decisions, rational conversation is the key to successful dispute negotiation.

2. Are construction professionals immune to bias?

Since rationality is a key quality for a successful negotiation, it is essential that construction negotiation parties make unbiased judgment and decisions. However, Bromiley and Papenhausen [13] claimed that complete rational human decisions cannot be expected in reality. Human decisions are not always made through deliberate analysis as the limitation of rationality was rooted in cognitive judgments [14, 15]. With the simplified application of heuristics, people tend to make systematic and predictable errors, resulting in bias in their decision making [16, 17]. Construction professionals are no exception. Unsatisfactory outcomes of decision-making bias, such as ineffective risk management, poor project planning and failure to respond to failing signals were reported [18, 19, 20]. Thus, the rationality of construction professionals is subject to the impact of biases.

Even in construction dispute resolution process, faulty judgements were often observed. Examples include the famous endowment effect phenomenon and reactive devaluation phenomenon [21, 22]. Endowment effect phenomenon was firstly raised by Richard Thaler, who was the 2017 Nobel laureate in economics science. This phenomenon captures the irrationality in human judgement when people tend to overvalue the items they own [23]. Because of this ownership attachment to their belongings, it will take much psychological effort for people to giving them up, as this would be considered as losing their identity [24, 25]. This happens in the context of construction dispute resolution when people consider their previous assessments, beliefs and positions as part of themselves, thus leading to the reluctance to step back and make changes [26]. Another renowned example demonstrating the partiality during construction dispute resolution is reactive devaluation phenomenon, representing an unjustified aversion on the proposals and suggestions raised by the opponents [27, 28, 29]. Undervaluing the counterpart's proposal would result in the rejection of a bona fide offer and make amicable settlement even more difficult [30]. Therefore, the occurrence of these faulty judgments forms psychological barriers against amicable dispute settlement.

3. Four main types of bias in construction dispute negotiation

Li and Cheung [31] first examined the types of bias that may happen in construction dispute negotiation. Four types of bias were identified and validated in the construction dispute negotiation process, namely preconception bias, self-affirmation bias, optimism bias and interest-oriented bias [32, 33, 34]. These four types of bias were found underlying and explaining other partiality and faulty judgements in construction dispute resolution [35].

Preconception bias, as the name indicated, describes the preconception disputing parties form before entering or at the commencement of negotiation. This preconception shapes how the disputing parties view their counterpart, their estimations to the counteroffer as well as their strategies in collecting and interpreting information. This unjustified preconception may lead to the under-evaluation and aversion against the arguments and evidence provided by the counterpart—reactive devaluation, and in turn the rejection of a proposal settlement which might derive the most utility. Moreover, inadequate effort was put to examine whether the preconception was grounded in truth or not [33]. Li and Cheung [35] found that this previously formed preconception is difficult to remove and could be even hardened along the negotiation process.

This preconception also links to the second type of bias: self-affirmation bias, meaning that disputing parties are inclined to confine their information collection scope and pay disproportionate attention to the information that confirms their points and arguments. Under self-affirmation bias, information interpretation serves the intention to prove already formed assessments, thus prevents the assimilation of information from the opposite perspective. This limited access to information and subjective interpretation drive the disputing parties away from complete and objective understanding of the situation.

The third type of bias, optimism bias depicts the emotion and attitudes of the disputing parties during the negotiation process. Overly optimistic negotiation parties tend to have overconfidence in their standing points and assessments, thus leading to unrealistic expectation to the negotiation outcome. This overconfidence on their assessments and position could strengthen their ownership attachment to their beliefs, leading to a stronger endowment effect on their formed assumptions. This wildly optimistic attitude makes it difficult for negotiating parties to calm down and communicate on potential compromises. Thus, the happening of optimism bias might result in strategic bargaining habits and opportunistic behaviours in negotiations, which would destroy mutual trust and impede acts of good faith [36, 37].

The fourth type of bias is interested-oriented bias, which stems from the human nature of self-serving [38]. Self-serving nature is a cognitive discourse whereby people tend to claim contribution while attribute negative outcomes to external factors, it is a manifestation of self-protection whereby self-esteem is maintained by shifting the responsibility of negative results [38, 39]. Affected by interested-oriented bias, negotiating parties put their own interests ahead of their counterpart. With the intention to fight for a larger piece of the "fixed pie", negotiating parties might consider making compromise as losing their proportion of interests [40]. When the dispute failed to reach an amicable settlement, with no rethink of their insistence on pursuing their interests that could have led to the impasse, they would believe the counterparts should be fully responsible for the unsatisfactory outcome.

4. The caveats in using third-party neutrals

When negotiating parties have difficulty in reaching a mutual agreeable settlement, third-party neutrals are the professionals to facilitate the settling process [41, 42]. The employment of third-party neutrals, such as dispute resolution advisors (DRA) and mediators, are a critical part of the dispute resolution service sector [5]. In order to better facilitate the communication, it is essential for the third-party neutrals to gain the trust from both parties by upholding their impartiality and neutrality. However, since the negotiating parties are subject to the effects of bias, could the third-party neutrals be possibly bias-free?

Cheung et al. [43] investigated the possibility of construction dispute resolution third-party neutrals being affected by judgmental biases. A simulation was designed to mimic a construction dispute resolution process and the behaviours of the third-party neutrals were recorded to examine their neutrality when receiving different sets of information. Notably, their study confirmed the possibility for biases to creep in third party neutrals' decisions, especially under the influence of opportunistic behaviours of the project participants. When third-party neutrals are under the effects of bias, they are unlikely to have a complete picture of the issues in dispute. Their credentials as well as their service in facilitating settlement would be seriously jeopardised. Thus, special caveats to guard against bias would be prudent as the impartiality of standing neutrals are the corner stones for the usefulness of alternative dispute resolution mechanisms. Moreover, Cheung et al. [43] highlighted in their study that third party neutrals of less experience are more inclined to practise biased behaviours. Thus, training and experience can be important determinants of third-party neutral's ability to remain rational. The above collectively highlight that the use of third-party neutrals could not eliminate bias from construction dispute negotiations.

5. Then, how to reduce bias?

Li and Cheung [33] proposed three categories of approach that could be deployed to mitigate biases in construction dispute negotiation, these are strategy-based, attitude-based and process-based. Strategy-based approach was designed for reducing preconception bias and self-affirmation bias with two types of measures: 1) suggesting the disputing parties to allow adequate time and effort in making decisions, review the cases multiple times before an assessment to be made, and be open to alternative assessments that might be different from the previous ones; 2) suggesting the disputing parties to deliberately consider the opposite, form the habit of readily questioning the soundness of previous positions and pay special attention to the information and evidence that might be opponent to their formed ideas.

Attitude-based approach was developed mainly to deal with interest-oriented bias and optimism bias [35]. The approach is to encourage the disputing parties to be rational and consider long-term business relationship. Under attitude-based approach, the disputing parties are suggested to avoid their heated emotions and calmly consider potential mutually beneficial gains and long-term development plan with cooperative partnership building. Besides, disputing parties are also to develop empathy, trying to put in the shoes of the counterpart and respectfully listen to their grievances.

Last but not least, process-based approach was found helpful in minimizing preconception bias and interest-oriented bias [44]. Dispute resolution mechanism design is to be optimized by conducting training sessions before the commencement of negotiations. The process of re-assessment, review and rec-construction of ideas should be required when starting a new round of negotiation, and the initial needs should be re-evaluated before negotiation offers to be made. The notion of re-thinking the design of dispute resolution mechanism also suggests new members to be added at each new round of negotiation to ensure fresh observations and remove preconception.

6. Embracing debiasing in reality testing

Notwithstanding that third party-neutrals are not immune to decision making biases, with adequate training on their techniques, they would be at better position to help the negotiating parties staying away from biases. Incorporating bias minimizing instruments into the third party-neutrals toolkit would prove invaluable in maintaining conducive dispute resolution environment.

Reality testing is a commonly used strategy for mediators to steer the disputing parties on rational courses [45,46]. In the process of reality testing, by raising previously designed questions, a mediator would guide the negotiating parties in a process of reviewing and re-estimating their assessments and decisions during the negotiations. Therefore, reality testing creates more opportunities and environment for the negotiating parties to chew on their assumptions and acknowledge their unrealistic beliefs and expectations [47]. Li and Cheung [44] explored ways to integrate bias mitigation measures into mediators' technique of reality testing. To cultivate a less-biased mindset of the negotiating parties, seventeen reality testing questions were developed in Li and Cheung [44]. By asking these reality testing questions, mediators could help the negotiating parties to be aware and self-realize their biased behaviours. The seventeen reality testing questions include three types: questions about decision making strategies, questions about attitudes during the dispute resolution process and questions about designing the dispute resolution mechanism. The usefulness and practicality of these reality testing questions were verified by construction dispute resolution professionals with more than twenty years' experience. Based on their assessment, questions about decision making strategies were rated as the most helpful as these could encourage the negotiating parties to re-visit their assumptions and decision-making process. Questions about attitudes were ranked as the second most helpful when emotion controlling was acknowledged as critical yet most challenging. Questions about designing the dispute resolution mechanism were rated as the third most useful. Therefore, it is instrumental for third party-neutrals to receive relevant training and develop necessary capabilities in enriching their toolkit to help the negotiating parties to make sound decisions.

7. Future of construction dispute negotiation

The emergence of industry 4.0 and relevant continually evolving technologies, such as machine learning, the Internet of things (IoT), blockchain, etc. are leading business in this digital journey across industries [48]. In the construction industry, as huge volumes of data are generated and considered as valuable assets, as well as the speedy embracement of digital technologies, the nature of construction dispute resolution would be dramatically changed in both dispute matters and resolution methods [49, 50]. One simple example could be the use of collaborative technology, like video conferencing in reducing the cost, streamlining the process and improving the efficiency of cross jurisdiction dispute resolution, where the disputing parties and expert witnesses will not have to commute to the other side of the world to attend hearing and interviews [51, 52]. Another example is before entering dispute resolution process, the negotiating parties could use artificial intelligence based data analytics applications to collect and synthesise data from past arbitrations and trials or gain the knowledge of the behaviours of a particular judge or arbitrator on historical cases [49]. More knowledge, understanding and preparation before entering the dispute resolution process could help achieve early settlement and boost efficiency. The diffusion and application of digital technology surely have brought many new opportunities, will the application of artificial intelligence and machine learning be able to avoid human cognitive limitations and eliminate biases in construction dispute negotiations? Stepping into the digital future of construction industry, to better harness the disruptive technologies, construction dispute resolution professionals need to explore and develop suitable capabilities for building a more rational and efficient dispute resolution community.

8. References

[1] Ribeirinho M J, Mischke J., Strube G., Sjödin E., Blanco J. L., Palter R., Biörck J., Rockhill D., and Andersson T. 2020 "The next normal in construction: How disruption is reshaping the world's largest ecosystem", Mckinsey

[2] Kumaraswamy M M 1997 "Conflicts, claims and disputes in construction", Eng., Const. Archit. Manage., 4 (2), 95-111

[3] Cheung S O and Yiu T W 2006 "Are construction disputes inevitable?", IEEE Trans. Eng. Manage., 53 (3), 456– 470.

[4] Stipanowich T J 2004 ADR and the "Vanishing Trial": the growth and impact of "Alternative Dispute Resolution". Journal of Empirical Legal Studies, 1(3), 843-912.

[5] Cheung S O 2010. "Construction mediation landscape in the civil justice system in Hong Kong.", J. Legal. Aff. Dispute. Resolut. Eng. Constr., 2(3), 169-174.

[6] Brandon M and Robertson L 2007 Conflict and dispute resolution: A guide for practice. South Melbourne, Vic., Australia: Oxford University Press.

[7] Cheung S O, Suen H C and Lam T I 2002. "Fundamentals of alternative dispute resolution processes in construction.", J. Constr. Eng. Manag., 10.1061/(ASCE)0733-9364(2002)128:5(409).

[8] Cheung S O 2014. "The effective use of ADR processes in construction.", In Construction Dispute Research, 299-317. Springer, Cham.

[9] Lee C K, Yiu T W and Cheung S O 2016. "Selection and use of Alternative Dispute Resolution (ADR) in construction projects—Past and future research." Int. J. Proj. Manage.,, 34 (3), 494-507.

[10] Flyvbjerg B 2017 "The Iron Law of Worldwide Mega Projects" in The Oxford Handbook of Megaproject Management, Flyvbjerg (Edr), Oxford University, 1-20.

[11] Brooks A W 2015 Emotion and the art of negotiation. Harvard Business Review, 93(12), 57-64.

[12] Lobel I B 1994 "Realities of interest based (win-win) bargaining.", Labor. Law. J, 45 (12), 771-778.

[13] Bromiley P and Papenhausen C 2003. "Assumptions of rationality and equilibrium in strategy research: The limits of traditional economic analysis.", Strat. Org., 1(4), 413-437.

[14] Simon H A 1955 "A behavioral model of rational choice.", Q. J. Econ., 69 (1), 99–118.

[15] Ariely D 2008 Predictably irrational: The hidden forces that shape our decisions (1st ed.). New York, NY: Harper.

[16] Keren G and Teigen K H 2004 "Yet another look at the heuristics and biases approach." in Koehler, D. J., & Harvey, N. (Eds.), Blackwell handbook of judgment and decision making, John Wiley & Sons, 89-109.

[17] Tversky A and Kahneman D 1974 "Judgment under Uncertainty: Heuristics and Biases." Sci., 185 (4175), 1124-1131.

[18] Kutsch E and Hall M 2010. "Deliberate ignorance in project risk management.", Int. J. Proj. Manage., 28 (3), 245-255.

[19] Pinto J K 2013 "Lies, damned lies, and project plans: Recurring human errors that can ruin the project planning process.", Bus. Horiz., 56 (5), 643-653.
[20] Haji-Kazemi S, Andersen B and Klakegg O J 2015 "Barriers against effective responses to early warning signs in projects.", Int. J. Proj. Manag., 33 (5), 1068-1083.
[21] Cheung S O, Li K and Ng Y H 2020 Study of endowment effect in construction project dispute resolution. Journal of Legal Affairs and Dispute Resolution in Engineering and Construction, 12(1), 04519041.
[22] Cheung S O, Li K and Chow O Y 2020 Reactive Devaluation as a Psychological Impediment to Construction Dispute Negotiation. Journal of Management in Engineering, 36(4), 04020025.
[23] Thaler R 1980 "Toward a positive theory of consumer choice.", J. Econ. Behav. Organ., 1(1), 39-60.
[24] De Dreu C K and van Knippenberg D 2005 "The possessive self as a barrier to conflict resolution: Effects of mere ownership, process accountability, and self-concept clarity on competitive cognitions and behavior.", J. Pers. Soc. Psychol., 89 (3), 345.
[25] Morewedge C K and Giblin C E 2015 "Explanations of the endowment effect: an integrative review." Trends. Cogn. Sci., 19(6), 339-348.
[26] Liedtka J 2015 "Perspective: Linking design thinking with innovation outcomes through cognitive bias reduction", J. Prod. Innov. Manag., 32(6), 925-938.
[27] Ross L and Stillinger C 1991 "Barriers to conflict resolution", Negotiation J., 7(4), 389-404.
[28] Ross L and Ward A 1995 "Psychological Barriers to Dispute Resolution", In Advances in Experimental Social Psychology, 255-304. San Diego, CA: Academic Press.
[29] Ross L 1995 "Reactive devaluation in negotiation and conflict resolution", In Arrow, K., Mnookin, R. H., Ross, L., Tversky, A., andWilson, R. (eds.), Barriers to Conflict Resolution, W.W. Norton, New York, 26–42.
[30] Ward A, Atkins D C, Lepper M R and Ross L 2011 "Affirming the self to promote agreement with another: lowering a psychological barrier to conflict resolution", Pers. Soc. Psychol. Bull., 37(9), 1216-1228.
[31] Li K and Cheung S O 2016 The potential of bias in multi-tier construction dispute resolution processes. In Proceedings of the 32nd Annual ARCOM Conference (Vol. 1, pp. 197-205).
[32] Cheung S O and Li K 2019 "Biases in construction project dispute resolution.", Eng., Constr. Archit. Manage., 26(2), 321-348.
[33] Li K and Cheung S O 2019 "Unveiling cognitive biases in construction project dispute resolution through the lenses of third-party neutrals". J. Constr. Eng. Manag., 145(11), 04019070.
[34] Li K and Cheung S O 2018 "Bias Measurement Scale for Repeated Dispute Evaluations.", J. Manag. Eng., 10.1061/(ASCE)ME.1943-5479.0000617
[35] Li K and Cheung S O 2020 Alleviating bias to enhance sustainable construction dispute management. Journal of Cleaner Production, 249, 119311.
[36] Troutman C M and Shanteau J 1977 "Inferences based on nondiagnostic information." Organ. Behav. Hum. Perform., 19 (1), 43-55.
[37] Peer E and Gamliel E 2013 "Heuristics and biases in judicial decisions." Ct. Rev., 49, 114.
[38] Sedikides C, Campbell W K, Reeder G D and Elliot A J 1998 "The self-serving bias in relational context.", J. Pers. Soc. Psychol., 74 (2), 378-386.
[39] Zuckerman M 1979 "Attribution of success and failure revisited, or: The motivational bias is alive and well in attribution theory.", J. Pers, 47 (2), 245–287.
[40] Caputo A 2013 A literature review of cognitive biases in negotiation processes. International Journal of Conflict Management.
[41] Regan P M 1996 "Conditions of successful third-party intervention in intrastate conflicts.", J. Conflict. Resolut., 40(2), 336-359.
[42] Saeb A, Mohamed O B, Danuri M S B M and Zakaria N B 2018 "Critical Factors for Selecting a Neutral to Support Alternative Dispute Resolution Methods in the Construction Industry.", Civ. Eng. J., 4(1), 11-23.

[43] Cheung S O, Li K and Levina B 2019 Paradox of Bias and Impartiality in Facilitating Construction Dispute Resolution. Journal of Legal Affairs and Dispute Resolution in Engineering and Construction, 11(3), 04519007.
[44] Li K and Cheung S O 2020 Embracing debiasing in mediator's tactic of reality testing. Journal of Legal Affairs and Dispute Resolution in Engineering and Construction, 12(1), 04519046.
[45] Holaday L C 2002 "Stage development theory: A natural framework for understanding the mediation process.", Negotiation J., 18(3), 191-210.
[46] Lande J 2002 "Using dispute system design methods to promote good-faith participation in court-connected mediation programs.", UCLA L. Rev., 50 (1), 69-141.
[47] Leung H M 2014 Hong Kong Mediation Handbook, 2nd edition, Sweet and Maxwell, Hong Kong.
[48] Xu L D, Xu E L and Li L 2018 Industry 4.0: state of the art and future trends. International Journal of Production Research, 56(8), 2941-2962.
[49] Morgan C and Reed R 2019 Dispute resolution in the era of big data and ai, https://www.herbertsmithfreehills.com/latest-thinking/dispute-resolution-in-the-era-of-big-data-and-ai
[50] Li J, Bao N J, Hu S, Hu W and Zerbino M 2020. Digitalization and International Tax Dispute Resolution: A Window of Opportunity for BRITACOM. Available at SSRN 3664039.
[51] Wilson-Evered E and Zeleznikow J 2021 Online Family Dispute Resolution: Evidence for Creating the Ideal People and Technology Interface (Vol. 45). Springer Nature.
[52] Shang C S and Guo W 2020 The Rise of Online Dispute Resolution-Led Justice in China: An Initial Look. ANU Journal of Law and Technology, 1(2).

Acknowledgement
The work described in this paper was fully supported by a HKSAR General Research Fund (Project no. 11209118).

Assessment of existing BIM implementation processes of a public organization to improve building assets management and maintenance

N Boufares[1], A Motamedi[2], I Iordanova[3] and D Forgues[4]

[1] M. Eng., Department of Construction Engineering, École de Technologie Supérieure, Montreal, Quebec, nouha.boufares.1@ens.etsmtl.ca

[2] PhD., Professor, Department of Construction Engineering, École de Technologie Supérieure, Montreal, Quebec, ali.motamedi@etsmtl.ca

[3] PhD., Professor, Department of Construction Engineering, École de Technologie Supérieure, Montreal, Quebec, ivanka.iordanova@etsmtl.ca

[4] PhD., Professor, Department of Construction Engineering, École de Technologie Supérieure, Montreal, Quebec, daniel.forgues@etsmtl.ca

Abstract. In the lifecycle of a building, the longest phase is generally that of operation and maintenance (O&M). The data needed to support O&M is mainly generated during the design, construction, and commissioning of the built asset phases. However, the extraction and transfer of relevant O&M data from the project documents remain a major issue in that it is time consuming and error prone. Nowadays, Building Information Modeling (BIM) allows different project team members to collaborate and share building data in real time. However, as-built models handed-over to the asset management team are quite voluminous and usually lack the necessary information for the O&M phase. This is due to an absence of O&M information requirements specification, and the lack of compliance monitoring and control during the project. This paper presents an action research focused on the issues encountered by a public building owner to manage their BIM processes. It demonstrates issues of BIM readiness and capabilities, both at the project and O&M levels, by mapping the gaps in the existing processes. Building owners rely on the design professionals' expertise to guide them but do not have the adequate resources, knowledge, and tools to ensure the quality of the models delivered regarding O&M requirements. Major issues in the management of information for O&M were identified, and some recommendations are proposed.

1. Introduction

BIM is a widely adopted solution to ensure the coordination between the various stakeholders involved in a construction project. Despite its potential for facility managers, BIM is primarily used during design and construction [1]. This paper uncovers some of the reasons for the insufficient use of BIM for operation and maintenance (O&M) and proposes initial steps asset managers can take to increase its adoption.

Regarding the collection of information generated and shared during the project lifecycle and its transfer for real estate asset management, the following two aspects were found to be problematic: (1) The contractual requirements at the design and construction stages lack the information required for asset management, resulting in additional costs and errors related to the manual retrieval of construction documents, missing information, or loss of information; and (2) Asset information is either unavailable or of poor-quality, leading to additional costs during the building operations stage.

The question that arises from these problems is: What issues exist in the present asset management processes and how can BIM help? This action research focuses on the issues faced by the construction and building asset management department of a Canadian university in the integration of BIM for the management of its assets. The main objective of the project is to propose a production and reception framework for the digital content for new building projects. This paper reports on the efforts to identify current issues in the process of production and transfer of asset data in a BIM-based project.

2. Literature review

The use of BIM technology has a strong potential to generate cost savings in facility management (FM) operations if the data is accurate and well maintained [2]. In the AEC sector, for each new project, the main challenge is in controlling both the design process and corresponding deadlines. In addition, in every project, there is an expectation that vital high-quality information be available and promptly ready throughout the construction process. However, there is generally either too much information to maintain, or information is insufficient to ensure excellent performance. Also, the data quality is often poor, and it is difficult to locate as it is stored in various formats. To tackle these challenges, an appropriate information flow must be integrated, accessible, and transparent to all parties [3]. BIM promises to solve this problem, both by greatly improving the inherent quality of building design information and by providing mechanisms and procedures so that the information can be communicated and shared among members of a project team [4]. However, BIM is not a miracle solution. It alone cannot solve the issue of the barriers created by the design and construction phases being separate. For instance, typically, project stakeholders have no interest in sharing their intermediate models due to context conditions such as contractual framework [5]. Another problem is that BIM models delivered at the commissioning phase cannot be easily used during the operation phase, mainly because designers do not consider the potential use of such models in the operation and maintenance phase. As a result, operation-related data is usually not included in the models [1].

Building managers suffer from a lack of data regarding the building's assets, which can lead to inefficient decisions [6]. The choice to adopt BIM for FM still requires facility managers to gain a level of awareness of its additional benefits over the status quo, namely through Computer-aided Facility Management (CAFM) [2]. There are problems observed in the adoption of BIM for FM, for example, the required O&M data is usually not present, and the file size of a model can become too large and cumbersome [7]. Additionally, the FM task requirements are not clearly included in the beginning of the project process. Therefore, the general project data transmitted from the design and construction phases should be filtered so that asset managers have access only to the data they need for their daily work [8].

Researchers innovating in the construction industry know that if their efforts are to be put into practice, they must build on existing technologies and processes, in addition to attempting to introduce new tools and workflows that require time for training and tightly allocated resources [9]. There is a demand to make processes more efficient for quick access to data and instant quality checks [10]. Data-centric organizations have learned that to have the greatest impact, data cannot be added to an organization's current practices, but rather it must be integrated into their pre-existing mindsets and processes [9]. This process involves the direct integration of data captured in the field into an asset database and the automatic generation of reports based on pre-programmed policies and/or templates [10]. Some professionals and their organizations have been reluctant to add data to their repositories, explaining that they were not prepared to deal with data on top of everything else, including new technologies, work processes, and workflows [9].

Rojas et al. [11] find that the use of BIM is the first step towards providing intelligent technical data regarding the state of construction. To prepare for the adoption of BIM for O&M, McArthur [7] recommends a stepwise modeling approach. This process begins with an initial analysis and data collection step, followed by five steps that will be repeated in the later implementation and development of each use case. This offers the flexibility needed to adapt the model to the changing needs of the building operations teams. A copy of the building model is saved to keep all the data, then simplified with families containing only the data needed for operations. The creation and adoption of BIM specifications and protocols can lead to a consistent and efficient method of data exchange with great benefits for construction projects. Several national standards and best practices have been developed around the world [3]. Computer design tools are growing in tandem with BIM, changing the game and changing the owners' perspective on the value of model-based design [9].

3. Methodology

This is an action research project studying the use of BIM by a Canadian university in its project to complete two buildings. The data collection was performed through various methods, while the main researcher was part of the Client's team as an intern. First, documents and models of the construction projects were analyzed. Then, a series of interviews were conducted with key stakeholders on the client and supplier sides, and a focus group consultation with representatives of the construction and the building asset management departments. Subsequently, current processes were mapped, analyzed, and validated by department representatives.

4. Case study

4.1. Context

The Construction Department of the Canadian university is undertaking a new BIM project (Project B). At the same time, this department and the Buildings Asset Management Department are completing the reception process of another project (Project A). Project A and B represent an interesting research opportunity: on the one hand, they allow to analyze the issues encountered in the past (Project A) in the production and transfer of digital data to the Buildings Asset Management Department for asset management. On the other hand, they offer the possibility to accompany the Construction Department and the Buildings Asset Management Department in the implementation and validation of a framework for the management of the digital data of projects (Project B and the subsequent construction projects).

This paper analyzes the practices surrounding the management of the BIM process for Project A. It aims to make a series of recommendations regarding the management and reuse of BIM model data. It is intended for the actors of the projects of the Construction Department and the Buildings Asset Management Department, to highlight the current deficiencies of the BIM practices in the management and operation phases of the project. The main objective is to help improve the efficiency and effectiveness of the realization and management of assets, to enhance the productivity and quality of future construction projects of the Canadian university.

4.2. Investigation

During the analysis of the current situation, different processes of BIM management were examined, starting with the contract and the creation of BIM management and execution plans, including the process of production and transfer of data, and the processes of reception and management of documents, and ending with an overview of the issues at stake.

4.2.1. BIM process (contract and BIM management/execution plans)

The goal is to digitize the asset data and information and to automate the capture and transfer of project information to the O&M management systems. For Project A, the client did not require BIM in a contractual manner. As a result, and without having qualified resources in BIM project management on the client's side, many BIM management aspects were missing: there was no monitoring process of the modeling (which was done on a voluntary basis by some project participants); the construction manager did not have a contractual requirement to deliver a federated As-built model at the end of the project; and there was no requirement for an O&M model to be delivered by the responsible stakeholders.

4.2.2. Production and transfer of data process

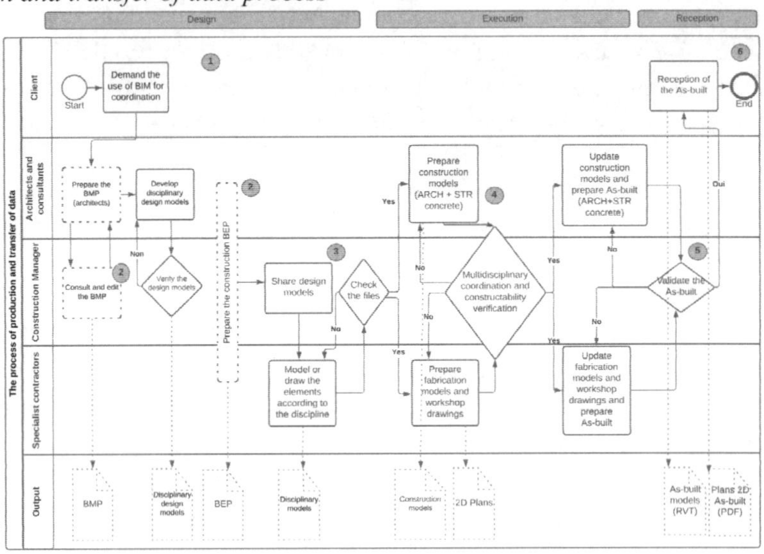

Figure 1. The process of production and transfer of BIM data

Following the analysis of the existing process of data production and transfer during the design, execution and acceptance phases of the project shown in Figure 1, some challenges are identified such as: (1) The lack of BIM-FM Client requirements in the contract; (2) Limited client involvement in the creation of the BIM management plan and the BIM execution plan; (3) Lack of quality control of the design BIM models by the client; (4) Little involvement by the client in monitoring the evolution of the project models; (5) No validation of information by the client about the transferred As-built models; and (6) Insufficient communication between the documents received and the O&M systems.

4.2.3. Document reception and management processes

The process of receiving documents for a new building is the responsibility of the Construction Department of the Client institution. The documents and data are then forwarded to the Buildings Asset Management Department. Figure 2 shows the multiple data storage platforms used by the latter. The representation of this process highlights once again the amount of manual data entry required due to the lack of interoperability between the systems. The main identified issue is the risk that the information on the systems is not updated automatically once the work is done. The fact that multiple manual entries are needed increases the risk of errors and the potential for inconsistent or missing information. The consequences of this could be serious, not to mention that any error will lead to a loss of user confidence in the validity of the information contained in the systems.

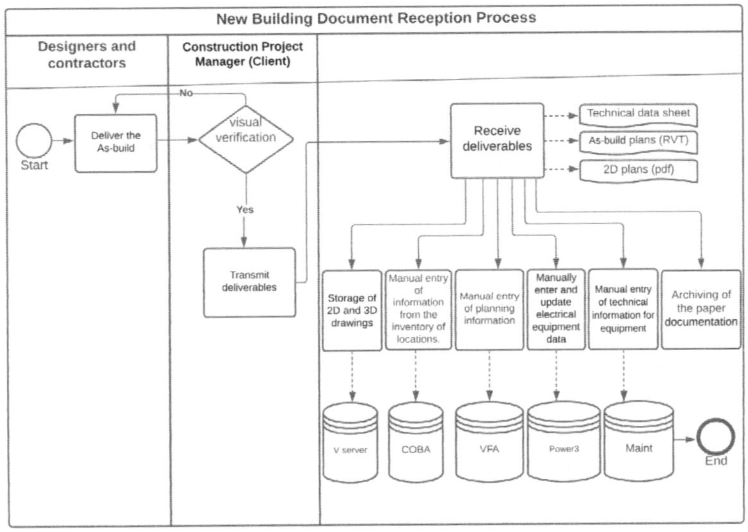

Figure 2. New building document reception process

4.2.4. Issues at hand-over

Following the analysis of the project documents and the communication with the different participants of the project through interviews, it was observed that there are several issues at the delivery level of this project that hinder the reuse of the As-built BIM models for the operation phase:

1. The client has not established a standard of information requirement to be included. There are project completion documents with all shop drawings. However, there is no quality control activity of the models by the Buildings Asset Management Department.
2. There is a lack of a standardized procedure to transfer the model data to the existing platforms.
3. The As-built models must be delivered at the final acceptance stage, which deprives the maintenance team of information sources for interventions on a building that has been in operation for several months (since provisional acceptance).
4. As there were no requirements or guidance processes from the client, the models that were delivered were not designed for reuse or to meet the information requirements for the computer-assisted maintenance management system (CMMS). The quality of these models has not been verified either.
5. Since there is no process for updating the models, new changes in the building are not reflected in the models. Soon, the models will become obsolete.

These issues stem from the fact that the models are primarily designed for 3D coordination purposes and not for use in the operational phase; no naming standard and information requirements were applied to the models; and no good modeling practices or modeling guidelines for use in the O&M phase were included. This is the result of lacking information requirements, naming convention for assets, systems, etc. and BIM-for O&M modeling best practices guide for Project A.

5. Discussion and Conclusion

In the AEC/FM industry, the use of BIM in the O&M phase is currently in progress, and it faces several challenges. This paper investigated the issues related to the production and transfer of data, the documents reception process, and the model update process. The analysis of the processes was based on a project in which the models were produced mainly for the purpose of 3D coordination, which explains the separation between the construction phase and the building management phase. The analysis confirmed the importance of specifying modelling requirements for assets management. This research aims to provide improved process maps related to the production and reception of digital content based on the structured definition of project information requirements and the asset information requirements. The improved process flow will also include: the process of planning and tracking of model production, the process of reception and transfer of models to the asset management department, the process of data extraction and transfer to the CMMS systems, and the process of data update for major renovation projects. The workflows will be discussed in future publications.

Following the observations and analysis of the different processes and documents, here are the recommendations drawn from this research: (1) The use of a common data environment (CDE) must be operationalized for the exchange of information of models between the project office, suppliers, and the building asset management department; (2) ISO 19650 and ISO 55000 standards should be adopted as references for the migration to the digitalization of the university's building asset management; and (3) A joint working group should be created between the major project management, the project office of Project B and the building assets management department to operationalize the recommendations in preparation for the reception of future projects.

Acknowledgements

This research was funded by the construction and the building asset management departments of the ÉTS. The authors would like to thank the technical staff of the construction and the building asset management departments of the ÉTS for their contributions in providing data for the project.

References

[1] A. Motamedi, I. Iordanova, D. Forgues, FM-BIM Preparation Method and Quality Assessment Measures (2018). https://www.researchgate.net/publication/325905180_FM-BIM_Preparation_Method_and_Quality_Assessment_Measures
[2] Y. Arayici, T. Onyenobi, C. Egbu. Building Information Modelling (BIM) for facilities Management (fM): the Mediacity case study Approach. (2012). http://usir.salford.ac.uk/id/eprint/19330/5/arayici_article_IJ3DIM_1%282%29.pdf
[3] A. K. Kiczak, Guidelines for preparation of OIR (Organizational Information Requirements) document for Miastoprojekt Wrocław following ISO 19650. (2020). https://bimaplus.org/wp-content/uploads/2020/10/2020-AnnaKiczak-Dissertation.pdf
[4] R. Crotty, The Impact of Building Information Modelling (2013).
[5] C. Preidel, A. Borrmann, C-H. Oberender, M. Tretheway, Seamless Integration of Common Data Environment Access into BIM Authoring Applications: the BIM Integration Framework (2016).
[6] L. Alileche, Use of BIM for the optimal management of existing building (2018). Université de Lille. https://pepite-depot.univ-lille.fr/LIBRE/EDSPI/2018/50376-2018-Alileche.pdf
[7] J.J. McArthur, A Building Information Management (BIM) Framework and Supporting Case Study for Existing Building Operations, Maintenance and Sustainability (2015). Procedia Engineering, vol. 118, p. 1104-1111.
[8] M. Munir, A. Kiviniemi, S. W. Jones, Business value of integrated BIM-based asset management (2019). Engineering, Construction and Architectural Management, vol. 26, no 6, p. 1171-1191. https://doi.org/10.1108/ECAM-03-2018-0105.
[9] R. Deutsch, Leveraging data across the building lifecycle (2015). International Conference on Sustainable Design, Engineering and Construction.
[10] A. S. Kachwalla, M. I. Hughes, Real Time Pavement Asset Management and Integration with Work Order Management (2019). https://ascelibrary.org/doi/pdf/10.1061/9780784482452.036
[11] E. M. Rojas, C. S. Dossick, J. Schaufelberger, B. A. Brucker, H. Juan, C. Rutz, Evaluating Alternative Methods for Capturing As-Built Data for Existing Facilities (2009). Computing in Civil Engineering (2009).

Automated extraction of attributes of IFC objects based on graph theory and SPARQL query

Xiang Xinglei[1], Ma Zhiliang[1], Huang Huanmin[2] and Xie Peng[2]

[1] Department of Civil Engineering, Tsinghua University, Beijing, China
[2] Qianhai Digital City Technology Co., Ltd, Shenzhen, China

E-mail: xiangxl1996@163.com

Abstract. Building Information Modelling (BIM) has been widely adopted as an effective means for supporting information exchange in Architectural, Engineering and Construction (AEC) industry based on unified and platform-independent standards. Among them, the Industry Foundation Classes (IFC), an ISO standard for BIM, is the most widely used one. However, due to the complexity and flexibility of IFC, the extraction of the attributes of IFC objects is a challenging process for practitioners and software developers in AEC industry, which impedes in-depth application of BIM in many cases. To solve the problem, an approach is proposed in this paper based on graph theory and SPARQL query to automate and simplify the process. The approach consists of four operators, i.e., IFC-to-OWL Convertor, P-Path Acquirer, SPARQL Query Generator, and Ontology Reasoner. IFC-to-OWL Convertor translates IFC instance model into ifcOWL instance model, which can be queried by using SPARQL. P-Path Acquirer obtains all possible Predicate Paths (P-Paths) from an object entity to an attribute entity in the graph created based on ifcOWL schema model. Then SPARQL Query Generator generates SPARQL queries for extracting attributes based on the P-Paths. Finally, the ifcOWL instance model and the SPARQL queries are input into Ontology Reasoner to query attributes of IFC objects. The approach is validated by conducting a case study. The approach contributes to the convenient application of IFC standards in AEC industry.

1. Introduction

The rapid development of Building Information Modelling (BIM) has a huge impact on Architectural, Engineering and Construction (AEC) industry over the past decades. BIM provides an effective means for information exchange and interoperability based on unified data standards. Among them, the Industry Foundation Classes (IFC) developed by buildingSMART International are the most widely used one and has been accepted as an ISO (International Standardization Organization) standard (ISO 10303-11). The standard is defined by using EXPRESS language, while the release also provides the version described in XSD language (ISO 10303-28) alternatively.

IFC model, as a dataset storing building information, is classified into schema model and instance model. IFC schema model is a formal and overall description of the definition and structure of BIM data. It defines a generic data model for storing and exchanging building information, instead of storing information for specific buildings. According to the data model defined in IFC schema model, IFC instance model stores the BIM data for a specific building. Each IFC instance model is composed of entities defined in IFC schema model. For example, IfcWall entity defined in IFC schema model is used to describe the concept of wall, and IfcWall_23543 can be an entity in a certain IFC instance model to represent a specific wall.

IFC schema model consists of four layers, i.e., resource layer, core layer, shared layer and domain layer [1]. The entities in a former layer can be referenced by a latter layer. Entities in the resource layer describe basic supporting concepts and data structures, for example, IfcPoint represents all points within a Cartesian coordinate. Resource layer entities cannot exist independently, but can only exist if referenced by entities in the other three layers. Except the entities in the resource layer, others all inherit from IfcRoot entity. IfcRoot has the following three sub-entities, i.e., IfcObjectDefinition, IfcPropertyDefinition and IfcRelationship. IfcObjectDefinition represents all objects or processes in AEC industry, specially including all physical product such as roofs, walls and slabs. IFCPropertyDefinition describes the characteristics, i.e., attributes that may attach to IFC entities. Further, attributes of IFC objects can also represented by IFC entities, e.g., IfcMaterial and IfcColourRgb. IfcRelationship summarizes all the relationships among entities. IFC objects in this paper are defined as the entities inherited from IfcObjectDefinition.

IFC objects link with their attribute entities in two ways, as shown in figure 1, i.e., the attributes that are directly attached to the objects (named as direct attributes), and the attributes that are connected with the objects through relationships (named as relationship attributes) [2]. Due to the complexity and flexibility of IFC standard, it is difficult to extract attributes of IFC objects. Indeed, IFC schema model defines a complex data model, consisting of 768 object entities and 1480 attributes (IFC4_ADD1). To make matters worse, for given attributes and IFC objects, the ways that the attributes attach to the objects are not unique and determined, but alternative and flexible [3]. Evidently, correctly and efficiently extracting attributes of IFC objects in IFC instance model is vital for many BIM applications, because it is often needed to obtain its attributes when an IFC object is focused. However, to extract attributes of IFC objects directly, an in-depth understanding of IFC schema model is required, which is hard for practitioners and software developers in AEC industry. To solve the problem, several approaches have been proposed.

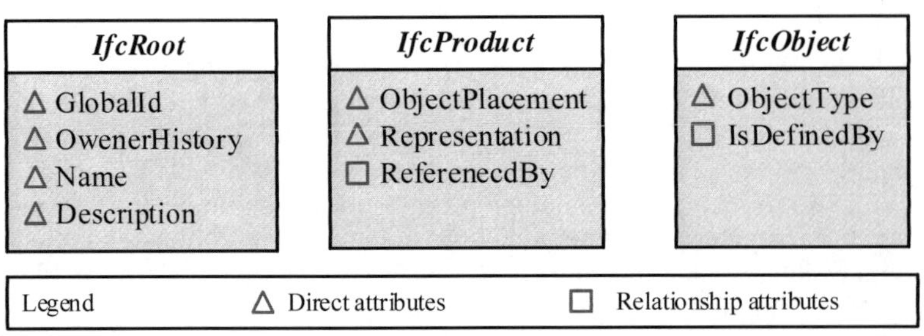

Figure 1. Examples of different types of attributes in IFC standard [2].

An intuitive approach is to develop a specialized query language to extract attributes of IFC objects. Partial Model Query Language (PMQL) is such an early effort [4]. PMQL is an XML-based language and provides functionality select, update, and delete IFC attributes. However, writing PMQL query still requires an in-depth knowledge about IFC standard and XML language. Besides, BIMQL is an ongoing development of IFC query language [5]. Compared with PMQL, BIMQL provides a few pre-defined shortcuts functions for handling commonly used cases to simplify the query writing process, such as retrieving information from property sets that are referred by IFC objects. However, these shortcuts are very limited and it is hard to extract information in the use cases that are not pre-defined.

Furthermore, semantic web technology was introduced to extract attributes of IFC objects. It provides another ontology language, i.e., Ontology Web Language (OWL), to formally represent IFC standards instead of EXPRESS and XSD language [6]. In OWL, a set of data is organized as triples, consisting of a subject, a predicate and an object, as shown in figure 2. Subject is the domain of predicate, and object is the range of predicate. The OWL-version of IFC standard, ifcOWL, has been developed as a counterpart of the EXPRESS version and recommended by buildingSMART [6]. In ifcOWL, IFC entities are linked with each other by using a range of predicates, named as P-Path in this paper. A structured query language, SPARQL, has been proposed to manipulate and query ifcOWL instance models, so that the process for extracting attributes is standardized and simplified greatly. Zhang et al. developed SPARQL extensions for querying ifcOWL instance model, called BimSPARQL [7], in which, a few new OWL vocabularies were pre-defined to represent the attributes of IFC objects and a set of queries were implemented to retrieve implicit attributes. Additionally, Liu et al. proposed an approach to extract construction-oriented quantity take-off (QTO) attributes based on ifcOWL [8]. However, although related researches indicate that ifcOWL and SPARQL has the great potential for extracting information, it is still challenging for the practitioners and software developers to program the SPARQL queries manually.

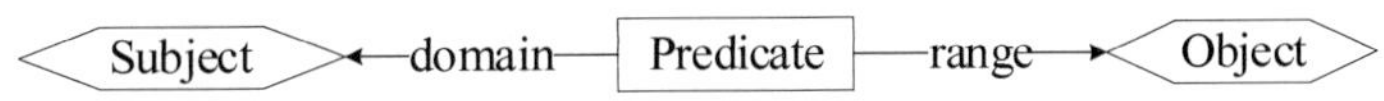

Figure 2. The triple form of an OWL statement [6]

Another approach that is based on data format conversion is to translate IFC instance model into graph and apply graph theory for extracting attributes. Ismail et al. demonstrated the potential of using graph theory concepts to manage the huge amount of information and complex relationships of IFC instance models [2]. Tauscher et al. presents an approach towards information retrieval from IFC instance model based on graph theory and shortest path search algorithm [3]. The graph-based method for extracting IFC information does not require the users to have in-depth knowledge about IFC standard. However, current graph-based approaches have the following problems. First, the conversion process from IFC to directed graph is still not uniform. Second, although related researches have provided several successful test cases, a systematic, standardized and generic extraction method has not been established. Third, it is time-consuming to generate a graph based on a heavy IFC instance model and to apply the graph algorithm on it.

This paper proposes an approach to extract attributes of IFC objects automatically based on the integrated application of graph theory and SPARQL query. In the approach, graph theory is used to facilitate generating SPARQL queries for extracting attributes automatically based on the graph created from ifcOWL schema model. Then after converting IFC instance model into ifcOWL instance model, the SPARQL queries are executed to actually extract the attributes. The integration of graph theory and SPARQL query overcomes the shortcomings of above-mentioned approaches from the following aspects. First, the SPARQL queries are generated automatically. Second, graph algorithm is applied on the light-weight schema model instead of heavy instance model to improve extraction efficiency greatly. Third, the whole process is easy to handle and efficient, without requiring in-depth knowledge about IFC standard.

The remainder of the paper is organized as follows. In Section 2, the workflow of the approach is introduced. Section 3 presents the key operators of the workflow. Section 4 verifies the approach by conducting a case study. Section 5 is the conclusion.

2. Workflow of the approach

Figure 3 shows the workflow of the proposed approach. As shown in the figure, the initial input for extracting attributes of IFC objects includes the following four items, i.e., ifcOWL schema model, object entity (such as IfcWall), attribute entity (such as IfcMaterial) and IFC instance model.

The workflow is characterized by four operators, i.e., P-Path Acquirer, SPARQL Query Generator, IFC-to-OWL Convertor and Ontology Reasoner. The former two operators are executed on the ifcOWL schema model to generate the SPARQL queries for extracting attributes automatically. Namely, P-Path Acquirer is used to obtain all possible P-Paths from the attribute entity to the object entity in ifcOWL schema model, and SPARQL Query Generator is used to generate SPARQL query for extracting attributes automatically based on the P-Paths. Then IFC-to-OWL Convertor is used to translate IFC instance model into ifcOWL instance model, which can be queried by using SPARQL. Finally, the ifcOWL instance model and the generated SPARQL queries are input into Ontology Reasoner. Consequently, the reasoning and query result in the attributes of the IFC objects.

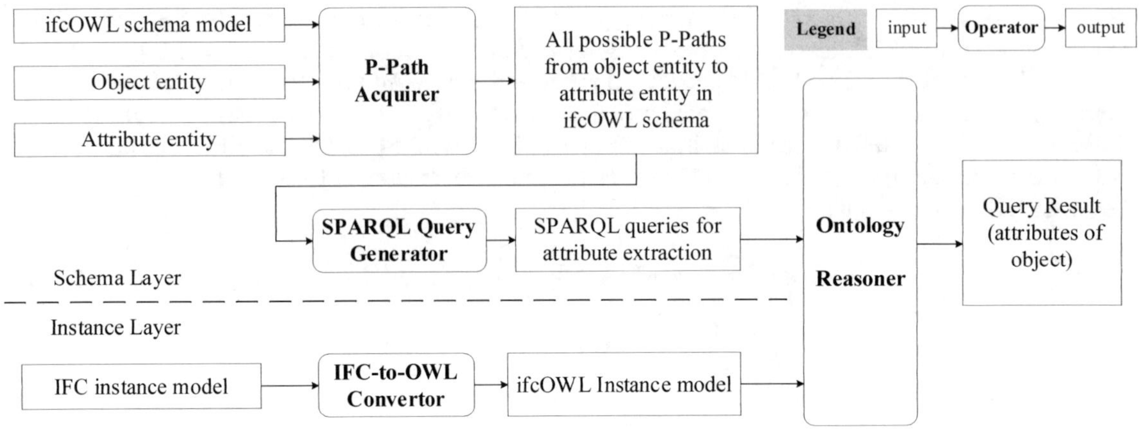

Figure 3. Workflow of the approach.

Note that IFC-to-OWL Convertor and Ontology Reasoner can be implemented by using existing and available tools. For IFC-to-OWL Convertor, BuildingSMART provides a tool, i.e., the UGent-Aalto IFC-to-RDF converter (https://github.com/pipauwel/IFCtoRDF), which is a widely-used converter to translate IFC instance model into ifcOWL instance model. For Ontology Reasoner, the reasoning engine Jena can parse ifcOWL instance model and execute SPARQL queries efficiently (https://jena.apache.org/). Therefore, this paper focuses on how to generate SPAQRL queries for extracting attributes of IFC objects, that is, the P-Path Acquirer and SPARQL Query Generator. P-Path Acquirer is named after the meaning of acquiring all P-Paths from an object entity to an attribute entity.

2.1. Basic concept of P-Path

As mentioned in Section 1, data is organized as a series of triples in OWL, consisting of subject (domain of predicate), predicate and object (range of predicate). Also, ifcOWL schema model organizes BIM data into triples. In each triple, the predicate links its domain entity and range entity. For example, <IfcDocumentInformation, hasDocumentReferences_IfcDocumentInformation, IfcDocumentReference> is one triple, and hasDocumentReferences_IfcDocumentInformation (predicate) links IfcDocumentInformation entity with IfcDocumentReference entity. IfcOWL4_ADD1 schema model contains 24097 such triples in total.

In ifcOWL schema model, an attribute entity often links to an object entity through a range of predicates. As an example, figure 4 shows how IfcProject links with IfcDocumentReference in ifcOWL schema model through several predicates.

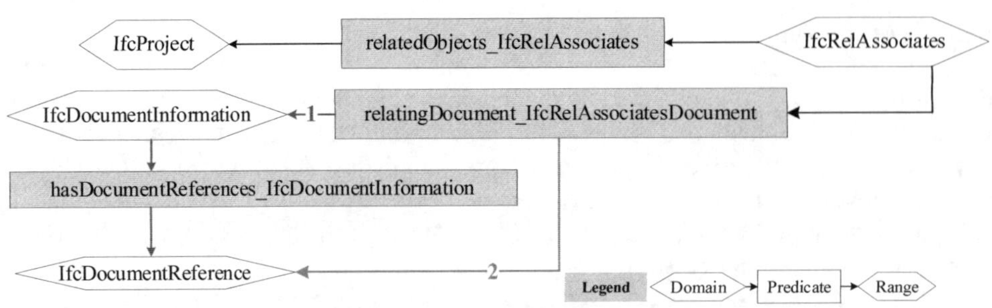

Figure 4. An example of P-Path (from IfcProject to IfcDocumentReference).

In this paper, a P-Path is defined as an ordered list of predicates to link an object entity with an attribute entity in ifcOWL schema model, e.g., as the coloured predicates in figure 4, there are two P-Paths which link IfcProject with IfcDocumentReference. The first P-Path is <relatedObjects_IfcRelAssociates, relatingDocument_IfcRelAssociatesDocument, hasDocumentReferences_IfcDocumentInformation>, and the second is <relatedObjects_IfcRelAssociates, relatingDocument_IfcRelAssociatesDocument >.

2.2. Create predicate graph based on ifcOWL schema model

To acquire all the P-Paths from an object entity to an attribute entity by using graph theory, the concept of predicate graph (P-Graph) is introduced at first in this paper. P-Graph is defined as a directed graph G (V, E), where the vertex set V consists of all the predicates in ifcOWL schema and the edge set E stands for the reachability between predicates. It is created based on ifcOWL schema model. The reachability between predicates is defined as follows. If the range of predicate I is the super entity of or the same as the domain of predicate II, predicate I and predicate II can be the before-after adjacent vertexes in a P-Path, meaning that it is reachable from predicate I to predicate II and an edge should be added from predicate I to predicate II in the P-Graph.

Figure 5 shows two example cases of reachable predicates in the P-Graph. In Case 1, predicate I is hasAssignments_IfcObjectDefinition, with the range of IfcRelAssigns. Predicate II is relatingGroup_IfcRelAssignsToGroup, with the domain of IfcRelAssignsToGroup. Because IfcRelAssigns is the super-entity of IfcRelAssignsToGroup, hasAssignments_IfcObjectDefinition and relatingGroup_IfcRelAssignsToGroup can be the before-after adjacent vertex in a P-Path. Then an edge between them will be added into the P-Graph. Similarly, Case 2 shows an example where the range of predicate I (hasProperties_IfcPropertySet) is the same as the domain of predicate II (name_IfcProperty).

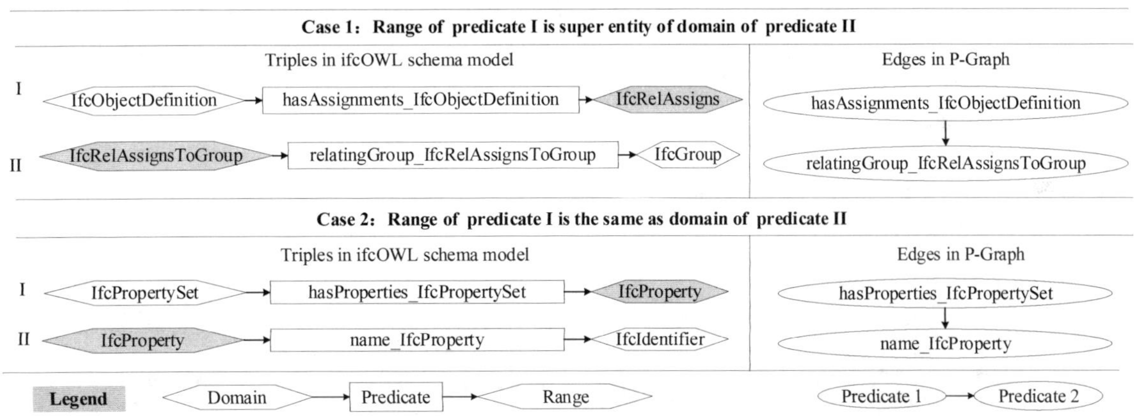

Figure 5. Example cases of reachable predicates in the P-Graph.

After adding the edges into edge set according to the above rules, in most cases, the result is complete. But there are some special situations that need be handled by using the following two operations to modify the edge set of P-Graph, as shown in figure 6.

Operation 1: Add edges between the predicates that have same domain of IfcRelationship and its sub-entities. The rationale for doing this can be explained by using figure 6. The figure shows an example of relatedObjects_IfcRelDefines and relatingPropertyDefinition_IfcRelAssociates, which shares the same domain IfcRelDefinesByProperties. In ifcOWL schema model, IfcPropertySetDefinitionSelect can be linked to IfcObjects through the two predicates, i.e., relatedObjects_IfcRelDefines and relatingPropertyDefinition_IfcRelAssociates, to represent the properties of IFC objects. it means that the two predicates can be the before-after adjacent vertexes in a P-Path and should be connected with an edge in the P-Graph. Therefore, an edge between such predicates needs to be added to modify the P-Graph.

Operation 2: Delete edges between inverse predicates. Inverse predicates in ifcOWL schema model are a pair of predicates with opposite meanings. The inverse predicates are used to describe the same semantics between entities from opposite direction, only one of the both is used to represent the meaning of the specific entities in ifcOWL instance model. It means that inverse predicates cannot be the before-after adjacent vertexes in a P-Path. However, they are considered incorrectly in the above-mentioned two cases and many redundant cycles are introduced into the P-Graph. Therefore, edges for inverse predicates need to be deleted to modify the P-Graph. As an example, in the figure 6, engagedIn_IfcPerson and thePerson_IfcPersonAndOrganization is a couple of inverse predicates. Such inverse predicates are declared explicitly in ifcOWL schema model and ifcOWL4_ADD1 declares 94 pairs of inverse predicates.

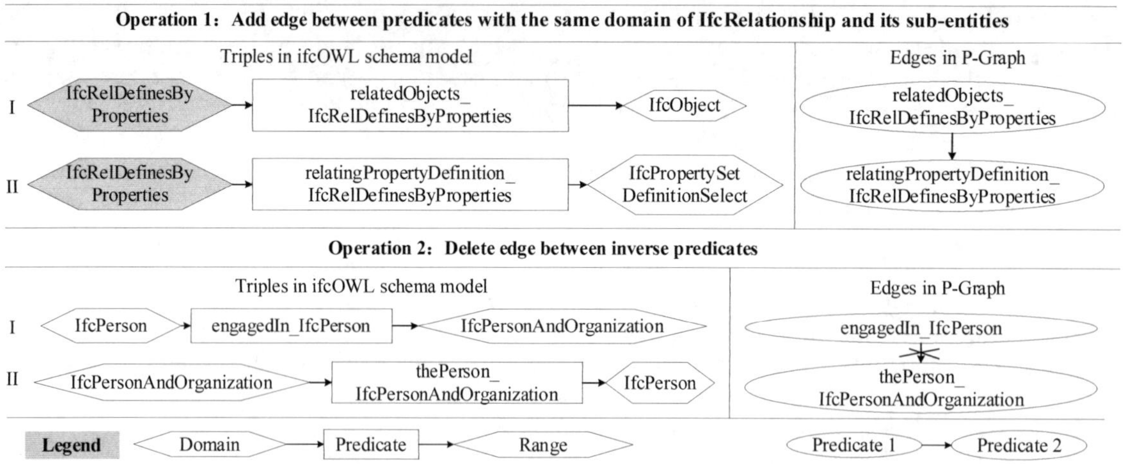

Figure 6. Examples of special operations to modify the edge set of P-Graph.

Finally, P-Graph based on ifcOWL4_ADD1 has been generated in this research, with 1556 vertex and 8794 edges. Figure 7 shows the partial P-Graph with the vertex relatedObjects_IfcRelAssociates.

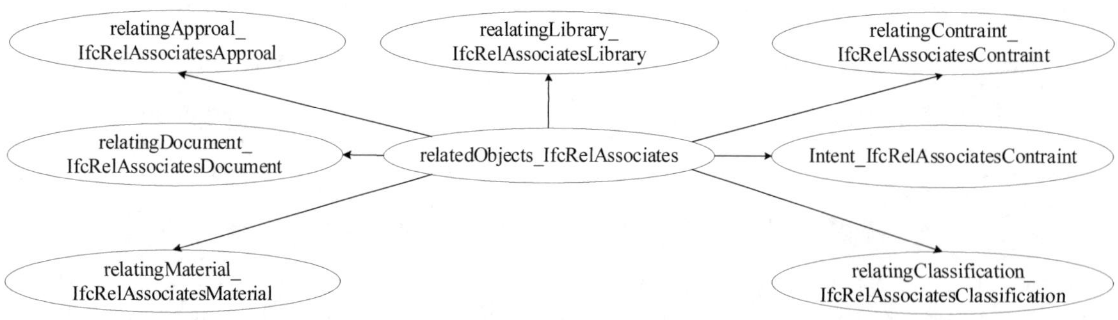

Figure 7. Partial P-Graph with the vertex relatedObjects_IfcRelAssociates.

2.3. Get all P-Paths on the P-Graph based on depth first search algorithm

Based on the P-Graph, all possible P-Paths from an object entity to an attribute entity can be obtained by using Depth-First Search (DFS) algorithm. DFS is an algorithm for traversing or searching graph data structures, which provides a systematic and effective way to find all the paths between reachable vertexes.

Given the P-Graph, all the P-Paths from an object entity to an attribute entity are obtained through the following two steps.

(1) Get Start Predicates set (SP-set) and End Predicates set (EP-set). SP-set is a set of predicates whose domain is the object entity. EP-set is a set of predicates whose range is the attribute entity. SP-set and EP-set are obtained from ifcOWL schema model directly.

(2) For each predicate-pair [sp, ep] (sp ∈ SP-set, ep ∈ EP-set), DFS algorithm is employed to find all the P-Paths from sp to ep in the P-Graph. The process is as follows.

Algorithm depth-first search

Input: start predicate (SP), end predicate (EP)

Output: All P-Paths

```
function DFS (SP, EP, Current_Path)
  if SP==EP:
     All_P_Paths ← All_P_Paths + Current_Path
     return
  end if
  SP. STATE ← ONGOING
  for P ∈ SP. Neighbours do
     if P. STATE == ONGOING then
        Current_Path ← Current_Path + P
        DFS (P, EP, Current_Path)
        Current_Path ← Current_Path - P
     end if
  end for
  P. STATE ← DONE
```

3. SPARQL Query Generator: Generate SPARQL queries based on P-Paths

Given the following three user input items, i.e., an object entity, an attribute entity, all P-Paths from the object entity to the attribute entity. SPARQL Query Generator is used to generate SPARQL queries for extracting attributes of IFC objects automatically. SPARQL Query Generator is implemented through two steps, i.e., to design SPARQL template for extracting attributes, and to make a program which generate queries based on the template and user input items.

3.1. Design SPARQL template for extracting attributes

Because SPARQL is a highly standardized and structured query language, as shown in Listing 1, SPARQL queries template for extracting attributes needs to be pre-designed. In this paper, two templates are designed to extract attribute for type and individual object. In Listing 1, the bold key words need to be filled out by using above mentioned input items. Specially, PREDICATE_x means a vertex within a P-Path from an object entity to an attribute entity and m is the length of the P-Path.

In Listing 1, Template A is used to extract all objects of specific type and their attribute values from ifcOWL instance model. For example, all walls and their material can be extracted by using Template A. Template B is used to extract the attributes of an individual object. In Template A, object_entity is the type entity of an object, such as IfcWall, while object_entity is the individual name of an object in Template B, such as IfcWall_54627.

3.2. Make a program which generate queries based on the template and user input items

Given SPARQL query template, the above-mentioned three user input items are used to fill out the template to generate the SPARQL queries for extracting attributes, i.e., an object entity, an attribute entity and the P-Paths from the object entity to the attribute entity. Listing 2 gives an example to show the SPARQL queries generation process by filling out the template using user input items. In Listing 2, the object entity is IfcWall, the attribute entity is IfcMaterial and the P-Path is <relatedObjects_IfcRelAssociates, relatingMaterial_IfcRelAssociatesMaterial, materials_IfcMaterialList, hasContents>. After filling out the Template A by using these input items, the generated SPARQL query is used to extract material for all walls in an ifcOWL instance model.

Listing 1. SPARQL query template for extracting attributes.

Template A for Type	**Template B for Individual**
SELECT ?varible_1, ?varible_m+3	SELECT ?varible_1, ?varible_m+2
WHERE {	WHERE {
?varible_1 rdf:type **object_entity**	? **object_entity PREDICATE_1** ?varible_1.
?varible_1 **PREDICATE_1** ?varible_2 .	?varible_1 **PREDICATE_2** ?varible_2 .
?varible_2 **PREDICATE_2** ?varible_3 .	……
……	?varible_m-1 **PREDICATE_m** ?varible_m .
?varible_m **PREDICATE_m** ?varible_m+1 .	?varible_m rdf:type **attribute_entity** .
?varible_m+1 rdf:type **attribute_entity** .	?varible_m ifc:name_**attribute_entity**
?varible_m+1 ifc:name_**attribute_entity**	?varible_m+1.
?varible_m+2 .	?varible_m+1 express:hasString ?varible_m+2
?varible_m+2 express:hasString ?varible_m+3	}
}	

Listing 2. An example of SPARQL query generation process by filling out the template (SPARQL Query to extract material attribute for walls).

Input items:

(1) object entity: IfcWall (2) attribute entity: IfcMaterial

(3) P-Path: <relatedObjects_IfcRelAssociates, relatingMaterial_IfcRelAssociatesMaterial, materials_IfcMaterialList, hasContents >

SPARQL Query Template A for type	**Generated SPARQL Query**
SELECT ?varible_1,	SELECT ?varible_1, ?varible_7
?varible_m+3	WHERE {
WHERE {	?varible_1 rdf:type **ifc:IfcWall**
?varible_1 rdf:type **object_entity** .	?varible_2 **ifc:relatedObjects_IfcRelAssociates**
?varible_1 **PREDICATE_1** ?varible_2 .	?varible_1.
?varible_2 **PREDICATE_2** ?varible_3 .	?varible_2
……	**ifc:relatingMaterial_IfcRelAssociatesMaterial**
?varible_m **PREDICATE_m**	?varible_3.
?varible_m+1.	?varible_3 **ifc:materials_IfcMaterialList** ?varible_4 .
?varible_m+1 rdf:type **attribute_entity** .	?varible_4 **list:hasContents** ?varible_5 .
?varible_m+1 ifc:name_**attribute_entity**	?varible_5 rdf:type **ifc:IfcMaterial** .
?varible_m+2 .	?varible_5 ifc:name_**IfcMaterial** ?varible_6 .
?varible_m+2 express:hasString	?varible_6 express:hasString ?varible_7 .
?varible_m+3 .	}
}	

4. Validation and Case Study

To validate the feasibility and effectiveness of the proposed approach, a prototype composed of P-Path Acquirer and SPARQL Generator is developed by using python, which was used to generate SPARQL queries for extracting attribute based the following input items automatically, i.e., ifcOWL schema model, an object entity and an attribute entity. Then the SPARQL queries and the ifcOWL instance model converted from IFC instance model by using IFC-to-OWL Convertor are input into Ontology Reasoner to extract attribute for IFC objects.

As a case study, the approach is tested by using a public IFC 2x3 instance model available online, which is a duplex apartment BIM model [9], as shown in figure 8. In the IFC instance model, there are 48 IfcWallStandardCase entities to represent walls. The proposed approach is used to extract material attribute for the 48 walls as validation. For this purpose, the object entity is set as IfcWallStandardCase and the attribute entity is set as IfcMaterial.

Consequently, 17 SPARQL queries are generated to query material attribute for walls. Listing 3 shows one of the generated SPARQL queries as an example. The query result indicates the approach extracts attribute for IFC objects completely and correctly, as figure 9 shows. The summary of query results is given in Table. 1.

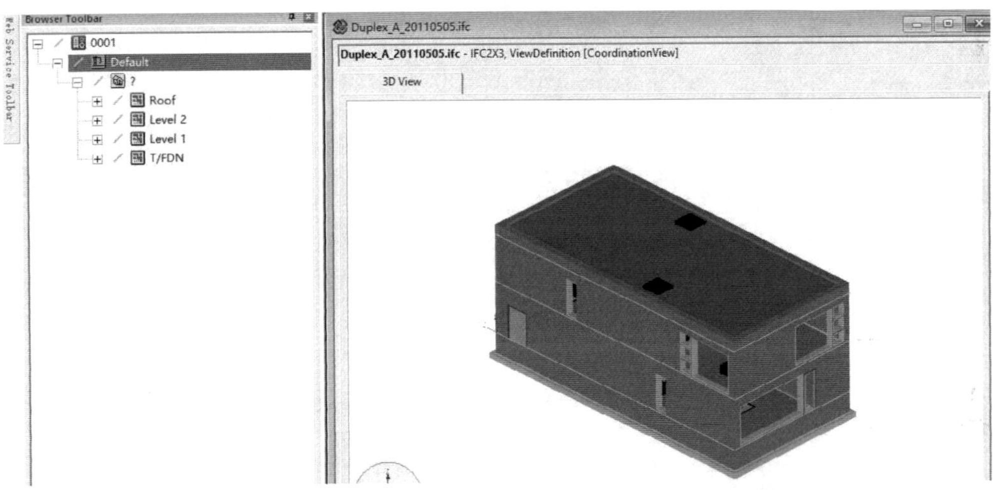

Figure 8. IFC instance model of a duplex apartment BIM model.

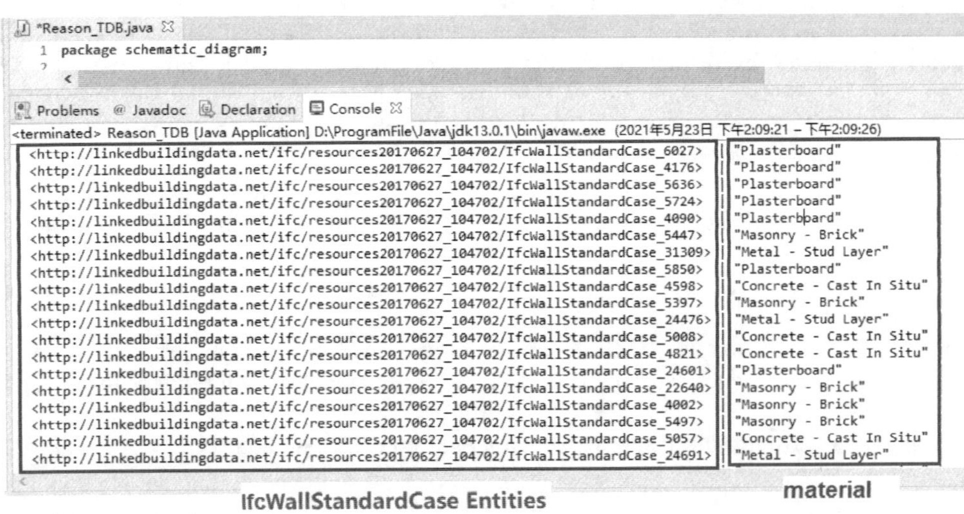

Figure 9. Jena query result interface for material attribute of all IfcWallStandardCase entities.

Listing 3. One P-Path from IfcWallStandardCase to IfcMaterial and corresponding SPARQL Query.

A P-Path from IfcWallStandardCase to IfcMaterial	A SPARQL query to extract material attribute for walls
	SELECT ?varible_1 ?varible_9
	WHERE {
ifc:relatedObjects_IfcRelAssociates	?varible_1 rdf:type ifc:IfcWallStandardCase.
↓	?varible_2 ifc:relatedObjects_IfcRelAssociates ?varible_1.
ifc:relatingMaterial_IfcRelAssociatesMaterial	?varible_2 ifc:relatingMaterial_IfcRelAssociatesMaterial
↓	?varible_3.
ifc:forLayerSet_IfcMaterialLayerSetUsage	?varible_3 ifc:forLayerSet_IfcMaterialLayerSetUsage
↓	?varible_4.
ifc:materialLayers_IfcMaterialLayerSet	?varible_4 ifc:materialLayers_IfcMaterialLayerSet
↓	?varible_5.
list:hasContents	?varible_5 list:hasContents ?varible_6.
↓	?varible_6 ifc:material_IfcMaterialLayer ?varible_7.
ifc:material_IfcMaterialLayer	?varible_7 rdf:type ifc:IfcMaterial.
	?varible_7 ifc:name_IfcMaterial ?varible_8.
	?varible_8 express:hasString ?varible_9.
	}

Table. 1. Summary of the query result.

Material	Masonry - Brick	Plasterboard	Metal – Stud Layer	Concrete - Cast in Situ
Number of IfcWallStandardCase	12	23	6	7

5. Conclusion

This paper proposed an automatic approach for extracting attributes of IFC objects, based on graph theory and SPARQL queries. In this approach, DFS algorithm are employed to find all the P-Paths from an object entity to an attribute entity in ifcOWL schema model. Then SPARQL queries for extracting attributes are generated by filling out pre-defined templates using the P-Path vertexes. Finally, after converting IFC instance model into ifcOWL format, Jena reasoner is used to parse the instance model and execute SPARQL queries to extract attribute for IFC objects. The correctness and effectiveness are validated by using a test case.

The proposed approach has the following advantages compared with existing researches.

(1) Automation. SPARQL generation process is automatic based on graph algorithm, instead of manual compilation.

(2) User-friendliness. The approach does not require users to have an in-depth knowledge about the underlying structure of IFC standard.

(3) High efficiency. Compared with the existing approaches to translate large-scale instance model into graph, the approach generates P-Path based on schema model, which reduces the computation and complexity.

The approach contributes to the convenient application of IFC standards in AEC industry.

Acknowledgements
The study has been supported by Shenzhen Qianhai Digital City Technology Co. Ltd.

References

[1] BuildingSMART International 2013 IFC4 Documentation Available online: https://standards.buildingsmart.org/IFC/RELEASE/IFC4/ADD2_TC1/HTML/ (accessed on 24 May 2021)

[2] Ismail A and Nahar A 2017 Application of graph databases and graph theory concepts for advanced analysing of BIM models based on IFC standard *24th Proc. Int. Workshop on Intelligent Computing in Engineering (Nottingham)* pp 146–157

[3] Tauscher E, Bargstädt H J and Smarsly K 2016 Generic BIM queries based on the IFC object model using graph theory, *Int. Conf. 16th on Computing in Civil and Building Engineering (Japan:Osaka)*

[4] Adachi Y 2003 Overview of partial model query language P*roc. of the 10th ISPE Inte. Conf. Concurrent Engineering (Portugal: Madeira)*

[5] Mazairac W and Beetz J 2013 BIMQL - An open query language for building information models *Adv. Eng. Inform.* **27** 444–456

[6] Pauwels P and Terkaj W 2016 EXPRESS to OWL for construction industry: Towards a recommendable and usable ifcOWL ontology *Autom. Constr.* **63** 100-133

[7] Zhang C, Beetz J and Vries B D 2018 BimSPARQL: Domain-specific functional SPARQL extensions for querying RDF building data *SWJ.* **9** 829–855

[8] Liu H, Lu M and Al-Hussein M 2016 Ontology-based semantic approach for construction-oriented quantity take-off from BIM models in the light-frame building industry *Adv. Eng. Inform.* **30** 190–207

[9] DuplexModel-IFC-2011-05-05 2011 Available online:http://smartlab1.elis.ugent.be:8889/IFC-repo/http.openifcmodel.cs.auckland.ac.nz/030811DuplexModel-IFC-2011-05-05/ (accessed on 24 May 2021)

Comparative Evaluation of Fleet Management Software in the Greek Construction Industry

Kleopatra Petroutsatou[1], Ilias Ladopoulos[1], Grigorios Vlachokostas[1]

1 Transport and Project Management Division, Department of Civil Engineering, School of Engineering Aristotle University of Thessaloniki (AUTh), Thessaloniki, Greece

Abstract. Nowadays, construction companies are facing multiple difficulties on how properly and effectively manage their fleet of construction equipment. Fleet management (FM) is a feature that allows companies to avoid or minimize the risks associated with investing in equipment, efficiency, productivity, overall transportation costs, impartial compliance in legislation. FM can include a variety of functions such as equipment acquisition, maintenance, licensing, accident management, replacement, equipment tracking, health and safety management, and many others. Internationally, there have been developed many Fleet Management Software (FMS) that offer a company with several functionalities. In this research the functions of the most globally used FMS are analysed and evaluated according to their qualitative characteristics. A questionnaire was distributed among significant large construction companies and public construction institutions in Greece, in order to evaluate those characteristics. This allowed to identify their extent of use and to highlight the most important abilities that the users desire. Finally, these characteristics were ranked using the Relative Importance Weight (RIW). By this hierarchy process users' needs are revealed, and further areas of software development are enlightened.

1. Introduction

Construction equipment FM constitutes a crucial, but also difficult part for the optimum and efficient operability for small, medium or large companies. Despite the fleet's size, the necessity of having a fleet management software to leverage their productivity is more than imperative. FM is considered more as an administrative approach, which allows the enterprises to organize their equipment, with the ultimate goal to improve their efficiency, to reduce their operating cost and to comply with their country's regulations and legislation. The advent of computing

FMS are very popular among companies across the world, when coming to execute several tasks, such as long-distance transports and earthworks. In the case of enterprises with a large equipment fleet, FM becomes a challenge and a very complex procedure. It can also be explained as the processes that fleet managers utilize to manage all fleet and asset information, from acquisition through to disposal (from cradle to grave). This enables companies to reduce costs, improve efficiency and ensure compliance across an entire fleet operation.

FMS for construction equipment appeared during the '70s, and at that time the software offered only a few and very basic capabilities and functions. Since then, their continuous development reached a level where they are now able to manage the operability of hundreds or even thousands pieces of equipment. Some of their main abilities are focusing on: a. Equipment tracking, b. Monitoring operator's behavior, c. Distance control, d. Main components and parts condition monitoring through sensors, e.

Damage and malfunction prevention, f. Live interaction between the supervisor and his operators, g. Preventive maintenance, h. Maintenance historical records, i. Manage spare parts reserves.

This study focuses on those unique characteristics of each FMS and offers a comparative analysis and software evaluation. Through a structured questionnaire, construction equipment experts, practitioners and Original Equipment Manufacturers (OEMs) were asked to evaluate each of their abilities, reflecting what the Greek construction industry demands. The analysis ranks the important characteristic according the users' needs and discuss the areas of future development in FMS.

2. Literature Review

Jin et al. (2000) examined the fleet replacement problem of a profit-maximizing manager, by using an optimal control model that captures both utilization and replacement decisions. Their results indicate that the optimal replacement schedule and fleet size are influenced by utilization schedules, and vice versa. Thus, replacement and utilization strategies should be determined jointly.

Fan et al. (2006) highlighted the need for the equipment managers to shift their attention from operational-level decision-making to corporate-level strategic decision-making regarding equipment management, which is often a challenging job with the current equipment management system. They presented an equipment data warehouse and a prototype decision support system (DSS), which enables equipment managers to visually analyze the equipment fleet data from different perspectives and at various level of details.

Sener and Iseley (2009) evaluated the use of the telematic technology in construction companies, determined its user acceptance, and assessed the differences it was making in FM. The difference made was to be assessed in terms of how this new technology had changed spatial equipment tracking, equipment utilization, equipment maintenance scheduling, operations analysis, job costing, and jobsite/project management related to construction equipment.

Dandwate (2015) distincts the three basic components of FM, which are Equipment assignment and optimization, Production monitoring, and Position and material monitoring, in order to analyze performance factors such as dump movement and haul road congestion to boost overall site productivity.

Gransberg (2015), in his final report for the Minnesota Department of Transportation, argues that Equipment Life-cycle Cost Analysis (LCCA) is typically used as one component of the equipment FM process and allows the fleet manager to make equipment repair, replacement, and retention decisions on the basis of a given piece of equipment's economic life. He managed to develop a robust method that permits equipment fleet managers to maximize the cost effectiveness of the fleet by optimizing the overall lifecycle value of each piece in the fleet. The research compared output using actual data from current software to the output from the new stochastic LCCA method using equipment deterioration curves and probabilistic input variables for capital costs, fuel, and other operating costs to demonstrate enhanced ability to optimize FM decisions. His report was the foundation of this current study.

Said (2015) presented the development of novel telematics-based computational methodologies to support two major equipment FM tasks: fleet use assessment and equipment health monitoring. He used the telematics data sets of large equipment fleets (dozers, excavators, backhoes, and dump trucks) from two companies and he managed to verify and validate his research developments by providing insightful FM information.

Avetisyan and Skibniewski (2017) stressed the need of large-scale companies to monitor the exact allocation of their construction equipment, by utilizing geographical allocation methods. Further development of their research identified the need also to use management tools and metrics for a company's equipment, to allow contractors to reduce their emissions, comply with local or federal regulations and to accurately assess the sustainability of their construction equipment fleet.

Vahdatikhaki et al. (2017) aimed to design a Multi-Agent System (MAS) architecture that combines Location based Guidance Systems (LGS) technology with advanced safety management methods to support the equipment operators on the construction site. A two-layer safety mechanism was proposed for the safety management in the MAS that enables proactive and reactive responses for the prevention of equipment-related collisions on the construction site. It was found that the proposed MAS structure was able to effectively address the fleet-level coordination between earthwork equipment and potentially improve the safety of earthwork projects.

Lee et al. (2018) developed a GPS only-based fleet telematics system for heavy earthwork equipment which can analyze time log information of each equipment using GPS location data without utilizing any other on-board sensors and Controller Area Network (CAN) bus data. They also stated that fleet telematics aims to comprehensively control equipment groups through location tracking, maintenance control, and productivity monitoring of individual equipment based on real-time location and status transmitted to fleet control center from the equipment.

3. Methodology

3.1. Software Search

An initial research was conducted, to identify the existing FMS available in the global market. According to Dandwate (2015), FM can be defined by three basic components: Equipment assignment and optimization, Production monitoring, and Position and material monitoring. At its most basic level, FM is about monitoring equipment location for an entire fleet but beyond that, it also helps to ensure that machines are in the right location and that the amount and type of material they are moving is accurate. In modern projects, the customer's needs are demanding higher expectations. In order to consider a software as FMS, certain abilities should be provided to the demanding users. Table 2 presents the main characteristics of the examined FLSs.

3.2. Questionnaire Dissemination

From the Greek construction industry sector, there were 16 expert evaluators selected and asked to weight each FLS ability, in a scale of 1 to 5 (Table 1). These experts were selected according their experience (more than 20 years in the field) in large infrastructure projects either as construction managers of superintendents of construction equipment on sites.

Table 1. Evaluation Scale

1	2	3	4	5
Irrelevant	Less Important	Important	Very Important	Extremely Important

3.3. Software Characteristics/Abilities

After conducting a holistic global market overview of the existing FMS, this study concludes to examine the main functionalities of 15 FMS with 30 different abilities, as depicted on Table 2.

Table 2. Fleet Management Software Main Characteristics

		Wialon	MyGeotab	Verizon Connect Reveal	GPSWOX	DIRECTOR Fleet Software (Teletrac Navman)	FleetLocate	Trackmatic	Fleet Complete	Fleet Commander	FleetFocus	FleetWave	Go Fleet GPS Tracker	ManagerPlus	Fleet Maintenance Pro	Fleetio
1	Equipment Cost Tracking	X	X	X	X	X	X			X	X	X	X	X	X	X
2	Customizable		X	X					X	X		X				X
3	Fuel Management	X	X	X	X	X	X	X	X	X	X	X	X	X	X	X
4	Risk Management					X		X	X	X	X	X				
5	Parts Inventory Management		X		X			X	X	X	X	X	X	X	X	X
6	Equipment Log					X							X	X	X	
7	Maintenance History	X		X		X	X		X	X	X	X		X	X	X
8	Tire Management	X	X	X	X		X	X	X	X	X	X	X	X	X	X
9	GPS Tracking	X	X	X	X	X	X	X	X	X	X	X	X			

10	Mobile Application	X			X		X	X	X					X	X	X
11	Network Support	X	X	X	X	X	X	X	X	X	X	X	X	X	X	X
12	Data Import/Export	X	X	X	X	X		X	X		X	X		X	X	
13	Auto e-mail	X	X	X	X	X	X	X	X	X	X	X	X			X
14	Route Tracking File	X	X	X	X	X										
15	Maintenance Management	X	X	X	X	X	X	X	X	X	X	X	X	X	X	X
16	Inspection Management		X		X	X	X	X	X	X	X	X	X		X	X
17	Work Order Management	X	X	X	X			X	X	X	X	X	X	X	X	X
18	Life Cycle Cost		X	X				X		X		X		X		
19	Accident Reports	X	X	X	X		X	X	X	X	X	X	X			X
20	Incident Reporting	X	X	X	X		X	X	X	X	X	X	X	X		X
21	Preventive Maintenance	X	X	X	X		X			X	X			X	X	X
22	Dispatch Management	X	X	X	X	X	X	X	X	X	X	X	X			
23	Routing	X	X	X	X	X	X	X	X				X			
24	Trip Logs	X	X	X	X	X	X	X	X	X	X	X	X	X	X	X
25	Real Time Tracking	X	X	X	X	X	X	X	X	X	X		X			
26	Driver Management	X	X	X	X	X	X	X	X	X	X	X	X		X	X
27	Alerts/Notifications	X	X	X	X	X	X	X	X	X	X	X	X			
28	Equipment Status	X	X	X	X	X	X		X	X	X	X	X	X	X	X
29	Repairs Tracking	X	X	X	X	X	X	X	X	X	X	X	X	X	X	X
30	Warranty Tracking							X		X	X	X		X	X	X

Equipment cost tracking refers to stay on budget, on time and on track. The construction industry is transforming with the ability to monitor costs related to materials, labor and equipment usage automatically and from anywhere with sensor-driven data taken directly from the field. Using this data, FM software provides businesses the insights to predict and overcome project delays and communicate with customers before they become unhappy. To make FMS more attractive, it should be also customizable, meaning to enable users to customize features for their organization by integrating rules feature that is configured and ready for use with a simple selection. Fuel Management is related to any mode of transportation, involves the maintenance, control & monitor over the storage and consumption of fuel. Effective Fuel Management ensures the best utilization of resources and allows to reduce costs to a large extent.

According to Sener and Iseley (2009), construction equipment FM should allow companies to remove or minimize the risks associated with equipment investments, improving efficiency, productivity and reducing their overall ownership costs through what it enables equipment owners to do. Furthermore, according to the American Association of International Certified Professional Accountants (AICPA), for the overall success of any business, it's in each company's best interest to invest in optimized spare parts inventory management, by following nine simple rules: a. Choose a Method for Labeling Critical Components, b. Balance Inventory Volume With Sawtooth Diagrams, c. Keep Your Bill of Materials up to Date, d. Calculate Optimal Economic Order Quantity, e. Develop a Standard Work Order Process, f. Focus on Inventory Control During Employee Training, g. Perform Cycle Counts on a Regular Basis, h. Make Spare Parts Easily Accessible and i. Invest in a Quality Computerized Maintenance Management System (CMMS).

Equipment Maintenance Logs/Records are recorded histories on all the actions that have been taken on each equipment. Keeping track of the equipment's condition helps keep the fleet at optimum working condition and minimizes the risk of unscheduled downtime or the equipment's unavailability. The tire management system is a specialized system that monitors and helps to maintain all the tires. It can track the tires in every step until the time of scrapping. It ensures that the tire is repaired, changed, and checked to increase its workspan. A GPS tracking system includes asset tracking devices combined with an asset tracking platform, allowing to gain complete asset visibility and control and maximize asset utilization. The mobile application refers to the ability of having the full potential of FM through a mobile phone application. Network Support, in a broad sense, has to do with everything that keeps the organization's computer network running, productive, and up-to-date. It assures that all devices are healthy, and that hardware, software, and personal devices are compatible and working. The import and export of data is the automated or semi-automated input and output of data sets between different software applications. True exports of data often contain data in raw formats otherwise unreadable to end-users without the user interface that was designed to render it.

Automated emails are carefully planned emails to be sent to subscribers at specific time intervals or as a response to the actions of users on a particular website. These emails can be sent individually or as part of a drip email campaign. Route Tracking File is the ability to store the executed equipment routes in a specific file system, for further exploitation. Maintenance management, when done properly, keeps assets in optimum operating condition. Less disruption in production or service leads to efficient operations, increases quality, and improves productivity. In addition, maintenance management lowers operational costs, protects the organization from liability, and improves environmental and personal safety. Inspection Management refers to inspect, collect, analyze, diagnose and monitor equipment. When combined with other information, like machine history and use, offers the ability to suggest repairs and maintenance for optimal safety and performance.

Equipment Life-cycle Cost Analysis (LCCA) is typically used as one component of the equipment FM process and allows the fleet manager to make equipment repair, replacement, and retention decisions on the basis of a given piece of equipment's economic life (Gransberg 2015). To be effective, an accident or incident report should be timely, specific and concise. It should avoid vague or easily misinterpreted language and phrases that may cause confusion and instead keep things on point. Preventive maintenance (PM) is the regular and routine maintenance of equipment and assets in order to keep them running and prevent any costly unplanned downtime from unexpected equipment failure. A successful maintenance strategy requires planning and scheduling maintenance of equipment before a problem occurs.

Dispatch management improves scheduling and dispatching efficiency by avoiding downtime and delays, reducing unnecessary rentals, and automating routine tasks. Routing includes routes planning, create a series of stops or waypoints where vehicles are expected to stop to provide a service or make a delivery or pickup. TripLog is a trusted enterprise class vehicle mileage tracking and FM solution that provides easy to use mobile and web software as a service to simplify compliance, bring transparency and reduce employees time related to business expense reporting.

Real-time locating systems (RTLS), also known as real-time tracking systems, are used to automatically identify and track the location of objects or people in real time. Implementing a driver or operator management system allows to get a comprehensive look at the drivers or the operatos, their productivity and the overall assets safety. A driver/operator management system can automatically upload vital information for the equipment fleet, such as fuel usage and telematics data. Alert notification is machine-to-person communication that is important or time sensitive. An alert may be a calendar reminder or a notification of a new message. The equipment status is the ability to transmit and record all the vital information coming from the equipment's current condition. Repair tracking means the specific ability to monitor the equipment's repairing condition. Warranty tracking is a specially designed for automating warranty claim process, which reduces the time consumed in processing the claims, by offering cross-browser compatibility as well as a responsive design.

3.4. FMS Characteristics/Abilities Evaluation

Table 9 presents the total weighting given for each FLS ability, where the response number of answers is represented, for each FMS ability. The related percentage reflects the desire rate of each ability.

3.5. FMS Overall Evaluation

Based on the questionnaire results, the use of the Relative Importance Weight (RIW) factor introduces the ranking for each software's qualitative characteristic, as on Table 3, by applying Eq. 1 (Doloi, 2013).

$$RIW = \frac{\sum_{i=1}^{5} a_i \cdot n_i}{\sum_{j=1}^{N} x_j} \cdot 100 \quad (1)$$

where: x_j=the sum of the jth factor;
j=the abilities 1, 2, 3, 4, ...N;
N=total number of abilities (30);
a_i=constant expressing the weight given to the ith response: i=1, 2, 3, 4, 5
n_i=the variable expressing the frequency of the ith response

for a response of 'extremely important': a_1=5
for a response of 'irrelevant': a_5=1
accordingly, n_1=frequency of 'extremely important' response and n_5=frequency of 'irrelevant' response

The summary of each ability's RIW for every FMS provides the total software's score. Table 4 demonstrates those aggregated results.

Table 3. FM Software Abilities Ranking

RANK	CHARACTERISTIC	RIW	RANK	CHARACTERISTIC	RIW	RANK	CHARACTERISTIC	RIW
1	**Equipment Status**	**3,76**	9	Real Time Tracking	3,39	21	Tire Management	3,23
2	Work Order Management	3,71	9	Maintenance History	3,39	21	Repair Tracking	3,23
3	Maintenance Management	3,61	9	Preventive Maintenance	3,39	21	Life-Cycle Cost	3,23
3	Risk Management	3,61	9	Network Support	3,39	24	Inspection Management	3,18
5	Incident Reporting	3,55	15	GPS Tracking	3,34	24	Equipment Cost Tracking	3,18
5	Accident Reports	3,55	15	Alerts/ Notifications	3,34	26	Dispatch Management	3,13
7	Fuel Management	3,45	15	Trip Logs	3,34	27	Warranty Tracking	3,08
7	Data Import/Export	3,45	18	Parts Inventory Management	3,29	28	Driver Management	3,02
9	Mileage Tracking	3,39	18	Equipment Log	3,29	28	Customizable	3,02
9	Routing	3,39	18	Mobile Application	3,29	**30**	**Auto Email**	**2,76**

Table 4. FML Aggregated Results

FMS	Wialon	MyGeotab	Verizon Connect Reveal	GPSWOX	DIRECTOR Fleet Software (Teletrac Navman)	FleetLocate	Trackmatic	Fleet Complete	Fleet Commander	FleetFocus	FleetWave	Go Fleet GPS Tracker	ManagerPlus	Fleet Maintenance Pro	Fleetio
Score	77,28	**83,32**	69,91	80,36	66,64	69,96	80,37	80,47	83,22	80,42	79,89	70,18	57,68	60,33	66,47

4. Findings and Recommendations

The results from Table 5 are representing the percentage evaluation of all the existing software abilities, where network support (68,75%), parts inventory management (56,25%) and maintenance management (56,25%) considered as the most precious. These rankings are demonstrating the evaluator's concerns and sensitivity on maintenance issues, recognizing the important role of maintenance regarding equipment's productivity and efficiency.

According to Table 3, most of the evaluators consider that the Auto Email ability is of less importance (2,76), while other characteristics are demonstrating higher importance weighting, like the equipment status (3,76), work order management (3,71) and maintenance management (3,61). Once again, monitoring the maintenance status and the equipment's constant working condition reveals to be of high priority for fleet managers.

Finally, the aggregated results of the above evaluation highlighted MyGeolab as the most promising and comprehensive software for FM, with a total score of 83,32/100. The worst FMS, according to the evaluators needs was Fleet Maintenance Pro, with a total score of 66,47/100, a fact that could be taken into consideration from the software developing company.

Although all the evaluators recognize the importance of utilizing this type of software, as a management multiplier that leverages each project's efficiency, they stated that they are not using any FMS for their projects. Two reasons supported their choice: a. They heavily rely on field superintendents experience and project managers rather than new technology, and b. construction industry is very reluctant to the adoption of new technology in construction processes and their resources (i.e. equipment).

5. Conclusion and Future Research

This study aims to comparatively highlight the FMS features internationally and through a structured survey in the Greek construction industry to explore the user's needs and recommend on future development in the field. Monitoring the maintenance status and the equipment's constant working condition reveals to be of high priority among Greek fleet managers. Although, there is a common consensus that FMS are necessary, their use is limited. They seem to incorporate a strategy of "operate to failure" rather than "fixed time-based maintenance" or "condition-based maintenance" (Kannan, 2011) coupled with a general reluctance of construction industry to follow new technology. The future in FMS should be able to retrieve information gathered by sensors concerning equipment's condition, combined with other specially) designed sensors for operating data in order to be able to: a. offer a smart real-time construction equipment monitoring system and b. a smart prescriptive maintenance system, for boosting productivity and investment return of construction equipment assets.

References

[1] Avetisyan H and Skibniewski M 2017 Web-based construction equipment fleet management system: cost-effective global and local allocation *Frontiers of Engineering Management* 4 76-83

[2] Avetisyan H, Skibniewski M and Mozaffarpour M 2017 Analyzing sustainability of construction equipment in the state of California *Frontiers of Engineering Management* 4 138-145

[3] Dandwate S 2015 Construction Equipment Fleet Management *International Journal of Modern Engineering Research* 5

[4] Doloi H 2013 Cost Overruns and Failure in Project Management: Understanding the Roles of Key Stakeholders in Construction Projects *Journal of Construction Engineering and Management* 139 267-279

[5] Fan H, Kim H and Zaïane O R 2006 Data warehousing for construction equipment management *Canadian Journal of Civil Engineering* 33

[6] Gransberg D 2015 *Major Equipment Life Cycle Cost Analysis* Minnesota Department of Transportation

[7] Jin D, Hauke L and Powell K 1999 Optimal Feet utilization and replacement. *Transportation Research Part E: Logistics and Transportation Review* 36 3-20

[8] Kannan G 2011 Field Studies in Construction Equipment Economics and Productivity *Journal of Construction Engineering and Management* 137 10 823-828

[9] Lee S, Park S and Seo J 2018 Utilization analysis methodology for fleet telematics of heavy earthwork equipment *Automation in Construction* 92 59-67
[10] Said H 2015 Utilizing Telematics Data to Support Effective Equipment Fleet-Management Decisions Utilization Rate and Hazard Functions *Journal of Computing in Civil Engineering* 30
[11] Sener E and Iseley T 2009 Construction Equipment Fleet Management Using Telematics Technology: Research and Resultant Educational Perspectives
[12] Vahdatikhakia F, Langarib S M, Taherc A, Ammarid K E and Hammad A 2017 Enhancing coordination and safety of earthwork equipment operations using Multi-Agent System *Automation in Construction* 81 267-285

Table 5. Qualitative Evaluation of the FMS Abilities

	Ability	Irrelevant (1)	Percentage	Less Important (2)	Percentage	Important (3)	Percentage	Very Important (4)	Percentage	Extremely Important (5)	Percentage	Total
1	Dispatch Management	1	6,25%	1	6,25%	3	18,75%	8	50,00%	3	18,75%	16
2	Fuel Management	1	6,25%	0	0,00%	3	18,75%	5	31,25%	7	43,75%	16
3	Incident Reporting	0	0,00%	1	6,25%	3	18,75%	4	25,00%	8	50,00%	16
4	**Parts Inventory Management**	0	0,00%	1	6,25%	3	18,75%	9	**56,25%**	3	18,75%	16
5	Equipment Log	0	0,00%	3	18,75%	3	18,75%	3	18,75%	7	43,75%	16
6	Mileage Tracking	0	0,00%	2	12,50%	2	12,50%	6	37,50%	6	37,50%	16
7	Routing	0	0,00%	2	12,50%	2	12,50%	6	37,50%	6	37,50%	16
8	Driver Management	1	6,25%	2	12,50%	4	25,00%	5	31,25%	4	25,00%	16
9	Inspection Management	0	0,00%	3	18,75%	3	18,75%	5	31,25%	5	31,25%	16
10	Tire Management	0	0,00%	2	12,50%	4	25,00%	5	31,25%	5	31,25%	16
11	Work Order Management	0	0,00%	0	0,00%	2	12,50%	6	37,50%	8	50,00%	16
12	GPS Tracking	0	0,00%	3	18,75%	0	0,00%	8	50,00%	5	31,25%	16
13	Alerts/Notifications	1	6,25%	1	6,25%	3	18,75%	4	25,00%	7	43,75%	16
14	Real Time Tracking	1	6,25%	1	6,25%	2	12,50%	5	31,25%	7	43,75%	16
15	Trip Logs	0	0,00%	2	12,50%	2	12,50%	7	43,75%	5	31,25%	16
16	**Maintenance Management**	0	0,00%	2	12,50%	1	6,25%	4	25,00%	9	**56,25%**	16
17	Maintenance History	0	0,00%	2	12,50%	3	18,75%	4	25,00%	7	43,75%	16
18	Preventive Maintenance	0	0,00%	1	6,25%	3	18,75%	7	43,75%	5	31,25%	16
19	Repair Tracking	0	0,00%	2	12,50%	4	25,00%	5	31,25%	5	31,25%	16
20	Vehicle Status	0	0,00%	0	0,00%	1	6,25%	7	43,75%	8	50,00%	16
21	Vehicle Cost Tracking	0	0,00%	3	20,00%	3	20,00%	4	26,67%	5	33,33%	15
22	Warranty Tracking	0	0,00%	2	12,50%	4	25,00%	8	50,00%	2	12,50%	16
23	**Network Support**	0	0,00%	1	6,25%	1	6,25%	11	**68,75%**	3	18,75%	16
24	Auto Email	2	12,50%	2	12,50%	5	31,25%	4	25,00%	3	18,75%	16
25	Life-Cycle Cost	0	0,00%	2	12,50%	5	31,25%	6	37,50%	3	18,75%	16
26	Accident Reports	0	0,00%	0	0,00%	3	18,75%	7	43,75%	6	37,50%	16
27	Customizable	0	0,00%	2	12,50%	5	31,25%	7	43,75%	2	12,50%	16
28	Risk Management	0	0,00%	0	0,00%	2	12,50%	8	50,00%	6	37,50%	16
29	Data Import/Export	0	0,00%	1	6,25%	3	18,75%	6	37,50%	6	37,50%	16
30	Available Mobile App	1	6,25%	1	6,25%	2	12,50%	7	43,75%	5	31,25%	16

IOP Conf. Series: Materials Science and Engineering 1218 (2022) 012026 doi:10.1088/1757-899X/1218/1/012026

Configuring BIM models to Support A Systems-Driven Visualization Approach: A Case Study

M Ensafi, A Harode and W Thabet

Department of Building Construction, Virginia Tech, Blacksburg, VA, USA

mensafi@vt.edu

Abstract. Models developed through design and construction are not intended to support facility management. They don't generally have the life-cycle data and information required or they are not configured to support visualization of model graphics from a system's perspective. Current implementations for BIM-FM focuses mainly on including facility life cycle data in BIM models to support space management or asset management. Models are not customized with necessary data and classification that allow for isolating and visualizing components of a specific building system (e.g. supply air). This is necessary in an emergency situation to allow facility staff to diagnose the problem associated with the system and its components, identify the problem and make necessary and informed actions. In this paper, we discuss requirements for preparing an as-built BIM model to be systems-centric using a case study approach of a two-story science building. A system classification hierarchy was developed to classify the systems and sub-systems of the building. Sixteen systems-centric properties were defined to search, filter, isolate, and display elements in a specific system or sub-system. The configured systems-centric model will allow facility managers understand the dependencies across different systems to accurately determine impacts across the facility and develop strategies to rapidly respond to any emergency. The case study analysed the mechanical and plumbing systems in the building. This paper will focus on results from the analysis of the mechanical system. An example of emergency scenario is discussed and presented.

1. Introduction

Facility management (FM) is an integrated approach for operating and maintaining the facility to meet the main objectives of the organization, facility managers, owners, and the end users [1]. Facility operators or owners are responsible for preparing reasonable plans to effectively operate and maintain their facilities. Building Information Modeling (BIM) is one effective tool that can be used as a repository for storing information, supporting maintenance management, and improving decision making. At the same time, BIM can be used for model visualization as well as for locating the assets within the facility, saving time and cost [2]. However, due to different challenges and barriers such as process or organizational complications, the BIM implementation in the FM field has been slower compared to the design and construction phases [3,4,5,6]. Furthermore, models developed during design and construction do not fully address facility operation and maintenance (O&M) needs and that is mainly due to lack of proper data requirements set by the owners [5]. BIM models configured for FM should contain critical information to fully support FM and the challenges faced by it [4]. Such models are referred to as FM-Capable BIM or FM-enabled BIM [3].

2. Methodology

An FM-Capable BIM should have four main features. First it should be data-centric meaning that it should act as a repository for O&M data. Second, the graphics should be accurate and complete to fully represent the building. Third, it should be systems-centric to allow to view and isolate model elements and components based on its different systems and subsystems within the building as well as their relationship and dependencies. Fourth, it should allow for direct linking to the FM system used by the facility owner to allow data exchange and update.

This paper proposes an approach for configuring BIM models to be systems-centric. During facility emergency, this will allow to filter and isolate model elements belonging to a system or subsystem from the entire model. A four-level classification containing multiple systems and subsystems was developed for the HVAC and plumbing systems. Additionally, fourteen specific systems-centric properties were defined. The proposed approach was then tested and discussed using a case-study. The combination of the proposed classification and properties will allow facility managers to search, filter, and display elements belonging to a specific system or sub-system.

2.1. System Classification

The system classification has four levels (Table 1). The first level comprises two main systems: mechanical and plumbing. The second level was developed by breaking the first level into seven System Types based on function. The third level (System Classification) includes further breakdown of the System Type. The fourth level breaks down the third level based on specific functions of the assets. Table 1 presents part of the developed system classification. The table can be further developed and adjusted based on building systems that are identified. The multi-level hierarchy helps linking elements belonging to the same sub-system or system.

Table 1. Level 1-4 System Classification

System Type (Level 2)	System Classification (Level 3)	System Name (Level 4)	Abbreviation
Mechanical System (Level 1)			
1. Cooling System	Hydronic Supply	Geothermal Well Supply	GWS
	Hydronic Return	Geothermal Well Return	GWR
	Hydronic Supply	Condenser Water Supply	CWS
	Hydronic Return	Condenser Water Return	CWR
2. Heating System	Hydronic Supply	Geothermal Well Supply	GWS
	Hydronic Return	Geothermal Well Return	GWR
	Hydronic Supply	Condenser Water Supply	CWS
	Hydronic Return	Condenser Water Return	CWR
3. Air System	Supply Air	AHU Supply Air	ASA
	Return Air	AHU Return Air	ARA
	Exhaust Air	AHU Exhaust Air	AEA
	Supply Air	DOAS Supply Air	DSA
	Return Air	DOAS Return Air	DRA
	Exhaust Air	FAN Exhaust Air	FEA
Plumbing System (Level 1)			
4. Domestic Water System	Domestic Cold Water Supply		DCW
	Domestic Hot Water Supply	BLR Water Supply	DHW
5. Storm System	Storm Water	Storm Water	SW

2.2. Systems-Centric Properties

To configure a systems-centric BIM model and allow users to search, filter, and visualize elements and systems, fourteen properties were defined. Some of the fourteen systems-centric properties were derived based on the system classification discussed above. Table 2 provides examples of the identified properties with their description.

Table 2. System-Centric Properties

	Property	Description
1	ASSET_TAG	A unique identifier for the asset based on a defined naming convention (e.g. "GR-AHU-RT-Roof-AHU-1")
2	CATEGORY	The standard Revit category for the asset (e.g. "Mechanical Equipment")
3	SYSTEM	Identifies the building system (Level-1) that the asset belongs to (e.g. "Mechanical System")
4	SYSTEM_TYPE	Identifies the system type (Level-2) the asset belongs to. (e.g. "Cooling Systems")
5	SYSTEM_CLASSIFICATION	Identifies the system classification (Level-3) the asset belongs to (e.g. "Supply Air")
6	SYSTEM_NAME	Identifies the system name (Level-4) the asset belongs to (e.g. "Geothermal Wells Supply System").
7	FLOOR	The asset floor number (e.g. Roof)
8	LOCATION_SERVED	The room or space number that the asset is serving (e.g. "R-108")
9	EQUIPMENT_SERVED	The asset ID for the equipment that the ducts/pipes are serving.

The full list of properties include ASSET_TAG, ID, EQUIPMENT_DESCRIPTION, SYSTEM, SYSTEM_TYPE, SYSTEM_CLASSIFICATION, SYSTEM_NAME, CATEGORY, ASSET_GROUP, FLOOR, ZONE, LOCATION, LOCATION_SERVED, and EQUIPMENT_SERVED. The properties are generalizable and can be adapted for other facilities.

2.3. Case Study

A case study of a science building on the campus of a 4-year academic institution was used to test the proposed approach while providing a proof of concept for the generalizability of the developed system classification and systems-centric properties. The building was divided into five HVAC zones based on its cooling, heating, and ventilation systems used. Table 3 presents the areas and floors of each identified zone.

Table 3. Zones 1 through 5

	Zone-1	Zone-2	Zone-3	Zone-4	Zone-5
Location	Event Space	Entire building except event space, IT rooms, electrical rooms, and aquatic center	Aquatic Center	IT and Electrical Rooms	Mechanical Penthouses rooms 320 and 310
Floor	First and Second Floors	First and Second Floors	First Floor	First and Second Floors	Roof

3. Systems-Driven Visualization

This section describes the process for configuring the BIM model as well as an emergency scenario example to present a use case for the configured model.

3.1. Configuring BIM Model

The Navistools plugin [7] for Navisworks was used to populate and configure the case study BIM model with the systems-centric properties. A user-defined tab was created which contained all the systems-centric properties identified.

Facility managers and experts can use the properties to filter and create various viewpoints presenting different systems and sub-systems to rapidly identify asset location, the dependencies between different elements and systems, and the impact of the situation and maintenance during emergency maintenances. For this paper, the authors have categorized the viewpoints based on room number and equipment. The room numbers are then broken down to specific assets serving that location. Figure 1 presents an example of viewpoint structure developed by the authors that has been also used for the following emergency scenario. Different facilities can have different folder structure for their viewpoints depending on their needs. For this case study the model was pre-configured with these necessary viewpoints before handover to the owner. It should be noted that additional viewpoints can be developed by the FM staff and saved either in the viewpoints hierarchy defined in this paper or in the viewpoint hierarchy that suits the need of the FM staff.

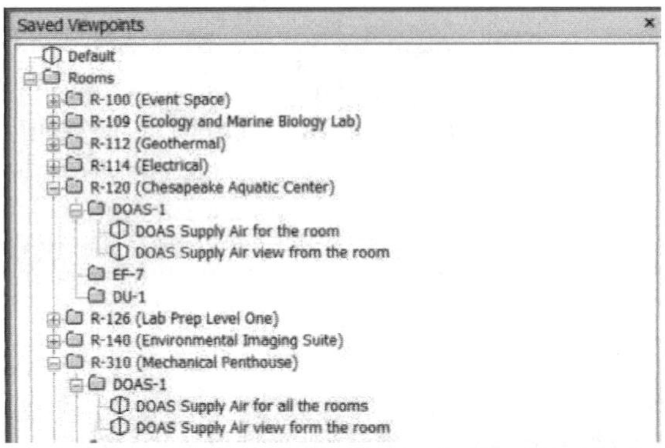

Figure 1. Example of the viewpoint structure

3.2. Emergency Scenario: High CO2 levels in Aquatic Center (R-120)

Room R-120 serves as the aquatic center and is dedicated to the study of marine ecology. Because of the nature of the lab, it is necessary to maintain adequate humidity, and CO_2 levels. To monitor these levels and to warn the researchers of any fluctuation in the pre-set levels, multiple sensors have been installed in the room. On a given day, the researchers noticed that the CO_2 sensor was showing a reading significantly higher than the set levels. As a result, a facility management staff was contacted to respond to the emergency issue. The facility staff proceeded to check the air quality and determine if there is proper air supply and exhaust in the lab.

Using a systems-centric model, a facility staff can use predefined viewpoints to identify the supply air and exhaust system connected to R-120 and determine the physical location of its different components to perform the needed inspections and determine the root of the problem. The facility staff can also visually determine if any other areas are impacted by the situation.

Using the viewpoints defined and saved under "R-120" folder (Figure 1), the Supply Air (SA) and Exhaust Air (EA) system servicing R-120 can be identified. Figure 2a shows all components of the Supply Air (SA) System supporting R-120. This includes: Dedicated Outside Air System (DOAS) coloured in orange, Supply Air Ducts coloured in blue, Diffusers coloured in orange, and Outside Air Valves (OAV) coloured in orange. While figure 2b shows all components of the Exhaust Air (EA)

System supporting R-120 which includes: Exhaust fan EF-7 coloured in orange, and Exhaust Air Ducts coloured in green.

Table 4 and Table 5 show the Navisworks search set functions that were used to create the viewpoint of Figure 2a and Figure 2b to isolate all components of the Supply Air System and Exhaust Air System respectively for lab R-120.

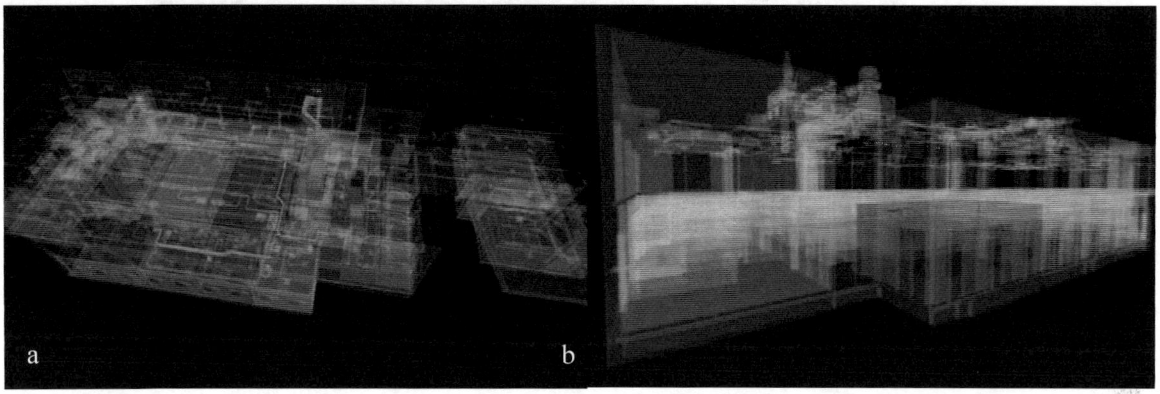

Figure 2. Components of supply (a) and exhaust (b) air systems supporting R-120

Table 3. Search sets for isolating supply system for room 120

Rule No.	Property	Condition	Value	Results
1	SYSTEM_NAME	Contains	DOAS Supply Air	Isolate supply air system associated with and serving lab 120 only
(and) 2	LOCATION_SERVED	Contains	R-120	

Table 4. Search sets for isolating exhaust air system for room 120

Rule No.	Property	Condition	Value	Results
1	SYSTEM_NAME	Contains	FAN Exhaust Air	Isolate exhaust air system associated with and serving lab 120 only
(and) 2	LOCATION_SERVED	Contains	R-120	

If the facility staff realizes that the problem is associated with the DOAS (e.g. DOAS is overly noisy indicating that it may not be functioning properly) and needs to shut it down, other areas impacted by the shutdown need to be identified to analyze the impact and inform affected individuals using these areas. To do so, the viewpoint structure (Figure 1) can be used to select the DOAS by its location (in this case R-310) and identify the areas that this equipment is serving (Figure 3). With reference to Figure 3 and the floor plan, rooms belonging to Zone 2 and Zone 3 will be impacted. Occupants of these rooms can be contacted if the DOAS must be shut down for maintenance.

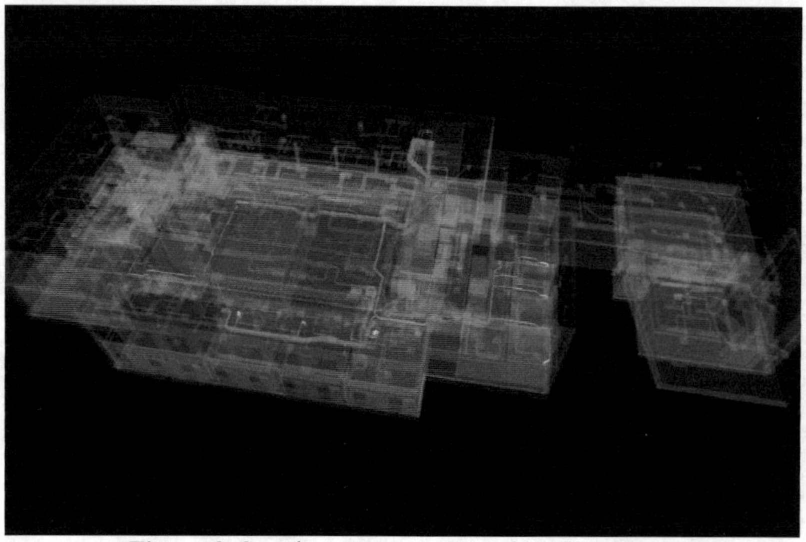

Figure 3. Supply system supported by the DOAS

4. Conclusion

In this paper, the authors presented a hierarchical system classification as well as a set of systems-centric properties to support their proposed approach in configuring BIM models for supporting FM needs during emergency maintenance. The proposed approach allows facility managers to search, filter, isolate, and display elements belonging to the same system or sub-system to understand the impact of an emergency situation or maintenance process and make timely decisions to respond. Additionally, preparing and configuring the models with viewpoints will allow facility staff to easily and rapidly navigate the model and visualize their desired elements, sub-systems, and systems without having a deep understanding of the software and its tools.

For future research, the authors are planning to conduct more case studies in order to determine the gaps in the proposed approach, and further adjust and develop the system classification, the systems-centric properties, as well as the workflow.

References

[1]
Chen, W., Chen, K., Cheng, J. C. P., Wang, Q., & Gan, V. J. L. (2018). BIM-based framework for automatic scheduling of facility maintenance work orders. Automation in Construction, 91, 15–30.

[2] Sadeghi Marjan, Mehany Mohammed, & Strong Kelly. (2018). Integrating Building Information Models and Building Operation Information Exchange Systems in a Decision Support Framework for Facilities Management. Construction Research Congress 2018, 770–779.

[3] Becerik-Gerber, B., Jazizadeh, F., Li, N., & Calis, G. (2012). Application areas and data requirements for BIM-enabled facilities management. Journal of construction engineering and management, 138(3), 431-442.

[4] Lavy Sarel, Saxena Nishaant, & Dixit Manish. (2019). Effects of BIM and COBie Database Facility Management on Work Order Processing Times: Case Study. Journal of Performance of Constructed Facilities, 33(6), 04019069.

[5] Thabet, W., & Lucas, J. (2017). Asset data handover for a large educational institution: Case-study approach. Journal of Construction Engineering and Management, 143(11), 05017017.

[6] Yang, X., & Ergan, S. (2017). BIM for FM: information requirements to support HVAC-related corrective maintenance. Journal of Architectural Engineering, 23(4), 04017023.

[7] Navistools (n.d.), Accessed February 7,2020. https://www.codemill.fi/.

Financing Public-Private Partnership Infrastructure Projects through Tokenization-enabled Project Finance on Blockchain

Y Tian[1], R E Minchin[2], C Petersen[3], E Moayed[4] and P Adriaens[5]

[1] School of Construction Management, University of Florida, Gainesville, FL 32603, USA
[2] School of Construction Management, University of Florida, Gainesville, FL 32603, USA
[3] North LNG Legal PLLC, Houston, TX 77075, USA
[4] Department of Project Management, University of Tehran, Tehran, Iran
[5] Department of Civil and Environmental Engineering, University of Michigan, Ann Arbor, MI 48109, USA

Corresponding author: Yifeng Tian. Email Address: yifeng.tian@ufl.edu

Abstract. Infrastructure is critical for enabling society to function and the economy to thrive. Unfortunately, there is an increasing mismatch between the need for infrastructure investment and available financing globally due to constraints on public resources and limited capacity to effectively leverage private sector co-financing in the current system. This research explores the integration of blockchain-enabled asset tokenization with public-private partnership (PPP) project finance to engage private sector resources and innovation to supplement limited public sector capacities and to leverage the inherent efficiencies in infrastructure financing. A conceptual framework of tokenization-enabled PPP project finance is proposed. The transactional and contractual structures of the blockchain-driven infrastructure financing system are illustrated and evaluated. It is found that the potential benefits can be summarized into three categories: better project management, improved bankability, and enhanced inclusivity.

1. Introduction

Infrastructure is critical for enabling society to function and the economy to thrive. Infrastructure investment helps alleviate poverty, combat climate change, and meet the rising aspirations of billions of people worldwide. However, the mismatch between the need for more and better infrastructure and the available financing continues to increase due to constraints on public resources and fiscal space. The contractual structure of public-private partnership (PPP) was conceived in the 1990s to engage private sector resources and innovation to supplement public sector capacities and to leverage the inherent efficiencies to narrow the infrastructure finance gap [1]. The majority of PPP projects are financed using project finance, a limited recourse financing mechanism, whereby the financing, construction and operation of a project are based on the project company's own operating cash flow and a detailed evaluation of the project's construction, operation and maintenance (O&M) risks. Project finance is a structure under which the public and private sectors can easily collaborate [2].

Under the current financing models, private investment in infrastructure, typically through loans and private equity, is limited to certain asset classes and a narrow set of institutional investors such as pension funds and insurance companies, thereby leaving a significant amount of capital and resources on the sidelines. The current models also face limitations in how liquidity risks can be mitigated, retail investors can be engaged, and social and environmental impacts can be integrated [3]. Pioneering thinking and a groundbreaking financing model will be required to ameliorate public sector engagement and mobilize broader private sources efficiently to catalyze the development of a broader mix of projects and a deeper financial base.

Blockchain is a type of distributed ledger for maintaining a permanent and tamper-proof record of transaction-based data. Improved security, transparency, and automation are features of a blockchain-driven system. Building on blockchain technology, asset tokenization enables the transmission of conventional assets in the conventional form (equity or debt) into cryptographic tokens existing in the crypto world, where efficiency is expected to be improved by orders of magnitude[4]. The integration of tokenization and project finance could introduce a more efficient financing system to facilitate infrastructure development.

Despite the aforementioned potential benefits and growing attention from scholars and practitioners, tokenization-enabled project finance has yet to be executed at scale and integrated into the PPP contractual context. To the best of the authors' knowledge, this research is the first to propose a conceptual framework of tokenization-enabled project finance in PPP infrastructure projects.

2. PPP project finance

There are two main entities in the PPP: the public sector entity, which organizes procurement, and the private sector entity, which is delegated to perform specific tasks on behalf of the public sector. The public entity can be public sector institutions at different administration levels (e.g., municipality or national government). Private entities are typically a syndicate of investors, comprised of equity sponsors and lenders organized in a special purpose vehicle (SPV), along with engineering, procurement and construction (EPC), and O&M contractors. Depending on the scheme of PPP, the private entity secures all or a portion of the financing to supplement potential public funding [5].

Private-sector project-finance debt is provided from two primary sources: loans and bonds. Commercial banks offer long-term loans to project companies. The majority of institutional interest in project-finance debt is found in the bond market. Bondholders are typically long-term institutional investors. Bonds are tradable instruments offering theoretical liquidity. However, many PPP bonds are sold on a private placement basis in an illiquid environment. Since the financial crisis in 2008, tighter financial regulations (e.g., Basel III) and reduction in long-term bank loans have limited the traditional debt capital to finance infrastructure [6].

Equity investors for PPP projects are divided into two main categories, strategic investors and financial investors. Strategic investors are companies for whom investment is part of a strategy for securing their business as subcontractors (e.g., construction and O&M). Financial investors are interested in the investment and the internal rate of return (IRR) (e.g., insurance companies, pension funds, and infrastructure funds) [7]. The generalized contractual structure of PPP project finance is illustrated in Figure 1. Equity investment in PPP, through the current financing system, is limited to a narrow set of large-scale institutional investors, thereby leaving a significant amount of capital and recourses on the sidelines, which exacerbates issues such as illiquidity, barriers to entry, higher cost of capital, and a narrower set of projects being developed [8].

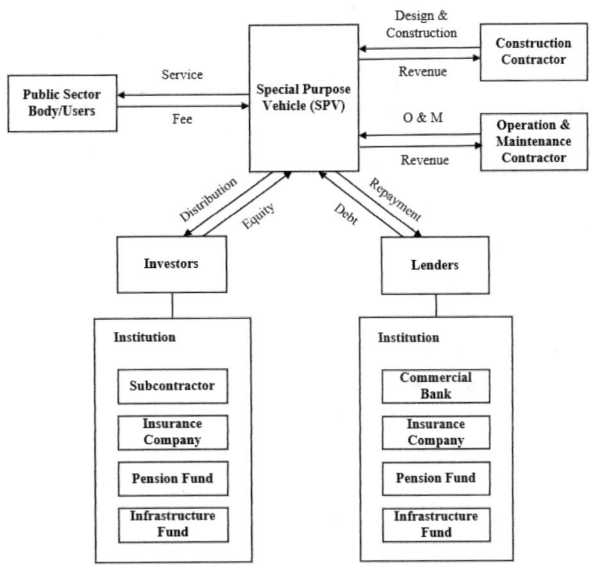

Figure 1. Generalized contractual structure of PPP project finance

3. Infrastructure asset tokenization

Tokenization is the process of digitally representing assets in the original form on a distributed ledger. Asset tokenization enables the representation of assets on the ledger by linking or embedding the economic value and rights derived from the underlying assets into cryptographic tokens created on the blockchain. Cryptographic tokens represent programmable assets or access rights, such as shares in a company, permissions to a platform, or electricity produced by energy plants. In theory, any asset and associated rights can be tokenized and represented on the blockchain. Tokens are categorized into three classes: payment tokens, utility tokens, and security tokens depending on their function and purpose.

Infrastructure asset tokenization refers to the process of tokenizing utilities (e.g., services or products provided by the facility) or ownership interests of the conventional securities (e.g., equity of infrastructure companies or funds, loans, and bonds) on the blockchain [9]. Given the increasing transition of infrastructure to intelligent and inclusive systems and the desire to unlock efficient financing, tokenization may support alternative financing models to overcome obstacles faced by the conventional financing system. Tokens backed by infrastructure assets are governed and executed through smart contracts, which are software algorithms integrated into a blockchain with trigger actions based on pre-defined parameters. Smart contracts are self-enforcing and self-executing. The automation reduces the administrative burden and intermediaries, which leads to lower transaction costs and faster execution. Besides the efficiency gains driven by automation and disintermediation, tokenization delivers other benefits, including enhancing transparency, improving liquidity to currently illiquid infrastructure assets, lowering barriers to small-scale projects, and engaging individuals in project investment and execution.

4. Tokenization-enabled PPP project finance

By integrating tokenization and PPP project finance, an innovative infrastructure financing system can be developed. As in the current project finance structure, an SPV is set up and is responsible for managing the PPP project. The functionality of the token, whether defined as a security (equity or debt) or utility (service or products), is determined by the project and the policy regime in which the project company operates. Theoretically, smart contracts can be generated to automate project governance and execution in a frictionless manner (e.g., regulatory requirements, policy on distributions, voting of shareholders, etc.). After the legal and deal terms are established, token issuance services providers, Know Your Customer/ Anti Money Laundering (KYC/AML) vendors, custodians, and primary/secondary marketplaces are selected and implemented by the project company. Tokenization could enable, aside from institutional investors, allow individuals, the community impacted by a project, and even the unbanked to be involved in infrastructure development at the project level. Potential investors need to pass KYC/AML checks and meet project requirements to become qualified before investing. Once these processes are completed, newly minted tokens are transferred to wallets of qualified investors or are listed on token exchanges. Qualified investors are able to transfer their tokens to other Qualified investors through peer-to-peer (P2P) trading or trade tokens in secondary markets if allowed by the project requirements. Future dividends and interest payouts generated from the underlying asset are sent out to wallets of token owners in the form of cryptocurrencies or equivalent fiat currency. The process is automated by smart contracts whenever pre-defined parameters are triggered. The transactional structure of tokenization-enabled PPP project finance is illustrated in Figure 2. The contractual structure is shown in Figure 3.

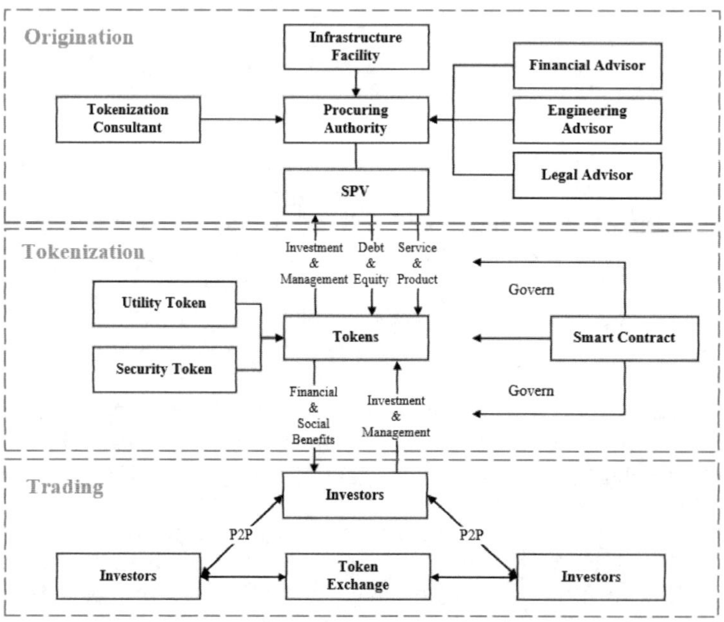

Figure 2. Transactional structure of tokenization-enabled PPP project finance

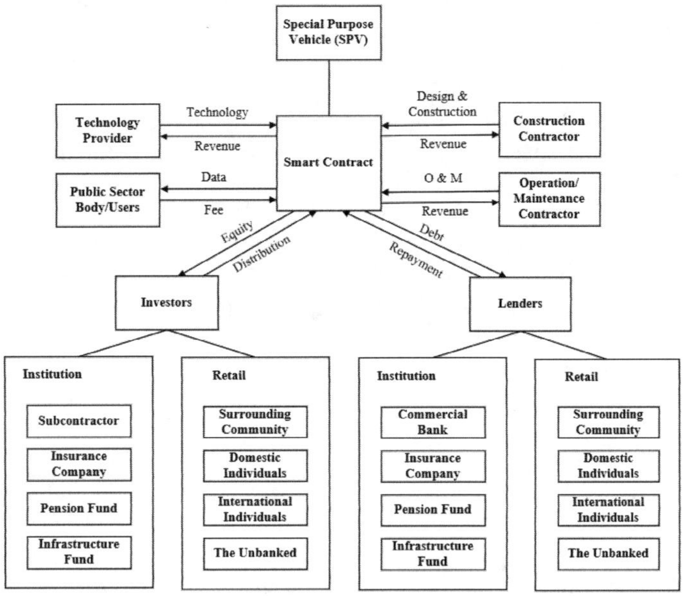

Figure 3. Proposed contractual structure of tokenization-enabled PPP project finance

Tokenization has the capacity to mitigate some of the limitations inherent to conventional PPP project finance. The potential benefits brought by tokenization can be summarized into three categories: transparent project management, improved bankability, and enhanced inclusivity.

- Management
  - Operational and financial information of the project can be automatically and immutably recorded on the blockchain [10]. Through tokenization and the smart contract functionality, the operational and financial information on the project can be immediately made available to project stakeholders. The improved quality and quantity of data will grant the project authority, investors, debt providers, and the surrounding community direct access to monitoring the project.
- Bankability
  - Smart contracts allow for bi-directional instant transfer of funds and tokens by removing intermediaries without the need for a separate settlement process [11]. The automation enabled by smart contracts reduces the cost of transaction and administration in project finance, which improves the bankability of PPP projects.
  - Tokens help stimulate the consumption of sustainable services and goods by providing financial incentives to users and further improving the profitability and bankability of high-impact projects. Non-financial values (e.g., social and environmental impacts) can be monetized through tokenization to serve as an additional cash flow resource [9].
- Inclusivity
  - By deploying tokenization, the scale of an infrastructure project is no longer a determining factor for financing cost efficiency due to the disintermediation enabled by smart contracts,

which results in a significant reduction in fixed costs. Tokenization makes investment in small-scale infrastructure at the project level possible [4].

- From the investors' perspective, tokenization enables fractional investment, promoting inclusiveness by offering financial access to individuals and small and medium-sized enterprises (SMEs) to PPP infrastructure investment. Tokens can also be distributed to the unbanked in the low-income community to address the inequality issue.

5. Conclusion

This research introduces a conceptual framework of tokenization-enabled PPP project finance. The transactional and contractual structures of the innovative blockchain-driven infrastructure financing system integrate tokenization into project finance. By integrating tokenization into project finance, the advantage of blockchain in data recording and storing assists different parties to better manage and monitor projects. The implementation of smart contracts reduces the cost of capital and makes the monetization of social and environmental impacts possible to provide the project with additional cash flow resources. Therefore, the projects' bankability can be improved. Tokenization-enabled PPP project finance can remove barriers to small projects and includes retail investors in infrastructure development. Tokenization is still in the early stage of development. Future research needs to focus on the financial modeling of real-world use cases to better understand the benefits of this financing approach. Once the potential risks and barriers for broader tokenization applications are carefully examined and mitigated, tokenization could play a critical role in revolutionizing infrastructure finance to provide efficiency gains to both the public and private sectors.

6. References

[1] Yescombe E R and Farquharson E 2018 Public-private partnerships for infrastructure: Principles of policy and finance *Butterworth-Heinemann*

[2] Yescombe E R 2013 Principles of project finance *Academic Press*

[3] Mostafavi A, Abraham D and Sinfield J 2014 Innovation in infrastructure project finance: A typology for conceptualization *International Journal of Innovation Science*

[4] Tian Y, Lu Z, Adriaens P, Minchin R E, Caithness A and Woo J 2020 Finance infrastructure through blockchain-based tokenization *Frontiers of Engineering Management* 7(4) pp 485-499

[5] Ke Y, Wang S, and Chan A P 2010 Risk allocation in public-private partnership infrastructure projects: comparative study *Journal of infrastructure systems* 16(4), 343-351

[6] Ma T 2016 Basel III and the future of project finance funding. *Mich. Bus. & Entrepreneurial L. Rev.* 6, 109

[7] Lu Z, Peña-Mora F, Wang S Q, Liu T and Wu D 2019 Assessment framework for financing public-private partnership infrastructure projects through asset-backed securitization. *Journal of Management in Engineering* 35(6), 04019027

[8] Tian Y, Zhang Y, Minchin R E, Asutosh A and Kan C 2020 An innovative infrastructure financing instrument: Blockchain-based tokenization. In *Construction Research Congress 2020: Infrastructure Systems and Sustainability* pp 731-740

[9] Tian Y, Adriaens P, Minchin R E, Chang C, Lu Z and Qi C 2020 Asset Tokenization: A blockchain Solution to Financing Infrastructure in Emerging Markets and Developing Economies. *Institute of Global Finance Working Paper*

[10] Hofmann F, Wurster S, Ron E, and Böhmecke-Schwafert M 2017 The immutability concept of blockchains and benefits of early standardization. *In 2017 ITU Kaleidoscope: Challenges for a Data-Driven Society (ITU K)* (pp. 1-8). IEEE

[11] Wang S, Yuan Y, Wang X, Li J, Qin R, and Wang F Y 2018 An overview of smart contract: architecture, applications, and future trends. *In 2018 IEEE Intelligent Vehicles Symposium (IV)*

Investigating Employee Performance for Improved Learning Curve in the South African Construction Industry

S Chauke, N Mashwama, C Aigbavboa and W Thwala

Department of Construction Management, University of Johannesburg, Doornfontein Campus, Johannesburg, South Africa

chaukesally1@gmail.com

Abstract. Employee performance is one of the most critical factors that have an impact on organisational success. The quality and characteristics of an employee in the organisation are important attributes for success. it is not possible for an organisation to attain its goals and objectives without employees. Most construction projects in South Africa face several problems, including poor project performance and poor-quality project delivery, among others. These problems are mainly a result of a failure in employee performance. Improving employee's performance is key to its success in delivering completed projects on schedule. The study is mainly a literature review/survey on the effects of employee turnover, the various ways to enhance employee performance, factors affecting employee performance, and the measures to improve employee performance.

Keywords: performance, employee, employee performance.

1. Introduction

The South African construction industry faces several problems and challenges, such as poor performance of construction projects (Omran, Abdalrahman and Pakir, 2012:55 and Sibiya, 2015:1). A construction project's success depends on its performance, which is measured based on timely completion, within the budget, required quality standard and customer satisfaction. Improving an employee's performance is key to its success in delivering completed projects on schedule (Sadraei, 2019:2). The organization attains its goals by combining the employee's expertise and skills when performing an action that results in the organization's outcomes and improving productivity. Improved organizational performance indicates all the employee's efforts, skills, and capabilities to attain organizational goals (Dakhoul, 2018:12). For employees to perform according to the required corporate standards, it is essential for management to understand the employees' needs (Francis, 2018:3). Hence, the study investigates employee performance in South Africa's construction projects to help critical players in the construction industry overcome performance problems and improve employee performance in the South African construction industry. The study also explored the causes of poor employee performance in the South African construction industry.

2. Literature Review

Zhang (2012:16) defines employee performance as the record of the results achieved by the employee.
Omran (2016:14) further describes employee performance as the level and quality of an effort. Cooperation, commitment, lateness, compliance with standard shown by an individual, and describes it as employee effort to do their best ability to achieve their work goals, resulting in good outcome and behaviour. Employee performance of great importance, and it is the core of the organisation's productivity. Companies' success relies on employee performance; hence employee performance is vital (Sadraei, 2019:3). Moreover, the organisation's goals are attained by combining employees' expertise and skills when performing an action that results in the organisation's outcomes and improving productivity. Improved organisational performance indicates all the employee's efforts, skills, and capabilities to attain organisational goals (Dakhoul, 2018:12). For employees to perform according to the required corporate standards, management needs to understand the employees' needs (Francis, 2018:3). Employees' performance indicates the organisation's level of competence and the individual's productivity towards organisational goal achievement. Organisations have realised that set goals cannot be achieved unless employees' performance is above board. Therefore, enhancing employee output is essential to accomplish organisational goals and increase organisational productivity (Tamunomiebi and Oyibo, 2020:2). Employee performance is measured through employee evaluations done annually, quarterly, semi-annually, depending on the organisation. (Francis, 2018:3 and Akinbowale, Lourens, and Jinabhai, 2014:345). Nandy (2016:2), Agrees that for an organisation to achieve its set goals, employee performance must be monitored or measured regularly.

2.1 Causes of poor employee performance

According to Amoah, Ahadzie, and Dansoh (2011:45), the causes of poor employee performance in the construction industry are many. Some of the elements are the unavailability of training proprietors and technicians, inability to delegate responsibility. Omran, Abdalrahman and Pakir (2012:61) outline the following as the factors affecting employee performance; adequacy of design and specification, cost progress monitoring, leadership skill of project team leader, adaptability to changes in a project plan, solving problems skill of project team leader, consultant's commitment to ensure construction work, according to specifications, team leader's responsibility to meet cost, time and quality, project team leader's working relationship with others, and unavailability of information.

2.1.1 *Communication*

Olanrewaju, Tan, and Kwan (2017:763) identified poor communication as the primary cause of poor employee performance in the Malaysian construction sector. Most construction disputes arise due to inadequate communication among the team members or stakeholders; therefore, communication is considered strategic. Adequate communication during the design stage is of utmost importance because poor communication leads to delays. Communication is one of the most important processes of transferring information in any construction project. Poor communication has become a common problem among construction stakeholders. Poor communication negatively affects aspects that are important for the success of a project, such as a timeframe, employee performance resulting in time and cost overruns and ultimately leading to the project's failure (Hussain, Othman, Gabr, and Aziz, 2018:2).

2.1.2 Motivation

Employee motivation is an essential factor for employee performance, lack of motivation results in poor performance, among other things. Lack of recognition of employee good performance is demotivating, and as a result, the employee performance is badly affected (Muze, 2014:2). Bala lawal (2018:11) further adds that employee motivation can be achieved by understanding and acknowledging employees' needs, feelings, wants, and boosting their work abilities. When adequate measures are taken towards motivating employees, there is an increase in performance.

2.1.3 Lack of Resources

When an employee lacks the time, money, personnel, or supplies to complete a task, it affects completing the task at hand. Lack of resource is a great contributor to poor employee performance (Barr 2015). According to Mashwama and Aigbavboa (2019:112), failure to plan for resources in time or failure to scrutinise the work tasks or activities contributes to poor employee performance. Improper equipment selection and faulty equipment affect the employee's performance.

2.1.4 Skills

Performance problems include the lack of required skills. This might be the result of an employee promoted or an employee assigned new job duties (Talyor 2015). Mashwama and Aigbavboa (2019:111) urge that low skill levels among employees cause poor performance in a construction project. Unskilled employees' employment on a site project hinders the proper execution of work. Which leads to error or mistakes resulting in poor project performance. Also affecting the project specification and standards (Prokopet 2017).

2.1.5 Working Environment

Bushiri (2014:1) states that the working environment plays a crucial role in employee performance. Furthermore, the low working environment contributes to poor employee performance immensely. It can either have a negative or a positive consequence. According to Abrey and Smallwood (2014:429), Poor site conditions contribute to time delays. Because poor site conditions are a source of complications and unsafe working sites, which impose Health &Safety risks, an increased likelihood of incident accidents occurring, causing time delays.

2.1.6 Job Satisfaction

It is found that employees with higher levels of job satisfaction are physically and mentally fit, but employees who are dissatisfied with their job are often frustrated and uncertain. Job satisfaction is one of the most crucial factors for organisational behaviour because each organisation has specific goals representing the existential philosophy (Abdulkhaliq and Mohammadali, 2019:164). In conclusion, a satisfied employee is a happy employee, and a happy employee is an excellent performing employee (Alromaihi, Alshomaly, and George, 2017:4).

3. Research Methodology

The study was conducted as a systematic/ narrative literature review on employee turnover's impact on employee performance. The literature review was conducted from a collection of existing resources the library, online journal articles, hard-copy journal articles, textbooks, and conference papers published and unpublished. The study is a literature review on the effects of employee turnover on employee performance. A qualitative research approach was adopted for this study. Mohajan (2018:2) state that qualitative research is suitable for a study that does not include statistical analysis and empirical calculation. Therefore, making it ideal for this study is part of a master's thesis at a questionnaire distribution stage.

4. Lesson Learned

There are many causes of poor employee performance in the construction industry. Proper communication in any project because inadequate communication is one of the contributing factors of poor employee performance. A Healthy &Safet working environment is crucial for employee performance. Job satisfaction is one of the most crucial factors for organisational success because a satisfied employee is a happy employee, and a happy employee is an excellent performing employee.

5. Conclusion

The South African construction industry faces several problems and challenges, such as poor performance of construction projects. Companies' success relies on employee performance; hence employee performance is vital. the organisation's goals are attained by combining employees' expertise and skills when performing an action that results in the organisation's outcomes and improving productivity. Improved organisational performance indicates all the employee's efforts, skills, and capabilities to attain organisational goals. the causes of poor employee performance in the construction industry are many. Some of the elements are the unavailability of training proprietors and technicians, inability to delegate responsibility. The causes of poor employee performance including lack of inadequate communication, lack of motivation and lack of resources. The work environment plays a crucial role in employee performance, poor site conditions contribute to time delays. Because poor site conditions are a source of complications and unsafe working sites, which impose Health &Safety risks, an increased likelihood of incident accidents occurring, causing time delays. Job satisfaction is another factor causing poor employee performance, an unsatisfied employee is an unhappy employee, which results in the employee performing poorly.

References

[1] Abdulkhaliq, S.S. and Mohammadali, Z.M., 2019. The Impact of Job Satisfaction on Employees' Performance: A Case Study of Al Hayat Company-Pepsi Employees in Erbil, Kurdistan Region-Iraq. Management and Economics Review, 4(2), pp.1-13.

[2] Abrey, M. and Smallwood, J.J., 2014. The effects of unsatisfactory working conditions on productivity in the construction industry. Procedia Engineering, 85(2014), pp.3-9.

[3] Akinbowale, M.A., Lourens, M.E. and Jinabhai, D.C., 2014. Employee performance measurement and performance appraisal policy in an organisation. Mediterranean Journal of Social Sciences, 5(9), p.342

[4] Amoah, P., Ahadzie, D.K. and Dansoh, A. (2011). The factors affecting construction performance in Ghana: the perspective of small-scale building contractors. The Ghana Surveyor.

[5] Bushiri, C.P., 2014. The impact of working environment on employees' performance: the case of institute of finance management in Dar Es Salaam region. Unpublished Doctoral Dissertation, The Open University of Tanzania.

[6] Dakhoul, Z.M., 2018. The Determinants of Employee Performance in Jordanian Organizations. Journal of Economics Finance and Accounting, 5(1), pp.137-143.

[7] Francis, J.S., 2018. Conflict Management Styles and Employee Performance in Public Hospitals in Nyeri County, Kenya (Doctoral dissertation, Master's Thesis, Kenyatta University, Kenya).

[8] Hussain, A.M.A., Othman, A.A., Gabr, H. and Aziz, T.A., 2018. CAUSES AND IMPACTS OF POOR Communication in the Construction Industry

[9] Mashwama, N. and Aigbavboa, C.O., 2019. Assessment of the performance of roads infrastructure in the Gauteng province of South Africa-stakeholders' perspective.

[10] Nandy, D.K., 2016. The Performance Management System-an Indian Perspective.

[11] Olanrewaju, A., Tan, S.Y. and Kwan, L.F., 2017. Roles of communication on performance of the construct. Assessment of factors affecting the performance of construction projects in South Africa (Doctoral dissertation, University of Johannesburg).

[12] Omran, A., Abdalrahman, S. and Pakir, A.H.K. (2012). Project Performance in Sudan Construction Industry: A Case Study. Academic Research Journals (India), pp. 55-78. GJAER Volume 1 • Number 1 • January-June 2012.

[13] Prokopets, E., 2017 Blastingnews https://uk.blastingnews.com/education/2017/05/what-are-thereasons-for-poor-performance-at-work-001605795.html [accessed online 04 February 2021]

[14] Sadraei, B., 2019. A Study of Performance Management system as a Way to Productivity and Achieving Organizational Goals in Small Business

[15] Sibiya, M., Aigbavboa, C. and Thwala, W., 2015. Construction projects' key performance indicators: a case of the South African construction industry. In ICCREM 2015 (pp. 954-960).

[16] Tamunomiebi, M.D. and Oyibo, C., 2020. Work-Life Balance and Employee Performance: A Literature Review. European Journal of Business and Management Research, 5(2).

[17] Taylor, T., 2015 Payscale https://www.payscale.com/compensation-today/2015/06/uncoveringthe-hidden-causes-of-poor-employee-performance [accessed online 04 February 2021]

[18] Zhang, Y., 2012. The impact of performance management system on employee perfomanceAnalysis with WERS 2004 (Master's thesis, University of Twente).

IOP Conf. Series: Materials Science and Engineering 1218 (2022) 012029 doi:10.1088/1757-899X/1218/1/012029

Platforms - A Novel Paradigm of Construction Collaboration

Žiga Turk, Mewail Giddey, Robert Klinc

All: University of Ljubljana, Faculty of Civil and Geodetic Engineering, Slovenia.

Corresponding author: ziga.turk@fgg.uni-lj.si

Abstract. Business and innovation models based on an idea of a digital platform have revolutionized information technology use in many consumer areas where uptake of IT would have otherwise been slow. Slow is also the uptake of information technology in construction. The hypothesis of this paper is that the delivery mechanism for technology has not been the most suitable one and that this can be improved upon – using the digital platforms. Term platform has been used before. In this paper we define the concept of a digital construction platform. We then analyse their potential in construction processes, focusing on construction design. The paper contributes to the understanding of an emerging new paradigm for the delivery, marketing, and innovation in digital technology in construction. Follow up work will examine the platforms from the structure-function-behaviour perspective and present their architectural foundations.

1. Introduction

In 1999, Nokia was the most valuable company in Europe. Over the next decade, Nokia dominated the global cell phone industry (Moazed and Johnson, 2016). By the beginning of 2014, Nokia's market share has dwindled. The reason for the downfall is the iPhone. It turned the cell phone industry giants like Nokia and BlackBerry into virtually irrelevant players within just a few short years. It was technically inferior, had a shorter battery life, it transferred a lot of data, and it wasn't secure (Moazed and Johnson, 2016). Its advantage was that it was not a product but a platform for other people's products.

Similar pattern can be observed in other industries. Air Bed & Breakfast (Airbnb) started out as a means of extra cash, by two friends who couldn't afford the rent. Today, Airbnb is a giant enterprise active in 220 countries, where it lists over seven million properties ranging from studio apartments to actual castles and has served over 750 million guests (Airbnb Inc., 2020). Airbnb did all that without owning a single hotel room of its own (Parker, Van Alstyne and Choudary, 2016; Cusumano, Gawer and Yoffie, 2019). It has achieved a scope and value that a traditional hotelier can hope to reach, only after decades of often risky investment and hard work, in just a few years (ibid.). It did so because it was a *platform*. Airbnb applied the platform model to the hotel business.

Both, iPhone and Airbnb, are just two examples from a list of disruptive platforms that includes Amazon, YouTube, eBay, Wikipedia, Upwork, Twitter, KAYAK, Instagram, Pinterest, Uber, Alibaba, Facebook, and dozens more (Parker, Van Alstyne and Choudary, 2016). Platform transforms industries, changes how value is created and results in creating new opportunities (Parker, Van Alstyne and Choudary, 2016).

1.1. Construction industry

The construction industry, with an estimated expenses of $14 trillion by 2025, has only an average of 1 percent labour-productivity growth per year over the past two decades, compared with a rate of 2.8 percent in the case of the total economy and 3.6 percent in manufacturing (McKinsey Global Institute, 2017). McKinsey Global Institute (2017) has identified infusing digital technology to be one of the ways to tackle the root causes that underline the poor productivity growth of the industry.

However, the construction industry has been slow to adopt process and technology innovations, in fact, it is one of the least digitized industries (Agarwal and Sridhar, 2016; Turk, 2021). This paper argues that the tool to do so could be the *digital platform*. That the process to achieve more and faster digitalization of construction industry is the process in which the innovation, work, and business would get on a platform. In short – platformization.

1.2. Hypothesis and problem statement

The hypothesis of this paper is that there are sufficient similarities between construction industry and the areas in which platforms have been successful to warrant the restructuring of the industry and its ICT technology towards platforms.

2. Platforms

In a nutshell, all platforms do two things: reduce transaction costs and enable complementary innovation (Moazed and Johnson, 2016). The platform model has been a success in "long tail" areas – where there are not only many different consumers but more importantly a multitude of producers and creators. Apple offered an opportunity to millions of app authors to offer their products, AirBnB to millions of property owners, YouTube to millions of movie producers, Uber to millions of car owners.

Oxford English dictionary (Oxford University Press, 2020) defines platform as a raised level surface on which people or things can stand. For example, the raised area alongside a rail track in a train station that allows passengers a convenient access to trains is called a railway platform (*Railway platform - Wikipedia*, 2020).

In the context of construction information technology, the term platform meant many things over the last decades. There exists, for example, a European Construction Technology Platform ECTP (Bourdeau, 2009). Platforms were understood as various tools for concurrent engineering (Katranuschov et al., 2003), the integration of the construction processes (Singh et al, 2011), cloud based BIM solutions (Goulding et al., 2014), collaboration environments (Ma, 2018) and also social BIM clouds (Das et al., 2015). It generally meant a place where people, information and technology would loosely come together.

This Section aims at defining an explaining the new meaning of the term.

2.1. Platform business models

The main concern of the paper are platforms that support a certain kind of business models. Platform business model is a business model based on connecting individuals and organizations for a common purpose or for sharing a common resource (Cusumano, Gawer and Yoffie, 2019). Parker, Van Alstyne and Choudary (2016) also defined it as a new business model based on enabling value-creating interactions between external producers and consumers.

Platforms deliver products or services by bringing together two or more market actors or "sides" (e.g., buyers and sellers, or an operating system maker with users, application developers, and hardware producers) that would otherwise not interact or easily connect (Cusumano, Gawer and Yoffie, 2019). For this reason, they are often called *multisided platforms* (Evans and Schmalensee, 2016).

Platforms usually act as intermediaries, developing and managing an aggregation platform for goods and services of the same type (Ruggieri *et al.*, 2018). The role of platform business model is to provide a space and a set of standards and protocols that facilitate interactions across a large number of participants (Kim and Yoo, 2019).

We can distinguish at least between three kinds of platforms. *Transaction platforms* serve as an intermediary for direct exchanges among market actors. *Innovation platforms* serve as a technological foundation upon which actors engage in complementary innovation – innovation performed by external complementors (Tsai, 2018). The third and most successful one is a combination of both – a *hybrid platform* (Cusumano, Gawer and Yoffie, 2019).

2.2. Where to apply Platform Business Model?

The reduction of substantial friction is a necessary but not sufficient condition for a multisided platform to succeed (Evans and Schmalensee, 2016). Parker, Van Alstyne and Choudary et al. (2016) identify the characteristics of industries ready for a platform revolution:

- *Information-intensive industry:* how does the business generate value, using physical asset or information asset?
- *Industries with non-scalable gatekeepers:* How is the business transaction characterized? Does it depend on gatekeepers? Are the gatekeepers scalable or non-scalable? Gatekeepers manage the flow of the business value. A good example is a receptionist. Receptionists are first in line in managing hotels' business transaction, i.e. the use of rooms by customers. A human receptionist can only accommodate one customer at a time, thus non-scalable. However, a virtual receptionist can accommodate as many customers as they are online, simultaneously, thus scalable.
- *Highly fragmented industries:* Is it possible to increase the efficiency?
- *Industries characterized by extreme information asymmetries.* Does one party have information advantage compared to other one?

2.3. Platform success factors

The success of a platform is determined by the connection, gravity and flow of the platform (Choudary and Bonchek, 2013).

- *Connection:* Does the platform focus on specific transaction? Is it easy to use? When Facebook started it focused on a simple core transaction. It was simple by design, unlike its' earlier competitors, who had many more feature (Moazed and Johnson, 2016).
- *Gravity*: Does the platform have a positive or a negative network effect? Network effect refers to the impact that the number of users of a platform has on the value created for each user (Parker, Van Alstyne and Choudary, 2016). A *Positive* network effect is characterized with the increase of significant value whereas *Negative* being with the reduction in the value produced for each user (Parker, Van Alstyne and Choudary, 2016). When they are positive, the results are nonlinear increases in utility and value, which enabled Facebook to grow from two users to over 2 billion users in just a few years (Cusumano, Gawer and Yoffie, 2019).
- *Flow*: Does the platform have the right core transaction? The core transaction is the set of actions consumers and producers must complete in order to exchange value (Moazed and Johnson, 2016). Getting the core transaction right is the most important piece of platform design (Moazed and Johnson, 2016).

3. Analysis of construction from platform perspective

The construction sector is one of the largest in the world economy, with about $10 trillion spent on construction-related goods and services every year (McKinsey Global Institute, 2017). It is suitable for the disruption though platforms.

3.1. Construction is Information Intensive

Many (Sommerville and Craig, 2006; Hua, 2013; Christodoulou, 2017; Perera *et al.*, 2017) have characterized the construction industry as information intensive, where huge number of documents are generated and exchanged between the individual parties contractually bound to the project (Sommerville and Craig, 2006; Hua, 2013). Successful and timely project completion depends on the accuracy and timeliness of information (Christodoulou, 2017).

The introduction of Building Information Modelling (BIM) presents the construction sector's moment of digitalization (EUBIM Task Group, 2016). A Building Information Model (BIM) is a comprehensive digital representation of a built facility with great information depth (Borrmann *et al.*, 2014).

3.2. Construction is dependent on Gatekeepers

Gatekeepers are market actors in a pipeline business model, responsible to manage the flow of value from the producer to the consumer (Parker, Van Alstyne and Choudary, 2016). In the construction industry, General Contractor have the role of gatekeeper (Laine *et al.*, 2017).

3.3. Construction is fragmented

The industry is also known for being fragmented (Hua, 2013; McKinsey Global Institute, 2017; Perera *et al.*, 2017). A major research direction is about computer integrated construction (Turk, 2020). The industry is highly fragmented in the sense of, (1) vertical fragmentation (between project phases), (2) horizontal fragmentation (between different specialists of disciplines at a given project phase), where communication and information exchange problems proliferate (Hua, 2013), (3) project fragmentation where each project is unique and starts from scratch, building on previous work only to limited extent and also leaving behind little but the work that was done specifically for this project; and finally (4) construction is fragmented across businesses, many of them being small or medium.

3.4. Construction is prone to information asymmetry

Communication develops the relationships necessary for successful project's outcomes (Project Management Institute, 2017). However, because of self-interest, the involved parties will not be willing to share all the information all the time (Ceric, 2013). This results in information asymmetry. Schieg (2008) describes information asymmetry as a situation in which one of the two co-operation partners is better informed than the other one.

4. Discussion of construction platformization

The current state of the art technology in construction digitalization is related to building information models (BIM) – creating and managing digital replicas of the future projects and current assets. Some elements of BIM technology are understood as platforms on which others can build. For example, Buildingsmart in its OpenBIM/phone analogy sees the IFC's as a platform on which others build applications (Buildingsmart, n.d.). However, this is not the entire relationship of BIM technology with platforms.

- *Information intensive:* BIM technology made construction even more information intensive.

- *Gatekeeping.* The BIM manager or the provider of the infrastructure for hosting BIM servers, BIM models, CDE's etc. are providing gatekeeping function to what information and tools are used when working on a project. Gatekeeping is seen as a feature, not bug, in projects which are considered properly organized.
- *Fragmentation.* While BIM addresses the vertical (1) and horizontal fragmentation (2) it does little or nothing about the fragmentation of the industry into multitude of projects (3) and multitude of companies (4).
- *Information Asymmetry.* While BIM at least theoretically improves the access to information within projects, this is not the case across projects and across the industry. A lot of information and knowledge is locked inside projects and companies without the opportunity to be found valuable across those contexts.

For the success of the construction platforms and as requirements towards its architecture the three success criteria need to be taken into account – connection, gravity and flow:

- *Connection:* The platform should focus on specific transaction such as designing, procurement or operation. It should be accessible and easy to use by all stakeholders, including the SMEs. It should grow out of a simple core that provides infrastructure for information, software tools, and people.
- *Gravity*: The network effect of a construction platform should be possible. Values should be created by users, the more the better. This is only possible if platforms cease to be project platforms (where more participants create more problems) but are industry platforms. Of course, mechanisms should be put in place, and this is the feature of the platform, that allow for intellectual property to be traded.
- *Flow*: The platform needs the right core transaction. This transaction in case of design platforms would be design information creation, at a request of a customer, done by the performer. This is the core element of the exchange of value of such platform and not the delivery of a project from lead designer to the client.

5. Conclusions

It was in the 1990s that Bill Gates quipped *"As we look ahead into the next century, leaders will be those who empower others." (Goodreads, n.d.).* Empowerment came in many forms and flavours, including more advanced services and tools. But the step change in this trend were services, digital infrastructures in fact, that allow others to build services and tools on top of it. This was then called digital platform. Providers of services, tools, software, information and content were empowered by the functionality and foundations a platform offered them.

We have argued, theoretically, that construction is suitable for the platform business model and listed some requirements towards success factors. To date, construction practitioners have been empowered by different kinds of software and tools. The industry was seeking optimizations in the use of these technologies withing projects. For example, the usage of BIM is advertised to generate savings of 20-50% (Kim et al., 2017). What has not been explored are the economics of switching of the industry from a project context to a platform context, from trying to save by using IT in a single project towards re-use of data, information, knowledge among projects and collaboration as well as specializations among the stakeholders outside of the project context. This is something that a platform could offer.
Future work is needed in several directions – architecture of such a construction platform, core functionality of such a platform, details of the business models, paradigm shifts in the industry and the changes to the overall value chains.

6. Acknowledgement

This research was conducted in part in the context of EU H2020 project Digiplace and Slovenian research program eConstruction (P2-0210).

7. References

[1] Agarwal, R. and Sridhar, S. C. M. (2016) 'Imagining construction's digital future | McKinsey'. Available at: https://www.mckinsey.com/industries/capital-projects-and-infrastructure/ourinsights/imagining-constructions-digital-future (Accessed: 1 June 2020).
[2] Airbnb Inc. (2020) Airbnb Newsroom - Fast Facts, Press.atairbnb.com. Available at: https://news.airbnb.com/fast-facts/ (Accessed: 1 June 2020).
[3] Borrmann, A. et al. (2014) Building information modeling, Building Information Modeling. doi: 10.4324/9781315797076.
[4] Bourdeau, Luc. "Presentation "The key role of ECTP and collaboration with other ETPs"." ECTP and E2BA Secretary General (2009).
[5] Buildingsmart. "How adopting the open standards approach to BIM will mean you are always a winner - buildingSMART International." Buildingsmart, n.d., https://www.buildingsmart.org/adopting-open-standards-approach-bim-will-mean-always-winner/.
[6] Ceric, A. (2013) 'The principal-agent theory and the role of project managers in construction: Guidelines for future research', Proceedings of the Joint CIB International Conference, Management of Construction: Research to Practice, pp. 766–776. Available at: https://www.irbnet.de/daten/iconda/CIB_DC25687.pdf.
[7] Choudary, P. S. and Bonchek, M. (2013) 'Three Elements of a Successful Platform Strategy', Harvard Business Review, pp. 1–4. Available at: https://hbr.org/2013/01/three-elements-of-a-successful-platform (Accessed: 29 March 2020).
[8] Choudary, S. P., Van Alstyne, M. and Parker, G. G. (2016) 'Pipelines, Platforms, and the New Rules of Strategy', Harvard Business Review, 94(4), pp. 54–62. Available at: https://hbr.org/2016/04/pipelinesplatforms-and-the-new-rules-of-strategy (Accessed: 28 March 2020).
[9] Christodoulou, S. (2017) eWork and eBusiness in Architecture, Engineering and Construction: ECPPM 2016, eWork and eBusiness in Architecture, Engineering and Construction: ECPPM 2016. doi: 10.1201/9781315386904.
[10] Cusumano, M. A., Gawer, A. and Yoffie, D. B. (2019) 'The Business of Platforms: Strategy in the Age of Digital Competition, Innovation, and Power'.
[11] Das, Moumita, Jack CP Cheng, and Srinath S. Kumar. "Social BIMCloud: a distributed cloud-based BIM platform for object-based lifecycle information exchange." Visualization in Engineering 3, no. 1 (2015): 1-20.
[12] EUBIM Task Group (2016) 'Handbook for the introduction of Building Information Modelling by the European Public Sector', EUBIM Task Group, p. 84.
[13] Evans, D. S. and Schmalensee, R. (2016) Matchmakers: The New Economics of Multisided Platforms., Harvard Business School Press Books.
[14] Goodreads, n.d. A quote by Bill Gates. Retrieved 4 August 2020, from https://www.goodreads.com/quotes/12422-as-we-look-ahead-into-the-next-century-leaders-will
[15] Goulding, Jack Steven, Farzad Pour Rahimian, and Xiangyu Wang. "Virtual reality-based cloud BIM platform for integrated AEC projects." Journal of Information Technology in Construction 19 (2014): 308-325.
[16] Hua, G. B. (2013) Implementing IT Business Strategy in the Construction Industry, Implementing IT Business Strategy in the Construction Industry. IGI Global. doi: 10.4018/978-1-4666-4185-3.
[17] Katranuschkov, Peter, Raimar J. Scherer, and Ziga Turk. "Intelligent services and tools for concurrent engineering? An approach towards the next generation of collaboration

platforms." Journal of Information Technology in Construction (ITcon) 6, no. 9 (2003): 111-128.

[18] Kim, J. and Yoo, J. (2019) 'Platform growth model: The four stages of growth model', Sustainability (Switzerland), 11(20). doi: 10.3390/su11205562.

[19] Kim, Seongah, Sangyoon Chin, Jintaek Han, and Cheol-Ho Choi. "Measurement of construction BIM value based on a case study of a large-scale building project." Journal of Management in Engineering 33, no. 6 (2017): 05017005.

[20] Laine, E. et al. (2017) 'Platform ecosystems: Unlocking the subcontractors' business model opportunities', IGLC 2017 - Proceedings of the 25th Annual Conference of the International Group for Lean Construction, II(July), pp. 177–184. doi: 10.24928/2017/0325.

[21] Ma, Zhiliang, Dongdong Zhang, and Jiulin Li. "A dedicated collaboration platform for Integrated Project Delivery." Automation in Construction 86 (2018): 199-209.

[22] McKinsey Global Institute (2017) 'Reinventing Construction: A Route To Higher Productivity', McKinsey & Company, (February), p. 20. doi: 10.1080/19320248.2010.527275.

[23] Moazed, A. and Johnson, N. (2016) Modern Monopolies : What It Takes To Dominate The 21st-Century Economy. New York, N.Y. : St. Martin's Press, [2016].

[24] Oxford University Press (2020) Platform | Definition of Platform by Oxford Dictionary on Lexico.com also meaning of Platform. Available at: https://www.lexico.com/en/definition/platform (Accessed: 12 July 2020).

[25] Parker, G., Van Alstyne, M. and Choudary, S. (2016) Platform revolution: How networked markets are transforming the economy and how to make them work for you, W.W. Norton & Company.

[26] Project Management Institute (2017) A guide to the project management body of knowledge (PMBOK guide) / Project Management Institute. 6th edn. Newtown Square, PA: Project Management Institute, 2017.

[27] Railway platform - Wikipedia (2020). Available at: https://en.wikipedia.org/wiki/Railway_platform (Accessed: 12 July 2020).

[28] Ruggieri, R. et al. (2018) 'The impact of Digital Platforms on Business Models: An empirical investigation on innovative start-ups', Management and Marketing, 13(4), pp. 1210–1225. doi: 10.2478/mmcks-2018-0032.

[29] Singh, Vishal, Ning Gu, and Xiangyu Wang. "A theoretical framework of a BIM-based multi-disciplinary collaboration platform." Automation in construction 20, no. 2 (2011): 134-144.

[30] Sommerville, J. and Craig, N. (2006) Implementing IT in construction, Implementing IT in Construction. doi: 10.4324/9780203030394.

[31] Turk, Ž. (2016) 'Ten questions concerning building information modelling', Building and Environment, 107, pp. 274–284. doi: 10.1016/j.buildenv.2016.08.001.

[32] Turk, Žiga. "Interoperability in construction–Mission impossible?." Developments in the Built Environment 4 (2020): 100018.

[33] Turk, Žiga. "Structured analysis of ICT adoption in the European construction industry." International Journal of Construction Management (2021): 1-7.

[34] Tsai, Chung-Lin (2018) The timing of fostering complementary innovation: exploring the antecedent of industry platform emergence, Technology Analysis & Strategic Management, 30:10, 1121-1135, DOI: 10.1080/09537325.2018.1442924

IOP Conf. Series: Materials Science and Engineering 1218 (2022) 012030 doi:10.1088/1757-899X/1218/1/012030

Process modelling for managing digitalised information exchanges in construction inspections and quality checks

S Comai[1], S Mastrolembo Ventura[2], S Costa[1] and A Ciribini[3]

[1] Graduate Research Assistant, University of Brescia, Brescia, Italy
[2] Post-doc Research Fellow, University of Brescia, Brescia, Italy
[3] Full Professor, University of Brescia, Brescia, Italy

E-mail: sara.comai@unibs.it

Abstract. Processes of the Architecture, Engineering and Construction (AEC) industry are known to be characterised by a high level of complexity in terms of activities and involved actors, which requires a consistent exchange of information through each phase of a project life cycle. A digital transition applied to information modelling and management may support the optimisation of such processes. To overcome the traditional paper-based and document-centric approach, the so-called Common Data Environment (CDE) has been introduced and its adoption through the project life cycle is now regulated by ISO 19650. A CDE is a platform, whose purpose is to collect, manage and disseminate sets of information through a managed, interoperable and collaborative process. The fullest adoption of CDEs requires a methodological change towards data-based information exchanges. As a preliminary step to such a goal, the formalisation of procedures by process mapping and modelling can allow a more coherent management of use cases and related information exchanges. With a focus on the construction stage, the formalisation of inspection procedures is proposed through the Business Process Model and Notation (BPMN 2.0) format. Quality checks are taken into account in relation to the supply and installation of an External Thermal Insulation Composite System (ETICS). Quality control inspections and related activities are analysed and discussed within the framework of public works. An open source BPMN modeller and its workflow engine have been adopted. The result is a process model that (1) represents the complexity of managing information exchanges in such a scenario; (2) can be used as a basis for the development of a CDE where all the information is shared and accessible to the various actors of the construction process; (3) automatically feed, as a future work, an inspection management support system.

1. Introduction

In recent years, the construction industry, and in particular the public sector [1], has begun to investigate new innovative strategies based on digital approaches to improve information management during the entire project life cycle. Digital innovations like Building Information Modelling (BIM), in fact, could offer considerable advantages in the construction industry over the conventional methods, document-centric and paper-based ones, that still largely prevail in current practice [2] [3]. The construction sector, being an industry which considers multiple players, requires a high degree of coordination to manage a complex construction process that involves a high exchange of information.

The information flows originated by the building process, are the result of the elaboration of multiple datasets. These data are generated in several phases and by different actors, which work to achieve a specific objective. The transition from different phases to another not only identifies a project development but also identifies changes of actors, objectives, profit margins, information requirements, levels of detail, methods and tools for processing information. These changes introduce a high level of difficulty in the management of information flows and make the interoperability among different actors inefficient.

In order to better organise information flows, methodological standards have been introduced:

- ISO 19650-1 / 2 [4] [5] standard defines concepts regarding the organisation and digitisation of regulatory flows;
- UNI 11337-5 [6] defines information management in the adoption of BIM methods and tools.

Within such a context, process mapping and process modelling may support better understanding of complex systems. In fact, the adoption of these approaches in construction projects, especially if combined with BIM methodologies, is more and more discussed in the industry as a method to provide valuable insights to optimise data flows, to gather information and to share knowledge during construction processes through the life cycle management of a project. Gardini et al. (2020), for example, proposed the process modelling approach for the management of the inspection of execution processes in construction [7]. Moreover, a clear overview of construction processes and workflows is also needed as a preliminary step to guarantee the efficiency behind the implementation of so-called Common Data Environments (CDE). A CDE is a platform whose purpose is to collect, manage and disseminate sets of information through a managed, interoperable and collaborative process [4]; its adoption is now managed by the EN ISO 19650 series.

ISO 29481-1:2017 [8] recommends the Business Process Modelling Notation (BPMN) as an approach for representing process maps. The purpose of process maps is "to describe the flow of activities within the boundary of a particular business process, the roles played by the actors involved together with the information required, consumed and produced". A process map includes the exchange requirements that are within the boundary of the process as well as an overview that provides a comprehensive description of the overall process. BPMN is an international standard [9] that is maintained by the Object Management Group. It "provides a means of graphically representing processes in a formal manner thanks to various standardised node and edge elements available" [2] and in the AEC context, BPMN is "often used to visualise the processes of model creation, collaboration and data exchange" [2].

With a focus on the construction stage, the paper evaluates the adoption of a process mapping and modelling approach as an adding value to the BIM-based management of a construction procedure. Moreover, the formalisation of inspection procedures is within the scope of this paper and it is proposed through the BPMN 2.0 format. In particular, quality checks are taken into account in relation to the supply and installation of an External Thermal Insulation Composite System (ETICS). Quality control inspections and related activities are analysed and discussed within the framework of public works. An open source BPMN modeller and its workflow engine have been adopted to execute the developed processes. Main findings and future works are finally discussed.

2. Proposed approach

The paper shows the preliminary results of an on-going research project on the adoption of a digital approach to the management of construction processes. In particular, with a focus on public works, the objective of the research is the development of an executable process to semi-automatically feed common data environments and manage information exchanges, providing support for the various stakeholders involved.

The proposed approach is based on four main steps: (1) analysis of normative texts, guidelines and other documentation materials useful to (2) map a construction process in terms of actors involved, activities to be carried out, data to be shared and documents to be produced. (3) Process modelling is then used to show the sequence of events, activities/tasks that occur in the process, as well as stakeholders involved and data to be exchanged. The modelling phase has been developed implementing the BPMN 2.0 notation tools and adopting an open source process modeller and workflow engine (i.e., Camunda Modeller). Finally, (4) in the process execution phase a preliminary attempt to translate the representation of the process into an operational one has been developed.

3. Sample use case: the supply and installation of an External Thermal Insulation Composite System (ETICS)

The formalisation of procedures by process mapping and process modelling can allow a more coherent management of use cases and related information exchanges. With a focus on the construction stage, the formalisation of inspection procedures is proposed and quality checks are taken into account. EN ISO 9000:2015 defines 'quality' as the 'degree to which a set of inherent characteristics of an object fulfils requirements", which are defined as a 'need or expectation that is stated, generally implied or obligatory' while 'a specified requirement is one that is stated, for example in documented information' [6]. If the specified requirements are not met, EN ISO 9000:2015 speaks of non-compliance or errors [2].

Within the scope of this paper, digitalised information exchanges in construction inspections and quality checks are managed in relation to the supply and installation of an External Thermal Insulation Composite System (ETICS).

3.1. Analysis of normative texts and guidelines

Normative texts and guidelines for the installation of ETICS in a workmanlike manner have been analysed in order to identify process phases and activities [10] [11] [13]. The Cortexa manual [11] and notebooks [12] have been taken as main references. Cortexa is a founding member of the European Association 'Manufacturers of Thermal Insulation Systems (EAE)'. The Cortexa manual is inspired by the best European practices, described in the EAE Manual, and has been adapted to Italian market needs by the Cortexa Technical Commission. The new 2020 edition of the Cortexa manual for the application of the External Thermal Insulation Systems (ETICS), specifying the design, the installation procedures and the certification needed by the installers. The goal of these documents is to describe the necessary conditions which have to be applied to carry out the activities in the best possible way. These considerations are even more important considering the fact that the benefits of this energy efficiency system are achieved only if the installation work is performed in a workmanlike manner.

Another standard that has been analysed is the UNI/TR 11715 'Thermal insulation products for buildings - Design and installation of thermal insulation systems for the exterior (ETICS)' which describes in a precise and rigorous way how to proceed in the correct design and installation of state-of-the-art thermal insulation systems.

3.2. Process mapping

A process map is characterised by a set of interrelated activities that arise from an input and provide an output. outputs and inputs can be represented by a product, documents or data. Along the execution of a process, many figures with different skills are required. Moreover, all these players are supposed to collaborate and interact in order to achieve the set goal. Thanks to the analysis of the specific texts indicated above, it has been possible to identify all the different actors involved in the executive process of installation of External Thermal Insulation Systems. Particularly, actors taken into consideration are three: the executor (the company or a subcontractor), the works management office (represents the client) and the supplier (manufacturer or distributor).

Subsequently, all the activities that each actor is supposed to carry out in order to lay a ETICS in a workmanlike manner were identified specifying, at the same time, all the checks points that must be accomplished along the process. Each check has been defined to have two different possible answers, a positive and a negative one. In case of a positive answer the transition to the next processing is allowed otherwise, a non-compliance has to be resolved before proceeding.

The activities that have been identified can be divided into activities that have to be carried out, communication activities, document creation activities and document/communication reception activities. Documents and data that are produced throughout the entire process are collected in the

IOP Conf. Series: Materials Science and Engineering 1218 (2022) 012030 doi:10.1088/1757-899X/1218/1/012030

Common Data Environment (CDE), which represents the central archive where information on the construction project are stored. The CDE is an environment constantly updated and its use improves collaboration, security and data verifiability.

3.3. Process modelling

The modelling of an ETICS supply and installation procedure has been developed using the BPMN 2.0 in order to represent the sequence of steps as a series of activities, events and sequence flows in a process model.

First, the entire process has been mapped at a high level (non-executable process), allowing the generation of a digital representation of the activities to be carried out chronologically. The map is made up of a macro and micro processes, such as the resolution of non-conformities.

The elements of BPMN used most frequently are sequence flow, pool, lane, task/activity, gateway, and message flow [14]. In this case, the notation tools that have been implemented are:

- pools: three pools have been created, one for each actor, and one for CDE;
- events: they represent the beginning and the end of each actor's process;
- activities: represents the activities in chronological order that must be carried out;
- gateway: was used to check the divergences and convergences of the sequence flow;
- data object: this tool has no effect on the sequence flow but represents data and documents produced by the activity.

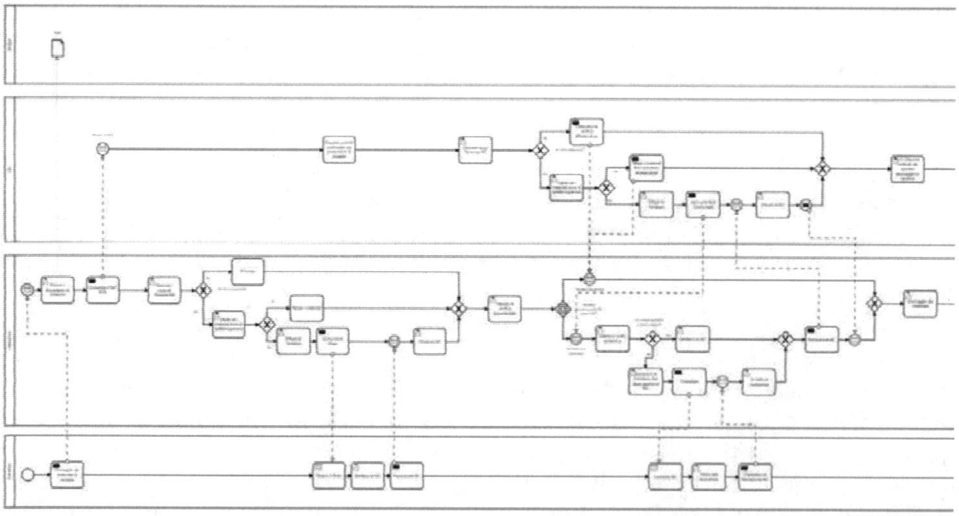

Figure 1. Part of the process map modelled in Camunda Modeler [15] with the BPMN 2.0 notation

3.4. Process execution

A process can be modelled to be formalised and visualised. Moreover, it can be also executed automatically once generated 'using a workflow engine that performs the process step by step and in accordance with the modelled logic' [2]. The workflow engine can also 'react to events during runtime as well as trigger events itself. Such an event might be a data input by the user, the calculation of mathematical formulae, a decision resulting from an if-then condition, or the execution of text-based source code' [2].

Once the modelling of the ETICS supply and installation process was obtained, it has been possible to verify the correctness of the modelling itself by implementing with some properties each notation tool used to make the process executable. Each notation tool can be customised using the properties panel; this panel can be used to assign the activities marked as 'User Task' to the various actors. Additionally, the properties panel allows the identification of the branches dictated by the gateway, and setting the sending and receiving of communications associated with the activities.

After completing the enrichment of the map, it has been possible to verify the correctness of the representation and automatically retrace the process from start to finish. The only tasks that required actors input to be complete were 'User Task'.

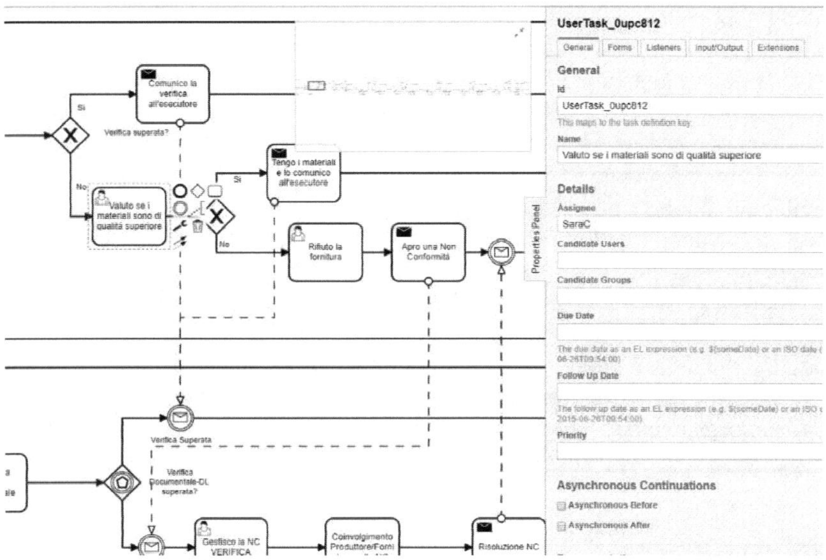

Figure 2. Customisation of "User task" activities by assigning the activity to the actor involved in the process

4. Results and conclusions

The research project has investigated the applicability of the BPMN 2.0 notation in the construction sector as a way to model processes with the aim to develop a process-oriented digital quality assessment procedure. Based on the results obtained, the potential and functionality of the BPMN format in process modelling and automation for the construction industry are confirmed.

Mapping and modelling quality control inspections and related activities for the supply and installation of an external thermal insulation composite system has allowed the identification of all the different layers of complexity of the process, highlighting the critical steps with a high probability of a non-conformity generation. The content validation of the process map is currently under development based on semi-structured interviews with experts in order to evaluate its comprehensiveness and to identify possible missing factors.

Moreover, the execution of the process model in an open source workflow engine is a first step towards the goal to automatically feed an inspection management support system.

Finally, as a future work, the process model could be also used as a basis for the development of a CDE where all the information is shared and accessible to the various stakeholders. Furthermore, future activities will include the further specification of the level of information need for information exchanges in the analysed process and the direct link of the process model with the related building information models and additional information containers within the framework of a common data environment.

Acknowledgements
The authors would like to acknowledge Alice Gardini and Marco Alvise Bragadin for the useful discussion about the topic and the collaboration to the research project. The research is funded by the Italian research grant PRIN 2017 "A distributed Digital collaboration framework for small and medium-sized engineering and construction enterprises".

References
[1] Directive 2014/24/EU of the European Parliament and of the council of 26 February 2014 on public procurement and repealing Directive 2004/18/EC
[2] Häußler M, Esser S and Borrmann A 2021 Code compliance checking of railway designs by integrating BIM, BPMN and DMN *Automation in Construction* 2020 doi: 10.1016/j.autcon.2020.103427
[3] Opitz F, Windisch R and Scherer R J, 2014 Integration of document – and model – based building information for project management support *Creative Construction Conference* 2014 pp. 403-411, doi: 10.1016/j.proeng.2014.10.566.
[4] EN ISO 19650-1:2018 Organization and digitization of information about buildings and civil engineering works, including building information modelling (BIM) - Information management using building information modelling - Part 1: Concepts and principles.
[5] EN ISO 19650-2:2018 Organization and digitization of information about buildings and civil engineering works, including building information modelling- Information management using building information modelling - Part 2: Delivery phase of the assets.
[6] UNI 11337-5:2017 Construction and civil engineering works - Digital management of information processes of construction - Part 5: Information flows in digitized processes.
[7] Gardini A, Bragadin M A, Naticchia B, Carbonari A and Corneli A 2020 BPMN 2.0 modelling for the management of the inspection of execution processes in construction Creative *Construction e-Conference* 2020 pp. 37-41 doi:10.3311/CCC2020-073.
[8] ISO 29481-1:2017 Building Information Models – Information delivey manual – Part 1: Methodology and format
[9] ISO/IEC 19510:2013 Information technology – Object management group Business Process Model and Notation
[10] UNI 11715:2018 Thermal insulation products for buildings – Design and implementation of thermal insulation system for the exterior (ETICS)
[11] Cortexa thermal insulation manual. https://www.cortexa.it/manuale-cappotto-termico-cortexa/ (last accessed: September 2020)
[12] Technical notebooks Cortexa on the thermal insulation for designers. https://www.cortexa.it/quaderni-tecnici-sul-cappotto-termico-per-progettisti/ (last accessed: September 2020)
[13] EN ISO 9000:2015 Quality management systems – Foundamentals and vocabulary
[14] Park Y, Cho C Y and Lee G, 2011 Identifying a subset of BPMN for IDM development Conference: CIB W78-W102 Joint Conference at France
[15] Camunda modeler and camunda engine. www.camunda.com (last accessed: February 2021)

Reflections on the Performance of South Africa's Construction Industry: Hope Beyond Covid-19 Effects

K Dithebe, W D Thwala, C Aigbavboa and B Madumelane

SARChl in Sustainable Construction Management and Leadership in the Built Environment, Faculty of Engineering and the Built Environment, University of Johannesburg, Johannesburg, South Africa

khdithebe@gmail.com

Abstract. South Africa's economic activity has declined in recent years owing its troubles to years of policy uncertainty, bad governance, corruption and limited budget and finance for the development of infrastructure. The woes are no different to the ones experienced by the construction industry. It must be categorically stated that the woes currently experienced by the industry did not begin with the effects of the coronavirus pandemic. The pandemic was simply the last nail to the coffin. The study thus provides a clear reflection on the dynamics and challenges that were encountered by the industry, with the hope of resuscitating it to what is used to be – a key sector in growing South Africa's economy. A quantitative approach was adopted. Reviewed literature formed basis of secondary data whilst the distribution of a close-ended questionnaire to construction professionals, economists and researchers formed the basis of primary data collection. Snowballing sampling technique was employed. A total of 77 questionnaires were returned. SPSS was used to analyse data using mean item scores to determine the level of agreement between respondents. From the results it was clear that corruption and fraudulent activities contributed significantly to the progression of South Africa's construction industry. Unregulated adoption of technology, political uncertainty and non-compliance to laws and safety standards emerged as key factors affecting the productivity of the industry. The adoption of advanced technologies such as robots and automated machines need to be formally regulated by government to be inclusive, this requires inclusive procurement practices to SMMEs and policy certainty and decisiveness from government. Government must be able to inform or direct business, firms who do not comply state departments must impose fines or sanctions (from bidding for government projects in the future).

Keywords: construction industry, corruption, economic decline, investor interest and confidence, policy certainty and decisiveness.

1. Introduction

The growth forecast for 2016 for South Africa's gross domestic product (GDP) was revised down to 0.9 percent from an estimated 1.7 percent during the Medium-term Budget Policy Statement (MTBPS) [1]. The weaker outlook resulted from lower commodity prices, higher borrowing costs, lower business and consumer trust, and drought. South Africa's economy is quite varied, questioned by poor financial development at 0.3 percent, elevated unemployment at nearly 27 percent, inflation at 6.1 percent, inequality and rates of poverty [1] at a high. In 2018 South Africa has further experienced negative economic growth in the first and second quarter, the consensus for the country's economic growth reached 0.7% [2]. Cook [3] and Watermeyer and Phillips [4] highlight that the country's performance has been damaged by factors such as poor mismanagement of key policy areas, state capture, corruption, policy uncertainty and the rise of business weakening the national competitiveness. These factors have worn off the confidence of the investors to invest in the country's new projects. From the abovementioned statistics South Africa's construction industry has also followed in the same footsteps with its own dynamics and challenges of recurring economic decline.

Statistics South Africa [5] indicates that the government of South Africa is the major construction client, covering between 40 to 50% of the domestic expenditure. StatsSA [6] indicates that the construction industry in 2014 had the largest turnover and employed 1.4 million people nationwide, contributing positively to lowering the unemployment rate and poverty. The success of the industry continued in the first quarter of 2015, with a contribution of 4% to the economic growth [7]. However, since then things have taken a sharp turn in recent years, adding to the economic decline of South Africa's economy in 2020, with its sixth consecutive quarter of economic decline [8]. South Africa Chamber of Commerce and Industry (SACCI) [9] opines the problems of the sector to limited budget from the national government. The study concurs, however asserts that good governance, policy certainty and decisiveness, business confidence, sustainable construction, restructuring and inclusive procurement practices as well as technology and innovation are attributes to a stable and productive construction industry, the study dissects and reflects on the dynamics and challenges that led the industry to the state it is today, before the coronavirus pandemic.

2. Dynamics and factors affecting the progression of the construction industry

The contribution of South Africa's construction industry in the economic growth of the country has not been positive and has affected the delivery and maintenance of infrastructure [5]. Construction firms are forced into filing for bankruptcy whilst others are operating under business rescue [10]. Amongst many other failures experienced in the industry is failure to complete projects within the agreed schedule, period and against the required quality standards. Contractors responsible for the failures are later required from time to time to rectify work done which in turn increases the cost of the project resulting both in cost and time overruns.

The industry's woes have been complicated further by the declining activity and productivity in the industry as a result of the coronavirus disease 2019 (Covid-19). Covid-19 pandemic is a respiratory disease that affects the lungs of individuals who have contracted it and can result in fatalities of victims mostly with underlying health issues [11], thus to limit the spread of the virus many sectors and industries have limited operations and are operating under lockdown regulations. The construction industry is one of the sectors that for some time ceased operations but is now also operating under strict lockdown regulations in South Africa.

2.1. Policy regulation and governance

Ofori [12] has defined laws and regulations as factors that guide the construction industry in any country and must include insurance requirements, specifications, defects liability duration and safety and health provisions. However, with so much laws to adhere to overregulating industries remains a detrimental issue in the construction sector. The Construction Development Industry Board report [13] suggests that more than 1,000 pieces of legislation have since 1994 been passed by the South African government, which in turn has created numerous regulatory requirements that give the impression of overregulation. These laws affect contractual and contracting procedures, labour and employment practices, permits to plan and supervise, training, and business practices. As a result, the development of approvals and zoning processes of local authorities are delayed and lead to unnecessary holding costs for developers [13].

Issues of overregulating and non-compliance in the construction industry remains a challenge for progression. In view of local laws and laws pertaining to corporate management, employment, taxes and health, and security conditions, there is a danger of non-compliance according to Price Water House Coopers [14]. Thus measures amongst many that addresses non-compliance include imposing fines and sanctioning companies from bidding for specific projects. It is also true that the issue of non-compliance in some instances is a result of policy uncertainty and indecisiveness from regulatory bodies.

According to the Construction Industry Development Board [13], public-sector capacity is a key constraint on the delivery of infrastructure and sustainable growth in South Africa' construction

IOP Conf. Series: Materials Science and Engineering 1218 (2022) 012031 doi:10.1088/1757-899X/1218/1/012031

industry. The public sector does not have the necessary capacity in terms of resources, skills and technology to undertake projects. Amongst the solutions to limited public sector capacity has been to approach the private sector and share responsibilities with them. However, Milford [15] posits that the public sector has continued to fail to fulfil its own mandated duties which has led to inefficient undertaking of processes of funding construction projects, leading also to corruption and backlogs of payments to contractors.

As another key responsibility of the public sector the limited availability of suitable land to commence with development has also indicated the inadequate capacity of the public sector to avail resources. Boshoff [16] further emphasizes that private land is not readily available in South Africa, adding that there are numerous land claim issues in the courts, zoning issues, and heritage sites, all of which combine to make the price of existing land inhibitive, in that way delaying development processes.The title is set 17 point Times Bold, flush left, unjustified. The first letter of the title should be capitalized with the rest in lower case. It should not be indented. Leave 28 mm of space above the title and 10 mm after the title.

2.2. Investor interest and confidence

Investor interest and confidence in any country is immensely determined by the stability of a country, this can range from policy certainty and good governance. Lack of decisiveness and implementation from government and multiple reshuffling of cabinet ministers also contributes to limited interest and confidence of investors towards directing funds in the growth of the economy. South Africa has over the years experienced years of uncertainty, which directly affects investor confidence [9], this is as a result of the existing political conditions and the poor decision-making behind the operations of many parastatals such as Eskom and South African Airways.

Dlamini [17] and Dithebe et al. [18] has alluded that the contribution of the construction industry in the declining economic growth of South Africa has immensely affected the confidence of the investors to finance and take interest into the country's infrastructure development. Reduced funding on the delivery of infrastructure has not only affected the business sectors of the country [19] but has also affected the social responsibility of the state to deliver the much-needed services to the public.

2.3 Productivity and challenges of construction firms

In reflecting on the dynamics of the construction industry it is consequential that the basic units of the industry are looked into, the productivity of the construction firms. Productivity is one of the most significant factors affecting the overall performance of any organization, big or small [20]. The productivity of an organisation can be realised in many ways, through restructuring, inclusive procurement practices and a labour workforce that is mostly skilled, as well as adopting advanced technological methods and equipment. Organisations restructures to protect the business – reducing costs, decreasing debt and improving competitive advantage [21]. Companies undergoing restructuring as a turnaround strategy may be forced to release employees from their duties [22]. In the current state, companies are confronted with either preserving the business through inevitable retrenchments or filing for bankruptcy, well aware that an increase in the unemployment rate also affects the growth of any economy. Employees or workers that are likely to face retrenchment are the ones without skills, referred to as general workers. Though there is a need for general workers the industry is also faced with a shortage of skilled labour. Windapo [23] reports that the skilled labor force in the construction sector of South Africa is in shortage. The mismatch or the gap between available skills and required skills is too wide to an extent that illegal/legal skilled immigrants are hired over unskilled nationals [24]. Makhene and Thwala [25] opines that the high quantity of industry participants who have no education, let alone a degree, are in a serious obstacle to the improvement of the construction industry, hence the need for skilled labourers.

The already employed workforce is influenced by industrial actions and community communication officers in private contracts and by council members in the public sector contract works, these two factors have affected labour and firm productivity [13]. Specifically, strikes by the none union members have greatly affected the profit margin of a project as a maximum number of working days in a project are lost due to labour strikes.

The CIDB [13] report observes that there is a shortage of skills not only in the construction sector but within the South African skills sector and in state-owned enterprises. The report further argues that mentors are required to provide developing contractors with support on-site to link skills gaps. The Department of Public Works [26] continues to contribute to high-level human capital development and to the creation of a pool of critical skills through the Expanded Public Works Programme (EPWP). The programme is aimed at changing the profile of the workforce by growing a representative workforce for the built and property environment, while growing the size and raising the quality of the human capital base needed to add to the skills pool and improve the quality of life of all South Africans [26].

In the process of uplifting and empowering the unskilled workers the inclusivity of small medium-micro enterprises (SMMES) must be considered, this can be achieved through improved and inclusive procurement practices that favours SMMES. According to Masterman [27] and Thwala and Mathonsi [28] the procurement system is an organizational structure or method of undertaking a project approved by the client for the application and at times eventual operation of a project. The current method of procurement used in South Africa has resulted in unhealthy levels of rivalry and hinders the development of small enterprises [29], the methods are also susceptible to manipulation and corruption. The public procurement legislation and the Construction Industry Development Board Act (CIDB Act) 32 of 2005 must be scrutinised to be more inclusive and less vulnerable to manipulation by big firms.

In the productivity of firms another important aspect is the use of advanced technology. The construction industry in the adoption of digitalised systems is still in its early stages, alludes Osunsanmi et al. [30]. However Oke et al. [24] argues that advanced technology in the form of digital collaboration is gaining traction in South Africa's construction industry. It is indeed true that advanced technology and innovation in South Africa's construction industry though not at an advanced stage is momentously being adopted and part of the firm's daily operations. South Africa has reasonable access to the latest technology [12], and this tends to limit the scope of the projects that can still be undertaken using traditional methods and human labour, increasing tensions between technology and labour relations. Construction firms are encouraged by government policy to employ more human labour to boost the economy and lessen poverty [19] than fully embracing the technological wave. However productivity is said to be slow without the usage of construction technology [31] and where projects involve and require relatively new technology, small contractors do not have the capability and resources to undertake them [32].

2.4 Investing in sustainable construction

According to Agenda 21 [33], South Africa is a developing country that experiences two significant problems namely limited natural resources such as water and the increasing pollution in the environment. These problems are caused by the construction activities employed to which the resultant is increased pollution and enormous consumption of energy. [34]. In this situation, and due to the increase in energy prices and shortage of water, the government set a vision to drive towards sustainable construction to ensure the importance of natural resources for the development plan and to improve the quality of life [35].

Huovila and Ritcheter [36] adds that the construction industry plays an important role as it is responsible for the infrastructure that economies require to maintain development. The development comes at a price as it exerts pressure on natural resources and this has had a serious impact on all living organisms and on the environment. The degradation of the environment has caught the world's attention and has been one of the most discussed subjects nationally and globally [37], the effects of the construction are felt throughout the projects' life cycle. The economic development of every nation has to focus more on the protection and sustainability of the environment. A lot of this attention has to be on the construction industry as it is a large consumer of resources and energy and its activities have a significant impact on the environment [38].

Another important aspect of sustainability which is ignored under the umbrella of sustainable construction is the health and safety of workers and the surrounding environment. The construction industry is alarmed with the health and safety not only of workers but also of the individuals who actually live or work around the construction project. Employee health and productivity are connected to the quality of the indoor environment, and that quality is largely determined by decisions made during project development and construction [39]. Smallwood and Haupt [40] states that though cost, quality and time are still seen as the significant indicators of a successful project, the health and safety aspect of the project is also a critical feature that must be prioritised. The safer the working environment, materials and equipment used, the better for the health of workers and the safety of the environment. The Department of Public Works report [26], through its statutory compliance, has continued with its health and safety campaign to inform and educate staff on the importance of health and safety in the workplace which is guided by the Occupational Health and Safety Act of 1993 (Act No. 85 of 1993). Please ensure that affiliations are as full and complete as possible and include the country. The addresses of the authors' affiliations follow the list of authors and should also be indented 25 mm to match the abstract. If the authors are at different addresses, numbered superscripts should be used after each surname to reference an author to his/her address. The numbered superscripts should *not* be inserted using Word's footnote command because this will place the reference in the wrong place—at the bottom of the page (or end of the document) rather than next to the address. Ensure that any numbered superscripts used to link author names and addresses start at 1 and continue on to the number of affiliations. Do not add any footnotes until all the author names are linked to the addresses. For example, to format

2.5 Construction corruption
Transparency International [41] states that the construction sector is the industry most affected and susceptible to corruption. Forms of corruption experienced are not limited to bribery, tender rigging, fronting, kickbacks and conflict of interests [42]. The nature of the construction industry, where the procurement of goods and services and the selection of contractors and suppliers on large-scale projects may be decided or influenced by individuals within an organization, provides a number of opportunities for manipulation and eventual corruption [43]. The continuation of corruption is amplified by a lack of confidence in the justice system to deal with corruption cases, lack of accountability and transparency and poor governance [43].

3. Methodology
The study assessed factors affecting the progression of South Africa's construction industry. The study adopted a quantitative survey approach to gather data from the nominated respondents through a structured close-ended questionnaire. The snowball technique was utilized because the researcher used a google form to distribute the questionnaire, a criteria was used to select the initial respondents. The questionnaire was divided into two parts. Part A deliberated on the background information of the respondents. Part B with the use of a 5-point Likert scale determining the level of agreement dealt with factors affecting the progression of South Africa's construction industry. The respondents were inclusive of economists, researchers in the Department of Construction Management and Quantity Surveying under the Sustainable Human Settlement and Construction Research Centre as well as the construction professionals in the construction industry, quantity surveyors, engineers, construction project managers and project managers. A total of 77 questionnaires were returned and fit for analysis. The reliability of the questionnaire (Part B) was tested and a Cronbach's alpha value of 0.939 was attained. A higher value, closer to +1 satisfies the reliability of the research instrument (Santos, 1999). In terms of data analysis a percentile approach in analysing Part A of the questionnaire was considered whilst in Part B mean item score (MIS) and standard deviation (SD) to rank factors affecting the progression of South Africa's construction industry were employed.

4. FINDINGS AND DISCUSSION
4.1 Background information
The results of the study on the background information of the respondents revealed that from a total of 77 participants 72.8% were between the ages of 25-39, whilst the remaining 27.2% were aged 40 and above. From the pool of respondents targeted it is evident that the target area has less professionals aged 40 and above which is good for the industry to invest more on young people. The respondents were also required to indicate their level of education, majority of the respondents with 42.9% had bachelor's degree, 28.6% had a diploma, 22.1% had a master's degree, 5.2% had a doctoral degree and 1.3% had

only a matric certificate and no formal higher education. The evidence that the respondents poses some form of education contributes to the study's reliability and ability to answer the questions in a manner that is acceptable and not skewed. More responses were obtained from quantity surveyors (36.4%), followed by construction project managers (26%), project managers (14.3%), researchers (10.3%), engineers (5.2%), architects and economists (each with 3.9%). The findings thus gravitates more on the perception of quantity surveyors and construction project managers on their view on the progression of South Africa's construction industry pre-covid-19 crisis. The background information of the study further revealed 48% of the respondents had experience in the construction industry ranging from 1-3 years, this finding is a direct reflection of a higher percentage of young people in the industry, which is also indicated under the section of respondents' age. Only 13% of the respondents had an experience in the range of 4-6 years, clearly revealing that as the age of participants increases so does the experience. It was also recorded that 21% of the respondents had experience in their respective fields ranging between 7-9 years, whilst 18% of the respondents had 10 years' experience and more in their respective fields. It is thus evident that the pool of respondents used for this study was a mixture of professionals with different levels of qualifications, experience, and age groups, with new entrees in different professions dominating. The dominance of new entrees to the industry can be attributed to sectors investing in the development and empowerment of young people.

4.2 Factors affecting the progression of South Africa's construction industry

The results of 'factors affecting the progression of South Africa's construction industry' are presented in Table 1. From the table it is evident that the factors have and average mean score of over 4.0. In the Likert scale employed in Part B of the close-ended questionnaire 4.0 represented 'Agree', clearly revealing that the respondents were in agreement on the detrimental impact of the factors affecting the progression of South Africa's construction industry. Prior to the agreement obtained it was further necessary to rank the factors in a descending order to determine their criticality. From the results corruption with a mean score value of 4.38 (SD=.744) was the factor above any other factor that has the most detrimental impact in the progression of the construction industry. The study agrees with Bowen et al. (2012) that construction corruption in the form of bribery, collusion, rigging, fronting and kickbacks affect the progression of the industry. Commissioned projects due to corruption suffer from cost and time overruns to a point where projects are abandoned, when the projects are not abandoned but complete inferior materials and methods are used, further affecting the sustainability (MIS=4.24; SD=.855) and maintenance of built-facilities. Transparency International [41] states that the construction industry is susceptible to corrupt activities, without measures regulating the susceptibility of the industry to corruption, corruption will always find expression with the industry.

Technology and innovation plays a significant role in the advancement and competitiveness of the industry locally and globally, this factor was ranked second with a mean score value of 4.35 (SD=.943) after corruption. The significance of technology cannot be over emphasised in the progression of any industry, however there are challenges delaying the full embracement of the aspect, fear to change, less use of human labour (MIS=4.31; SD=.979) and digital divide are amongst the issues. Though embracing technology is followed by productivity changes including restructuring (MIS=4.09; SD=.955) in the construction firms may affect progression, and fundamental questions remain unanswered whether these firms have the capacity and resources to embrace the change, whether there are plans to cater for the replaced human labour and most importantly the inclusivity of procurement practices to include small medium-micro enterprises (SMMEs) in the new wave. SMMEs are susceptible to digital divide due to limited resources and are likely to be excluded from the progression of the industry thus affecting the overall output of the industry.

Any form of development and innovation must be guided by the government of that country through political certainty and decisiveness. But the industry is subjected to indecisiveness and overregulation which then leads to non-compliance by the construction firms. The inability to show decisiveness and certainty has made observers to question the capacity of the state to carry out its responsibilities [15; 13]. Changes and indecisiveness on policies and regulatory frameworks contribute to the uncertainty and instability of any country's infrastructure development. The ability and capability to take strategic decisions and implement them is also informed by the capacity of the government institutions to handle responsibilities and be accountable. The study concurs with the Price Water House Coopers [7] and World Economic Forum [39] that changes in health and safety regulations (MIS=4.35; SD=.774), financial and labour policy and environmental standards interrupt planning and operational responsibilities of construction professionals. Indecisiveness and poor governance which has resulted in the country's instability has also affected investor interest and confidence. Lack of investor confidence in the country's ability to return investments limits any form of investment that can be made towards infrastructure development [9]. The study concurs with Dithebe et al. [18] and Dlamini [17] that limited project finance has had a negative impact on the performance of the construction industry and has thus negatively impacted South Africa's economic growth.

TABLE I. FACTORS AFFECTING THE PROGRESSION OF SOUTH AFRICA'S CONSTRUCTION INDUSTRY

Factors	$\bar{x}$	**σX**	**R**
Corruption	4.38	.744	1
Technology and Innovation	4.35	.943	2
Political uncertainty and indecisiveness	4.35	.900	2
Health and safety	4.35	.774	2
Inclusive procurement practices	4.31	.765	3
Investor interest and confidence	4.31	.782	3
Public- sector capacity	4.26	.834	4
Sustainable construction	4.24	.855	5
Regulatory and legislative framework	4.23	.841	6
Labour workforce and available skills	4.14	.979	7
Restructuring and productivity of construction firms	4.09	.955	8

5. Conclusion and recommendations

The study presented a case that the woes currently experienced by South Africa's construction industry must not be solely associated with the impact covid-19 pandemic has had in the industry due to limited movement and closure of firms in recent months. It must be categorically stated that the industry already and has consecutively made negative contributions to the country's economic growth, the pandemic just shined light to the existing crisis. But the question then remains, is there hope for the industry, even beyond the covid-19 crisis? The answer is yes, whilst the need for infrastructure development exists the need for a prosperous construction industry indeed still exists.

The study reflected on the economic decline of South Africa's construction industry with a clear view on the factors affecting the progression of the industry. The study revealed that any attempt on the resuscitation of South Africa's construction industry, even beyond the covid-19 crisis must pay particular attention to the industry's structural challenges including restructuring of firms and susceptibility to corruption, it cannot be that the culture of corruption is treated as a secondary if not a third contributor to economic decline if things are to change. The study further concurs that though the contribution of advanced technology cannot be disputed inclusive measures on the side of SMMEs are required to regulate its full adoption. Political certainty and decisiveness from the public sector cannot be emphasised enough, this has a direct impact to investor interest and confidence. An industry that is futuristic and protective of its future generations and environment is an industry that is developmental, sustainable and not afraid to change, South Africa's construction industry has potential in abundance to be that industry.

IOP Conf. Series: Materials Science and Engineering 1218 (2022) 012031 doi:10.1088/1757-899X/1218/1/012031

References

[1] National Treasury. (2016). Chapter 2: Economic Overview, 2016 Budget Review. Available at: http://www.treasury.gov.za/documents/national%20budget/2016/review/ [Accessed 19 July 2019].

[2] World Bank. (2016). Data, South Africa. The World Bank. Available at: http://data.worldbank.org/country/south-africa. [Accessed 18 July 2019].

[3] Cook, A. (2019). South Africa: Current Issues, Economy and U.S Relations. Congressional Research Service. Available at: https://fas.org/sgp/crs/row/R45687.pdf [Accessed July 19 2020]

[4] Watermeyer, R and Phillips, S. (2020). Public infrastructure delivery and construction sector dynamism in the South African economy. Department of planning, monitoring and evaluation, Pretoria, South Africa

[5] Statistics South Africa. (2017). Construction industry, 2017. Department: Statistics South Africa, Pretoria, South Africa

[6] Statistics South Africa. (2014). Construction industry, 2014. Department: Statistics South Africa, Pretoria, South Africa

[7] Price Water House Coopers. (2016). South African Construction Industry, 4th edition. Available at: https://www.pwc.co.za/en/publications/sa-construction.html [Accessed April 2019]

[8] Statistics South Africa. (2020). Category Archives: Economic growth. Department: Statistics South Africa, Pretoria, South Africa. Available at: http://www.statssa.gov.za/?cat=30 [Accessed July 19 2020]

[9] South African Chamber of Commerce and Industry (SACCI). (2016). Updated and Revised SACCI Business Confidence Index. Available at. http://www.sacci.org.za/images/stories/BCI/MEMO_BCI_UPDATE_2.pdf [Accessed 9 July 2019]

[10] Windapo, A.O and Cattell, K. (2013). The South African Construction Industry: Perceptions of key challenges facing its performance, development and growth. Journal of Construction in Developing Countries, 18(2): 65-79.

[11] Anjorin, A.A. (2020). The coronavirus disease 2019 (COVID-19) pandemic: A review and an update on cases in Africa. Asian Pacific Journal of Tropical Medicine, 13(5): 199-203

[12] Ofori, G. (1990). The construction industry aspects of its economics and management. Singapore University Press, Singapore City

[13] Construction Industry Development Board. (2014). Development through partnership, concrete, issue 1, July 2014. Available at: http://www.cidb.org.za/publications/Documents/Concrete%20Newsletter%20-%20Issue%201.pdf [Accessed July 19 2020]

[14] Price Water House Coopers. (2013). South African Construction Industry: Economic uncertainty levels have hit an all-time high. Available from: www.pwc.co.za [Accessed July 19 2020]

[15] Milford, R. (2010). Public capacity payment and procurement issues should be a challenge to the operations of contractors in South Africa. Paper presented at the Fifth Built Environment Conference of the Association of Schools of Construction of Southern Africa. Durban, South Africa, 18–20 July.

[16] Boshoff, S. (2010). The Cape Town 2030 initiative: A contested vision for the future. In Counter-Currents: Experiments in Sustainability in the Cape Town Region. Cape Town: Jacana Media (Pty) Ltd and African Centre for Cities, 70–79

[17] Dlamini, S. (2012). Relationship of construction sector to economic growth. School of Construction management and engineering, University of Reading, UK

[18] Dithebe, K., Aigbavboa, C., Oke, A and Muyambu, M.A. (2018). Factors influencing the performance of the South African construction industry: A case of Limpopo province. International Conference on Industrial Engineering and Operations Management, Pretoria/ Johannesburg, South Africa

[19] Construction Industry Development Board. (2004). SA Construction Industry Status Report; Pretoria, South Africa

[20] Kazaz, A and Ulubeyli, S. (2007). Drivers of productivity among construction workers: A study in a developing country. Building and Environment, 42(5): 2132-2140

[21] Deloitte. (2017). South Africa restructuring outlook survey results 2017: Seeing through the fog. Deloitte Touche Tohmatsu Limited, Johannesburg, South Africa

[22] Groenewald, D., Le Roux, S.J. and Rossouw, D. (2007). Strategic management: An applied South African perspective. Cape Town: New Africa Books

[23] Windapo, A.O. (2016). Skilled labour supply in the South African construction industry: The nexus between certification, quality of work output and shortages. SA Journal of Human Resource Management, 14(1):1-8

[24] Oke, A.E., Aghimien, D. O., Aigbavboa, C.O and Koloko, N. (2018). Challenges of Digital Collaboration in the South African construction industry. International conference on industrial engineering and operations management, Bandung, Indonesia

[25] Makhene, D and Thwala, W.D. (2009). Skilled labour shortages in construction contractors: A Literature Review. Construction Industry Development Board. Available at: http://hdl.handle.net/10210/5303 [Accessed July 11 2020]
[26] Department of Public Works report (2016). 2016/17 Department of Public Works Annual Performance Plan. Department: Public Works, Pretoria, South Africa
[27] Masterman, J. W. E. (2002). An introduction to Building Procurement Systems. 2nd edn. London: E & FN Spon.
[28] Thwala, W.D and Mathonsi, M.D. (2012). Selection of procurement systems in the South African construction industry: An exploratory study. Acta Commercii. Available at: https://www.researchgate.net/publication/307848017_Selection_of_procurement_systems_in_the_South_African_construction_industry_An_exploratory_study [Accessed July 13 2020]
[29] Construction Industry Development Board. (2019). Construction monitor: contractor development. Available at: http://www.cidb.org.za/publications/Documents/Construction%20monitor%20-%20July%202019.pdf [Accessed July 15 2020]
[30] Osunsanmi, T.O., Aigbavboa, C and Oke, A. (2018). Construction 4.0: The future of the construction industry in South Africa. International Journal of Civil and Environmental, 12(3): 206-212.
[31] Valverde-Gascuena, N., Astor, E.N., Fuentes-Del-Burgo, J and Ruiz-Fernandez, J. (2011). Factors that affect the productivity of construction projects in small and medium companies: Analysis of its impact on planning. Procs 27th Annual ARCOM Conference, Bristol, UK
[32] Kamar K. A. M and Hamid Z. A. (2011). Sustainable construction and green building: the case of Malaysia
[33] Agenda 21 (2002). Johannesburg Summit Full text. "Protecting and Promoting Human Health" Chapter 6, Summary. Available at: https://www.un.org/esa/dsd/agenda21/res_agenda21_00.shtml [Accessed July 20 2020]
[34] Ali, H. and S. F. Nsairat. (2009). Developing a Green Building Assessment Tool for Developing Countries-Case Study of Jordan. Journal of Building and Environment, 1053-1064
[35] Visser, F. (2009). Sustainable house in Aqaba, Jordan. Aqaba Residence Energy Efficiency (AREE). Online architectural magazine 2009. Available at: http://www.archimag.com/eng/aqaba_eng.php. [Accessed: May 25 2019]
[36] Huovila, P. and C. Richter. (2010). Life cycle building design in 2010. Schriftenreihe WDK
[37] Ametepey, S.O and Ansah, S.K. (2014). Impacts of construction activities on the environment: the case of Ghana. Journal of Construction Project Management and Innovation, 4(1):934-948
[38] Ge, Z. and Gao, Z. (2008). Applications of nanotechnology and nanomaterials in construction. First Inter. Confer. Construct. Develop. Countries, pp.235-240.
[39] World Economic Forum. (2016). Shaping the Future of Construction report. Available at: http://www3.weforum.org/docs/WEF_Shaping_the_Future_of_Construction_full_report.pdf [Accessed July 19 2020]
[40] Smallwood, J. J and Haupt, T.C. (2007). Impact of the South African Construction Regulations on construction health and safety. Journal of Engineering, Design and Technology, 5(1):23-34
[41] Transparency International. (2005). Transparency international annual report 2005. Available at: https://www.transparency.org/en/publications/transparency-international-annual-report-2005# [Accessed July 20 2020]
[42] Hartley, R. (2009). Fighting corruption in the Australian construction industry: The national code of practice. Leadership Management Engineering, 9(3):131-135
[43] Bowen, P., Edwards, P. and Cattell, K. (2012). Corruption in the South African construction industry: A mixed methods study In: Smith, S.D (Ed) Procs 28th Annual ARCOM Conference, 3-5 September 2012, Edinburgh, UK, Association of Researchers in Construction Management, 521-531

IOP Conf. Series: Materials Science and Engineering 1218 (2022) 012032 doi:10.1088/1757-899X/1218/1/012032

Relationship between real estate and stock market during the coronavirus crises in the Czech Republic

J Holcman and Z Prostějovská

Department of Economics and Management in Construction, Czech Technical University in Prague, Faculty of Civil Engineering, Thákurova 2077/7, 160 00 Prague 6, Czech republic

jakub.holcman@fsv.cvut.cz

Abstract. This paper deals with the relationship between the residential house price index and the stock markets at the time of the coronavirus pandemic in 2020. The research focuses primarily on the real estate market in the Czech Republic and the development of the Prague Stock Exchange index. The period from 2007 to the end of 2020 is monitored. The subject of the examination is not the assessment of the severity or type of crisis, only the development of markets and their shocks. House price index (HPI) and stock market indices are used for consistent analysis. To find a broader context the mortgage interest rate and the unemployment rate are used. With the help of correlation and analysis of implemented government measures, the basic conditions for the development of indices and measures are sought. Government measures are not only aimed at local conditions, but often also affect cooperation in international trade and traveling. The aim is to evaluate the degree of dependence of the development of real estate and stock markets on measures implemented to reduce the effects of the crisis. The reason is also the fact that there are many common variables (tradable banking companies, local and international investors, etc.).

1. Introduction

The current situation in the world is increasingly proving that people can adapt to challenging conditions. This is not only due to modern technology and the pace of its development. It is primarily the ability to acquire and evaluate information. In an instant, anyone can find out what is happening in a neighbouring city or on another continent, in a variety of sectors and topics. Also for this reason, in this article, seemingly unrelated sectors and shocks in the Czech Republic for 2020 are put in context. These are house prices (House Price Index), the stock market (PX Index) and information on the progress of the fight against the COVID-19 epidemic. Data on unemployment and interest rates are also added. The data are presented for illustrative purposes, including historical developments. The impacts of the spread of historical infectious diseases are compared, and COVID-19 has a greater impact than it did for SARS [1].

Before COVID-19 became a global problem and issue, it was difficult to imagine the impact on individuals, let alone national and international economies. The change in the behavior of the basic economic actors (investors and households) in the Czech Republic is already demonstrable with the data available and presented in this paper. In 2020 and beyond, the behavior of these actors, in the context of real estate and equity investment, could primarily be guided in two basic ways:

IOP Conf. Series: Materials Science and Engineering 1218 (2022) 012032 doi:10.1088/1757-899X/1218/1/012032

a) Psychological - fear of the unknown, the social aspect and the uncertain evolution of the economy (uncertain behavior of other subjects)

b) Legislative - compliance with government regulations and anti-epidemiological measures

By analyzing and comparing with recent results from abroad, the opposite trend was found, i.e. the increase in residential property prices. The development of the Prague Stock Exchange is approaching the world trends.

2. Literature Review

There are many studies on COVID-19 from a medical point of view and they are still growing. On the other hand, papers dealing with the impact on economies and the real estate market have been slower in coming. Studies in 5 major cities in Australian states have presented results of a 1.26-1.77% p.a. decline in property prices when the number of infected people doubles over a period of time [2]. A possible substitution effect is also discussed. The thesis of declining real estate prices in regions with higher rates of promotion or less developed medical care is confirmed by a Chinese study [1]. The decline in residential property prices is quantified at 2.47% over the next three months, further depending on the situation [1]. One of the earliest papers dealing with COVID-19 and house prices [3] describes high liquidity constraints and credit generation as a tool to counteract the rise in house prices due to the large amount of money in the economy, due to possible quantitative easing. Based on the calculations, the relationships between different actors in the industries in China have also changed in 2020 [4]. This fact also affects the real estate market, as it is dependent on industry. Construction output is transformed by industrial output into products tradable in the real estate market. Furthermore, changes in transport, storage and production constraints are described [4]. These changes are also associated with an acceleration of development in the commercial sector and a shift of customers to the online environment. Changes in the use of existing properties and land are addressed as the impact of COVID-19 on commercial buildings is significant [5].

The COVID-19 shock has a demonstrable negative impact on the high sensitivity of long-term bond yields to index volatility in the run-up to the pandemic [6]. Increasing liquidity through quantitative easing has a greater effect on stock performance in sectors that are more constrained due to the epidemic [7]. Within [7], the performance of U.S. industries are analyzed with the conclusions that the performance impact was caused by the announcement of quantitative easing and the restoration of investor confidence.

The literature contains analyses of the real estate market and COVID-19 or financial markets and COVID-19. The link between the house price index and the stock market is presented below.

3. Methodology and Data

The research used quarterly HPI and Unemployment data from the Czech Statistical Office [8] [9]. Furthermore, monthly values of the PX index of the Prague Stock Exchange [10] and monthly average interest rate of FINCENTRUM HYPOINDEX [11], which are presented for the entire banking system in the Czech Republic. HPI values are presented as a national average. In these cases, these are hard data that can be objectively counted on. In order to illustrate the epidemiological situation, information from the official websites of the Ministry of Health [12] [13], the Ministry of Industry and Trade [14] and the Ministry of Finance [15] is further used.

The Prague Stock Exchange is not a well-known international market. However, international companies are also present here. It is possible to invest in the banking sector, the energy sector, the arms industry, IT companies or the food industry. All tradable securities are included in the PX index, which is the subject of this paper, as a representative descriptive sample of the Prague Stock Exchange. The current legal form of the exchange was established in 1993, although this year it celebrates 150 years since its foundation [10]. The volatility of stock markets is generally higher and faster than that of real estate prices, as the actual securities trades are more quickly settled. The volume per transaction tends to be higher for real estate and for many investors (households) one investment for life.

Among the government measures, the following were selected as major measures that had an impact on the behavior of economic agents. There were, of course, more measures taken. However, an analysis of the adequacy of their use in combating the epidemic is beyond the scope of this paper. The list of selected measures is recorded in Table 1.

Table 1. List of government measures taken [12]

Date	*Description of Government intervention*
24.02.2020	1. statement by the Minister of Health on the new situation in the world
12.03.2020	Declaration of a state of emergency
16.03.2020	Restriction of population movement - until 1 April
06.04.2020	Slight loosening of retail and outdoor sports
17.05.2020	Ending the state of emergency
17.08.2020	All districts in the country remain free from risk of community transmission
28.08.2020	In Prague, there is incipient community transmission of COVID-19
05.10.2020	Declaration of a state of emergency
22.10.2020	Restricted retail and services, office activities and free movement of residents
20.11.2020	The government has decided to extend the state of emergency until 12 December
10.12.2020	The government has decided to extend the state of emergency until 23 December
23.12.2020	Government extends state of emergency for another 30 days

The data were analyzed using the CORREL function in MS Excel. These are the pairs of Stock vs. HPI, Interest rate vs. HPI and Unemployment vs. HPI. Correlation was first done for the period 2008 - 2020 (available data) and then for 2020 only. The change in trend of these values is compared. By combining the time series of the indices, measures and dates of government measures, the impact of the measures taken on the evolution of these markets is monitored.

4. Results

The key insight is that the uncertainty in the form of the global COVID-19 epidemic has been reflected in the decline of the PX index. Immediate action and handling of the situation by government officials and the provision of assurances of priorities on protection of the population and economic activity returned confidence to economic agents. This fact caused the declines to be no higher. The development of the PX index after the start of the epidemic is copied by foreign markets, such as the NASDAQ-100 Index [16]. However, the PX index did not return to its original level as it did abroad. However, in contrast to other studies from abroad, residential property prices have started to increase. Real estate acts as a certain investment, especially in the case of acquiring a home of one's own and storing equity as a protection against inflation. This is supported by falling mortgage interest rates and abolished the property acquisition tax (4%) of the purchase price with effect from 1 January 2021 [15].

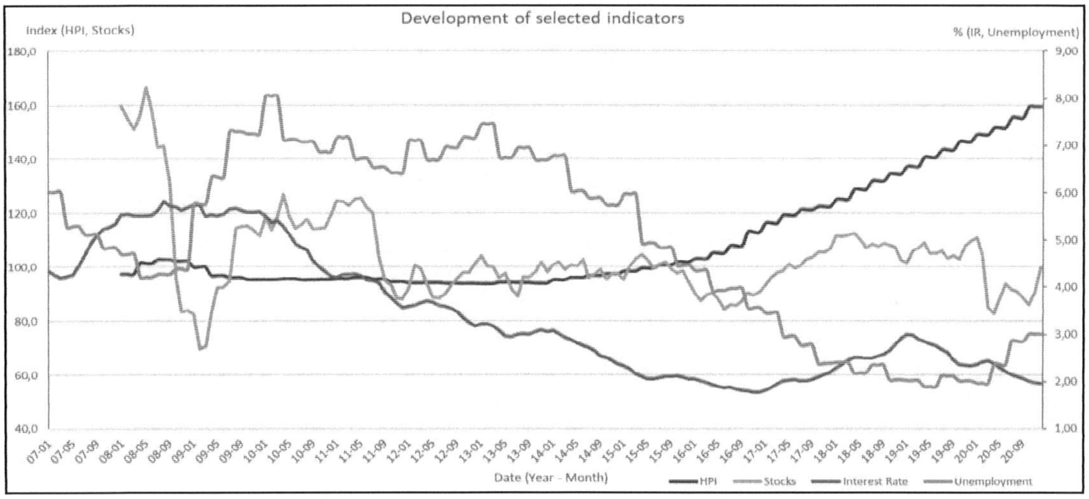

Figure 1. Development of selected indicators [8] [9] [10] [11].

The Ministry of Finance has also granted extended tax filing deadlines to economic entities in 2020, waived advance payments to businesses in closed industries, and is preparing a subsidy program for affected industries [14] [15]. Deferment of loan repayments was also discussed [15]. The curtailment of economic activity in some sectors is reflected in rising unemployment rates. We cannot speak only of seasonal fluctuations.

House prices (HPI) have not seen a decline even with the announcement of the presence of an epidemic or a high increase in the number of infected or stricter government measures.

The relationship between house prices (HPI) and the stock market (PX index) is such that financial volumes shift from stock markets to property markets. In terms of the dependence of these two markets, the correlation coefficient over the period 2008-2020 comes out at -0.124. The value indicates independence to slight inverse dependence, which supports the thesis of a transfer of financial transaction volumes between other investment options including Real Estate Investment. The indirect dependence increases to -0.258 in the 2020 assessment, see Figure 2.

There is a high inverse dependence between Unemployment and HPI, at -0.866. However, this coefficient subsequently changes to a value of 0.965. This is due, among other things, to support from the state to prevent people from becoming insolvent.

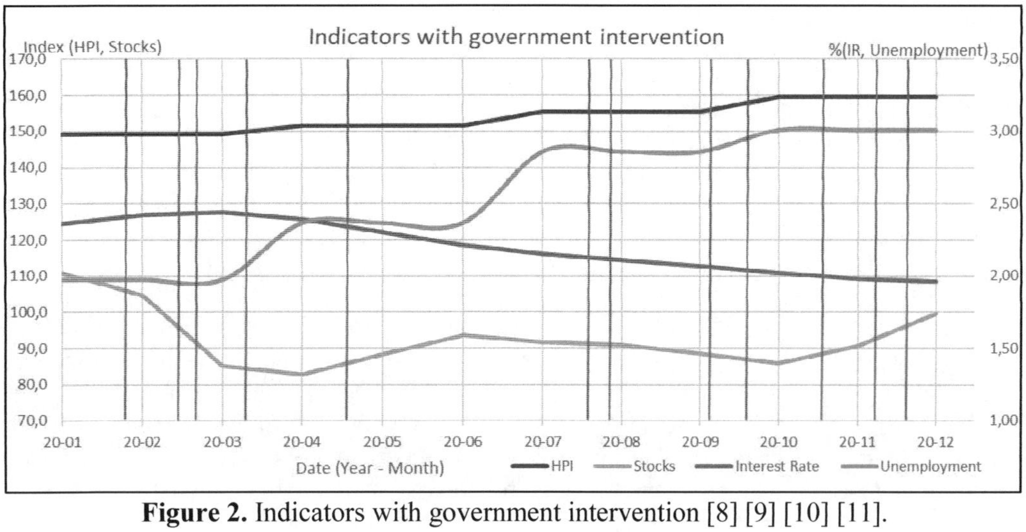

Figure 2. Indicators with government intervention [8] [9] [10] [11].

There is a high degree of inverse dependence for Interest Rates relative to HPI. It is -0.479 for the whole period under review. This changes rapidly in 2020 as interest rates continue to fall, encouraging new loan arranging and house prices rise. The value of the correlation coefficient is at -0.958.

During states of emergency and times of restricted movement of people, property inspections were also restricted, thus delaying the processing of property transactions. However, due to this situation, virtual tours and online interviews with real estate agents' clients flourished.

Due to the development of the number of infections in 2020, only the last quarter of the year can be considered extreme. However, the Czech health sector waited until the second quarter of 2021 for the highest extreme. Here, however, HPI data is not yet available, and therefore the graphs in Figures 1 and 2 end with December 2020.

Figure 3 shows the epidemiological waves. The initial shock in March 2020 had a much higher impact on equity markets than the subsequent three very aggressive waves. From a social point of view, it could be said that economic actors adapted to the situation and sought solutions to move their economic activities to the virtual space (e-commerce, home office, development of delivery companies, etc.).

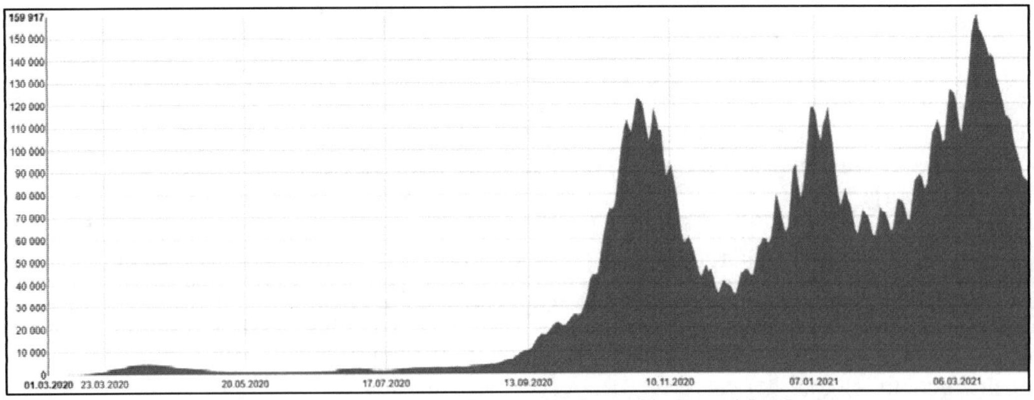

Figure 3. Evolution of the number of infected [13].

The impacts of COVID-19 on the data under consideration are undisputable. The economic closure of the Czech Republic [12] [13] took longer compared to literature solutions from other countries. Nevertheless, the evolution of real estate prices and stock markets is different. The reasons for this can probably be: a developed health system, a smaller population, the geographical location of the country and a very low unemployment rate.

5. Conclusion

The article dealt with the relationships between real estate prices, the value of securities on the Prague Stock Exchange and other aspects of unemployment and interest rates, all in the context of the spread of COVID-19 in the Czech Republic. The relationship between house price index and PX index has been found in the research of house prices and other related aspects. This is a correlation with an inverse relationship of -0.258 in the year 2020. Government anti-epidemic interventions during 2020 did not cause such significant fluctuations in these markets as the initial shock of the announcement of the presence of COVID-19 in the Czech Republic. However, the government intervention did not allow the PX index to return to its pre-Coronavirus form in this market. Given the degree of dependence, part of the volume of financial transactions has shifted specifically to the real estate market, namely residential real estate. Market rules of supply and demand also apply which currently corresponds to the rise in prices. Real estate prices need to be constantly monitored because of the potential for price bubbles and wider financial crises, as a large number of economic operators are involved in the real estate markets.

References

[1] Qian X, Qiu S and ZHANG G. 2021. *The impact of COVID-19 on housing price: Evidence from China. Finance Research Letters.* ISSN 15446123. doi:10.1016/j.frl.2021.101944

[2] Hu M R, Lee A D and Zou D. 2021. *COVID-19 and Housing Prices: Australian Evidence with Daily Hedonic Returns. Finance Research Letters.* ISSN 15446123. doi:10.1016/j.frl.2021.101960

[3] Apergis N. 2021. *The role of housing market in the effectiveness of monetary policy over the Covid-19 era. Economics Letters.* ISSN 01651765. https://linkinghub.elsevier.com/retrieve/pii/S0165176521000264 p 200

[4] Li Z, Zhou Q, Chen M and Liu Q. 2021. *The impact of COVID-19 on industry-related characteristics and risk contagion: Australian Evidence with Daily Hedonic Returns. Finance Research Letters.* ISSN 15446123. doi:10.1016/j.frl.2021.101931 p 39

[5] Nanda A, Xu Y, Zhang F and Liu Q. 2021. *How would the COVID-19 pandemic reshape retail real estate and high streets through acceleration of E-commerce and digitalization?: Australian Evidence with Daily Hedonic Returns. Journal of Urban Management.* ISSN 22265856. doi:10.1016/j.jum.2021.04.001 p 110-124

[6] Janus J, Xu Y, Zhang F and Liu Q. 2021. *The COVID-19 shock and long-term interest rates in emerging market economies: Australian Evidence with Daily Hedonic Returns. Finance Research Letters.* 110-124. ISSN 15446123. doi:10.1016/j.frl.2021.101976 p 110-124

[7] Chen H, Yeh Ch, Zhang F and Liu Q. 2021. *Global financial crisis and COVID-19: Industrial reactions. Finance Research Letters.* ISSN 15446123. doi:10.1016/j.frl.2021.101940 p 110-124

[8] *Český statistický úřad.* Český statistický úřad [online]. 2013 [cit. 2019-5-5]. https://www.czso.cz/csu/czso/index-cen-bytovych-nemovitosti-4-ctvrtleti-2020

[9] *Český statistický úřad.* Český statistický úřad [online]. 2013 [cit. 2019-5-5]. https://www.czso.cz/csu/czso/zam_cr

[10] *Prague Stock Exchange* [online]. Praha: Burza cenných papírů Praha, © 2021 [cit. 2021-6-26]. https://www.pse.cz/udaje-o-trhu/statistika/statisticke-soubory

[11] *Fincentrum Hypoindex: Hypoindex.cz* [online]. Prague, © 2008 [cit. 2021-2-24]. https://www.hypoindex.cz/

[12] *Ministry of Health Czech republic:* Tiskové zprávy - aktuální informace o COVID-19 [online]. Praha: Ministry of Health Czech republic, © 2021 [cit. 2021-6-26]. https://koronavirus.mzcr.cz/category/tiskove-zpravy/

[13] Komenda M, Karolyi M, Bulhart V, Žofka J, Brauner T, Hak J, Jarkovský J, Mužík J, Blaha M, Kubát J, Klimeš D, Langhammer P, Daňková Š, Májek O, Bartůňková M and Dušek L. 2020. *COVID-19: Přehled aktuální situace v ČR. Onemocnění aktuálně* [online]. Praha: Ministerstvo zdravotnictví ČR, [cit. 2021-5-25]. https://onemocneni-aktualne.mzcr.cz/covid-19.

[14] *Information about Coronavirus. Ministry of Industry and Trade* [online]. Praha, © 2005 - 2020 [cit. 2021-6-26]. https://www.mpo.cz/cz/rozcestnik/koronavirus/

[15] *Ministry of Finance Czech republic: Koronavirus (COVID-19)* [online]. Praha, © 2021 [cit. 2021-5-25]. https://www.mfcr.cz/cs/aktualne/koronavirus-covid-19

[16] *Daily Stock Market Overview, Data Updates, Reports & News: Nasdaq* [online]. Rockville, MD 20850: Nasdaq, © 2021 [cit. 2021-6-26]. https://www.nasdaq.com/

Acknowledgments

This work was supported by the Grant Agency of the Czech Technical University in Prague, grant No. SGS21/100/OHK1/2T/11 – Analysis of price-forming factors and entities influencing the real estate market with the included assessment of interdependence.

IOP Conf. Series: Materials Science and Engineering 1218 (2022) 012033 doi:10.1088/1757-899X/1218/1/012033

Stigmatized Properties: Filter Theory

I Hajnal

BUTE, Budapest, Hungary

istvan.hajnal@hu.gt.com

Abstract: International literature does not treat stigmatized properties in a uniform manner. In the proposed article, the Author employs the following definition: "A stigmatized property is one assigned with a real or perceived negative external effect on the users. That external effect modifies or reduces property market value through a specific, multilayered filter." Many property-related stigma cases are mentioned in the literature, such as high-voltage lines, noise, air pollution, airports, haunted properties, and many other similar situations. The various stigmas, as well as identical stigmas on different properties, have been assessed individually so far. Knowing the depreciating mechanism of a stigma allows for valuers the uniform handling of stigmas on different properties, as well as understanding the depreciating mechanisms of new stigmas. Using extensive literature research on various stigma cases, the Author built up a framework that explains the major variables and their interdependencies. As the result of the research, the new theory describes the change in value caused by the stigma takes place through a filter that combines the distance between the stigmatizing factor and the stigmatized property, the environmental conditions and community perceptions and interprets them differently over time.

1. Introduction

The problem of stigmatized properties is widely discussed in the international literature. However, there is no uniformly accepted definition of this concept, and the literature most often cites the definitions of Patchin [1], Mundy [2], Morgan [3], Roddewig [4], Greenberg et al. [5] or Slovic [6]. In this article, we use the most recently published definition produced on the basis of the above, namely: "Stigmatized property is property assigned with a real or perceived negative external effect on the users. That external effect modifies or reduces property market value through a specific, multi-layered filter." [7]

The impact of depreciation on stigmatized properties has been examined in numerous case types and locations, and the relevant literature now contains hundreds of articles. The case types studied range from the stigmatizing effect of airports [8,9,10] to the topic of so-called "haunted properties" [11]. A typical stigma case type is properties under high-voltage power lines [12] and the case of properties near mobile telephone relay towers [13]. Many studies have also examined the depreciating effects of spills of chemical pollutants or landfill sites [2,14,15][1].

The depreciation of stigma may be calculated using traditional valuation methods or new, so-called non-traditional methods. These are the Revealed Preference (RP) and Stated Preference (SP) procedures

[1] The referenced literature is just an illustration from a vast set of publications

[16]. Most case studies were carried out using the hedonic model within the SP methodology. The hedonic pricing method (HPM) essentially breaks down the value of the property into value components, to which multiplication factors are assigned, expressing the weight of each valuer in the total value as a "shadow price" [17]. This analysis starts out from the value of transactions (or ad data) and, by breaking them down into components, it quantifies the stigma effect. At the same time, the SP procedures examine the expected pricing preferences of the interviewed future consumers, which is often referred to under the umbrella term "contingent valuation" [18]. In both methodological groups, the usual research practice is to include variables based on empirical experience, while explanatory variables with stigmatizing effects do not cover the full range of factors causing the stigma. There is no guiding principle that systematically defines the range of explanatory variables.

The objective of this article is to present a uniform theory, using the same characteristics of the different case types, that provides the same approach for all stigmatised property scenarios.

After this introductory part, the article discusses the mode of action of the stigma in a separate chapter, and then presents the filter theory developed by the author, which is suitable for combining case types.

2. Mode of action of stigma in the case of properties

In literature research on the case types of stigmatized properties, researchers basically group the entire range of stigmatized properties according to the stigmatizing effect and the nature of the stigmatizing property. Most of these publications are related to a geographical location, as the most important variable determining the value is location. Case types have a lot in common; they have the same mechanism of a depreciation effect.

Based on the itemized case descriptions and literature, the following common characteristics for each stigma case type can be identified [7]:

- Each case has a negative charge and a negative judgement;
- Most of the cases listed are triggered by human activity (investment, emergency, regulation);
- This human activity is often related to construction and investment;
- The development of the effect is usually linked to a property or a group of properties (this property is the stigmatizing property);
- In cases discussed in the literature, the surrounding (stigmatized) properties are generally not affected by any actual physical effect;
- However, the persons using the property are generally affected by or presumed to experience an adverse effect;
- This adverse effect is difficult to objectify and difficult to measure;
- For this adverse effect, users list additional risks that are or considered to be serious;
- There was a lively social and professional discourse about these cases.

The approach described in this article is irrespective of the specific trigger effect and should be applied uniformly to the whole range of stigmatized properties.

In the literature, researchers studied the stigmatizing effect – depreciation function and gave a quantified opinion on the decrease in the expected value of the Market Value. As an example, the general rule of the effect of the proximity of an airport is that an increase in noise level of 1 dB reduces the value of the property by 1% [16]. Of course, user and residential perception is also present in the mapping of this function, but the extent and components of this are very difficult to filter out from the total depreciation effect. The literature often mentions and indicates the presence of this residential perception, but it fails to provide a method to show the extent and proportion of the total effect.

Similarly, most authors have recognized that the distance from the source of the stigmatized property has a significant effect on the extent of the stigma. Many of the researchers not only described the location effect by distance, but also by regional location and spatial coordinates (based on GIS or LIDAR systems). The so-called 'Space Syntax' theory has been used in urban planning since the 1970s, which serves to analyse the density of the urban fabric and the relationship between urban values. Giannopoulou et al. [19] combine this theory with the analysis of the evolution of spatial market values, thus relativizing "distance" depending on the urban location.

The literature did not point to an important issue related to the topic of distance: the presence of intermediate elements in the space between the stigmatizing effect (stigmatizing property) and the stigmatized property. If a mobile relay tower is 200 metres away from a residential building and there is a nice lawn between the two properties, it obviously has a greater impact than as if a busy road ran along the same place. The impact of the proximity to the airport can also be modified by a nearby, noisy highway, and even the proximity of a haunted house can override that if there is, say, a cemetery between the two properties. It should also be taken into account that, in the urban environment, a property can be affected by several stigmatizing effects at once. The effect may be single, but the effects of several stigmas may be cumulative (or they may even cancel each other out). As a variable, we must therefore also evaluate the environment, its structures and its effects in the model of depreciation of stigmatized properties.

It is important to note that buildings with different functions but under the same effect suffer different depreciation effects, and similarly, buildings with the same function but in different conditions show different depreciation levels even under the same effects.

The stigma's depreciating effect changes as time passes. Timeliness plays an important role in determining the market value; according to the valuation standards, the market value is valid for a given date (the valuation record date). We also try to learn the effect of the stigma by valuation date; however, from the appearance of the stigmatizing effect, the value modification effect changes – it typically decreases [4]. Regardless of the date, the effect is also modified by social opinion. Hence, it is not enough for the appraiser to record the fact of stigma; they must also examine the specific stage of the stigma. Therefore, determining the depreciation of the stigmatized property is also a particularly complicated task, because the existence of this variable, as well as its evolution over time, requires analysis. The vast majority of the literature did not include the temporal cycle of the stigma among the variables when creating models. This, of course, is not the case for a single analysis for a single valuation issued, but if multiple valuations are analysed (for example in a sample space test), this may become a relevant criterion.

In all cases, the fear of the specific effect is greater, and the risk associated with the effect is greater than what the objective extent of the effect would justify. The subjective perception of the person, and even more so the community, plays an important role in the mode of action of the market depreciation of stigmatized properties. Social inclusion must be interpreted through a social filter of environmental risks. The 1992 Royal Society report on risks states that risk acceptance can be interpreted through people's beliefs, attitudes, perceptions and feelings, as well as through the broad cultural and social position that determines their behavior towards hazards [20]. In fact, this is about accepting 'technical risks'. According to Bickerstaff's literature review [21] , this filter is of a societal and social nature. He proves that using the example of air pollution. According to him, the attitude of the people around him, the place of residence, the method of perception (e.g. the sight of dead animals) and the trust in the source of communication should be interpreted together. The author thus identifies social risk as a common perception of the recipients. The relationship between the actual degree of risk and the perception is not a simple "mistake". The community's opinion at any time is an essential element of the depreciating effect of the stigma.

So far, we have talked about the value of the stigmatized property in general, as the authors in the literature generally use the term "value" without an adjective. This is, by some consensus, nothing other than the market value [22]. Market value is a defined form of value that quantifies the opinions of property market's players about the property in a single sum. Research is typically aimed at including this generalization in the amount of depreciation due to stigma, i.e. we are looking for the average market player's opinion of the stigma. However, there are certain defined forms of value (such as investment value or special value) that can be attributed to a particular market player. Obviously, a special stigmatizing effect affects different individuals in different ways. It is therefore important to point out that the "common perception", i.e. the community opinion referred to in the previous paragraph, relates to the market value, and other forms of individual value should not be determined by the value judgement of the average market player but by the specific, individual value judgement.

3. Filter theory

Taking into account the aspects described in the previous section together, the interpretative theory called the filter theory develops step by step. Figure 1 below shows the market mechanism of the general depreciation effect: some external (or internal) circumstance directly modifies the market value of the property in a measurable, objective manner. The number of such objective factors is infinite, but in real estate valuation practice 10-20 factors are examined that have a significant impact on property value.

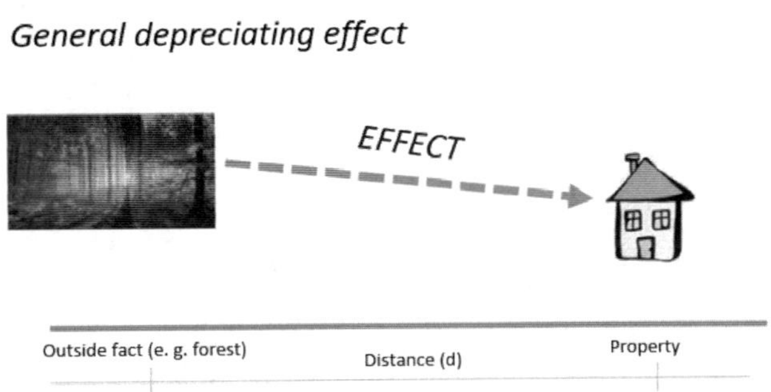

Figure 1: General value-adjusting effect

The stigmatizing element, in most cases the stigmatizing property, is one of the important value-adjusting effects. According to the usual literary approach, in the case of the depreciation of stigmatized properties, the community's opinion is a common element that changes the depreciation rate of the effect. This community opinion can be seen as a lens, and even more as a filter, which often magnifies or even reduces the objective effect. (Figure 2)

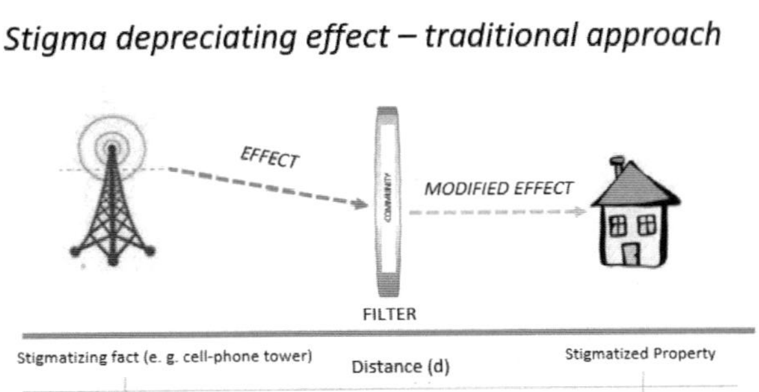

Figure 2: Traditional approach

Effects starting from the stigmatizing element (the stigmatizing property) weaken with distance. The effect of distance on value is also a complex relationship. There may also be environmental elements in the space between the two objects that positively or negatively affect depreciation. The depreciating effect of the relative location of the two objects is also influenced by the community's opinion. Because of the above, the distance between the point of impact and the stigmatised property cannot be interpreted without the environmental elements, so the filter is not single, but includes the characteristics of the environment, among them any other stigmatizing effects – and therefore, mutatis mutandis, the distance between the two objects themselves, as shown in Figure 3. So we are talking about three filters: distance, environment and community.

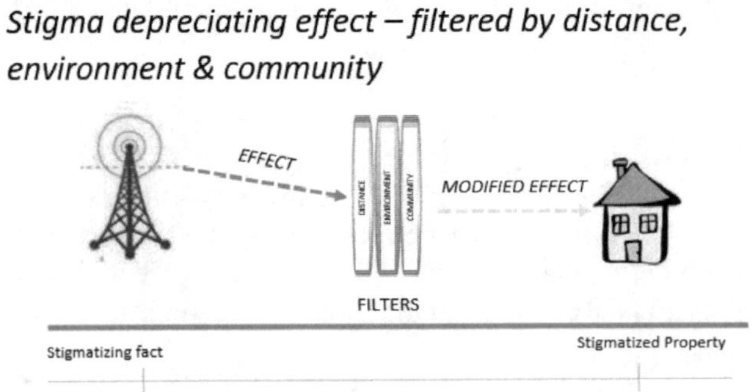

Figure 3: The combination of the three filters

Stigmas emerge and disappear. Something that was not a stigma 20 years ago might be one today and might not be one in 15 years' time. As mentioned earlier, the filters of the stigmatizing effect are constantly changing over time: based on the community opinion, each of the three filters is given or can be given a new meaning. The filters are joined together by a common variable: time. Thus, the change in value caused by the stigma takes place through a filter that combines the distance between the stigmatizing factor and the stigmatized property, the environmental conditions and the community perception and interprets them differently over time. The combined concept of the Filter Theory is illustrated in Figure 4.

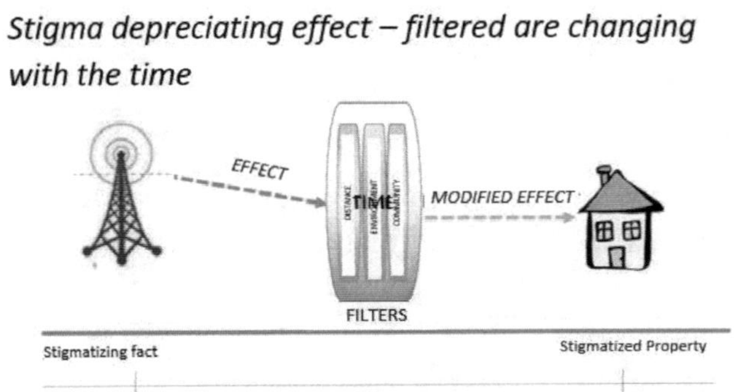

Figure 4: The concept of the filter theory

4. Conclusion

The library of literature on stigmatized properties mainly deals with the study of individual cases and locations, and so far the synthesis of these cases has been missing. This synthesising theory is necessary in order to include the full range of explanatory variables for further studies.

The filter theory described in the previous chapter explains the depreciation mechanism of the stigma. Accordingly, the change in value caused by the stigmatization effect takes place through a filter that combines the distance between the stigmatizing factor and the stigmatized property, the environmental conditions and the community perception and interprets them differently over time.

Knowing the depreciating mechanism of a stigma allows for the uniform handling of stigmas on different buildings, as well as understanding the depreciating mechanisms of new stigmas. The filter theory can be used in any research or expert opinion where, in a specific stigma's case, the decrease in the market value is the subject of the study, especially if the researcher wishes to quantify the general and temporal changes in market value caused by the stigmatizing effect.

References

[1] Patchin, P. J. (1988). Valuation of Contaminated Properties. Appraisal Journal, 56(1).
[2] Mundy, B. (1992). The impact of hazardous materials on property value, revisited; The Appraisal Journal, vol. 60, no. 4, pp 463-471.
[3] Morgan, R. M. (1994). The Expansion of the Common Law Duty of Disclosure in Real Estate Transactions: It's Not Just for Sellers Anymore. FLA. BJ, 68, 28.
[4] Roddewig, R. (1996). Stigma, environmental risk and property values: 10 critical inquiries. Appraisal Journal (Oct.) 375-387.
[5] Greenberg, M., & Hollander, J. (2006). Neighbourhood Stigma Twenty Years Later: Revisiting Superfund Sites in Suburban New Jersey. Appraisal Journal, 74(2).
[6] Slovic, P. (2009). Talking About Recycled Water—and Stigmatizing It. Decision Research Report No.15-01 Eugene, Oregon: Decision Research March, 2009
[7] Hajnal, I. (2017). Evaluation of stigmatized properties. Organization, Technology and Management in Construction: an International Journal, 9(1), 1615-1626.
[8] Bell, R.: (2001). The Impact of Airport Noise on Residential Real Estate. The Appraisal Journal, July 2001. pp 312-321.
[9] Nelson, J.P. (2004). Meta-analysis of airport noise and hedonic property values: Problems and prospects. Journal of Transport Economics and Policy, 38(1), pp 1–28.
[10] Lazic, A., Golaszewski, R. (2006): A Technical Note on Aircraft Noise and its Cost to Society. GRA, Incorporated, Corporate White Paper
[11] Chapman, D., & Ludlum, M. (2014). Teaching stigmatized property: You don't have a ghost of a chance. Journal of Business Cases and Applications, 11, 1.
[12] Chalmers, J. A., & Voorvaart, F. A. (2009). High-Voltage Transmission Lines: Proximity, Visibility, and Encumbrance Effects. Appraisal Journal, 77(3).
[13] McDonough, C. C. (2003). The impact of wireless towers on residential property values. Assessment Journal, 10(3), 25-32.
[14] Gamper-Rabindran, S., Timmins, C. (2013). Does cleanup of hazardous waste sites raise housing values? Evidence of spatially localized benefits, Journal of Environmental Economics and Management, Volume 65, Issue 3, May 2013, Pages 345–360
[15] Hurd, B. H. (2002). Valuing superfund site cleanup: Evidence of recovering stigmatized property values. The Appraisal Journal, 70(4), 426.
[16] Nelson, J.P. (2008). Hedonic Property Value Studies of Transportation Noise: Aircraft and Road Traffic. In: Baranzini, A. Ramirez, J., Schaerer, C. (Ed) „Hedonic Methods in Housing Markets, Pricing Environmental Amenities and Segregation" ISBN: 978-0-387-76814-4, Springer, 2008
[17] Ridker, R. G., & Henning, J. A. (1967). The determinants of residential property values with special reference to air pollution. The Review of Economics and Statistics, 246-257.
[18] Parkhill, K. A., Butler, C., & Pidgeon, N. F. (2014). Landscapes of threat? Exploring discourses of stigma around large energy developments. Landscape Research, 39(5), 566-582.
[19] Giannopoulou, M., Vavatsikos, A. P., & Lykostratis, K. (2016). A process for defining relations between urban integration and residential market prices. Procedia-Social and Behavioral Sciences, 223, 153-159.
[20] Royal Society (1992). Risk: Analysis, perception and management. Report of a Royal Society Study Group: Pp 201. The Royal Society. 1992.
[21] Bickerstaff, K. (2004). Risk perception research: socio-cultural perspectives on the public experience of air pollution. Environment international, 30(6), 827-840.
[22] RICS Valuation Standards; (2010); RICS, London, 6th ed. (Red Book)

Success Factors in Projects

Bilal Ktaish* and Miklós Hajdu

Budapest University of Technology and Economics, Doctoral School of Engineering, 1111 Budapest, Műegyetem rkp. 3, Hungary

*Email: Bilalktaish@outlook.com,

Abstract. The literature of project success is very wide one. Various articles focus on the evolution of the understanding of project success. Most of the projects we hear of in media are either over budget, late or are simply not good enough and still different lobbies of people claim that those projects have been successful. Neither the practitioners nor the academicians seem to agree on what constitutes project success. It seems to be a rather elusive concept. This paper attempts to put forth the points of views of different researchers in this field and focuses primarily on the factors that affect the project outcome regardless of the type of project. Two sets of factors contributed to successful outcome of a project: the macro and micro factors. A set of macro includes the realm of the owner, while micro factors fall in construction domain. Other factors were grouped into four areas: factors related to: project, project manager and team, organization, and the external environment. More factors were investigated by other researchers, factors including applied methods, individuals in projects, and organizational context. Therefore, every party in the project has its own perception thus forming different criteria of project success. These differences of perception explain the reason why the same project could be considered a success by one and unsuccessful by another.

1. Introduction

The word project is derived from Latin where "pro" means "forward" and "jacere" means "throw". Thus, the original meaning of project is something that in a figurative sense has been thrown forward [1], the term project is defined: 'A unique set of co-ordinated activities, with definite starting and finishing points, undertaken by an individual or organization to meet specific objectives within defined schedule, cost and performance parameters. Since 1960s, the project management analysts have been attempting to discover which factors lead to project success. A construction project is usually considered successful if it's achieved on time, within the budgeted cost and at prescribed standards However, problems may arise leading to more objectives other than cost, time, and quality; where there are some projects which have been completed in time and withing budget but wasn't considered as successful

The world's leading information technology research and advisory company, organizations should accept the increased risk to achieve the desired return and to accept failure and move on [22]. Since the organizations have increased focus on cost management, they should embrace the failure. In addition, the necessary causes of failure: Unsuitable definition of project scope, lack of communication, and inappropriate management of project. The study also examined the significance

of organizational characteristics such as applied project management methodology and tools. Considering the defined aims of the paper, there is a need for providing a review of the literature on the understanding of project, the phenomenon of project success.
Success in construction projects is a global phenomenon. Factors leading to project success have been investigated in many regions and have been extensively explained in the literature over the years. Part of these studies will be described below. This paper focuses primarily on the factors that affect the project outcome regardless of the type of project.

2. Literature Review

In this area, there is a broad divergence of views; the only consensus seems to be the disagreement on what form 'project progress'. An interesting case of the various perspective has been provided: An architect may consider success in terms of aesthetic appearance, an engineer in terms of technical competence, an accountant in terms of dollars spent under budget, a human resources manager in terms of employee satisfaction, and chief executive officers rate their success in the stock market [9]. Hence, There's no exact success in project and there's only a perceived one.
Project success is the evaluation of overall goals of the project; however, project management success is the evaluation of the classical measures of achievement against budget, time, and quality [23] the Project success handle with concerns for performance internally or externally for short and long period. A project can be defined as the accomplishment of a certain target, which includes a series of tasks that require resources [20]. However, project management can be considered as the procedure of controlling the achievement of the project targets. The objective of project management involves determining the requirement of work, preparing the execution of the work, controlling the progress of the activities, and adjusting the deviations from the plan. Originally these two descriptions may find to be overlap. In fact, its barely exaggerate that project management success is one of the aspects of project success since the latter is unattainable without the former [20].
Other distinction is between success criteria (the standards by which failure or success of business will be evaluated) and success factors (those inputs that contribute directly or indirectly to success). Where the criteria of project success are the group of standards by which the success of project can be reviewed. However, the factors for project success are the group of states, or effects which lead to project outcomes [20]. These effective forces can facilitate or delay the success of project. They lead to failure or achievement, but don't create the basis of judgement. An example to explain the differences between criteria and factors is by using a high school student who want to achieve a degree in a university. To get the degree, the student must get accepted in registration and graduate at the end. Obviously, these two sets of conditions are the criteria. Each criterion has its condition, for admission criteria: entrance examination, policy, fees, duration, etc. For awarding the degree: passing marks, projects, completing these two criteria mean success in awarding the degree. However, failure to satisfy any of the criteria would result in losing the degree. Conversely, factors affecting the admission and graduation criteria could consist of age, academic ability, distractions and so on. These factors could be important, yet they don't decide the failure or success of achieving the degree. The critical success factors were introduced and described as key factors required to reach goals and it was highlighted in the book benchmarking in construction by Mccabe, Steven, that they are crucial for managers to enhance their organization [2].
Project success became a complicated phenomenon, and can be treated from input and output perspective [8].The input-oriented perspective evaluates the factors contributing to success using critical success factors, while the output-oriented perspective assess the success by means of project criteria. the evaluation of success, cost and quality is for sure an essential part of project control, and when trying to evaluate the success, one must distinguish between project success and the success of project management effort as both are related but may be very different [23]. Most project management literature advocates that project management has three major objectives: a project must be managed on time, within budget and to quality/performance specifications. There are examples of projects which didn't met the required criteria (time, budget, quality,) and are nevertheless considered

successful projects. A large project was held in Kuala Lumpur and it has to be completed in 12 months however what happened next is that there was extension of time and additional budget [13]. Since opening of this project, it has shown to be very popular. This case study shows doubts: the developer and the contractor have lost and from their point of view the project has failed. On the other hand, the stakeholders' perception is different where the project was a big success. Therefore, every party in the project has its own perception thus forming different criteria of project success. These differences of perception explain the reason why the same project could be considered a success by one and unsuccessful by another.

There are two sets of factors that conclude the successful outcome of a project: the macro and micro factors [22]. A set of macro includes the realm of the owner, while micro factors fall in construction domain. A framework was carried out for the macro viewpoint of project success [13]. And two set of conditions were illustrated: completion criteria and satisfaction criteria. Those criteria are also affected by sets of factors. Normally, the stakeholders, the owner, and public will look at project success from macro viewpoint. The first criteria: The completion is influenced by the factors like weather, supervision etc... while the second criteria: satisfaction is affected by location, prestige, cost. The users are more demanding on the satisfaction criterion over the completion one. An example is an evidence about this user's point of view is the famous Sydney Opera house. This construction project needed 15 years (from 1958 to 1973) and additional budget to build, today it represents the symbol of Sydney and engineering masterpiece. In addition, another framework for micro viewpoint of project success is conducted in the study where the completion criteria are the set for deciding the project success. Generally, the contractor will determine the success of project from the micro viewpoint. In this type of viewpoint, the risk environment, finance, and human are the factors affecting the completion criteria [13].

The primary criteria of project success are determined by cost, time and quality that called iron triangle. The variation of project success has results into developed user's necessity, customers' requirement, and environment. There were four suggested dimensions of success: business success, planning for future, project effectiveness and impact on customer [20].

The reasons affecting the performance of the project were also studied: the factors of iron triangle, management of both risk and human resource, in addition to communication management [23]. Conceptual framework was recommended for the success of the project by recognizing schedule accomplishment, budget, satisfaction of owner and public, and profitability.

A research was prepared to identify the critical success/failure factors in projects. The factors were grouped into four areas: factors related to: project, project manager and team, organization, and the external environment [5]. The first area was characterized by: the size of project, value of it, and the life cycle. The factors related to project manager and team can affect the project outcome where the project manager's commitment become most critical during different stages of the project. Selecting a project manager is significant for successful project, where he/she should have the necessary technical and administrative skills. Moreover, the authors found that factors like manager and team members have impact on client satisfaction where skills like marketing can affect the stakeholder's attitude towards the project outcome [5]. The third area studied by the same article includes factors related to the organization; the top management support usually supervises the project manager. And the full support from the top management facilitates and achieve the strategies for the completion of the project. This mainly happen in projects with matrix organizational forms and requires negotiating skills. The last factor considered by the authors is linked to external environment such as political, economical and social for example: the weather, government restrictions and guidelines. The curiosity in project success factors was growing until was a hypothesis that the factors of success must stand alone because of the variety of projects in prioritization of project goals, the nature of organization and the environmental changes.

The idea of CFS were carried out and proposed four factors: teamwork enhancing policy, comprehensive facility team, adequate knowledge in handling different aspects of facilities, and information optimization in the planning and design stage [21]. The identified factors that considered critical to a large-scale construction project in Thailand are comprehensive, competence, commitment, and communication.

Having argued the factors alone, the project wouldn't be successful without satisfying the success criteria. Hence, the project success tends to be more accomplished with both project success factors, and the success criteria considered as whole.

A designed survey was involved in Australian institute of project management (AIPM), representing managers and users in Australia. Fifteen critical success factors were determined by the respondents [3]. The most identified CSF was the project understanding which includes understanding the needs of clients and the objective of the project, the clarity in the requirements, documentations, and scope. The other factor was the competent project team, some respondents indicated that the team members should concentrate on the project. Communication was also a factor commonly identified in the results, and it should exist in all phases of project and between the stakeholders [3].

Questionnaire was forward to project managers, owners, contractors, and other concerned parties. The population were requested to select five factors out of many factors that previously determined by Pinto, Slevin and Davis [6]. Based on the results, the respondents chose clearly defined goal and directions as a critical success factor following with competent project team members. The third ranked factor was clearly defined roles, while the fourth one was communication and consultation with stakeholders. The last factor that respondent select is the compliance with planned time, cost, and quality [3].

Other Factors affecting the success of project were investigated and listed as follow[19]:

- Applied Methods

The methods applied in project management are widely discussed by several authors. And in general, the influence level of this applied methods relies on the type of the project. The first step to increase the success of project is by identifying the risk [18], and those methods can be applied in technological aspects.

- Individuals in project

The significance of so-called agents was highlighted, including team members, project leaders, senior management, customers and suppliers in the process of NPD. Those groups would currently be named Project Stakeholders where stakeholders are assumed to be any kind of individual or institution with an interest in a project's outcome [18].

- Organizational context

This type of factor became more considered especially after companies managed high number of projects thus increasing of complexity in the process [4]. The idea of collaborative project environment concluding that the correlations between stakeholders are crucial for the success of failure of project [4].

The factors influencing the success of projects are identified and presented in the following areas: (1) applied methods, (2) people in projects, (3) and organizational context.

3. Aim

This paper intends to investigate the factors affecting the success of the projects. The methodology can be used to broaden the understanding of the success of the projects regardless the type of the project.

4. Research Methodologies

A secondary type of methodology was done. Data, results were gathered from different papers and journals. The process of quantitative research follows standard procedures, methods, forms of analysis and reporting the results of the research undertaken. This standardization is intended to maximize objectivity.

5. Results and Discussion

A comprehensive literature review on the success factor in projects was carried out in this study. In this context, studies focusing on the factors in different types of projects in different countries were reviewed. Most of these studies used the questionnaire survey and relative importance index to identify the most important factors causing success. In addition, almost all these studies have identified the factors based on previous studies.
According to the results of the literature review, the most cited factors of success are:

Authors	Project Success Factors
Pinto and Slevin (1987)	Defined goal and directions, competent project team members, defined roles, compliance with planned time, cost and quality
Sanvido, Grobler, Parfitt, Guvenis & Coyle (1992)	Teamwork enhancing policy, comprehensive facility team, adequate knowledge in handling different aspects of facilities, and information optimization in the planning and design stage
Belassi and Tukel (1996)	Project size and value, uniqueness of project activities, political environment, economic environment
Standish Group (2000)	Definition of project scope, lack of communication, and inappropriate management of project
(Cooke-Davies (2002)	Instruction on risk management ideas
(Baccarini & Collins, 2004)	Understanding the needs of clients and the objective of the project, the clarity in the requirements, documentations and scope
(Yean & Ling, 2009)	Management of both risk and human resource

Table 1 Success Factors from different studies

Establishing critical success factors will provide company/organization a competitive edge and a baseline of success in achieving the responsibility of a project management companies. This will lead to a rise to satisfied investors and make the project management company succeed. Several Investigations have presented the accent of the necessity of project planning and control. Large scale construction projects require a cautious planning before the execution. Social and economic expectations are presented in the large-scale projects. Hence, late finishing can affect the expected outcomes leading to enormous financial issue, performance, and the credibility of project itself. The backbone of a construction project is the adequate planning and active control mechanism. Having many resources with weak planning and control will lead to outcome issues. This mechanism has done be done in the early initial stages of the project. Inadequate planning causes project to finish behind agenda and with surprising delays. Therefore, specifying clear, realistic, recognizable goals by project team is vital. It permits the team to figure out where they stand, how far are they from the goal, and what they still require achieving the success. Once it has been done with fitting project planning and control, the accomplishment of the project tasks is far easier.
The competence of the personnel is another theme included in the rating of success factors. A team possessing experience, and knowledge is necessary for the success accomplishment of project goals. In a field where the project manager doesn't know how to solve a problem, face a challenge, and needs expert skills, his role would be crucial. Moreover, it will cause excessive delays, and failed negotiations.
The unsatisfactory project efficiency can be due to the client itself. It should be recognized that client interest has may dimensions. Clients must be straight forward, quick in response and aware of the

consequences. Construction projects include modifications in design. Thus, client responsibility is necessary to prevent any possible action later. Project manager and the team must deal with changes on daily basis. These changes should be communicated to the client and involve them.

6. Acknowledgments

The author of this paper wish to acknowledge the support from supervisor Dr. Miklos Hajdu for his support, encouragement, and patience. The meetings and conversations were vital in inspiring me to think outside the box, from multiple perspectives to form a comprehensive and objective critique.

7. Conclusion

Project success without any doubt is one of the most debated issues in the modern academic field of project management. However, an agreed description of project success hasn't been noticed yet. Project success criteria and critical success factors includes the two crucial components of project success. The criteria are the variables that evaluate the project success while factors are the independent variables that can affect the project outcome.

Project successes rely on numerous features that can involve factors related to human, project, management, and external environment. Based on their requirements, the project participants have different benefits; however, they must possess an agreement about the objectives of the project. A great number of studies which focus on the factors leading to success of the projects have been carried out by many researchers. The main objective of this study is to identify and better understand the factors affecting the outcome of the projects. The present paper tries to investigate the project success factors through review of the previous articles in the project management literature. Time, cost, budget are the three most cited and important criteria followed by criterion of client satisfaction. Having argued the factors alone, the project wouldn't be successful without satisfying the success criteria. Hence, the project success tends to be more accomplished with both project success factors, and the success criteria considered as whole.

References

[1] Adams, S., & Around, W. (n.d.). 2 . What is a project ?, 27–38.

[2] Alias, Z., Zawawi, E. M. A., Yusof, K., & Abra, A. (2014). Determining Critical Success Factors of Project Management Practice : A conceptual framework. *Procedia - Social and Behavioral Sciences*, *153*, 61–69. https://doi.org/10.1016/j.sbspro.2014.10.041

[3] Baccarini, D., & Collins, A. (n.d.). Critical success factors for projects Stream Z . Other (Project Management) * = corresponding author and Presenter Critical success factors for projects.

[4] Bai, C., & Sarkis, J. (2013). Int . J . Production Economics A grey-based DEMATEL model for evaluating business process management critical success factors. *Intern. Journal of Production Economics*, *146*(1), 281–292. https://doi.org/10.1016/j.ijpe.2013.07.011

[5] Belassi, W. (1996). A new framework for determining critical success / failure factors in projects, *14*(3), 141–151.

[6] Beleiu, I., Crisan, E., & Nistor, R. (2015). MAIN FACTORS INFLUENCING, 59–72.

[7] Bryde, D. J. (2005). Methods for managing different perspectives of project success. British Journal of Management 16(2), pp. 119-131. [11] Henard, D. H., and Szymanski, D. M. (2001). Why some new products are morel, successful than others. Journal of Marketing Research 38(3), pp. 362-375.

[8] Cooke-davies, T. (2017). The "' real "' success factors on projects, *7863*(October). https://doi.org/10.1016/S0263-7863(01)00067-9

[9] Freeman, M. A. (1992). Measuring Project Success. *Project Management Journal*, (March 1992).

[10] Hargreaves, N., and Endlar, L. (2011). The Impact of Cultural Dimensions on the Project Management Triple Constraint Model. Proceedings of the 10th European Conference on Research Methodology for Business and Management Studies, pp. 232-240.

[11] Huang, Z., Poli, M., and Mithiborwala, H. S. (2009). Project Strategy: Success Themes for Strategic Projects. Proceedings of Picmet 09 - Technology Management in the Age of Fundamental Change, Vols 1-5, pp. 1255-1262.
[12] Ika, L. A. (2009). Project Success as a Topic in Project Management Journals. Project Management Journal 40(4), pp. 6-19.
[13] Lim, C. S., & Mohamed, M. Z. (1999). Criteria of project success : an exploratory re-examination, *17*(4), 243–248.
[14] Pinto JK, Slevin DP. Critical success factors in R&D projects. Research Technology Management 1989;32(1):31-35
[15] Radolph, W.A., & Posner, B.Z. (1994). Effective Project Planning and Management. Prentice Hall International. Richardson, T. (1995). Project Management pitfalls. Business Communications review, 25(8), 49.
[16] Rickards, T., Chen, M. &, Moger, S. (2001). Development of a self-report instrument for exploring team factor, leadership and performance relationships. British Journal of Management, 12(3), 243-250.
[17] Rockart, J.F. (1979). "Chief Executives Define Their Own Information Needs," Harvard Business Review. Rowe, A. J., Mason, R. O. and Dickel, K. (1982). Strategic management & business policy, A methodological approach, Addison-wesley publishing company, Philippines.
[18] Schmidt, R., Lyytinen, K., Keil, M., Cule, P., Schmidt, R. O. Y., Lyytinen, K., … Cule, P. (2015). Identifying Software Project Risks : An International Delphi Study Identifying Software Project Risks : An International Delphi Study, *1222*. https://doi.org/10.1080/07421222.2001.11045662
[19] Seweryn Spalek. (2014). SUCCESS FACTORS IN PROJECT MANAGEMENT . LITERATURE REVIEW Seweryn Spalek, 4828–4835.
[20] Shenhar, A. J., Dvir, D., Levy, O., & Maltz, A. C. (2002). Project Success : A Multidimensional Strategic Concept, *34*(2001), 699–725.
[21] Standish, T., & Report, G. (2013). The Standish Group Report.
[22] Toor, S. (2009). Construction professionals ' perception of critical success factors for large-scale construction projects, (October 2015). https://doi.org/10.1108/14714170910950803
[23] Wit, A. De. (1988). Measurement of project success, 164–170.
[24] Yean, F., & Ling, Y. (2009). Key project management practices affecting Singaporean firms ' project performance in China. *International Journal of Project Management*, *27*(1), 59–71. https://doi.org/10.1016/j.ijproman.2007.10.004

System-theoretical Approach to Fundamental Limits of Controllability in Complex Organization Networks

Wolfgang Eber

Technische Universität München, Lehrstuhl für Bauprozessmanagement, München, Germany

eb@bv.tum.de

Abstract. Organisation is fundamentally based on elaborating the equilibrium of two contradicting forces. An increasing degree of separation of work leads to specialisation and therefore significant increase of productivity in terms of time and cost. This advantage is compensated for by the therewith rising effort required to define, shape and delegate the subtasks to the different players and have them well-informed, coordinated and motivated to successfully contribute to the overall project. The well-known and common approaches to optimize division of work are mainly based on strict hierarchical structures like work breakdown structures in order to perfectly identify work-packages and their respective interfaces. Control loops are then established maintaining the certainty to achieve the previously defined results of the subtasks so that they will perfectly fit and will be ready to flawlessly form the total product. However, this traditional approach presupposes perfectly defined stable and separable systems as well as perfectly operating controlling mechanisms. As soon as some imperfection of either of these is given, which can be safely assumed in reality, the method is bound to fail principally. With this paper, we propose a system-theoretical framework modelling in particular a local imperfect however controlled situation and providing the principal limits on a mathematical basis as well as allowing for means for a practical approach. Organisations are represented by slightly contradicting systems while interactions as well as controlling mechanisms are given by first and second order differential equations according to the Theory of Systems. The resulting long-term behavior of the model, optimally avoiding oscillating and probably escalating development, indicates the principal limits of controllability. We find that the concepts of Lean Construction mainly address exactly these requirements and therefore find their formal justification including some quantitative framework.

1. Introduction

Organization is all about improving efficiency of production by implementing division of work [1, 2]. The common structural approach to developing appropriately matching work packages is based on hierarchical tree-structures using reduced but nonetheless non-ambivalent interfaces to superior as well as sub-ordered structural elements [3, 4, 5]. Not only recent development of understanding implies that this strict concept is bound to fail principally. Meanwhile, alternative approaches like self-deterministic management [6, 7], agile management [8, 9] and to some degree concepts of lean management [10, 11, 12], where definitions are less strict and interfaces more cooperative, are gaining more and more importance.

Understanding the principle of incomplete contracts [13, 14, 15, 16] already states that no result nor consumption of resources can be perfectly determined in advance which would be absolutely required by a hierarchical approach to division of work [17].

The well-known Principal-Agent Problem [13] points out that a perfect description of work and respective supervision can only be gained at a price of an at least equivalent effort as if not outsourcing the package. Thus, the goal of organization is finding the second-best solution instead of hoping for a never occurring first-best solution. The second-best solution is then represented by the balance between control and tolerance [18].

Part of the concept of coordination is doubtless the implementation of control mechanisms to adjust processes to required results. However, even then, the definition of perfect results to compare with is required. Thus, coordination not only focusses on the agents' side in order to fulfil the given requirements but also needs to deal with the principals' inability to formulate the perfect determination of work package as a solution to a given problem. This situation resembles the V-model [6] controlling not only the quality of a solution, measured as deviation from then given plan, but also the applicability of the result to solving the given problem appropriately. The therewith demanded extension of the principle of control, being absolutely compatible with the traditional approach of governing, however, not only requires a significantly higher provision of controlling resources but higher controlling delays as well, since the goal comparison procedure resides on a much more abstract level than just comparing facts to a plan.

On this background, we state that perfect determination, e.g. based on strictly hierarchical approaches, is principally not possible, neither on the agents' side nor on the principals' side [19, 20, 21, 22, 23]. Therefore, the question is to be brought up, to which degree within an inevitably inconsistent system local contradictories can be ruled out by respective controlling mechanisms and where perfection reaches its limits given by fundamental rules.

2. Organisation Modelled as a System

In order to understand the behavior of an organization [24], modelling as a system is required, forming the system as a set of elements and respective (systemic) interactions [25, 26, 27, 28, 29]. This comprises not only physical elements but all imaginable elements to be processed in order to complete the project.

Broken down to the utmost level of detail, each element is represented by a single variable q_i forming the state vector $\vec{Q}$ while all interactions are given by linear differential equations as the second term of a respective Taylor series. The constant terms can be eliminated by linear transformation of the reference system to the state of equilibrium $q_0 = 0 \quad \partial\vec{Q} / \partial t = 0$. Higher order terms are neglected for this approach, linear superposition property of impact is assumed.

$$\vec{Q}_{t+dt} = \vec{Q}_t + \vec{F}(\vec{Q}) \rightarrow \frac{\partial\vec{Q}}{\partial t} = \vec{Q}_0 + A\vec{Q} + \ldots\text{(higher orders)} \tag{1}$$

where A is the adjacency matrix and $\vec{Q}$ the state vector.
Written as components we obtain:

$$\frac{\partial q_j}{\partial t} = \sum c_{j,i} q_i \tag{2}$$

3. Unruled Inconsistent Systems/Organisations

As long as the system is not contradictory [27], the differential equations lead to the trivial solution and the system stabilizes at a state $\vec{Q} = \vec{0}$ (zero vector). Inconsistency is given if multiple influencing nodes are driving a common target towards different values. Formally, within a linear system, the differential equations are superposed. However, if the influencing parameters $c_{i,j}$ are strong and the impact is not performed simultaneously but e.g. alternatingly, a nonlinear system is built and instabilities show up on the time axis.

Example (Principal-Agent Problem) [13]*:* Two partners (1 and 2) are not agreeing on a certain value q_j. The principal (1) expects and formulates a certain quality q_j for a sub-product, while the agent (2) provides only a different value, for whatever reason. He probably knows better due to his expertise or cannot do better due to the given circumstances. Then, over time this absolute value is determined by each of the participants according to their personal (valid, however inconsistent) definition somehow alternatingly and induces the respective consequences to the adjacent elements.

In this context, the temporary $q_j(t)$ value is simply set by the locally determining party. Since the differential equation refers to only a modification of the previously set value, this needs to be understood as a very strong and very fast adaption to the given value (i.e. notifying the inadequate value and immediately setting it).

Remark: If the adjustment process is slow compared to the alternating determination of the contradicting elements, the resulting value represents the average opinion. This situation is equivalent to both parties accepting the other side's value to some degree over time.

Such inconsistencies are represented by the average resulting fuzziness of system variables. As long as factually inconsistencies are existing, the system models the behavior correctly. This includes averaging as well as oscillating or even escalating development.

Obviously, the two determining elements (players) maintain no direct interaction, namely nothing in order to clarify the situation. Only if one of the players is officially declared to be wrong, or both agree on a common value, the discrepancy is solved. Such procedure, however, is based on both players to realize each other's position or at least resulting value and start an attempt to solve, which means establish a corrective interaction.

4. Introduction of Controlling Structures

If means of controlling are established, one or both of the disagreeing parties or possibly a third party is observing the value in question and comparing this to an expected value. Based on this knowledge, forces are implied to bring the value back to expectation. In terms of systems theory, this is the introduction of a loop, where some impact is derived from an observation of a value [30].

The mathematical representation of any short loop (one member only, the controller returns only proportional reaction) controlling loop is given by the most simple structure:

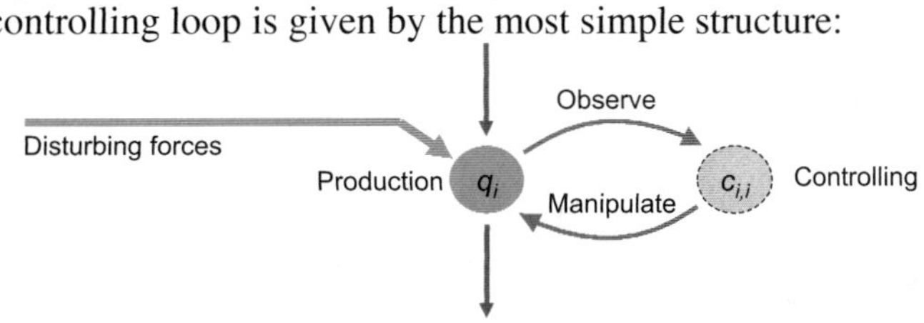

Figure 1. Basic System of a Theoretical Controlling Structure.

The general solution of this kind of equations is either oscillating, exponentially escalating or dampened, which is the preferred controlling behavior [31, 32].

$$\frac{\partial q_i}{\partial t} = c_i q_i \Rightarrow q_i \sim e^{c_i t} \tag{3}$$

The most general approach to understanding the time behavior of a starkly simplified subsystem of this kind is given by the differential equations for the harmonic oscillator. Any deviation of a value Q is answered by an automatized force leading back $-\beta$ to the desired value 0 working against the inertness μ towards any change and additionally taking some retarding forces ρ proportional to the rate of change into account.

$$\frac{\partial^2 Q}{\partial t^2} = -\frac{\beta}{\mu} Q - \frac{\rho}{\mu} \rightarrow \frac{\partial Q}{\partial t} - \omega^2 Q - k\frac{\partial Q}{\partial t} \tag{4}$$

using the damping factor $k = \rho/\mu$ and the frequency $\omega = \sqrt{\beta/\mu}$. The general solution is given by an approach of dampened oscillations as a complex function

$$Q(t) = Q_0 \cdot \exp(\lambda t) \quad \text{with } \lambda = -\frac{\rho}{2\mu} \pm \sqrt{\left(\frac{\rho}{2\mu}\right)^2 - \frac{\beta}{\mu}} := -\frac{k}{2} \pm \sqrt{\left(\frac{k}{2}\right)^2 - \omega^2} \tag{5}$$

Depending on the relationship between the damping factor $k = \rho/\mu$ and the frequency $\omega = \sqrt{\beta/\mu}$ this system is capable of performing dampened or escalating oscillations, as well as exponential approximating characteristics.

The situation of *weakly dampened oscillations* is determined by $k/2 << \omega$ where the root becomes negative and the solution therefore complex-valued. The frequency is different from the undamped frequency $\omega_\Lambda{}^2 = \omega^2 - k^2/4$ while the relaxation time is $\tau_R = 2/k = 2\mu/\rho$

Critical dampening refers to the situation where the root-term vanishes as no oscillations occur $k/2 = \omega \rightarrow \rho/2\mu = \sqrt{\beta/\mu}$. The solution reduces to a single exponential descent with an identical relaxation time constant $\tau_R = 2/k = 2\mu/\rho$:

Finally, the *overdamped case* is given if the root yields a real solution, i.e. $k/2 >> \omega$. The characteristic of the time development is also exponential, however, the time constant of relaxation is somewhat different.

$$\tau_R = \left[\frac{k}{2} \mp \sqrt{\left(\frac{k}{2}\right)^2 - \omega^2}\right]^{-1} \tag{6}$$

The primary solution (negative sign) indicates the damping factor $k/2$ reduced by a term depending on the relationship between frequency and damping factor and, thus, an increased relaxation time τ_R.

5. Theoretical Approach Applied on Delayed Integral Governors

Approximation of a Delayed Integral Controller

Understanding the characteristics of a harmonic oscillator obviously shows no direct connection to controlling mechanisms since terms like "inertia", "friction" and "retarding forces" have only symbolic meaning. However, from governing theory we know the principles of the integral controller as the most fundamental and stable concept.

$$\frac{\partial Q}{\partial t} = -k_C Q \tag{7}$$

The most influential parameter taken from real controlling systems which is not manifested here would be a significant time delay Δt resulting from finite detection patterns, lengthy considering and discussion procedures and, finally, from durations of initiation activities. Considering this parameter in particular, the DE takes on a different shape:

$$\frac{\partial Q(t)}{\partial t} = -k_C Q(t - \Delta t) \tag{8}$$

As a second order approach we substitute and develop $Q(t)$ in close proximity of t:

$$Q(t - \Delta t) \simeq Q(t) - \Delta t \frac{dQ}{dt} + \frac{\Delta t^2}{2} \frac{d^2 Q}{dt^2} + \ldots \tag{9}$$

Inserting this into the differential equation of the delayed integral controller yields finally

$$\frac{d^2 Q}{dt^2} = \frac{-2k_C}{k_C \Delta t^2} Q + \frac{2(k_C \Delta t - 1)}{k_C \Delta t^2} \frac{dQ}{dt} \tag{10}$$

This expression again needs to be compared to the harmonic equation

$$\frac{\partial^2 Q}{\partial t^2} = -\frac{\beta}{\mu} Q - \frac{\rho}{\mu} \frac{\partial Q}{\partial t} \tag{11}$$

Thus, we identify as a useful approximation using $\tau_C = 1/k_C$ as the time constant of the original governor:

$$\mu = \frac{\Delta t^2}{\tau_C} \quad \beta = \frac{2}{\tau_C} \quad \rho = 2\left(1 - \frac{\Delta t}{\tau_C}\right) \tag{12}$$

Characteristics of Delayed Integral Controllers

On this background, the behavioral cases of an integral controller with controlling strength $k_C = 1/\tau_C$ and subjected to some delay Δt can be formulated:

The *overdamped controller* is well capable to rule out any deviation, however not in the shortest possible time. The relaxation time is given by:

$$\tau_R = \left[\frac{k}{2} \mp \sqrt{\left(\frac{k}{2}\right)^2 - \omega^2}\right]^{-1} = \left[\frac{\rho}{2\mu} \mp \sqrt{\left(\frac{\rho}{2\mu}\right)^2 - \frac{\beta}{\mu}}\right]^{-1} = \Delta t^2\left[\left(\tau_C - \Delta t\right) \mp \sqrt{\tau_C^{\,2} - 2\tau_C\Delta t - \Delta t^2}\right]^{-1} \tag{13}$$

Developing the full term for $\Delta t^2/\tau_R << \tau_C - \Delta t$, i.e. a controller time constant τ_C far off the time delay Δt in comparison to Δt itself and it's relation to the relaxation time τ_R, we obtain with $\vartheta := \left(\tau_C - \Delta t\right)$

$$\mp\sqrt{\vartheta^2 - 2\Delta t^2} = \frac{\Delta t^2}{\tau_R} - \vartheta \quad \Rightarrow \quad 1 = \frac{2\vartheta\tau_R - \Delta t^2}{2\tau_R^{\,2}} \tag{14}$$

With $\Delta t^2 << \vartheta\tau_R$ the second term can be neglected and the remaining term yields $\tau_R = \vartheta = \left(\tau_C - \Delta t\right)$.

The *critical setting* is given by the condition $k/2 = \omega \rightarrow \rho/2\mu = \sqrt{\beta/\mu}$ and therewith $\tau_C = \Delta t(1+\sqrt{2})$ which leads to the optimal relaxation time constant:

$$\tau_R = 2/k = \frac{2\mu}{\rho} = \frac{\Delta t^2}{\tau_C - \Delta t} \tag{15}$$

Finally, the attempt to rule out deviations within shorter times (*weakly dampened*) invokes the oscillating solution where the overall relaxation time is:

$$\tau_R = 2/k = \frac{2\mu}{\rho} = \frac{\Delta t^2}{\tau_C - \Delta t} \tag{16}$$

Remark: As long as $\tau_R > 0$, the oscillations are dampened and finally run out. However, with $\tau_R > 0$ the exponential function changes its character to an escalating behavior. With $\tau_R = \Delta t^2/\left(\tau_C - \Delta t\right)$ this transition occurs at $\tau_C = \Delta t$ where the exponent changes sign.

Figure 2 shows the characteristic behavior of such a controlling structure over a wide range of controlling strength $k_C = 1/\tau_C$ for given unity time delay $\Delta t = 1$:

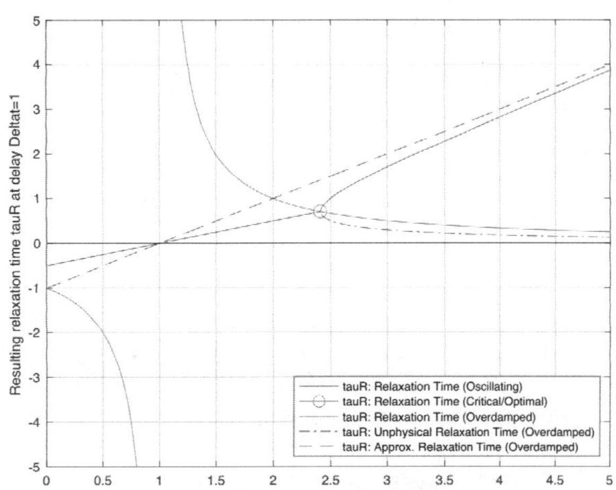

Figure 2. Theoretical development of relaxation time τ_R ranging from oscillating over the critical/optimal setting to the overdamped situation.

We clearly observe the critical setting as the optimal selection of controller time constants for obtaining the shortest possible governing time. Increasing given time constants τ_C, i.e. weaker governing strengths $k_C = 1/\tau_C$, lead to increasingly slower relaxation times. From the mathematically point of view, two branches are possible, where the correct one approaches a linear function with gradient 1, intersecting the abscissa at $\tau_C = \Delta t$.

Smaller time constants corresponding to applying a stronger controller (raise k_C) produces instable behavior performing increasingly oscillations, however damped and therefore still stabilizing after some time. Only if the controller time constant τ_C reaches the value of the given time delay Δt, the systems behavior changes from dampened to escalating oscillations with the change of sign.

6. Explicit Evaluation of Required Tolerance Margins

These theoretical results can easily be applied on a singular presumably linear production process. Let the rate of production, respectively the invested production resources be $R_{Prod} = dQ/dt$, leading to the product determined as quality Q_{Fin} after the production duration t_{Fin}.

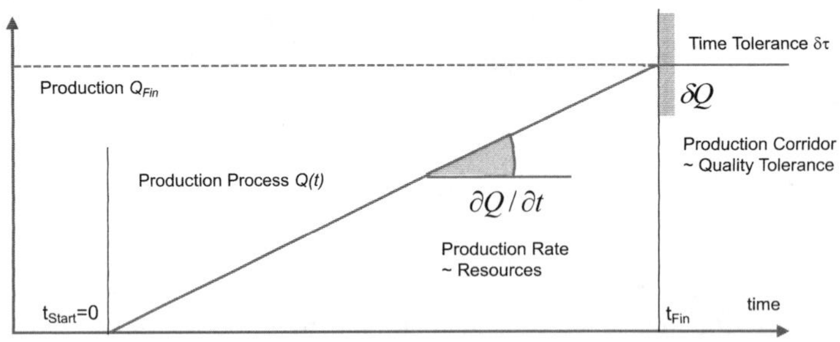

Figure 3. Linear process subjected to controlling.

Time Tolerance of a Controlled Linear Production Process

Using an implemented controlling process over the whole process where the delay Δt is given, the optimal controlling strength is about $k_C \simeq 0{,}41/\Delta t$ and the time constant $\tau_C \simeq 2{,}41 \cdot \Delta t$. From this we derive the optimal relaxation time $\tau_R \simeq 0.707 \cdot \Delta t$. However, since τ_R represents the time to bring a deviation down to $1/e$, we conclude a sensible required *time tolerance* to settle a possible deviation e.g. to 1% as: $\delta\tau = \tau_R \cdot \ln 0.01 \simeq 3{,}26\Delta t$, which is valid during the process and, thus, as well at the end of the process.

Quality Tolerance of a Controlled Linear Production Process

In order to derive a term helpful as quality tolerance we make use of [33] where we understand the meaning of the differential equation for an integral controller:

$$\frac{\partial Q_I}{\partial t} = [Q_I(t) - Q_{I0}(t)]k_I \tag{17}$$

as k_I corresponds to the percentage of the actual deviation of value Q_I that is invested efficiently in the production speed, thus in the resources ready exclusively for controlling purposes per time unit. Accordingly, we understand and name the ressources used for controlling purposes in this context $R_{Contr} = dQ_I/dt$.

Therein, we insert the known optimal controlling strength $k_C \simeq 0{,}41/\Delta t$ and transform to the equilibrium system $Q_{I0} = 0$: $R_{Contr} = 0{,}41 \cdot \delta Q_{Contr} / \Delta t$. Rearranging gives a measure for the deviation δQ_{Contr} which can be mastered within the time tolerance $\delta\tau$ if the controlling ressources are given:

$$\delta Q_{Contr} \simeq 2{,}41 \cdot \Delta t \cdot R_{Contr} \tag{18}$$

Dividing by the resources available for production $R_{Prod} = dQ/dt$ and integrating over the production time we obtain:

$$\delta Q_{Contr} = 2{,}41 \cdot \Delta t \cdot R_{Contr} \quad \Rightarrow \frac{\delta Q_{Contr}}{R_{Prod}} = 2{,}41 \cdot \Delta t \cdot \frac{R_{Contr}}{R_{Prod}} \quad \Rightarrow \frac{\delta Q_{Contr}}{Q_{Fin}} = 2{,}41 \cdot \frac{\Delta t}{t_{Fin}} \cdot \frac{R_{Contr}}{R_{Prod}} \tag{19}$$

Thus, we obtain the manageable relative deviation $\delta \tilde{Q}_{Contr} = \delta Q_{Contr} / Q_{Fin}$ requiring relative controlling ressources $\tilde{R}_{Contr} = R_{Contr} / R_{Prod}$ where the ratio is besides a factor of order 1 mainly determined by the time delay in comparison the process completion time.

$$\delta \tilde{Q}_{Contr} \simeq 2{,}41 \cdot \frac{\Delta t}{t_{Fin}} \cdot \tilde{R}_{Contr} \tag{20}$$

Remark: This value represents, what can be managed given the controlling ressources and the time tolerance derived from the controller during the process and, thus, at the end of the process as well.

In this context the time delay Δt represents the responsiveness of the controlling process. Thus, we replace Δt by the previously elaborated absolute time tolerance $\delta\tau = \tau_R \cdot \ln 0{,}01 \simeq 3{,}26\Delta t$:

$$\frac{\delta Q_{Contr}}{Q_{Fin}} = 2{,}41 \cdot \frac{\Delta t}{t_{Fin}} \cdot \frac{R_{Contr}}{R_{Prod}} \Rightarrow \frac{\delta Q_{Contr}}{Q_{Fin}} = \frac{2{,}41}{3{,}26} \cdot \frac{\delta\tau}{t_{Fin}} \cdot \frac{R_{Contr}}{R_{Prod}} \Rightarrow \frac{\delta Q_{Contr}}{Q_{Fin}} = 0{,}74 \cdot \frac{\delta\tau}{t_{Fin}} \cdot \frac{R_{Contr}}{R_{Prod}} \tag{21}$$

Referring to a normalized process ($Q_{Fin} = 1, t_{Fin} = 1, R_{Prod} = 1$), we obtain the relation between manageable quality and time tolerance:

$$\delta \tilde{Q}_{Contr} \simeq 0{,}74 \cdot \delta\tilde{\tau} \cdot \tilde{R}_{Contr} \text{ and thus } \frac{\delta \tilde{Q}_{Contr}}{\delta\tilde{\tau}} \simeq 0{,}74 \cdot \tilde{R}_{Contr} \tag{22}$$

So far and very obviously we state the ratio of relative manageable quality to relative time tolerance given by the continuously available controlling ressources held ready. This relationship is widely valid, however, since the time tolerance $\delta\tilde{\tau}$ is principally predetermined (and limited) by the underlying controlling process, the resulting manageable quality $\delta\tilde{Q}_{Contr}$ is now tightly bound to $\tilde{R}_{Contr}$ and limited as well.

On this basis we estimate the general ability of controlling mechanisms to enforce a given and predetermined quality over the production process. Assuming given controlling resources $\tilde{R}_{Contr}$, quality deviation incidents up to $\delta\tilde{Q}_{Contr}$ can be compensated for if the time-tolerance $\delta\tilde{\tau}$ is allowed for. This is certainly true during the run of the process as well as at the end t_{Fin} where the controlling needs to continue operation using $\tilde{R}_{Contr}$ for another period of about $\delta\tilde{\tau}$. However, this limits the manageable quality deviation principally and all expected deviation exceeding this value need to go into a definite quality tolerance margin.

$$\delta\tilde{Q} = \left|\delta\tilde{Q}_{Dev}\right| - \left|\delta\tilde{Q}_{Contr}\right| = \left|\delta\tilde{Q}_{Dev}\right| - \left|0{,}74 \cdot \tilde{R}_{Contr}\delta\tilde{\tau}\right| \tag{23}$$

7. Conclusion

Based on these theoretical considerations we state that any organizational structure can principally not be set up consistently and therefore will be in fundamental need of concepts of not only controlling overhead but also of tolerance. Any inconsistent system, i.e. not only far off the equilibrium state, but where the equilibrium state is only dynamically determined, cannot be stabilized totally.

If ruling mechanisms are at hand which principally allow inducing forces strong enough to compensate for inconsistencies, the system may become controllable. The optimal time constant of the balancing process as the central controlling parameter is widely independent of the degree of inconsistency or controlling force. The main parameter turns out to be the controlling delay, i.e. the reaction time of the controlling mechanism which cannot be assumed to vanish. Therewith, a principal minimum balancing time is given. At any attempt to reduce this by employing stronger controlling mechanisms, the system becomes increasingly unstable and leads to escalating behavior.

Thus, the rules for an acceptably controllable system are

- Clear definition of work-packages generated by probably strictly hierarchical structures.
- Understanding of the principally inherent inconsistency of this concept.
- Reducing complexity by avoiding long loops which increase the reaction time.

- Implementing local controlling structures, very close to all creative processes and independent of other structures (Complexity).
- Not only comparing quality to the pre-set plans but also to the capability of the production process and the suitability as a solution to the given problem.
- Determining the controlling strength according to the inherent controlling delay.
- And last and most importantly, introduce quantitatively sufficient tolerance on interfaces to deal with the unavoidable discrepancies of quality.

The consequences of this approach are proposed e.g. by concepts of lean management, however not exclusively. According to VDI 2553 [10] or the SCRUM manifest [8, 9] in particular the acceptance of inherent inconsistencies of organization and the short loop approach to tackle these is pointed out. After constructing the organizational system based on highly hierarchical structures with resulting inconsistent declarations, long communication paths and missing feed-back-loops, self-organizing teams with short paths and fast loops are implemented to solve discrepancies efficiently.

References

[1] Smith A 1776 *An Inquiry into the Nature and Causes of the Wealth of Nations* vol 1, Nachdruck von 1981, Indianapolis, Indiana, USA, S. 14f., ISBN 0-86597-006-8.

[2] *Taylor F 1911 The principles of scientific management.* London: Harper & Brothers, 1911 (Cosimo, New York 2006, ISBN 1-59605-889-7).

[3] Eber W and Zimmermann J 2018 *Evaluating and Retrieving Parameters for Optimizing Organizational Structures in Real Estate and Construction Management*, https://doi.org/10.3311/PPar.12709, Periodica Polytechnica Architecture, Budapest University of Technology and Economics, Budapest, Hungary, 2018. 49 pp 155–164

[4] Kerzner H 2003 *Project Management: A Systems Approach to Planning, Scheduling, and Controlling* (8th ed.), ISBN 0884222245414, Wiley, Berlin.

[5] Schulte-Zurhausen M 2002 *Organisation,* 3rd ed., Verlag Franz Vahlen, München.

[6] Schelle, Ottmann and Pfeiffer 2005 *Project Manager*, GPM Deutsche Gesellschaft für Projektmanagement, ISBN 9783800637362, Nürnberg

[7] Malik F 2003 *Systemisches Management, Evolution, Selbstorganisation*, 4th ed., ISBN 3834964409, Haupt Verlag, Bern

[8] Beck K et al. 2001 *The Agile Manifesto* downloaded April 2021 28th from https://agilemanifesto.org/iso/de/principles.html

[9] Schwaber K and Sutherland J 2020 *Scrumguide,* downloaded April 2021 28th from https://scrumguides.org/docs/scrumguide/v2020/2020-Scrum-Guide-US.pdf

[10] Verein Deutscher Ingenieure e.V. 2019 *VDI 2553 Lean Construction,* Beuth-Verlag, Düsseldorf

[11] Koskela L 2000 *An exploration towards a production theory and its application to construction*, ISBN 951-38-5565-1, VTT-Publications, Kivimiehentie, Finland.

[12] Koskela L, Ballard G, Howell G and Tommelein I 2002 *The foundations of lean construction,* Design and Construction: Building in Value https://doi.org/10.4324/9780080491080-23

[13] Picot A, Dietl H and Franck E 2008 *Organisation - Eine ökonomische Perspektive*, 5. rev. ed., ISBN 978-379-102371-7, Schäffer-Poeschel, Stuttgart

[14] Hoffmann W. and Körkemeyer K 2018 *Zum Umgang mit der Komplexität von Bauvorhaben – Ergebnisse einer* Expertenbefragung Bauingenieur 93 Springer, VDI-Verlag, Düsseldorf.

[15] Ebeling W, Freund J and Schweitzer F 1998 *Komplexe Strukturen: Entropie und Information*, ISBN 9783815430323, Teubner Verlag, Stuttgart

[16] Liening A 2017 *Komplexität und Entrepreneurship*, Springer Gabler , ISBN 978-3-658-13173-9, pp186ff, 431ff.

[17] Caldarelli G and Vespignani A 2007 *Complex Systems and Interdisciplinary Science*. Large Scale structure and dynamics of complex networks, https://doi.org/10.1142/6455 World Scientific Publishing Co. Pte. Ltd. Vol.2., pp. 5-16.

[18] Coase R H 1937 *The Nature of the Firm* Economica. https://doi.org/10.1111/j.1468-0335.1937.tb00002.x, New Series. vol 4 16
[19] Strogatz S H 2001 *Exploring complex networks*, https://doi.org/10.1038/35065725, Nature 410, p.268
[20] Wassermann S and Faust K 1994 *Social Network Analysis*. ISBN-10: 0521387078, Cambridge University Press, Cambridge.
[21] White D R, Owen-Smith J, Moody J and Powell W W 2004 *Networks, Fields and Organizations*, https://doi.org/10.1023/B: CMOT.0000032581.34436.7b, Computational and Mathematical Organization Theory 10:95-117
[22] Winch G 2006 *Towards a theory of construction as production by projects* In: Building Research & Information (2006) 34(2), 164–174, https://doi.org/10.1080/09613210500491472
[23] Eber W 2020 *Potentials of Artificial Intelligence in Construction Management, OTMC Organization, Technology and Management* in Construction: an International Journal, DeGruyter and The University of Zagreb, Faculty of Civil Engineering, Zagreb, 2020.
[24] Booch G et al. 2007 *Objektorientierted Analysis and Design*, 3rd ed. ISBN-13: 978-8131722879, Addison-Wesley, Bonn, 2007
[25] Bertalaffny *L* 1969 *General Systems Theory,* ISBN 780807604533 George Braziller Inc. New York 1969, P 54 ff.
[26] Luhmann N 2001 *Soziale Systeme. Grundriß einer allgemeinen Theorie*. ISBN 3-518-28266-2, Frankfurt am Main 1984. (ISBN 3-518-28266-2)
[27] Zimmermann J and Eber W 2017 *Criteria on the Value of Expert's Opinions for Analyzing Complex Structures in Construction and Real Estate Management,* Creative Construction Conference 2017, CCC 2017, 19-22 June 2017, Primosten, Croatia, https://doi.org/10.1016/j.proeng.2017.07.208, Procedia Engineering, Volume 196, 2017, Pages 335-342
[28] Haken H 1983 *Synergetik*, ISBN 978-3540110507, Springer Verlag, Berlin, Heidelberg, New York, Tokyo.
[29] Newman M E J 2003 *The Structure and Function of Complex Networks*, https://doi.org/10.1137/S003614450342480, SIAM Review 45, pp. 167-256
[30] Wiener N 1992 *Kybernetik*, ISBN 978-3430196529, Econ Verlag, Düsseldorf, Wien, New York, Moskau.
[31] Eber W 2019 *Managing Construction Projects: Developing Complexity into Complicatedness*, Sustainable Built Environment D-A-CH Conference 2019, Graz University of Technology, Graz, Austria, 2019
[32] Zimmermann J and Eber W 2014 *Mathematical Background of Key Performance Indicators for Organizational Structures in Construction and Real Estate Management,* https://doi.org/10.1016/j.proeng.2014.10.585, Procedia Engineering, Volume 85, 2014, Pages 571-580
[33] Eber W 2019 *Bauprozessmanagement und und Immobilienentwicklung - Von den Grundlagen zur Anwendung* (Vortrag), Kolloquium Investor - Hochschule Bauindustrie 2019, LBI-Lehrstuhl Bauprozessmanagement, München, 2019.

Teaching Design Management tools during a pandemic: A Peruvian case study

X Brioso[*] and C Calderon-Hernandez

GETEC Research Group, Department of Engineering, Pontifical Catholic University of Peru, Av. Universitaria 1801, Lima 32, Lima, Peru

[*] Corresponding author's e-mail address: xbrioso@pucp.edu.pe

Abstract. During the year 2020, the sanitary emergency norm forced universities to teach under the remote learning method, representing a challenge of designing courses that require high effectiveness. The purpose of this work is to present the strategies that were successfully applied during the teaching of design management tools for a building project, implemented in a course part of the Civil Engineering master of the Pontifical Catholic University of Peru. The methodology considered the effect of the Covid-19 Pandemic and had as objective to develop soft skills and employability, for which a balanced use of virtual platforms as well as software was proposed for videoconference and laboratories. On the other hand, during this course, students developed a proposal for integrated planning and management of the design phase using tools of norms and guidelines of project management systems such as ISO 21500, Lean Construction, Building Information Modeling, PMBOK of the Project Management Institute, Integrated Project Delivery, among others. Under this framework, the participation of stakeholders, multicriteria decision making, development of project design deliverables and project sequence was analyzed, which presented an important change under the collaborative contracts. The success of this methodology is reflected in the high quality of final papers and expositions, the result of surveys taken - effectiveness of 99%-, and in the improvement opportunities detected. This paper presents important information that may be adapted by academic and practitioners during the phases of the pandemic and post-pandemic.

Keywords: Teaching, New tools in education, Design Management, Decision making, Visualization in design.

1. Introduction

At the beginning of 2020, COVID-19 virus started to spread all over the world, affecting a large number of citizens. For this reason, on March 13th, the World Health Organization (WHO) declared a global pandemic. [1]. The WHO directives were taken into consideration by most countries' regulations, including Peru, with the objective of controlling the spread of the virus with diverse measures, such as stoppage of many sectors, including education and construction [2]. Before this scenario, the design, construction and commissioning phases of construction had to adapt to protocols that imposed strong restrictions for the restart of projects, maximum number of workers in an area, staggered entries, among others [3]. All these measures had to be embodied in the courses of architecture and construction at universities. Indeed, the sanitary emergency norm forced universities to apply remote teaching as an alternative to give students the opportunity to continue and successfully finish their semester [4]. Most of these students were not used to this, so strategies needed to be developed for both teachers and students to adapt. The switch starts with the lead professor who needs to involve the interested parties to assume the responsibility and act rapidly to develop a joint plan to face these challenges, trusting their leadership, vision, culture and capacity to create the needed relationships with innovation and creativity [5, 6].

The main purpose of this article is to describe the strategies implemented during the pandemic in the course "Planning and Integrated Management of the design of a Building Project", a course part of the master's degree of the civil engineering program at the Pontifical Catholic University of Peru. This teaching methodology had to be implemented in a very short period, so that the academic semester wouldn't be delayed any longer. Also, as a part of the course, students were taught tools such as multicriteria decision making methods, value stream mapping, and technology like Augmented Reality and BIM.

2. Course Description

During the times before the sanitary emergency, the teaching methodology was face to face and consisted of the development of classes, laboratories, and visits of professionals specialized in design and construction management, where students observed and discussed the design, constructability, and planning processes of a project. On the other hand, since 2019, the research from GETEC research group has been incorporated in courses, especially the ones generated by tools and technologies applied in design, planning and construction. However, as a consequence of COVID-19, the norm enforced that those classes, laboratories and evaluations were made virtual. [2, 4].

The proposal was a flexible integration of tools and technologies of construction management that allowed a work team to operate efficiently on the design and planning phases, according to Lean Construction [7, 8, 9], Project Management Institute [10], Building Information Modeling (BIM) [11, 12, 13], ISO 21500 [14], among others. Flexible learning methods were included to prevent eventual damage to the learning trajectory of students, providing resources that are useful during the pandemic and post-pandemic times [15]. Through these strategies competencies and soft skills were generated in our students, which are very valued by the industry, contributing to their employability.

3. Tools and Technologies

The real conditions of design and planning stages of a building Project were simulated, these accessible technologies were selected for the student body.

3.1. Building Information Modeling (BIM)

BIM is a system that promotes innovation and allows the management of digital representations of physical and functional characteristics of construction projects, creating long term value [16]. BIM system is a generic term used to name the software BIM, which includes tools, platforms or servers used within a BIM environment to admit the information modeling of a project [11]. During, the pandemic active methodologies are being generated based on real projects though BIM virtual models. With the implementation, students learn through the virtual models without the restriction of online classes, learning to collaborate from anywhere, acquiring skills that help them to face situations like COVID-19 efficiently [17].

3.2. Virtual Reality (VR)

Virtual reality (VR) is a visualization technique that is applied in different areas, including building construction, and may be immersive or non-immersive [18]. The use of virtual reality (VR) and the creation of digital prototypes (digital mock-up) for the design revision is being implemented more during the design and construction phases [18]. Moreover, there is a software of virtual reality rendering available for BIM tools of common use [11, 19].

3.3. Augmented Reality (AR)

Augmented Reality (AR) is a tool that enhances the user perception over a real environment with the use of mobile devices like tablets or cell phones [20]. There are several prototypes and framework proposals that allow the integration of BIM and AR, and there are several applications available for the design and construction phases of projects [21]. In our proposal we used the application of new visualizations techniques like AR integrated with BIM in the workflow.

3.4. Drones

Drones or UASs (Unmanned Aerial Systems) are equipped systems with video cameras that may be used for data surveys. The use of these systems for project monitoring is expanding and even though it still has its limitations, it's important to consider and identify the opportunities it brings [22].

3.5. Multicriteria Decision Making Methods

Tools such as Choosing By Advantages (CBA), which consists of listing factors and criteria, to then determine the advantage of each alternative by assigning value to each one [23]. As well as value stream mapping. CBA is based exclusively on advantages instead of advantages and disadvantages, avoiding double counting or sums and subtractions [24]. In addition, it has been proven that the use of CBA and BIM improves the collaboration among stakeholders [25], optimizes project planning [26], and reduces the decision times and improves the transparency of the system [23, 27].

3.6. Value Stream Mapping (VSM)

A value stream map is composed of the value-added actions and non-value-added actions needed to make a production Flow and manufacture products from the raw material to the client delivery [28]. VSM is a tool of the Lean Production that allows a systemic view of the value in the process of production, identifying problems and real loses, lessons learned and continuous improvement [29]. There are several adaptations to this Lean tool collaboratively implemented by work teams in construction projects [30].

3.7. Delphi Method

The Delphi method is an iterative decision-making method performed anonymously by experts leaded by a moderator who analyses the results each round and communicates them to the participants until consensus is reached [31]. This method spread rapidly since it offers a flexible and simple methodology while also reducing the influence of unwanted psychological effects among participants (inhibition, dominant personalities, etc.) [32]. The methodology of the Delphi study are: (1) Identify the research objective; (2) Identify and select the expert panel according to predefined criteria; (3) Developing questionaries; (4) Transmitting questionaries to the panel; (5) Analyzing responses; (6) Evaluating consensus and, if needed, repeating the process [33].

4. Methodology

Classes and evaluations needed to be done remotely, which was, as previously indicated, a challenge both for students, and teachers. For this, the university's Moodle platform named Paideia, as well as the Zoom Platform were used. In Paideia teachers uploaded all the content of the course, the folders where students turned in their assignments and the links for the zoom meetings. Zoom, allowed students to connect to a session either with their computer or mobile device. Teachers and teaching assistants were aware that given the emergency, not all students may be able to connect at the time of the session for classes, so these were recorded and later uploaded. Students were expected to assist all 5 evaluations, however. Case studies from our research group were shared with students for a more in-depth vision of the application of the tools and technologies imparted.

This course had 5 bi-weekly evaluations, which students needed to complete in 3 hours each. These were done in groups of 5 members, encouraging coordination and soft skills in students. These evaluations were divided in 2 sections, a theoretic part, and a laboratory part. First, the teacher would present a brief case study and explain the theory and tools that needed to be applied. Then, students were divided in break rooms, to work on the assignment. During this period, the teacher was able to enter and exit a breakroom to monitor their progress and to answer any questions.

5. Evaluations

5.1 Evaluation 1: Use of Scoring to select the best design alternative

Evaluation 1 consisted of the selection of the best out of 3 design alternatives for the design of an apartment complex with the use of the scoring method. Here, students were given tables, like table 1 shown below, with the score for the purposes of both the investor and the final user, and they had to ponder these with the score assigned to each one of the alternatives, that presented a different type of structure, finish, and installations. At the end of the assignment, students were asked to discuss the results and analyze the risks and mitigation actions associated with the selected alternative.

Table 1. Needs and Values and corresponding score for the investor point of view.

	Needs and Values	Indicator	Metric	1 – 5 score
INVESTOR	Cost Effectiveness	Investment	US $ 17,000,000.00	5
		Investment Period	18 meses	
		Minimum return rate	24%	
		Minimum profit	15%	
		Risk level	Tolerable	
	Image	Positioning	No	3
		Social Responsibility	Yes	
		Reputation	Yes	

5.2 Evaluation 2: CBA, VR and AR to select the best design alternative

For the second evaluation, the Nativity Church intervention Project was presented to students. This project consisted of a structural reinforcement of a foundation in the Saint Jerome Hall. For the stakeholders of this Project, the CBA method was used to choose between 2 design alternatives. For this case study, the technology applied was laser scanning, point cloud important, virtual reality and augmented reality. The app "Kubity" was used to share the 3D model of the two design alternatives proposed, so that they could use Augmented Reality to visualize the environment and the design for each one. Then, in groups, they discussed the alternatives and used the CBA method to choose the best alternative, they also had to discuss the results and the decision-making process, analyzing whether they considered it was transparent or not. Figure 1 shows virtual reality captures on the Kubity app for both intervention design alternatives.

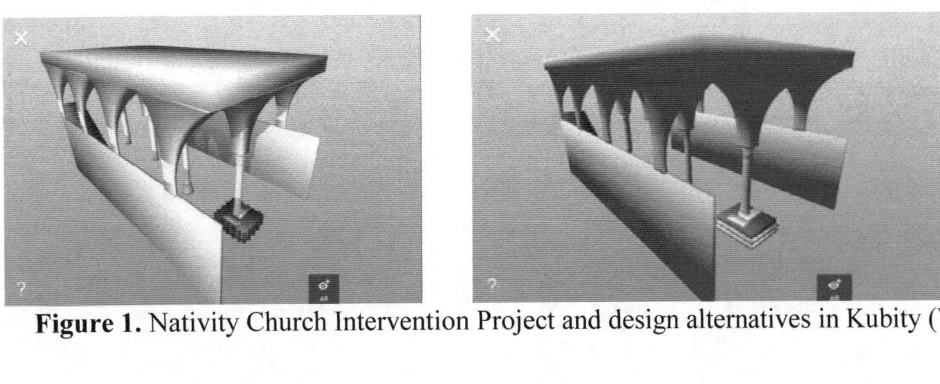

Figure 1. Nativity Church Intervention Project and design alternatives in Kubity (VR).

5.3. Evaluation 3: Using DELPHI and CBA method to select the best construction sequence

On the third evaluation, students used a combination of CBA and Delphi, to select the best construction process design. In this case, there were 3 alternatives, for a building with the same number of stories and construction area, each one had variable modulation, standardization, and industrialization. During the session, each group chose a stakeholder, to perform the analysis from that specific point of view. After using the CBA method in groups to select the best alternative, all students joined the main session to use the Delphi method and evaluate the consensus. A member of each group presented their winning alternative and their most important advantage. After 2 rounds, consensus was reached, and students understood the method and its applicability.

5.4. Evaluation 4: Using Value Stream Mapping for the design phase of a building project.

At the beginning of the session, students learned Value Stream mapping theory and were given examples of how to use this tool to improve processes and reduce waste. They were then asked to elaborate the Flow chart for the design process of a building Project, including all systems: architectural, structural, mechanical, electrical, and plumbing. For this, they used the app Miro, which allowed students to collaboratively work in the same board simultaneously. Figure 2 is a partial capture of the board prepared by one of the groups. Here, the flow, tasks and decisions are observed.

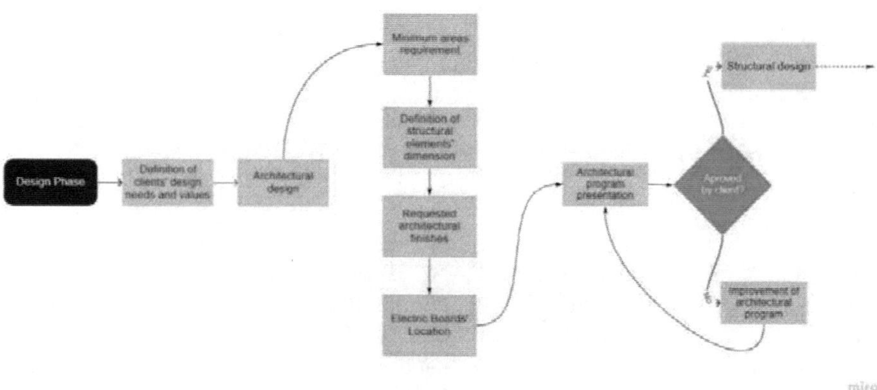

Figure 2. Real example of a group flow chart

5.5. Evaluation 5: Integration of Tools and Technologies applied to a Construction Project

For the final group Project, students had to develop a proposal for an integrated planning and management design phase using tools of norms and guidelines of Project management systems like PMI, ISO21500, BIM and Lean Construction. Here they also had the opportunity to apply all the tools and skills developed during the course for a Project of their choice. Figure 3 is a capture of a group Project presentation. Here, this group applied BIM, more specifically 3D modelling with Autodesk Revit, to integrate the information of the project. In the image we can observe the different systems of a building being integrated.

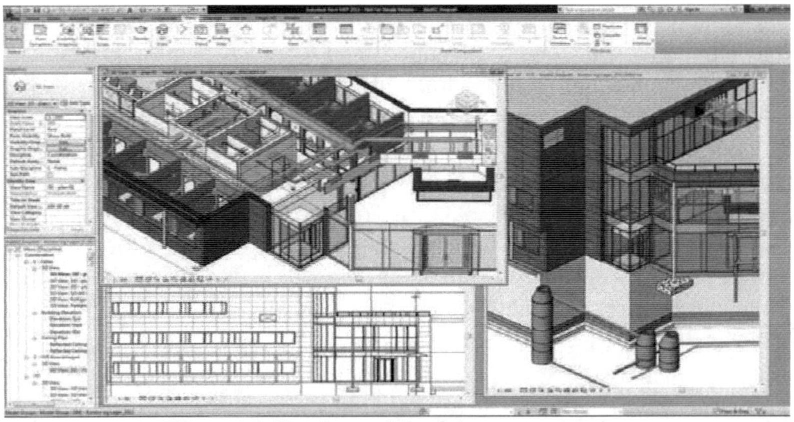

Figure 3. Revit model of the group project

In addition, the group used the software Revizto for collaborative design process and identifying the most important interferences between the different systems of a project. Here, they were able to apply this tool and identify the strengths of a collaborative design process.
Here are some of the interferences identified. In this case an unfinished Shell connection for a Steel structure. Finally, with this technology, the group made their simulations and final design proposal.

6. Results

The efficiency of the course is reflected on the excellent average of the class. Moreover, after every evaluation the students gave feedback regarding the methodology and the tools and technology applied. The feedback was positive and given the students previous experience, more opportunities to apply the tools and technology were given. For example, on the second evaluation, students indicated that the use of virtual reality and augmented reality made the process of using CBA even more transparent, since it increased the understanding of the design proposal, they also felt that it simulated the use of these tools in a real project. Something similar occurred after evaluation 4 where students shared the opportunities where they could also apply value stream mapping on their day to day professional lives.
The final course survey indicated that students benefited greatly from this methodology, as it is shown in Table 2, the course efficiency exceeded the efficiency in all categories, in comparison to the average efficiency of all courses of the engineering department in that semester.

Table 2. Course survey results.

Survey Category	Average efficiency of courses of the Engineering Department (%)	Course efficiency (%)
A. Learning strategy	87.0	97.6
B. Methodology and resources	84.6	100.0
C. Learning assessment	84.4	97.5
D. Communication and interaction	88.0	100.0
GENERAL QUALIFICATION	**85.8**	**99.0**

7. Conclusions

It has been observed that virtual courses can be just as efficient and beneficial for students as regular face to face courses when the proper methodology is applied. In this case, being students of a master's degree, they already had experience in building construction, so this had to be enhanced with case studies and examples that resonated with them and complemented their knowledge. Also, the group work was enriching, since they came from different backgrounds and work experience. Dynamism and permanent interaction between students, teachers and teaching assistants was key, so questions during classes and evaluations were encouraged, moreover, students used the forum outside sessions in the Paideia Platform, to share their personal experiences, doubts, and any resource they considered worth sharing with their classmates. According to students, classes and laboratories successfully simulated real work conditions of the pandemic times and may be also applied post-pandemic.

Finally, the use of lean tools and technology gave students a more open vision of the design management process. Also, they developed skills that they will be able to apply in their respective work areas, improving the design process of all projects they are involved with. This is very important considering this stage is often neglected.

References

[1] World Health Organization 2020 *Coronavirus disease (COVID-19).*

[2] Ministerio de Salud 2020 *Resolución Ministerial N° 448-2020/MINSA Lineamientos para la vigilancia, prevención y control de la salud por exposición al SARS-CoV-2.*

[3] Ministerio de Vivienda, Construcción y Saneamiento 2020 *Resolución Ministerial N° 087-2020* Protocolo Sanitario Sectorial.

[4] Ministerio de Educación 2020 *Resolución Viceministerial N° 157-2020/MINEDU Orientaciones para el desarrollo del servicio educativo en los Centros de Educación Técnico-Productiva e Institutos y Escuelas de Educación Superior, durante la Emergencia Sanitaria causada por el COVID-19.*

[5] Quezada R L Talbot C and Quezada-Parker K B 2020 *From Bricks and Mortar to Remote Teaching: A Teacher Education Programs Response to COVID-19* Journal of Education for Teaching, 46:4, 472-483

[6] National Policy Board for Educational Administration 2015 *Professional Standards for Educational Leaders* Reston, VA: Author

[7] Gao Z Ezekwem K and Aslam M 2020 *An Integration of BIM and GIS for Integrated Project Delivery* Construction Research Congress. March 8–10, 2020 | Tempe, Arizona.

[8] Brioso X 2015 *Teaching Lean Construction: Pontifical Catholic University of Peru Training Course in Lean Project & Construction Management.* Procedia Engineering, 123 (2015) 85 – 93.

[9] Brioso X. Humero A and Calampa S 2016 *Comparing Point-to-Point Precedence Relations and Location-Based Management System in Last Planner System: A Housing Project of Highly Repetitive Processes Case Study* Procedia Engineering, 164 (2016), pp. 12–19.

[10] Project Management Institute 2017 *A Guide to the Project Management Body of Knowledge (PMBOK® Guide)* 6th Ed Institute, 6th Ed.

[11] Sacks R Eastman C Lee G and Teicholz, P. 2018 *BIM Handbook: A Guide to Building Information Modeling for Owners, Designers, Engineers, Contractors, and Facility Managers*. Wiley.

[12] Brioso X Murguia D and Urbina A 2017 *Teaching Takt-Time, Flowline, and Point-to-Point Precedence Relations: A Peruvian Case Study* Procedia Engineering, 196, pp. 666-673.

[13] Aslam M Gao Z and Smith G 2021 *Integrated implementation of Virtual Design and Construction (VDC) and Lean Project Delivery System (LPDS)* Journal of Building Engineering, Volume 39, 2021, 102252.

[14] Brioso X 2015 *Integrating ISO 21500 Guidance on Project Management, Lean Construction, and PMBOK* Procedia Engineering, 123 pp. 76 – 84.

[15] Daniel S J 2020 *Education and the COVID-19 pandemic*. Prospects 49, 91–96.

[16] Project Management Institute 2016 *Construction Extension to the PMBOK® Guide* Newtown Square, PA, USA: Project Management Institute, Third Ed.

[17] Leon I. Sagarna M. Mora F Otaduy J P 2021 *BIM Application for Sustainable Teaching Environment and Solutions in the Context of COVID-19.* Sustainability 2021, 13, 4746.

[18] Paes D Arantes E Irizarry J 2017 *Immersive environment for improving the understanding of architectural 3D models: Comparing user spatial perception between immersive and traditional virtual reality systems*. Automation in Construction, 84, 292-303.

[19] Brioso X Calderon-Hernandez C Aguilar R and Pando M A 2019 *Preliminary Methodology for the Integration of Lean Construction, BIM and Virtual Reality in the Planning Phase of Structural Intervention in Heritage Structures* RILEM Book Series, Volume 18, 2019, Pages 484-492.

[20] Meza S Turk Z and Dolenc M 2015 *Measuring the potential of augmented reality in civil engineering* Advances in Engineering Software, vol. 90, pp. 1-10.

[21] Wang X Love P E D Kim M J Park Ch Sing Ch and Hou L 2013 *A conceptual framework for integrating building information modelling with augmented reality* Automation in Construction, vol. 34, pp. 37-44.

[22] Irizarry J and Bastos D 2016 *Exploratory study of potential applications of unmanned aerial systems for construction management tasks* Journal of Management in Engineering. Volume 32 Issue 3. ASCE.

[23] Brioso X Calderon-Hernandez C Irizarry J and Paes D 2019 *Using Immersive Virtual Reality to Improve Choosing by Advantages System for the Selection of Fall Protection Measures* Comput. Civ. Eng. 2019 Vis. Inf. Model. Simul. - Sel. Pap. from ASCE Int. Conf. Comput. Civ. Eng. 2019, no. June. 146–153.

[24] Suhr J 1999 *The Choosing by Advantages Decision making System*, Quorum, Westport, CT, 293 pp.

[25] Santa-Cruz S Córdova-Arias C Brioso X Vázquez-Rowe I 2021. *Transparency-based protocol for decision-making regarding seismic rehabilitation projects of public buildings*. International Journal of Disaster Risk Reduction 55, 102-116

[26] Suarez J C Zapata J Brioso X 2020 *Using 5D Models and Cba for Planning the Foundations and Concrete Structure Stages of a Complex Office Building*. In: Proc. 28th Annual Conference of the International Group for Lean Construction (IGLC). Berkeley, California, USA, 6-10 Jul pp 913-924.

[27] Arroyo P Mourgues C Flager F Correa M G 2018. *A new method for applying choosing by advantages (CBA) multicriteria decision to a large number of design alternatives* Energy and Buildings, 167, 30-37.

[28] Rother M and Shook J 2003 *Learning to see: value stream mapping to add value and eliminate muda*. Lean Enterprise Institute.

[29] Pasqualini F and Zawislak P A 2005 *Value stream mapping in construction: A case study in a Brazilian construction company* 13th Annual Conference of the International Group for Lean Construction. Sydney, Australia.

[30] Murguia D Brioso X and Pimentel A 2016 *Applying Lean Techniques to Improve Performance in the Finishing Phase of a Residential Building* 24th Annual Conference of the International Group for Lean Construction. Boston, Massachusetts, USA, 20-22 Jul 2016.

[31] Cruzado-Ramos F and Brioso X 2020 *Sustainability Performance Evaluation in Building Projects by Integrating Lean and Sustainable Management Using the Delphi Method* Proc. 28th Annual Conference of the International Group for Lean Construction (IGLC). Berkeley, California, USA, pp 805-816.

[32] Landeta J 2006 *Current validity of the Delphi method in social sciences* Technological Forecasting and Social Change, 73(5), 467–482.

[33] Hallowell M R and Gambatese J A 2010 *Qualitative research: application of the Delphi method to CEM research* Journal of Construction Engineering and Management, 136(1), 99-107.

The effectiveness of agile management on traditional projects within public organizations

J J Al Maamzi[1] and T Tawfik[2]

[1]Manager Projects (400kV Trans Projs), Dubai Electricity and Water Authority, Dubai, United Arab Emirates.
[2]Specialist - Trans Project Mgmt, Dubai Electricity and Water Authority, Dubai, United Arab Emirates.

E-mail: Jamila.juma@dewa.gov.ae and Tarek.Mohamed@dewa.gov.ae

Abstract. After the coronavirus pandemic, we must look for the notion of an agile methodology as one of the ways that supports the restoration of our businesses to what they were and even better; this is critical in order to remedy and grow the global economy once again. The notion of agility has been traditionally present, however, it was engaged only for IT and software projects. As this concept has been validated further during the pandemic, this methodology can be increasingly applied in different types of projects, such as construction projects and other infrastructure projects. With the growing adoption of agile methods, project managers at public organizations increasingly need to realize its applicability to their projects and the aspects that drive key project performance features. On the one hand, many firms confirmed that agile methods solved their difficulties during the pandemic, particularly in addressing changes due to unstable and unpredictable needs. On the other hand, some other firms have continued to struggle, arguably because of not practicing agile approaches. Consequently, this paper attempts to examine the effect of agile methodologies on the processes involved in traditional project management within public organizations, which it does through applying a qualitative method. Furthermore, this paper will explore how this methodology can allow project managers to evaluate the applicability of using an agile method within public organizations, especially for construction projects and other infrastructure projects.

Keywords. Agile methodologies, Combined project management, Kanban, Scrum, Traditional project management.

1. Introduction

As business processes are very complicated, opportunities are very dynamic and competition is global and intense, traditional approaches in project management need to be more advanced and new methods need to be developed, such as agile management. This is particularly relevant in terms of agile concepts being practiced in new fields, rather than just the software area. [7]

Though agile approaches are gaining recognition amongst traditional systems of development in different organizations, most of these organizations seem to indicate a preference to adopt both methods of development. [8]

Thus, this paper aims to demonstrate that the traditional approaches can be combined with agile approaches in any type of projects, such as those typically implemented by public organizations.

1.1. Aim of the research
The objective of this research is to examine the effectiveness of agile project management within the traditional projects, such as construction or infrastructure projects, within public utilities.

1.2. Motivation of the research
In order to accomplish this research aim, there were some questions that had to be explored, as per the following:

- What is the best way to apply agile methodologies for traditional projects, such as construction or infrastructure projects within public utilities?
- What is the impact of agile methodologies on the processes involved in traditional project management within public organizations?
- What are the roles of project managers within the agile project management?
- How are the agile manager roles different from the traditional manger roles? [1]

2. Theoretical Background
This paper is focused on two main project management approaches, which are the traditional waterfall methodologies and agile methodologies including Scrum and Kanban. Therefore, we need to know more about the traditional and the agile approaches (Scrum and Kanban); to select the best framework that can be used in this research and is most likely to be successful in the public project context. Thus, in the coming section there will be a more detailed discussion about these concepts.

2.1. Project Management
According to PMBOK Guide, project management is the application of knowledge, skills, tools and techniques to project activities to meet the project requirements. [2] pg (10)

2.2. Traditional Project Management (Waterfall) Practice
The traditional approach of project management (Waterfall) is the predictive life cycle of the project, with the bulk of planning occurring upfront, then executing in a single phase; a sequential process (e.g., analyze, design, build, test, and deliver). [3] As a result, project activities are executed in a serial way, as shown in figure 1.

Figure 1. Predictive life cycle [3]

2.3. Agile Methodologies
Agile methods and approaches are umbrella terms that shield a range of frameworks, practices, techniques or methods, in which all of those approaches are considered to be lean, since lean thinking concepts are involved, (e.g., small batch sizes, eliminating waste, focused on value). Moreover, agile life cycle can be defined as an approach that is both iterative and incremental to refine work items and deliver frequently, where the iterative life cycle is an approach that allows feedback for unfinished work

to improve and modify that work, and the incremental life cycle is an approach that provides finished deliverables that the customer may be able to use immediately. [3]
According to the Agile Practice Guide Book, there are four values of the agile manifesto, [3] which are:

- Individuals and interactions over processes and tools
- Working software over comprehensive documentation
- Customer collaboration over contract negotiation
- Responding to change over following a plan

Example of agile methodologies include SCRUM, Lean, XP and KANBAN. There are many differences amongst the traditional method of working and the agile style of working. Table 1 below shows the main differences between them. [1]

Table 1. Comparison table between traditional and agile methods [1]

Categories	Traditional (waterfall)	Agile
Development model	Traditional (Phases) Predictive	Iterative
Management	Controlling	Facilitating
Focus	Process	People
Developers	Work individually within team	Collaborative or in pairs
Customer involvement	Requirements gathering and delivery phases	On site and constantly involved
Product features	All included or nothing	Most important first
Technology	Any	Mostly object oriented
Testing	End of development cycle	Iterative and / or drives code
Documentation	Thorough	Only when needed

2.3.1. Characteristics of hybrid life cycle
A hybrid method is a combination of predictive, iterative, incremental, and agile methods. It is very important to know which approach to utilize for achieving certain goals. [3]
Therefore, if the waterfall approach is too inflexible, and the agile too liberated, maybe the hybrid method can work better for the project team. Especially as the hybrid method made it easier for the project team to identify the critical development path, and still complete tasks in 'sprints' while maintaining product quality. [3] [4] [5]
A hybrid approach in project management with both traditional and agile practices may be the most valid approach as concluded by Fernandez and Fernandez [6] in their study; moreover, the agility is the way to improve productivity, reduce production costs, customer satisfaction and product quality. [6]

2.3.2. Combined agile and predictive approaches
Another approach to consider is the combined agile and predictive approaches during the project life cycle. In which, a combined agile and predictive approach practiced simultaneously, or a small agile approach is used within a largely predictive approach, or vice versa as shown in figure 2.

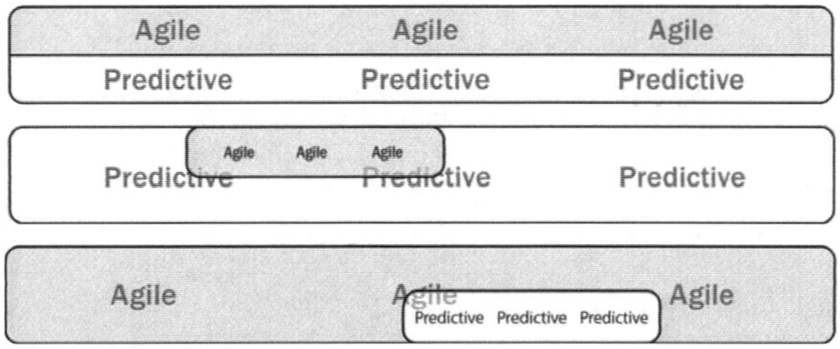

Figure 2. Combined agile and predictive approaches [3]

2.3.3. Scrum method

The Scrum method is one of the agile methodologies developed by Jeff Sutherland and formalized by Ken Schwaber in 1995. [1]

Scrum is an agile framework that supports providing the desired business benefits in the shortest time. It repeatedly and rapidly reviews actual work of the product. It emphasizes iterative progress of the product and highlights teamwork. Its aim is to provide a new product every 2-4 weeks. [10]

The product owner, scrum master and the team are involved in the scrum process. The product owner's role is to maintain the right business perspective; the scrum master works with the product owner and facilitates the team, and the team should contain five to 11 members (+- two). [1] [10]

In the scrum approach, the activities include sprint planning, sprint meetings and scrum review. A sprint normally lasts two to four weeks. The scrum master leads a daily 15 minute meeting where each team member briefs about their progress and challenges. The sprint review meeting reviews the previous sprint in terms of achieved activities and the coming sprint details are defined. [1]

The artifacts developed are called product backlog, sprint backlog and burndown chart, where the product backlog is a list of product features and it is maintained by the product owner. On the other hand, the sprint backlog, is about the developed tasks, which are required to implement a feature and is a subset of the product backlog. Finally, the burndown chart is a chart used for showing the total work remaining in the sprint. [1] [3]

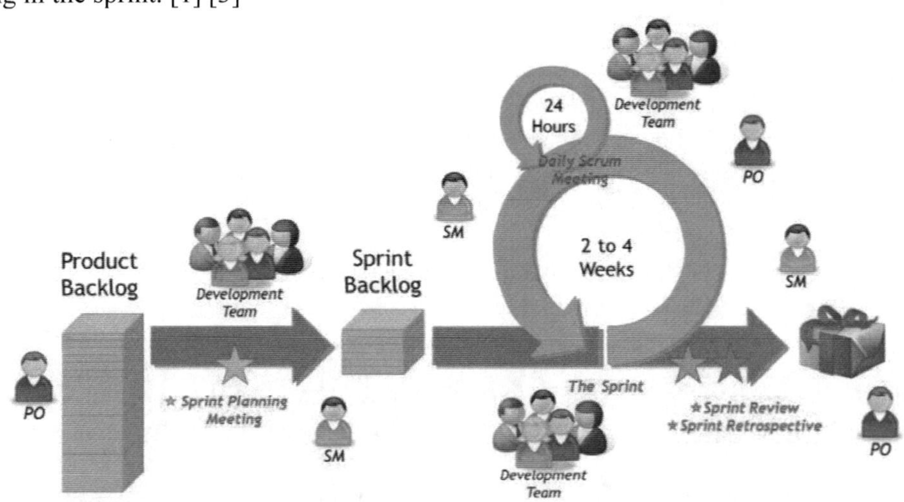

Figure 3. Scrum Framework [9]

2.3.4. Kanban method

Kanban is a visual system for managing work which was introduced by David J. Anderson in 2004. It pictures both the actual work and the process for small IT teams. The key goal of applying Kanban is to identify possible struggles in the process and fix them. Kanban aims for a workflow that should progress well and at optimal efficiency [10]. The core practices of Kanban are the following [9] [10]:

- **Visualize the flow** of work via using cards or software, as shown in the Kanban board in figure 4.
- **Limit work in progress** (WIP) by encouraging the team to finish current work first, before taking up new tasks.
- **Manage and measure flow** by noticing the work as it flows through the swim lanes and addressing any bottlenecks.
- **Make process policies explicit** via diagraming the process rules and guidelines.
- **Apply feedback** loops during the work process and reviewing team and customer feedback.
- **Continuous improvement** by expanding collaboratively and experimenting, as a team joins enhancement initiatives.

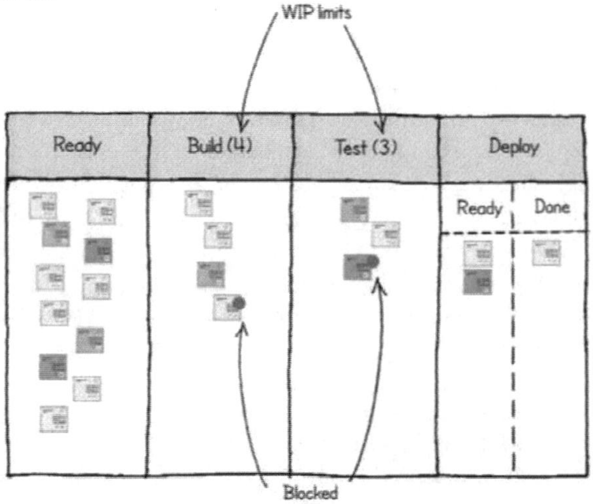

Figure 4. Kanban Board [9]

2.3.5. Comparison between Scrum and Kanban methods

The comparison between the Scrum and Kanban methods can support the agile team members in their choice for a multitude of purposes. Therefore, the threat of project delays can be reduced. The same can be tacked via some major factors like the method prescription, roles and responsibilities, team size, adoption time, feature size, batch, prioritization requirements, lead time, technical practices, quality and cost. In sum, the key points are as per the following [10]:

- Scrum is an agile process that allows us to focus on delivering business value in the shortest time, while Kanban is a visual system for managing software development work.
- Kanban method raises continuous enhancement, efficiency and productivity.
- Scrum is focused on the backlog, while Kanban on a dashboard.
- The scrum master acts as a problem solver, while Kanban encourages every team member as a leader and sharing responsibility amongst them all.
- Scrum prescribes time-boxed iterations, while Kanban focuses on planning a different duration for individual iteration.

- Both Scrum and Kanban have no technical practices.
- Scrum avoids cost saving, but focuses on knowledge, decision making and experiences, unlike Kanban, which focuses on cost cutting.
- Predefined roles and responsibilities are required for Scrum, but Kanban does not need predefined roles and responsibilities.

Table 2. Comparison table between Scrum and Kanban methods [9]

	Scrum	Kanban
Origin	Software development	Lean manufacturing
Ideology	Solve complex problems while delivering valuable products	Use visuals to improve work flows and processes
Practices	Sprint planning Sprint Daily scrum Sprint review Sprint retrospective	Virtualize the flow of work Limit work in progress Manage flow Make process policies explicit Implement feedback loops Improve, experiment
Roles	Product owner Scrum master Development team	No formal roles
Metrics	Velocity	Cycle time Throughput

2.4. Introducing agile in traditional project management
A study done by Danijela Ciric *et al.* [7] addressed all the strategies and actions needed to implement agile project management into a traditional method from multiple studies. The study concluded them as followings [7]:

- How Agile really works: agility methods do not necessarily fit every project, thus we need to understand it very clearly, and to learn the common software projects characteristics.
- Tailor the agile approach wisely: not every firm is similar in its nature, they have their own challenges during deploying the agile method. Accordingly, every firm needs to reflect upon its own environment, when choosing an agile method to apply.
- Build up processes instead of tailoring them down: via defining the project's needs and choosing of only those processes that appear crucial.
- Piloting: it is recommended to start small when introducing an agile method and gradually introducing it into small processes or departments rather than making a complete change; to be more effective.
- Commitment to changes and team autonomy: the right people, dedication and commitment and acceptable authority are the most important factors to be an agile team and should be defined from the start.

- Leadership success: a leader's personality profile is essential for agile project management; since it needs to focus on people and collaboration, especially with embracing changes.
- Roles and responsibilities: define all the roles and responsibilities that will be addressed with agile approaches.
- Set architectures: architectures shall support classification of traditional and agile teams.
- Redefine and realign traditional milestones: traditional milestones to be redefined and realigned with the iterative approach.
- Risk evaluation: risk is a key element to be considered during the project implementation, especially the risks of too much agile and to little traditional or vice versa.
- Flexibility: flexibility is a very important factor when change is frequent. Therefore, it needs to be applied to projects selectively, to deal with uncertainty.
- Keep the looping back: for projects beyond software development, it is very difficult to deal with changes, thus, it is required to find a way to modularize the tasks and work iteratively.
- Negotiation and setting expectations: in agile methods the team along with the client need to test the product after each iteration and gather feedback; thus, negotiation and setting expectations are very important elements in the agile approach.
- Critical options: the 'last responsible moment' technique is very important in helping make a better decision by the time the last responsible moment arrives, as this tool allows project managers to define a decision that is uncertain at the moment and that might change later on.
- Agile top-management: top senior and executive support is a very important element to successfully implementing the agile approach.
- The barriers: it is recommended to destroy the barriers for better agile behaviors, since agile needs changes as all stakeholders to be on the same page.
- Learning mode: coaching and training teams during agile method implementation is a crucial factor in change.

2.5. Agile and traditional project management reasons and challenges
According to a study done by Danijela Ciric et al. [7], a number of facts are defined; for instance, there are a number of industries utilizing agile project management beyond software applications. Moreover, there were a number of reasons to implement agile methods in and beyond software projects, where the most common reasons were accelerating project/product delivery and enhancing ability to manage changing priorities. While most significant challenges faced by the utilization of agile methods were work prioritization, long feedback loops, alignment among stakeholders on what to build next, incompatibility of agile methods within organizational functions and insufficient time for testing. In sum, there were a limited number of different reasons and challenges in and beyond software projects, as shown in figures 6 and 7. Thus, the study recommended to do further empirical research on the use of agility methods outside IT projects. [7]

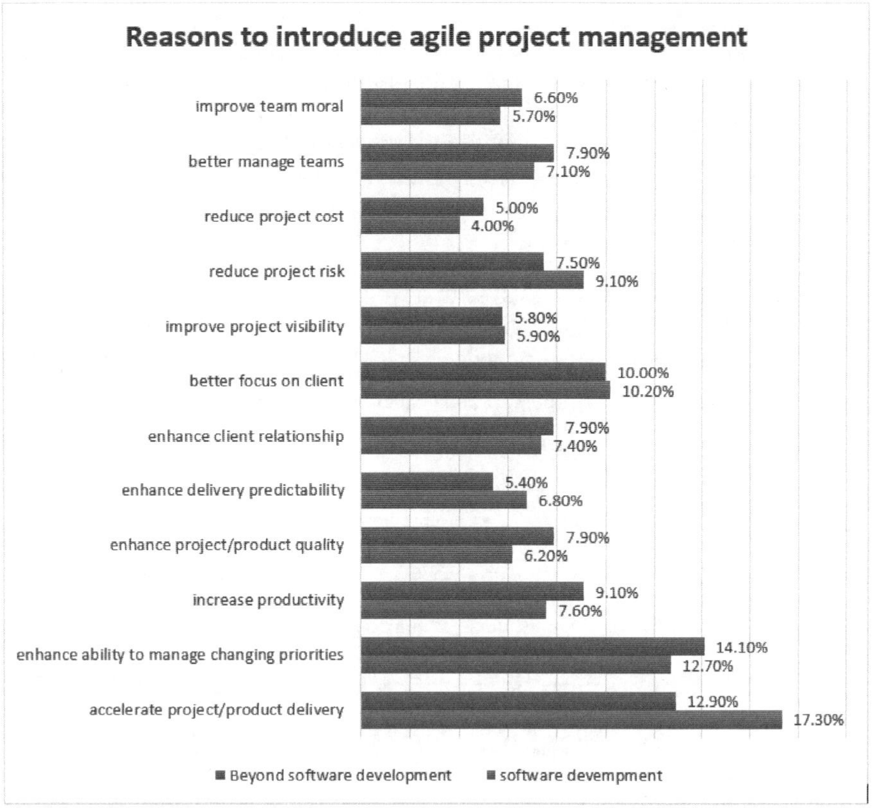

Figure 5. Reasons to introduce agile project management [7]

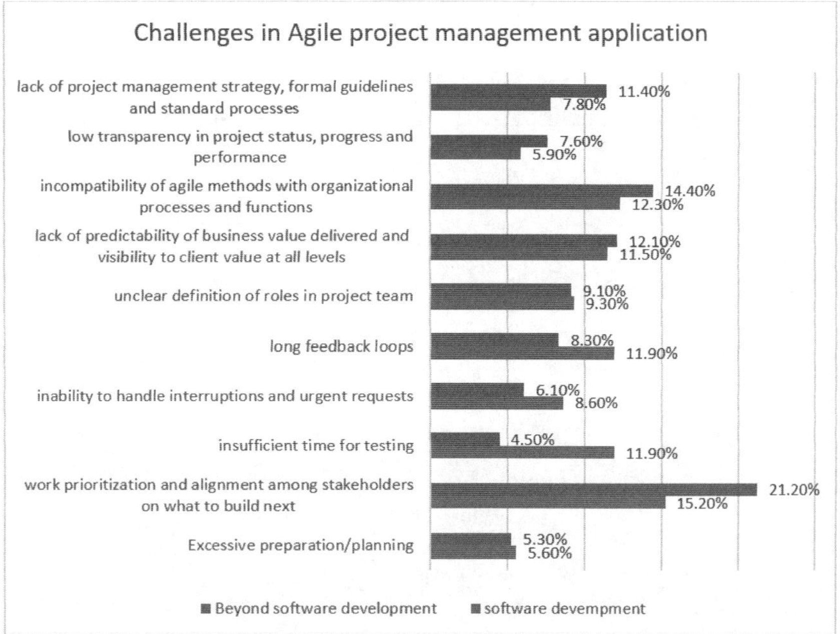

Figure 6. Challenges in agile project management [7]

In additional, another study done by Vinekar *et al.* [8] mentioned that there were obstacles in using both agile and traditional methods; this is because client features and project features vary, and it is hard to change the systems development of a firm's own features from project to project. The main challenges can be addressed at four levels: the management and organizational level, people level, process level and finally at the technology level. Table 3 shows the summarization of the characteristics of these two methods. They concluded that organizations could earn more benefits via utilizing both agile and traditional systems development. Moreover, the explorative and exploitative ability are very high in case of use of both agile and traditional approaches in project management, as shown in figure 7. [8]

Table 3. The ambidextrous system development organizations [8]

	Agile Subunit	**Traditional/Stable Subunit**
Management and organizational	Leadership and collaboration Cooperative Flexible Manager as facilitator Tacit knowledge Team reward systems	Command and control Autonomous Disciplined Manager as planner Explicit knowledge Individual reward systems
People	Collaborative work Multidisciplinary skills Pluralist decision making High customer involvement Small teams	Individual work Specialized skills Managerial decision making Low customer involvement Large teams
Process	People centric Speculative Assess progress Evolutionary development Write tests prior to code Individual approach to projects Adaptable Iterative Short durations	Process centric Standardized Measure progress Life-cycle development Write code prior to tests Unified approach to projects Preplanned Linear Long durations
Technology	Object oriented Tools for iteration	Structured or object oriented Standardized tools

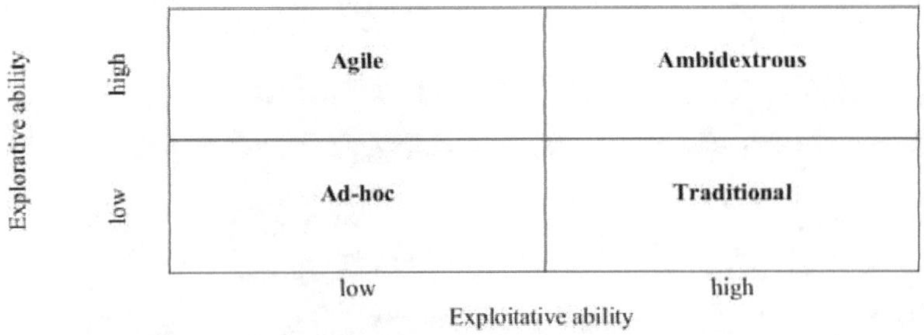

Figure 7. The association between explorative and exploitative abilities for systems development organizations [8]

3. Research Framework

According to the literature background and after knowing most of the possible types of methodologies in detail that can serve project management, and whether to select agile or flow like waterfall approaches in managing projects, all depends on the objectives and goals of a certain project. As per the literature that no methodology is perfect and they have their advantages and disadvantages. Therefore, advanced organizations are increasingly leaning toward the combined approach, which combines the benefits of both.

However, the agile-waterfall framework aims to maintain the dependency tracking and systematic approach of waterfall, whereas at the same time integrating the powers of the 'agile' methodology – flexibility and transparency are required for regulating the customer's fast-changing desires. [4] [5]

Therefore, this study examines the impact of the combined agile and traditional waterfall approaches on the ultimate project gained outcomes.

Moreover, the framework proposed for this study includes Scrum and Kanban that can be used together along with the combined agile and waterfall approach, as shown in figure 8. Where, both the Scrum and Kanban (Scrumban) frameworks have emerged. Adding Scrumban influences both frameworks can complement each other to better embrace agility, to enhance what is missing in each method, to facilitate higher performance in teams and enabling continuous process improvement. This is particularly relevant as Scrum provides a way for iterative and incremental development process, whereas Kanban ensures more visibility of the workflow and the quick identification of possible bottlenecks. Furthermore, by applying both frameworks many benefits can be secured, mainly from a people perception (e.g., fostering improvement and collaboration via feedback and concentrating on delivering business value). [11]

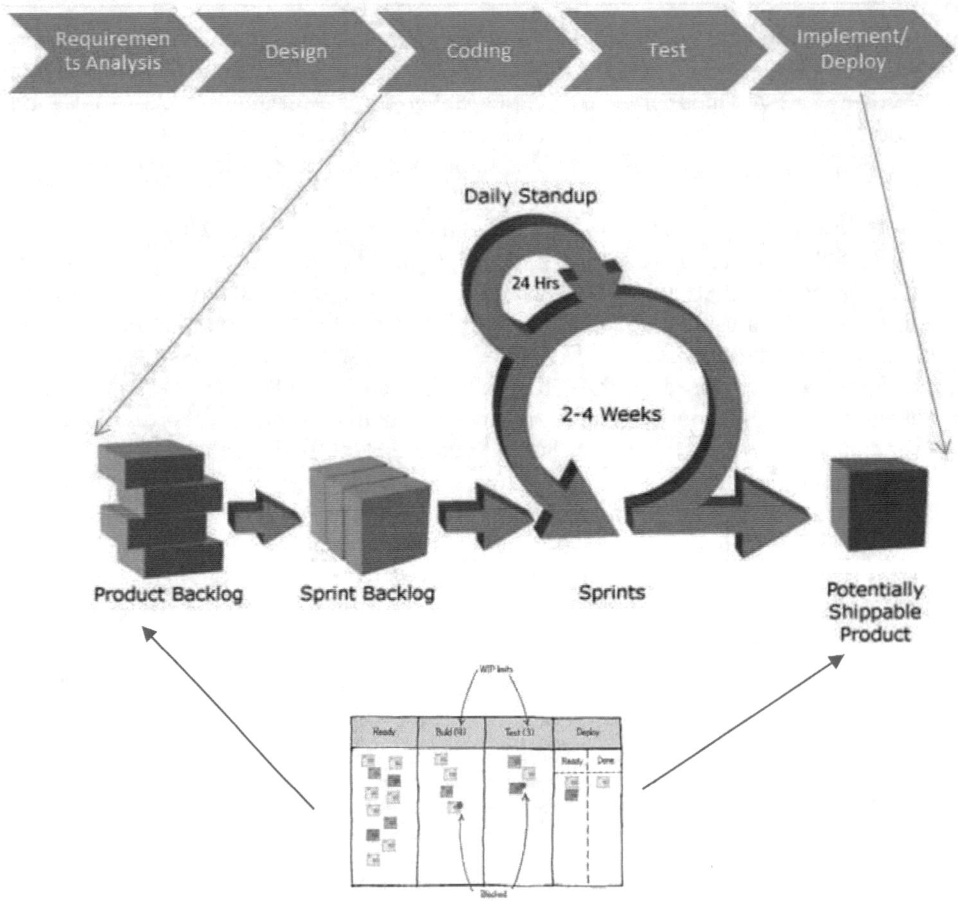

Figure 8. Collaborative (Hybrid) model (source: http://www.softwaretestinghelp.com)

4. Research Methodology Outline

This paper intends to make use of qualitative research method with the following major elements [12]:

1. The data can be taken from various sources as well as interviews and the most common observations.
2. Using different analytic and interpretation procedures can be used to reach theories.
3. Written and verbal reports.

4.1. Agile project management experts

In this part, interested parties will be identified (e.g., agile experts) for interviews and observations, and for companies, project teams, and firms that work in agile communities.

4.2. Interviews

For the interviews section, we need to gather information about the research topic, wherein we will conduct interviews with them. The research plan is to follow projects with 6 to 24 months, and embrace interviews at important phases/milestones of the projects; in order to capture the holistic view of agile project management during the entire project life cycle.

4.3. Data collection
The data collection part will be done after collecting the data and analysis of this data will begin to set the grounded theory for the study. All the data collected will be dealt confidentially.

4.4. Developing the theory
The study was built based on a qualitative method; to develop the grounded theory. Accordingly, there will be in depth analysis of all the information gathered to reach the study theories. All the research methods are to be applied (e.g., validating data).

Conclusions and future work
According to many research projects conducted on agile approaches and from growing practices of agile methods in many organizations, all that leads us toward thinking about the importance of the agile approach and the needs to explore agile project management.
This study wishes to examine the proposed combined framework and check the impact of the agile methodologies on the processes involved in traditional project management within public organizations, and if there are benefits of using the agile approach within traditional methodology, beyond software projects.
Finally, as many organizations are adopting agile methods, the authors of this paper have to develop a deeper knowledge and a comprehension awareness about the benefits of using a combined method of agile and traditional project management and serious issues - if any - to be addressed during utilization of the similar approaches.

Acknowledgements

We thank all those anonymous referees for their useful suggestions and helping to publish this work.

References

[1] Hoda R, Noble J, and Marshall S 2008 Agile project management In New Zealand *Computer Science Research Student Conference*, NZCSRC, URL: https://researchspace.auckland.ac.nz/docs/uoa-docs/rights.htm

[2] PMI 2017 *A Guide to the project management body of knowledge* 6, ed (USA: 14 Campus Boulevard, Newtown Square, Pennsylvania 19073-3299 USA: PMI, Inc.)

[3] PMI 2017 *Agile practice guide* (USA: 14 Campus Boulevard, Newtown Square, Pennsylvania 19073-3299 USA: PMI, Inc.)

[4] Kusters R J, van de Leur Y, Rutten W G, and Trienekens J J 2017 When agile meets waterfall-investigating risks and problems on the interface between agile and traditional software development in a hybrid development organization. In *International Conference on Enterprise Information Systems* (Vol. 2) 271-278 SCITEPRESS, DOI: 10.5220/0006292502710278

[5] Van Waardenburg G and Van Vliet H 2013 When agile meets the enterprise *Information and software technology*, ***55*(12)**, 2154-2171, DOI: https://doi.org/10.1016/j.infsof.2013.07.012

[6] Fernandez D J and Fernandez J D 2008 Agile project management—agilism versus traditional approaches *Journal of Computer Information Systems* ***49*(2)** 10-17, DOI: 10.1080/08874417.2009.11646044

[7] Ciric D, Lalic B, Gracanin D, Tasic N, Delic M, and Medic N 2019 Agile vs. Traditional Approach in Project Management: Strategies, challenges and reasons to introduce agile *Procedia Manufacturing* ***39*** 1407-14, DOI: https://doi.org/10.1016/j.promfg.2020.01.314

[8] Vinekar V, Slinkman C W & Nerur S 2006 Can agile and traditional systems development approaches coexist? An ambidextrous view *Information systems management* **23(3)** 31-42, DOI: https://doi.org/10.1201/1078.10580530/46108.23.3.20060601/93705.4

[9] AXELOS 2015 *PRINCE2 Agile Guidance* (UK: TSO).

[10] Alqudah M, and Razali R 2017 A comparison of scrum and Kanban for identifying their selection factors. *In 2017 6th International Conference on Electrical Engineering and Informatics (ICEEI)* 1-6 IEEE, DOI: 10.1109/ICEEI.2017.8312434

[11] Mahnic V 2014 Improving software development through combination of scrum and kanban. *Recent Advances in Computer Engineering Communications and Information Technology, Espanha*, 281-288, ISBN: 978-960-474-361-2

[12] Strauss, Anselm, Glaser and Barney 1967 The discovery of grounded theory: Strategies for qualitative research Chicago: Adline.

Impact of Environmental Sustainability on the Market Value of Office Buildings

J Kupec, K Eklová and Z Prostějovská

Czech Technical University in Prague, Faculty of Civil Engineering, Praque, Czech Republic

josef.kupec@seznam.cz

Abstract. The issue of environmental sustainability of buildings is relatively new in many areas and has not yet been sufficiently explored. One of the little-explored areas is the relationship between environmental sustainability and the properties' market value. In practice, LEED or BREEAM sustainability certificates are often used to determine the degree of environmental sustainability. These certificates are well known to international investors and tenants in domestic markets. Sustainability has evolved into a competitive factor due to its positive economic, ecological and social impact. Certified buildings are preferred by major corporate tenants with an international field of activity who often have ethical rules for sustainable development. This increases the attractiveness of the building, which has a positive effect on its market value.

1. Introduction

In recent years, sustainability in construction has become very important. Almost all newly built office buildings have a sustainability certificate. Nonetheless, Czech valuers usually do not examine in detail the impact of the object's sustainability certificate on its market value. Most commercial properties are valued using the income approach. Operating costs are generally transferred to the lessee and therefore not directly entered into the calculation of a property's market value. The importance of this issue lies in the impact of a building's certification on the value of real estate not having been sufficiently quantified.

The benefits of certified buildings are addressed in many published articles. Today, sustainability in real estate is firmly anchored in society, policy and in the minds of people. Green buildings have added value to real estate in several areas; this has an impact on the market value. In many markets, green buildings are perceived to be of better quality, and many of today's large tenants are willing to pay extra for green premises. For these tenants, renting green spaces is an opportunity to demonstrate their commitment to sustainability, thereby attracting the best employees and increase productivity [1]. For this reason, green buildings can generate higher rents per unit and thus increase the revenue from its operations [2].

Saving energy is not the only benefit of the green buildings. The added value of a green building is also reflected in risk mitigation for owners and banks providing financing. In the risk assessment process, it is necessary to take into consideration that green buildings can have added value in a more stable yield, higher resilience to consumer preferences, ever tightening requirements for sustainable construction and the impact of rising energy costs. By reducing the risk of investment, the yield is compressed and the value increases. Another effect of less risk is better bank financing when green buildings are constructed [2].

The two the most commonly used certification systems for the sustainability certification of office properties are LEED and BREEAM, both well known to international investors and tenants in domestic markets. The purpose of this proposed research is to quantify the impact of LEED and BREEAM sustainability certificates on the market value of office properties in Prague and to assess whether a higher sustainability certification grade increases market value.

2. Market value of office properties and its key valuation inputs

According to international valuation standards, three basic and generally accepted valuation approaches are used for the valuation of real estate: the market (comparative) approach, the income approach, and the cost approach [3]. Each office building is unique with individual elements affecting its market value. The income approach is most often used for the valuation of office buildings. During the income approach the valuation is performed based on the capitalisation of the potential net income obtained from the rent of the property based on the investment risks relating to the ownership of the property. When this approach is properly applied, it is considered a solid indication of the value of the property for its capacity to produce an income. All valuation approaches involve a valuation method whereby the capital value is found by capitalising or discounting the estimated future income to be derived from the property. The most common methods for establishing yield are the calculation of eternal fixed income (constant yield over a long period of time), establishing yield value by means of appraisal norm, and calculations for variable yield [3].

Simply put, the value of an office building is calculated as the ratio of rental income and the capitalisation rate [11]. Unit rent and yield are two key valuation inputs used for office property valuation and market value determination. For this reason, the unit rent and the yield are important indicators of the condition of the market segment and are analysed by real estate consulting companies, banks, appraisers and other real estate specialists [12].

Individual parameters have a positive or negative effect on these two main key inputs and are closely interconnected. The basic parameters that affect the market value of office buildings include location, age, technical condition, standard, occupancy, tenant structure, competition in the area, lettable area structure, supply and demand, and other individual characteristic.

3. Impact of environmental sustainability on the key inputs effecting market value

According to USGBC, LEED-certified buildings command the highest rents, while lease-up rates typically range from average to 20% above average. Vacancy rates for green buildings are an estimated 4% lower than non-green properties [9]. In the application of the income approach for real estate valuation, several valuation inputs such as structural vacancy, nonrecoverable costs, capital expenditures, voids on expiry, initial void, etc. are used. However, the inputs with the most significant impact on market value are level of yield and rental level. For these reasons, this article focuses mainly on these two key inputs.

3.1. Impact on the level of yield

The best way to determine the impact of sustainability certificates on the level of yield is by analysing direct market evidence with known initial yields. The initial yield is a particular type of income yield and is the net income received in the first year divided by purchase price, and is a common market measure of investment performance [5].

As described in Chapter 2, several parameters affect the value of office buildings. Restrictive conditions should be set to obtain a sufficiently meaningful sample and thus eliminate the impact of the parameters affecting the value.

Publicly available information sources, such as the real estate cadastre, business newspapers and press releases of investment funds were used to obtain information on realised transactions of office buildings corresponding to the selected parameters. For all properties, the sustainability certificate and its grade was investigated. The sample consist of 32 office properties which meet the selected parameters, 20 have obtained LEED or BREEAM certification at any level, and 12 properties are non-certified.

During the examined period (January 2017 – December 2020) movement in the market level of office yields could be observed. Therefore, it was necessary to adjust the obtained data to reflect market condition at the transaction date. Data adjustment by using coefficients through the indirect comparison method is commonly used in real estate valuation practice to determine market level [6]. The initial yield of all transactions was converted to market level as of December 2020. The adjustment was based on the movement in Prague prime office yields as reported by Colliers International [7]. The prime office yield in Prague in December 2020 was 4,25%. If the initial yield at the date of the transaction was higher (in 1Q/2017 – 1Q/2019) than the level in December 2020, the initial yield was reduced. The opposite procedure was used if the prime yield level was lower (in 3Q/2019 – 1Q/2020).

The aim of my research was not only to determine the impact of sustainability certification on market value but also the impact of a building's certification grade. To reflect the sustainability certificate grade, the properties were divided in for groups:

- sustainability certificate with highest grade – LEED Platinum/BREEAM Outstanding
- sustainability certificate with high grade – LEED Gold/BREEAM Excellent
- sustainability certificate – LEED Silver/BREEAM Very good - Good
- no sustainability certificate.

Subsequently, the initial yield was determined for each group. To determine the impact on market value, the properties without certification were used as a base.

Table 1: Adjusted average initial yield reflecting the sustainability certificate grade (source: author)

Certificate grade	Number of properties	Adjusted average initial yield	Impact on market value
LEED Platinum/BREEAM Outstanding	7	4,90%	128,78%

LEED Gold/BREEAM Excellent	7	5,62%	112,28%
LEED Silver/BREEAM Very good - Good	6	6,25%	100,96%
None	12	6,31%	100,00%

Table 1 shows that the certification of the office properties already has had an impact on the level of yield which effect market value.

For sustainability certification to have an impact, the property must at least achieve a LEED Gold or BREEAM Excellent certificate, so that due to yield reduction the market value could increase by 10-15%. If a building obtains the highest grade of sustainability certification, its market value can increase by up to 25 - 30% compared to a building with a lower grade or without certification.

3.2. Impact on the level of rent

The second most important key input is the level of rent. Certified buildings are preferred by major corporate tenants with an international field of activity who often have ethical rules for sustainability. One of the most obvious benefit of green buildings should be energy savings. Saving energy reduces operating costs and thus helps to increase headline rent, which has a positive effect on the value of the property. To verify this assumption, an analysis of headline rents and service charges on the Prague office market was performed. As a source of information, the CBRE office advertising server was used [8].

As described in Chapter 2, several parameters affect the value of office buildings. Restrictive conditions should be set to obtain a sufficiently meaningful sample and thus eliminate the impact of the parameters affecting the value. The selected sample consisted of 43 office properties. Subsequently, the properties were divided into the same groups used in the initial yield analysis. For these groups, the average headline rent and average service charges was determined.

Table 2: Average headline rent and service charge fee reflecting the sustainability certificate grade (source: author)

Certificate grade	Number of properties	Average headline rent	Average service charges
LEED Platinum/BREEAM Outstanding	6	15,08 €	3,65 €
LEED Gold/BREEAM Excellent	12	14,74 €	3,60 €
LEED Silver/BREEAM Very good - Good	5	14,60 €	3,51 €
None	20	14,43 €	3,78 €

Table 2 shows that the certification of the office properties has had a very low impact on the level of headline rent and service charges. The identified difference in headline rent of the property with the highest grade of sustainability certificate is lower than 5%. Also, the difference in service charges is almost negligible. Green properties may have lower costs for energy supplies, however, these costs are balanced by higher maintenance costs which are reflected in service charges.

The assumption that energy savings reduce operating costs and thus help to increase headline rent was not confirmed. Also, the assumption that tenants are willing to pay higher rent for properties with a higher sustainability certificate grade was not confirmed.

It is necessary to take into consideration that headline rent is only one component of net effective rent calculation. The net effective rent which is used for property valuation also includes rental incentives such as rent free or fit-out contributions. However. the impact of these components is significantly lower than the impact of headline rent and author team is not aware of any publication supporting the impact of sustainability certification grades on rental incentives.

4. Conclusion

Real estate is a specific type of asset and every real estate is unique. There may be significant differences between individual office buildings, especially in location, size, design standard, and other key factors that have a decisive impact on the resulting market value. The Czech real estate market is becoming increasingly interesting for international investors who want to invest in premium real estate. Foreign investors and tenants often do not have the know-how of local investors and therefore prefer certified properties they know from their markets. For them, certification is a sufficient indication of quality that ensures higher and stable income and lower risk levels. These factors contribute positively to the transaction prices of certified buildings. From the perspective of an international investor, the real estate market in the Czech Republic consists mainly of Prague. Regional cities are not as interesting for foreign investors, and therefore certification is not so widespread.

The income approach is most often used for the valuation of office buildings. When using the income approach, the unit rent and the yield are two key valuation inputs used for office property valuation and market value determination.

To determine the impact of sustainability certificates on yield, information on realised transactions of office buildings corresponding to pre-selected parameters was obtained from publicly available information sources. For sustainability certifications to have an impact on yield, the property must at least achieve a LEED Gold or BREEAM Excellent certificate, then the market value could increase by 10-15% due to a yield reduction. If a building obtains the highest grade of sustainability certification, its market value can increase by up to 25 - 30% compared to a building with a lower grade or without certification.

The analysis of real estate advertising shows that the certification of the office properties has a very low impact on the level of headline rent and service charges. The identified difference in headline rent of the property with the highest grade of sustainability certificate is lower than 5%. The difference in service charges is also almost negligible. Green properties may have lower costs for energy supplies, however this costs are balanced by higher maintenance costs which are reflected in service charges. The assumption that energy savings reduce operating costs and thus help to increase headline rent was not confirmed. The assumption that tenants are willing to pay higher rent for properties with a higher grade of sustainability certificate was also not confirmed.

This research shows that higher grades of sustainability certificates have an impact on the market value of office properties. A higher certificate grade causes yield compression, which increases the market value. This corresponds with the assumption that sustainability certificates mitigate risks for owners and banks providing financing.

Acknowledgements

This work was supported by the Grant Agency of the Czech Technical University in Prague, grant No. **SGS21/011/OHK1/1T/11.**

IOP Conf. Series: Materials Science and Engineering 1218 (2022) 012038 doi:10.1088/1757-899X/1218/1/012038

References

[1] Royal Institution of Chartered Surveyors (RICS) (2013), Sustainability and Commercial Property Valuation, second edition, ISBN 978-1-78321-029-9

[2] Appraisal Institute (2013), Green Building and Property Value, [online], [see 16.12.2019]. available at: https://www.appraisalinstitute.org/assets/1/7/Green-Building-and-Property-Value.pdf

[3] International Valuation Standards Council (2018), International Valuation Standard 2017. EKOPRESS, s.r.o., ISBN 978-0-9931513-0-9.

[4] THE EUROPEAN GROUP OF VALUERS ASSOCIATIONS, European Valuation Standards 2016, eighth edition. TEGoVA 2016. ISBN 978-90-819060-1-2

[5] PETER WYATT, Property Valuation in an Economic Context, first edition. Blackwell Publishing Ltd, 2007, ISBN 978-1-4051-3045-5

[6] Sayce, S. et al. (2006), Real Estate Appraisal: From Value to Worth, 1st ed. Oxford: Blackwell, 340 p. ISBN 1-4051-0001-X. pp. 150-170

[7] Colliers International (2021), Office Overview – Czech Republic [online], [see 27.4.2021]. available at: https://www.colliers.com/cs-cz/research/q4_2020_office-market-overview

[8] CBRE (2021), Offices to lease [online], [see 27.4.2021]. available at: https://www.cbreproperties.cz/kancelare/hledani/Praha

[9] U.S. Green Building, [online], [see 27.4.2021]. available at: https://www.usgbc.org/leed/why-leed

[10] BREEAM International New Construction 2016 Technical Manual, BRE Global Ltd. 2017, [online], [see 16.1.2020]. available at: https://www.breeam.com/discover/technical-standards/

[11] BAUM, Andrew, Nick NUNNINGTON and David MACKMIN. The Income Approach to Property Valuation. 5th ed. London: EG Books, 2006. xiii, 334 p. ISBN 0-7282-0464-9.

[12] ENEVER, Nigel, David ISAAC and Mark DALEY. The Valuation of Property Investments. Seventh edition. London: Routledge, Taylor & Francis Group, 2010. xii, 220 p. ISBN 978-0-7282-0550-5.

The Next Era of IPD Research: A Systematic Literature Review of The IPD Research Trends 2017-2020

A J Arar and E Poirier

École de technologie supérieure ÉTS, Montreal, Canada,

ahmad-jamal-harb.arar.1@ens.etsmtl.ca

Abstract. Questions around integration, innovation, and collaboration within project teams have been the focus of increasing research in recent years. This is in response to the notorious fragmentation and the inherent performance issues that face the construction industry. Among the solutions identified, new contractual approaches, namely, integrated project delivery (IPD), have been investigated to understand how they can help overcome the industry's issues. This paper aims to contribute to the previous efforts that studied the research trends of the IPD from 2001 to 2016 by synthesizing the last four years from 2017 to 2020. This period accounts for more than 70% of research on IPD compared to the previous 16 years combined. A systematic literature review was conducted to understand the trends in IPD research over the last four years. The results allowed the research team to understand better where such research was being conducted, what type of projects were being studied, and most importantly, the topics being covered within this research and its results. This paper proposes an initial review that will be further developed in a full-length journal paper.

1. Introduction

The past two decades have seen the emergence of integrated project delivery (IPD) as an alternative to traditional project delivery approaches. IPD relies on bringing the key participants of a project into a multiparty contract where risks and rewards are shared, and the project is managed in a way that enables collaborative decision-making. This approach has emerged as highly effective instead of other, more traditional project delivery systems. It has demonstrated distinct advantages in directing the attention of project parties towered meeting a project's objectives.

Many studies have been conducted on this innovative and collaborative approach to enrich the collective understanding of the principles and promises of IPD while investigating the factors that affect its adoption since its emergence at the beginning of the 2000'S. Researchers have conducted some reviews to synthesize this knowledge, namely Kahvandi et al. [4], who investigated research between 2001 and 2016. As IPD grows in popularity, its investigation through research also intensifies. This paper summarizes findings from a systematic literature review that builds on Kahvandi et al.'s [4] work to investigate research conducted between 2017 and 2020. The aim is to identify and discuss the research trends of the past four years concerning IPD. A full-length follow-up paper will present the results of this investigation in depth.

2. Literature review

The construction industry has been described as a broken industry by many. Indeed, it has failed to take full advantage of the technological advances in many areas that serve the construction process such as, material science, building technology, and equipment, to name but a few. Low productivity levels were among the prominent leading indicators of this missed opportunity to maximize the value generated. The construction industry has systematically experienced a decline in productivity levels in stark contrast to other Non-farm industries [1].

It is widely believed that fragmentation between project parties is one of the main reasons for this situation [2]. This deprives project parties of getting the most out of their respective and collective capabilities. These missed opportunities include early involvement of critical knowledge and experience early in the design phase, ensuring alignment to project goals across the project team, and incentivizing collaboration throughout the project's lifecycle [3]. Therefore, a call for change in the way projects are managed, contracts are developed, and teams are formed has been heard around the globe.

IPD has been hailed as a solution to many of the industry's current issues. The beginning of research on IPD can be traced back to 2001, with its roots being tied to relational contracting strategies and approaches developed in the 1980s and 1990s. Since then, IPD has rapidly grown in popularity and has been the subject of much research. For instance, the authors of [4] Investigated the developments of IPD over 15 years and displayed the research trends of IPD from 2001 to 2016. The results of their research can be summarized as follows:

- The USA leads in IPD from 2001 to 2016, taking advantage of the fact that most of the initiating studies on IPD were conducted there; however, IPD studies were conducted in many countries, including many developing countries.
- The most significant percentage of the studies in the period in question used some sort of conceptual data of IPD more than quantitative and qualitative data.
- Two different periods can be noticed; in the first, the most significant percentage of IPD research addressed the principles of IPD, while in the second, it investigated the challenges and barriers of IPD adoption. However, many studies that looked through other aspects of this promising delivery system, such as the integration between IPD and BIM, the implementation of IPD, and solutions to the barriers, were conducted.
- In terms of the application domain, most IPD research has studied the application of IPD on infrastructure projects in that period.

3. Methodology

In the last four years, research on IPD has continued to grow. Researchers from different countries around the world have scrutinized the different aspects of this promising approach. This study builds on the previous efforts in [4] that studied the research trends of IPD from 2001 to 2016 by covering the period from 2017 to 2020. It synthesizes the research done on IPD to form and understand the current research trends and the gaps that still need more contribution. Moreover, this study tried to enhance some aspects of the previous study by following a rigorous review process and providing the details of the review process, including the inclusion and exclusion criteria, and reform some used categories and classifications to represent the research trends better.

This study applied a systematic literature review methodology to discover the research trends around IPD from 2017 to 2020. The intent behind choosing this methodology is to provide more reliable results through a rigorous process containing predetermined inclusion and exclusion criteria. In addition, to provide the full path of the review process [5], where the systematic literature review "is designed to reduce the effect of the reviewers' own bias, and a full protocol should be written to define and guide the process" [6]. Lastly, the extracted data from the systematic review were combined or compared in many places with the data from [4] to understand the evolution of IPD research trends over the last two decades.

3.1. Review process

The process started by searching specific academic databases such as (ASCE, EI Compendex, Inspec, Knovel, IEEE, Science Direct, Springer Link, Wiley online library, Emerald, and ASME), for articles that contain any of the two terms, integrated project delivery and IPD, in any part of the article. The inclusion criteria were set to include all papers about IPD within the construction industry, published between 01-01-2017 and 31-12-2020, published in peer-reviewed journals or conferences. The exclusion criteria were established to exclude all papers that fall under other types of documents, addressing industries other than Construction, Studies that we could not get access to its full paper., and papers in languages other than English.

The first round of the review process resulted in collecting 1044 papers carried forward to the next step of the research. Then, those papers were first sorted and filtered to remove all duplicated papers. This process left a total of 615 papers, which were then scanned by title and abstract to remove irrelevant papers. Next, papers that do not have open access and those in a language other than English were excluded. As a result, the final number of publications considered in this review was 112 articles, as shown in Figure 1.

3.2. Validation
The review process and results were validated through two different stages; the first stage was to validate the review protocol and the review process, which was done through two specialist librarians at the École de technologies supérieure's library. The review protocol and a detailed research process were submitted to them for review. A meeting was held with them where the details of the process were discussed. As a result of this step, a revised version of the research protocol has been generated.

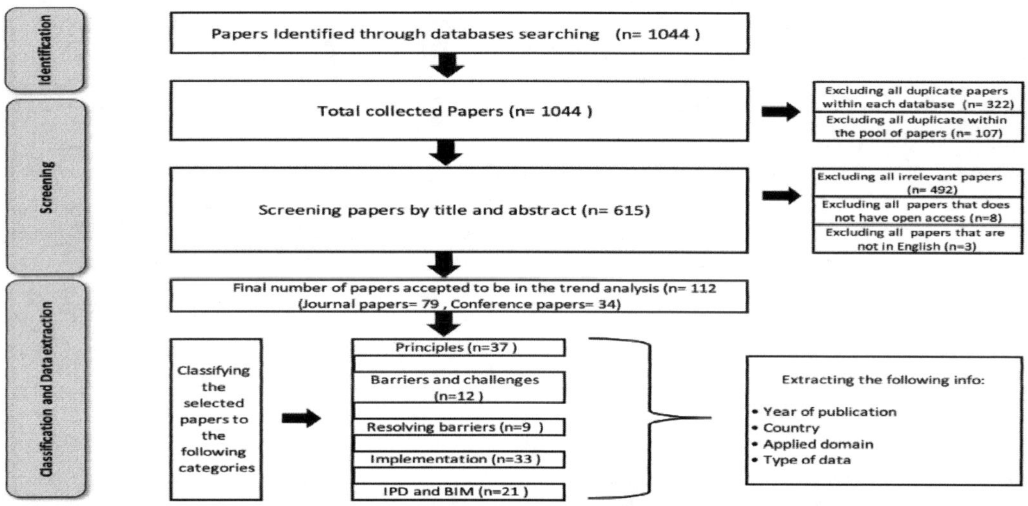

Figure 1. Flowchart of the review process

The second stage was used to validate the review results, which were done with the help of an external validator. This external validation process first explained the review protocol, review process, and the pool of papers to the researcher validating the approach. The person randomly selected 15% of the papers, extracted the required information, and classified the papers. The results from this round of validation process were compared with the researcher's results. The comparison revealed an 85% match of the data extracted. Next, the researcher added his explanation to the points of differences with the validator, and the sheet was sent back again to the validator for the second round of validation. Again, the validator responded to the researcher's arguments, and the percent of the match after this round reached 98%. Eventually, the logic concluded from this exercise on 15% of the articles was applied to all the other articles. In addition, the classification categories were reviewed by the validator, and a new application domain category was suggested that ensure better classification for the project types under each domain.

4. Data analysis and discussion
As illustrated in the previous section, the search process resulted in 112 papers, of which 78 were journaled papers and 34 were conference papers. These selected papers were read, and information on the publishing year, country of publishing, type of data used, the domain of application, and the subject area were extracted to form and understand the research trends around IPD research.

4.1. Publishing year
The first indicator developed was the year of publication. The year of publication was determined as the year the paper was made available, not the year of submission or approval. The results show that the last four years from 2017 to 2020 represent a continuation of the growing trend in IPD research, where the amount of the published research in the last for years equals more than 70% of the total published research in the period from 2001 to 2016. In particular. The year 2019 saw the highest number of IPD publications between the last four years with 32 papers, while both 2017 and 2018 have witnessed 30 and 29 publications respectively, while only 21 papers were published in 2020.

4.2. Country of publishing
Another aspect that was considered is the global reach of IPD research by identifying the country of publishing. In the cases where multiple authors with different affiliations were indicated, the first author's affiliation was used to determine the country of publication. The reviewed papers show that in the period from 2017-2020, 28 different countries have contributed to the efforts of investigating the different aspects of IPD. As shown in Figure 2, the USA, China, and Norway are leading in terms of the number of IPD research published in the period in question. The USA alone contributed to the diffusion of 31 research, while China diffused 22 research and Norway put in 9 research. Next in the list, UK and Canada arose with 5 and 4 research on IPD, respectively. It is interesting to note that Asia as a continent diffused 45 IPD research produced by 12 different countries. North America came next with 35 publications, then Europe with 26 research by ten different countries. In contrast, only two countries in Africa and one country in South America and Oceania have shown up on the list.

By looking back to the period from 2001 to 2016, remarkably, the result in [4] did not provide the total number of countries that diffused IPD research; instead, they mentioned that 13 countries and the European group have contributed to IPD research. Therefore, to be able to compare with that period, the 28 countries that diffused IPD research from 2017 to 2020 can be sorted in the same logic to 18 countries and the European group, and that is indicating the notable increase in the number of countries that are participating in the IPD research efforts.

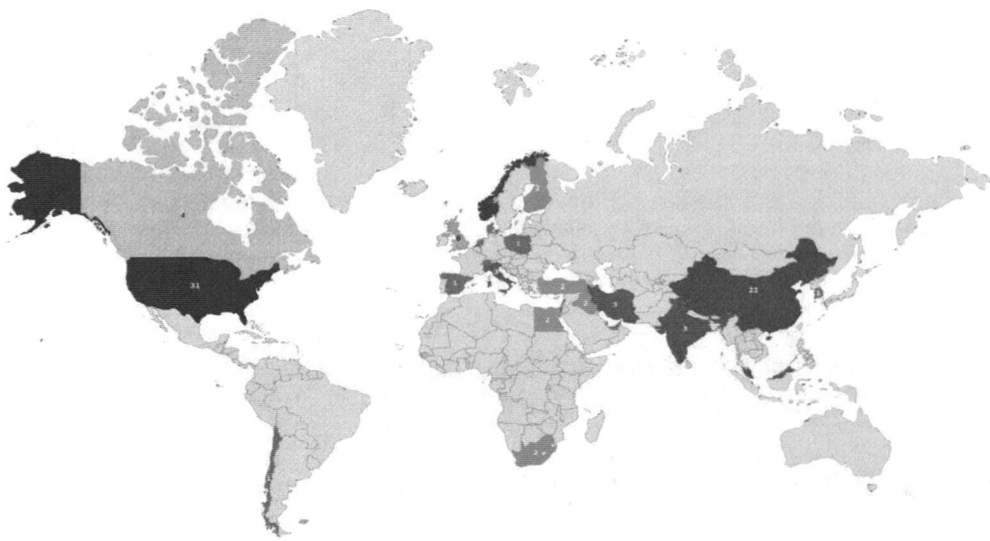

Figure 2. Distribution map of IPD research from 2017-2020

4.3. Type of research conducted
As indicated by the data collected and used to analyze and interpret the results, the type of research constitutes another avenue for evaluation. Throughout the papers reviewed in this study, and as shown in Figure 3, most of the papers employed qualitative methods to understand IPD. Indeed, almost 57% of the studies were qualitative. Conversely, studies using quantitative data amounted to 27%, while only 12% of the studies used the mixed method where both qualitative and quantitative data were collected. In addition, a small fraction that did not exceed 4% was for studies that used some sort of conceptual data.

4.4. Application domains
Many studies attempted to investigate IPD using a case study approach. That being said, this approach was not utilized homogenously across the sample. Indeed, case studies were rarely used in many countries where IPD is still in the early stages of adoption. Thus, finding cases to study that utilized IPD as a project delivery system was a barrier to this research approach. Therefore, only 46% of the selected research were case studies investigating an IPD or collecting data from persons directly involved in the IPD project. Those papers were

classified into six categories based on the application domain, as shown in Figure 4. Institutional (i.e., federal building, hospital, university), which is the most represented domain with 37% of the cases, Infrastructure (i.e., bridges, highways, dams), which the second most represented domain with 20 % of the cases. Next, the Residential and commercial (i.e., office buildings, housing, commercial centers) and the Industrial (i.e., factories, oil, and gas plants) appeared in 12% and 4% of the cases, respectively.

Lastly, two more categorizations are provided to serve the studies that do not fall under the categories mentioned above. Miscellaneous research studied project types not included in the previous categories, e.g., renovation projects, and Multiple for research that studied projects from more than one domain.

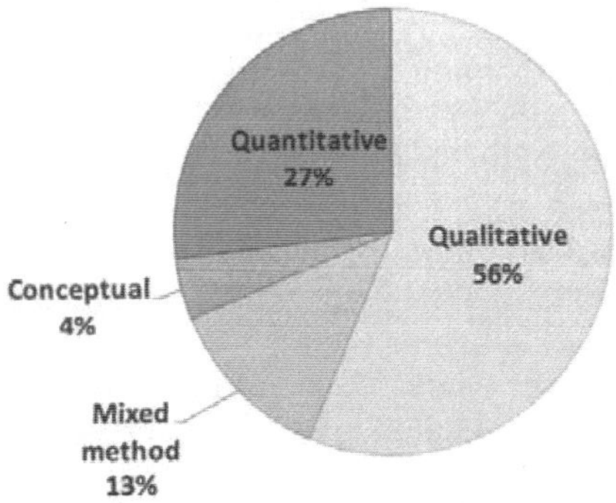

Figure 3. Research publication per data type.

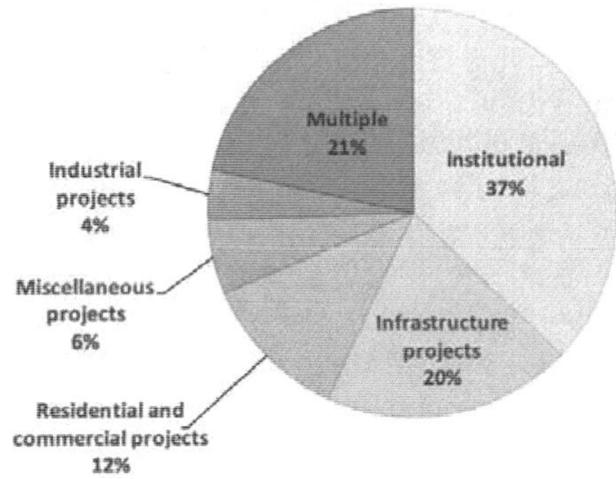

Figure 4. Research publication per application domain.

4.5. Subject area

Finally, and potentially most importantly, the research trends on IPD were investigated to identify the topics and the addressed subjects, which can help frame the needs and gaps, which would shape future trends in IPD research. The selected studies were categorized into five different groups, as shown in Table 1. Nonetheless, it is worth mentioning that the same study could be classified under more than one category in many cases. There is no clear dividing line between those categories, resulting in some research under two or even three categories. However, in this study, to simplify and clarify the research trends around IPD, the researcher meant to classify each study under one category based on the most addressed topic in the study.

As illustrated in Table 1, most of the research reviewed in this study can be classified under two subject areas, principles of IPD with 37 articles (33%) and implementation of IPD with 33 articles (30%). The next subject area that made up a considerable share of the research topics is the integration between IPD and building information modeling (BIM), which were addressed in 21 different studies representing 19% of the total reviewed studies. The next in line are the barriers and challenges; this topic was addressed in 12 research representing 11% of the total studies. Moreover, lastly, the subject that brought the least attention is solving the barriers of IPD adoption, which were addressed in 9 studies that speak for only 8% of the total reviewed studies.

5. Conclusions

This study aims to discover the research trends of IPD over the last four years from 2017 to 2020, driven by the increase in the popularity of this innovated and promising approach and the rise in the number of studies that addressed IPD in recent years, which make it paramount to understand what has been accomplished and identify the current trends on IPD research. Therefore, determine the needs and gaps that should be addressed in future research while the IPD is gaining more popularity and is expected to spread in more regions around the world due to the general dissatisfaction with the construction industry's performance.

Table 1. Classification based on the subject area

Subject area	Description	Number of articles
Principles	Papers around the principles of IPD, an introduction to IPD, the need for IPD.	37

Implementation	Papers investigated the circumstances of IPD implementation through a case study or the implementation attempts within a specific region or country.	33
Barriers and challenges	Papers around the barriers and challenges of IPD adoption and implementation.	12
Solving barriers	Papers that are dedicated to understand and propose solutions to the barriers of IPD adoption.	9
IPD and BIM	Papers that looked into integrating the IPD and BIM or addresses IPD and BIM together as innovative tools and approaches to the construction industry.	21

The last four years from 2017 to 2020 represent a continuation of the growing trend in IPD research, where 112 papers were published in the form of a journal or conference articles. This trend still can grow, as more and more projects are starting to adopt IPD and more successful stories of projects that employed IPD is spreading. This deployment would drive both researchers to reveal any vague understanding of IPD and industry parties who will be more willing to take steps towered adapting IPD.

Although 28 different countries have contributed to the efforts of IPD research in the last four years, more local research and case studies are needed to promote the adoption of IPD in more construction markets. The concepts of IPD still unknown in many construction markets around the world. There is a need for pioneers who could act as early adopters of IPD hand by hand with conducting local research on IPD to promote this innovative and collaborative delivery system.

The principles and the implementation of IPD were the most two addressed topics in the last four years, 2017 to 2020; however, as more new construction markets in different countries would enter and practice the IPD system, the need for studies that address the challenges and the barriers and ways to solve that barrier in the new markets are paramount. In addition, most case studies in the published researches are around institutional and infrastructure projects, which usually involve a high level of complexity. However, the problems of the construction industry are not limited to those project types only. Therefore, there is a need for more (low-medium) complexity-level case studies to test the effectiveness of IPD in managing different projects sizes.

Eventually, it should be noted that this paper proposes an initial review that will be further developed in a follow-up full-length journal paper where the results of this investigation will be presented in more depth.

Acknowledgments

This research is funded by the GRIDD research group. Special thanks of gratitude to Mr. Félix Langevin-Harnois and Mr. Prasun Lala, from the ETS library, who helped us validate the review process, and special thanks to Ms. Audrey Provost. She helped us invalidating the results of this systematic review.

References

[1] Teicholz P, Goodrum PM, Haas CT. US construction labor productivity trends, 1970–1998. Journal of Construction Engineering and Management. 2001;127(5):427–9.
[2] Ebrahimi G, Dowlatabadi H. Perceived challenges in implementing integrated project delivery (IPD): insights from stakeholders in the US and Canada for a path forward. International Journal of Construction Education and Research. 2019;15(4):291–314.
[3] Barbosa F. Reinventing construction–A route to higher productivity. February 2017. 2017.
[4] Kahvandi Z, Saghatforoush E, Alinezhad M, Noghli F. Integrated Project Delivery (IPD) Research Trends. EPPM-Journal. 2017 Jul 31;7(2):99–114.
[5] Crowther M, Lim W, Crowther MA. Systematic review and meta-analysis methodology. Blood. 2010 Oct 28;116(17):3140–6.
[6] White A, Schmidt K. Systematic literature reviews. Complementary Therapies in Medicine. 2005 Mar 1;13(1):54–60.
[7] Guide AIA. Integrated project delivery: A guide. American Institute of Architects, California. 2007;

IOP Conf. Series: Materials Science and Engineering 1218 (2022) 012040 doi:10.1088/1757-899X/1218/1/012040

Towards Inclusive and Sustainable Infrastructure Development through Blockchain-enabled Asset Tokenization: An Exploratory Case Study

Y Tian[1], R E Minchin[1], K Chung[2], J Woo[1] and P Adriaens[2]

[1] School of Construction Management, University of Florida, Gainesville, FL 32603, USA
[2] Department of Civil and Environmental Engineering, University of Michigan, Ann Arbor, MI 48109, USA

Corresponding author: Yifeng Tian. Email Address: yifeng.tian@ufl.edu

Abstract. Infrastructure development is a key to supporting the economy and building social resilience. Unfortunately, existing infrastructure financing models struggle with multiple issues to close the widening financial gap and integrate environmental, social, and governance (ESG) factors to improve resilience and achieve sustainable growth. With the emergence of innovative technology, blockchain-enabled asset tokenization shows the potential to create a new economic model to integrate non-financial values, such as positive social and environmental impacts, into tradable cryptographic tokens. Tokenization offers opportunities to generate long-term positive social and environmental impacts, along with better financial returns, which further improves the profitability and bankability of infrastructure projects. SolarCoin, WePower Token, and ZiyenCoin are analyzed in this exploratory case study research to demonstrate how blockchain-enabled asset tokenization can be applied to infrastructure development. It is identified that tokenization promotes inclusiveness and sustainability through shareholder empowerment, incentive monetization, and finance optimization. Obstacles hindering the broader adoption of tokenization and policy implications are also discussed.

1. Introduction

Infrastructure is fundamental to social development and economic growth through creating connections, enabling communications, powering communities, and creating opportunities for every member of society. From power generation facilities to transport systems and water networks, infrastructure provides essential services enabling society to function and the economy to thrive. Infrastructure plays a vital role in making positive environmental impacts and strengthening inclusiveness and sustainability [1]. Quality investment in infrastructure is one of the main drivers for contributing to the United Nations Sustainable Development Goals (SDGs). However, the mismatch between the need for more and better infrastructure and the available financing continues to increase. An analysis by the McKinsey Global Institute forecasted that the world was facing a $3.7 trillion gap between projected investment and the amount needed to provide adequate global infrastructure through 2035 [2]. Under the current financing system, public sector finance through direct grants, subsidies, and concessionary loans are subject to public budget constraints and political influence, which is unsustainable in the long run. Private investment in infrastructure is limited to a narrow set of institutional investors such as pension funds and insurance companies, thereby leaving a significant amount of capital and recourses on the sidelines. The current debt-dominated infrastructure financing system struggles with integrating environmental, social, and governance (ESG) considerations or providing incentives to promote SDGs. The COVID-19 pandemic has made the already difficult situation even more challenging. Pioneering thinking and a groundbreaking financing system are required to ameliorate public sector engagement and mobilize private sources from a broad spectrum to bridge the widening infrastructure gap and realize inclusive and sustainable growth to impact society and the environment positively.

Innovations are emerging at a swift pace in the information age, which offers the potential to transform the infrastructure system and its financing options more intelligent, efficient, and resilient. Blockchain is considered the most significant technological breakthrough since the internet. Blockchain is a type of distributed ledger for maintaining a permanent and tamper-proof record of transaction-based data. With the advancements in the blockchain, a decentralized, immutable, and trusted system can be built, which brings improved security, transparency, and automation. Building on the blockchain, asset tokenization enables the transition of assets in conventional forms into cryptographic tokens. Tokenization can introduce a new system to finance infrastructure in a more efficient, inclusive, and sustainable way [3].

Despite the aforementioned potential benefits and growing attention from media sources and practitioners, the field of infrastructure asset tokenization has not attracted comparable academic interest. It lacks a deeper grounding in theoretical and empirical analysis. This research aims to broaden current knowledge of infrastructure asset tokenization to deepen the understanding of how tokenization is applied to finance infrastructure while promoting inclusiveness and sustainability. So that infrastructure can better serve the surrounding community and the broader society. A detailed explanation of the potential benefits brought by infrastructure asset tokenization is outlined in Section 2. Use cases are analyzed in Section 3. Key research findings and recommendations for future research are included in the final section.

2. Infrastructure asset tokenization

Tokenization is the process of digitally representing existing assets on a distributed ledger, such as the blockchain. Asset tokenization enables the representation of assets on the ledger by linking or embedding the economic value and rights derived from the underlying assets into cryptographic tokens created on the blockchain [4]. Cryptographic tokens represent programmable assets or access rights, such as shares in a company, ownership of a piece of real estate, permissions to a platform, or electricity produced by energy plants. In theory, any asset and associated rights can be tokenized and represented on the blockchain. Tokens are categorized into three classes depending on functions and purposes [5].

- Payment tokens are a decentralized currency for making and receiving payments on the blockchain, such as Bitcoin. Payment tokens are used as an alternative means of exchange, supplement to fiat currency.
- Utility tokens grant access to products or services offered by the issuer of the tokens. This type of token is meant to be primarily used within the blockchain-based ecosystem. Utility tokens are issued through Initial Coin Offerings without the governance of existing security regulations in most jurisdictions.
- Security tokens are an investment instrument representing the legal ownership of a physical (e.g., real estate properties, gold, etc.) or digital asset (equity, bonds, loans, etc.) that has been verified within the blockchain. Issuance and transactions of security tokens are under the protection of existing security regulations.

Cryptographic tokens are governed and executed through smart contracts, which are software algorithms integrated into a blockchain with trigger actions based on pre-defined parameters. Smart contracts are self-enforcing and self-executing. The automation reduces the administrative burden and the number of intermediaries required in the process, which leads to cost reduction and faster execution. Besides the efficiency gains brought by automation and disintermediation, tokenization delivers other benefits, including enhancing transparency, improving liquidity to currently illiquid assets (e.g., infrastructure assets), and enabling more efficient clearing and settlement [6].

Infrastructure asset tokenization refers to the process of tokenizing utilities (e.g., services or products provided by the underlying facility) or ownership interests of the conventional securities (e.g., equity of infrastructure companies or funds, loans, and bonds) onto the blockchain. Tokenization allows for the use of tokenized digital records in the infrastructure life cycle, from planning and financing to construction and operations. The crucial benefits related to the enhancement of inclusiveness and sustainability in infrastructure development by implementing tokenization are summarized into three categories: stakeholder empowerment, incentive monetization, and finance optimization [7].

- Stakeholder empowerment
  - Tokenization provides communities with a new tool of governance. Tokens embedded with certain rights of the infrastructure facility can be distributed to neighbouring residents, who serve as special shareholders to influence the planning and operations of the project through voting enabled by smart contacts. Consumerization through tokenization thus has the capacity to create a sense of local ownership of the public infrastructure facilities and provides a platform to galvanize social acceptance. By owning tokenized revenue-generating infrastructure assets, neighbouring residents can enjoy the social spill-over effects brought by the new infrastructure facility as well as receive an additional source of income.
  - Through tokenization, operational and financial information of the underlying infrastructure facility is automatically recorded on the blockchain immutably. The improved quality and quantity of operational and financial data will grant stakeholders and investors direct access to monitoring the project.
  - From investors' perspective, tokenization democratizes infrastructure assets and allows for fractional ownership, which lowers barriers to investment and promotes inclusiveness by offering financial access to individuals and small and medium-sized enterprises (SMEs). Tokens can also be distributed to the unbanked in the low-income community to address the inequality issue.
- Incentive monetization
  - Tokenization shows the capability to completely reinvent business models, create new economies and change society's view on earning income by converting non-financial values that can be monetized and traded across a blockchain, which incentivizes positive behaviors and efforts for sustainability. Tokens help stimulate the consumption of sustainable services and goods by providing financial incentives to users and customers and further improving the profitability and bankability of high-impact projects. Transactions on the blockchain are verified instantaneously. The instant incentives increase motivation even more.
- Finance Optimization
  - By deploying tokenization, the size of infrastructure projects is no longer a determining factor for financing cost efficiency. The disintermediation (removal of intermediaries) enabled by smart contracts, results in a significant fixed financing cost reduction and improved bankability of projects. Tokenization makes investment in small-scale infrastructure at the project level possible.
  - Smart contracts allow for bi-directional instant transfer of funds and tokens by removing intermediaries without the need for a separate settlement process. The automation enabled by smart contracts reduces the cost of transaction and administration, increases

the speed of execution, and facilitates dividend distribution, escrow management, and collateral management.

3. Case Study Analysis

This research employs an exploratory case study approach. Through case studies, a novel phenomenon and concept are better understood [8]. Given the novelty of infrastructure asset tokenization, multiple cases were studied to indicate how asset tokenization can be applied to promote inclusiveness and sustainability in the context of infrastructure development. Cases were retrieved from desk research. The selected cases depict the common features of infrastructure asset tokenization and also include adequate contextual and structural diversity. The use of multiple cases is based on the theoretical replication logic for producing contrasts in a cross-case manner [9].

3.1. Case A: SolarCoin

SolarCoin is a type of payment token designed to facilitate solar photovoltaic facility installation [10]. It was an open community project created in 2014 operated by the SolarCoin Foundation. The Foundation distributes SolarCoins to generators of solar electricity using verified solar facilities at the rate of one SolarCoin per one megawatt-hour (MWh) of solar energy produced. The SolarCoin blockchain is a trusted data layer reporting solar power production data, which provides a Know-Your-Device (KYD) anchor data provenance layer for providing more security when monitoring solar power produced. SolarCoins can be used as a means of rewards and payment. The value of SolarCoins is determined by the faith of its users in utility across the network, similar to Bitcoin. SolarCoins can be traded in the secondary market (e.g., Lykke) in exchange for other cryptocurrencies or fiat currencies.

In terms of shareholder empowerment, the immutable operation and performance data recorded on the SolarCoin blockchain grant stakeholders and investors direct access and visibility in the project's performance. Furthermore, the broader SolarCoin community can provide essential support to assist smaller projects initiated by SMEs and solar energy producers in developing countries to facilitate business growth in this sector. From the perspective of incentive monetization, SolarCoins, serving as direct incentives to reward solar energy producers, encourages the development of solar infrastructure facilities and the transition to clean and renewable energy. However, according to the Coinmarketcap, the market capitalization of SolarCoin fluctuated within the range between one million to five million dollars in the past year. As a currency with small market capitalization, its capability to realize the proposed goal is limited. The most significant obstacle faced by SolarCoins is to gain mainstream acceptance, incentivize communities to transition to solar energy, and convince power producers to share generation data. The promotion of SolarCoins depends on its liquidity and consumer attractiveness as compared to other cryptocurrencies.

3.2. Case B. WePower

WePower Smart Energy Tokens are utility tokens issued by WePower, a renewable energy procurement and trading platform that aims at facilitating the global shift to renewable energy by democratizing the energy procurement process [11]. Smart Energy Tokens represent fractional ownership rights and obligations obtained under power purchasing agreement (PPA) contracts. The tokens are exchanged among owners through auctions. The WePower platform applies smart contracts to execute transactions and connects energy producers and purchasers directly in a frictionless and secure manner. Information is stored on the Ethereum blockchain and available for participants to monitor auctions, transactions, and executions.

By applying blockchain-enabled tokenization, WePower Smart Energy Tokens backed by the factional ownership rights of PPA simplifies the contracting process and enables individuals and companies, regardless of the size, to be engaged in renewable transactions transparently. In terms of financial optimization, the tokens allow energy producers to access a new segment of energy purchasers at a lower cost. Energy producers can also contract with multiple purchasers at the same time, thereby diversifying their counterparty risk. For projects seeking finance, these features make new renewable infrastructure projects more bankable, thus potentially leading to the delivery of more renewable projects. The more cost-effective and transparent contracting and transaction process would encourage more energy producers and purchasers to transit to renewables and further facilitate sustainable infrastructure development. While bypassing the traditional project financing process in securing funding for projects is attractive, the disclosure of consumption and production data in the energy markets may be limited depending on the jurisdiction.

3.3. Case C. ZiyenCoin

ZiyenCoins are security tokens issued by Ziyen Energy, a pioneering tokenization company offering a total of 500 million equity-based security tokens backed by energy assets in its portfolio. The price of one ZiyenCoin is $0.01. A maximum of $5 million is planned to be raised with the security token offering (STO) to both U.S. and non-U.S. investors. The initial investment requirement is as low as 50,000 ZiyenCoins which is equivalent to $500. The rights to the company's Class A securities, incorporation certificates, and bylaws are embedded into the smart contract of each ZiyenCoin. These rights are transferable in peer-to-peer (P2P) and secondary market trading. The company is acquiring more green energy projects facilitating the renewable energy transition [12].

The $500 minimum initial investment requirement removes small investors' barriers to invest directly through small-value tokens to become shareholders of Ziyen Energy. Investors can invest in energy infrastructure assets in the U.S. and are able to build up a diversified portfolio based on their specific financial expectations. The advantage of blockchain in the information recording and storing grants investors and project stakeholders access to improved monitoring and management of their investments. The implementation of small contracts makes raising funds from the public markets possible for small-scale companies like Ziyen Energy, currently valued at $5 million. This could significantly facilitate the development of SMEs and expedite the renewable energy transition. However, like many other early tokenization applications, ZiyenCoin is still evaluating its business model to scale market adoption and gain mainstream acceptance.

4. Conclusion

Three infrastructure-related tokenization cases are reviewed in this research. Cases were selected based on the type of cryptographic tokens. SolarCoins are payment tokens. ZiyenCoins and WePower tokens are security tokens and utility tokens, respectively. These cases indicate that various token models can be applied to facilitate infrastructure development and financing. The promotion of inclusiveness and sustainability is through shareholder empowerment (e.g., engaging the surrounding community, integrating high-quality data, and granting access to individual and SMEs), incentive monetization (e.g., monetizing non-financial values), and finance optimization (e.g., removing barriers to small projects and improving liquidity), relying on the blockchain and smart contracts. Three cases are all initiated by private entities in the energy sector, mainly owned by the private sector. No tokenization of public infrastructure cases was uncovered, indicating that public entities may be lagging in the early adoption of asset tokenization.

Three tokenization projects in this research are facing a common obstacle, which is high uncertainty at their early stage of development. SolarCoin was introduced in 2014, but it has not gained mainstream acceptance as planned when initiated. The structure of the WePower tokens was revised in 2019 after its initial launch in 2018. Information on these cases was extracted from whitepapers, which only demonstrates the business plans instead of actual operations. Due to the relative infancy of tokenization, there is minimal operational data available for in-depth analysis at present. Future research should conduct in-depth analysis based on data from operations to explicitly examine how tokenization

IOP Conf. Series: Materials Science and Engineering 1218 (2022) 012040 doi:10.1088/1757-899X/1218/1/012040

promotes inclusiveness and sustainability in the context of infrastructure development. The application of tokenization in other infrastructure sectors besides energy should also be studied.

Infrastructure asset tokenization shows the potential for facilitating infrastructure development while promoting inclusiveness and sustainability. A coordinated, proactive, and long-term approach that builds on existing practices by the public and private sectors is required to maximize the potential benefits brought by the emerging blockchain-enabled technology.

5. References

[1] Lenferink S, Tillema, T and Arts J 2013 Towards sustainable infrastructure development through integrated contracts: Experiences with inclusiveness in Dutch infrastructure projects *International journal of project management* 31(4) pp 615-627

[2] Woetzel J, Garemo N, Mischke J, Kamra P, Palter R 2017 Discussion Paper in Collaboration with McKinsey's Capital Projects and Infrastructure Practice *Mckinsey Global Institute*

[3] Tian Y, Lu Z, Adriaens P, Minchin R E, Caithness A and Woo J 2020 Finance infrastructure through blockchain-based tokenization *Frontiers of Engineering Management* 7(4) pp 485-499

[4] Uzsoki D 2019 Tokenization of Infrastructure. A blockchain-based solution to financing sustainable infrastructure *IISD*

[5] Gan J, Tsoukalas G and Netessine S 2021 Initial coin offerings, speculation, and asset tokenization *Management Science* 67(2) pp 914-931

[6] Tian Y, Zhang Y, Minchin R E, Asutosh A and Kan C 2020 An innovative infrastructure financing instrument: Blockchain-based tokenization. In *Construction Research Congress 2020: Infrastructure Systems and Sustainability* pp 731-740

[7] Tian Y, Adriaens P, Minchin R E, Lu Z and Qi C 2020 Asset Tokenization: A Blockchain Solution to Financing Infrastructure in Emerging Markets and Developing Economies. *Institute of Global Finance Working Paper*

[8] Yin R K 2017 Case study research and applications: Design and methods *Sage publications*

[9] Ragin C C 1987 The Comparative Method: Moving Beyond Qualitative and Quantitative Strategies *University of California Press*

[10] Johnson L, Isam A, Gogerty N and Zitoli J 2015 Connecting the Blockchain to the Sun to Save the Planet *SSRN 2702639*

[11] WePower 2019 WePower Whitepaper, https://wepower.com/ Accessed April 28 202

[12] Ziyen 2020 ZiyenCoin Whitepaper, https://www.ziyen.com/ Accessed May 2 2021

IOP Conf. Series: Materials Science and Engineering 1218 (2022) 012041 doi:10.1088/1757-899X/1218/1/012041

Tracking Productivity in Real-time Using Computer Vision

A Kadoura and E P Small

Department of Civil and Environmental Engineering, University of Delaware, Newark, Delaware, USA

akadoura@udel.edu

Abstract. The construction industry is lagging behind other industries in terms of productivity gains with stagnant growth over several decades. The reasons for the lack of growth are complex and multifaceted, yet all causes and the resulting effects are realized on the output of the workers and machinery on-site. Often, typical site management practices do not identify productivity issues in a timely fashion when corrective actions may impact construction activity progress. Better management tools are required to optimize the productivity of on-site crews. This promises to positively impact the completion of individual tasks, work packages, projects, and ultimately the industry with respect to productivity, which is the driving force behind this research effort. The approach pursued focuses on utilizing commercially available cameras together with computer vision and artificial intelligence. With these tools, the main objective is to develop an algorithm capable of providing real-time feedback on the productivity of workers. Ideally, the algorithm will further identify each type of activity being conducted, thereby capturing useful productivity data. The initial model is a simple classifier that checks whether a worker performs work or stands idle. This is performed through algorithms that identify and track a person's pose and joints and translate them to data points that can be evaluated and translated into helpful productivity measures. Finally, successfully developing a model capable of providing real-time productivity data will allow project managers and planners to better manage and utilize on-site resources. Additionally, since a large amount of data is collected and saved, trends in productivity levels can be tracked and studied further to optimize and improve them.

1. Introduction

The construction industry is regarded as one of the most important economic sectors globally, with an estimated market spending of 10.4 trillion USD. According to McKinsey research, the construction industry accounts for "13% of [global] GDP... [and]... employs 7% of world's working population." [1]. It is expected that with this amount of spending, productivity and total output will improve significantly. This is not the case, as McKinsey estimated that the average growth in construction productivity over the last two decades was 1 percent, compared to a global average of 2.8 percent [1]. This demonstrates that the construction industry lags behind other industries, and steps must be taken to realize the potential for increased productivity growth.

The study will look into the possibilities of leveraging existing tools to track labor productivity in real-time. It is exploratory in nature; thus, it is not meant to be comprehensive or all-encompassing for all construction industries or sorts of construction operations. Computer vision models will be investigated and used on sample imagery/videography rather than a comprehensive, representative set of construction activities. The ultimate, long-term goal is to get productivity data to management

IOP Conf. Series: Materials Science and Engineering 1218 (2022) 012041 doi:10.1088/1757-899X/1218/1/012041

decision-makers so they can intervene in a timely fashion and make required adjustments to resources and activities. Once proof of concept has been achieved, more studies can be done to expand on this effort.

There are several advantages to exploring the topic of real-time productivity tracking utilizing existing tools. The first benefit is that identifying and tracking productivity in real-time provides quick feedback on the construction operations and crews' performance. If productivity is reduced unexpectedly owing to a site fault, it could be quickly discovered and fixed. On the other hand, current ways to assess productivity rely on limited site observations, making it difficult to promptly address issues if there are delays or productivity reductions.

2. Literature Review

Several factors contributing to the low productivity were identified. Regulatory/contractual issues, design issues, procurement and supply-chain, and on-site execution have been identified as areas where significant differences can be made [1]. The main concern is with the approach taken to improve and correct these productivity problems. Typical improvement approaches rely on forensic analysis, where the productivity is calculated at a later stage of construction and then compared with the planned actions. With this approach, results are obtained too late to implement any corrective actions. Therefore, a significant improvement can be made by reducing the time between observing and calculating the productivity and comparing it to the planned value enabling the opportunity for corrective actions.

2.1. Productivity

The first step required to improve productivity is to be able to define it and measure it. This is not very easy because measuring productivity involves and depends on many factors that vary from project to project. In general, productivity can be quantified as a ratio of input to output costs or daily work output. The exact metrics and units used to calculate productivity typically depend on the type of data being analyzed by the project management team. According to Dozzi and AbouRizk, there is no universal industry standard available to measure or compare the productivity values [2]. To complicate matters further, productivity values are often measured on different scales. The most common levels used in the industry are macro-level and micro-level productivity analysis [2]. Macro-level productivity analysis compares productivity levels across countries or industries, while micro-level productivity analysis is more detailed and can be used to compare the productivity levels between different projects, work packages, or activities. Calvetti et al. introduced a third more detailed level of analysis: nano-level analysis. The nano-level analysis is based on the basic motion and work elements of laborers [3]. Basic motion elements build into work elements which then collectively complete the tasks. This identifies the productivity of each worker and allows for analysis between workers or crews.

2.1.1. Measuring and Collecting Productivity Data. Other industries have made strides toward automating the process of measuring productivity with new sensor technologies. Fortunately, these new technologies are gradually making their way into the construction industry. However, the current widely used methods rely on manual observations of workers by assigning a person to monitor workers on-site and producing a productivity measurement using hand calculations. Dozzi and AbouRizk presented three methods for measuring productivity on the job site: field rating, work sampling, and the five-minute rating method [2]. Each of these methods calculates productivity based on human-based observation and can be burdensome. Therefore, a better approach is to introduce new technologies to automate the process.

The observational limits can be overcome by employing approaches that rely on computer vision and data processing. Calculations can be completed in minutes, and results can be delivered quickly, which is required for using the data to make beneficial changes to site operations. With this methodology, any of Dozzi and AbouRizk's three approaches for calculating worker efficiency in performing tasks can be easily modified to provide an immediate alarm in the event of a work halt or other critical data.

Because the study aims to measure and track productivity in real-time, a significant portion of the literature review will be devoted to examining the various approaches that can be used to establish a measurement. Numerous approaches and equations have been identified to collect productivity-related field data. The majority of the approaches are conceptually similar but appear to differ in the procedure for collecting data or calculating productivity; thus, only a few approaches will be examined to better understand estimating field productivity.

Factors Impacting Productivity. While researching the factors that influence productivity in the construction industry, it was discovered that the factors are not constant but rather constantly change, even within the same project. However, these factors can be classified into major categories that have an impact on productivity levels. Although hundreds of factors have been identified in the literature, some of these factors include, but are not limited to, construction management and planning issues, improper or insufficient material and tools, incorrect construction methods, a lack of supervision, and reworks [4].

2.2. Construction 4.0

Construction 4.0 is the fourth construction revolution, sparked by Industry 4.0, the fourth industrial revolution. Industry 4.0 was launched in Germany in 2011 to create a fully integrated industry based on newly accessible technologies [5]. Based on the literature, Construction 4.0 can be defined as the process of digitizing and automating the construction industry by utilizing technologies and techniques to enhance performance. This, in turn, improves collaboration and information flow, enhances triple bottom line attributes, preserves safety, and increases productivity.

Construction 4.0 adopted new technologies, which are introduced as pillars. Examples of these pillars include artificial intelligence, big data analysis, the internet of things (IoT), field data-capturing technologies, and so on. The two most appropriate pillars used in this research are artificial intelligence (AI) and field data-capturing technologies (FDCT). These pillars are most related because computer vision is part of AI, and the cameras used to collect the data are part of FDCT. The research will use FDCT to capture and digitize the information from the construction site, while the AI model will predict whether the worker is working or not.

2.2.1. Computer Vision. Computer vision is an interdisciplinary field in computer data science that deals with how computers perceive data from images and videos. Computer vision attempts to imitate how humans use their visual system to perceive and understand their surroundings by using cameras and convolution neural network (CNN) algorithms [6]. The CNN model comprises of four main layers, the input layer, the convolution layer, the pooling layer, and the output layer. Depending on the complexity of the model, more convolution and pooling layers can be supplemented. It should be noted that the performance of CNN or any other AI model is primarily determined by the data used to train the model. Since a construction project is considered unique, it is usually challenging to achieve high accuracy due to the constantly changing nature of the project. Furthermore, there is limited available video of construction operations to use in training the models. Ideally, videos could be captured of actual site operations; however, this proved difficult as construction companies are often reluctant to share data due to confidentiality agreements [7], legal concerns, or site worker disinclination.

2.2.2. Field Data-Capturing Technologies. Field Data-Capturing Technologies (FDCT) are a group of technologies that capture and record information from the field and digitize it to raw data that can be used for further analysis, as the name implies. These technologies are intended to supplement or replace traditional manual methods, increasing efficiency and providing more reliable data. Productivity studies with surveillance cameras enable data collection with fewer direct observers to measure productivity. This method removes the observers from the work site, which has no effect on the crew's work performance. Workers, in effect, do not behave as if they are being observed. Because the process of gathering information still requires a significant amount of manual labor, using FDCT has significant advantages. Using FDCT to automate data collection and measurement of productivity is thus very appealing for this process.

3. Research Methodology

The goal of this research is to determine whether labor productivity can be assessed in real-time using computer vision, with the results being used to influence positive change in the construction process. This expands on the cutting-edge techniques summarized in the literature review and contributes to the body of knowledge in Construction 4.0 applications and labor productivity optimization.

3.1. Computer Vision Models

Several available computer vision models were tested for the applicability of determining productivity. Some of these models include YOLO and DEEPSort. The main issue with these models is that the model identifies and tracks the worker, but with no further details to the type of work being conducted. A better identified model is TF-Pose-Estimation, which uses a previously developed library (OpenPose) and builds upon it. The primary purpose of OpenPose, and TF-Pose-Estimation, is to provide real-time multi-person key-point detection [8]. It should be noted that there are several other models that are also capable of identifying joints and body parts with no significant difference between them and this model.

The benefits provided in TF-Pose-Estimation and similar models compared to other models is the ability to identify joints and body parts. Since the body movements and joint locations can be identified, the model can be capable of determining whether a worker is working or standing idle. Therefore, TF-Pose-Estimation was chosen to move forward with the research. The algorithms rely on supervised training, and a pretrained dataset was made available for use and application in this study. The model will not need to be re-trained because the developers made available a pre-trained algorithm, saving time and effort in the research process. The models' training recognizes individuals but does not distinguish between working and non-working individuals. This is accomplished through the use of a classifier variable that has been trained to indicate work positions and differentiate whether the labor is productive or idle based on the personnel's joint positions and poses. If necessary, the pre-trained classifier can be retrained using a dataset of construction workers in the future.

The model must be capable of translating and storing the FDCT input (i.e. imagery and videography) into useful data that can be analyzed for additional insightful and useful data. Ideally, the approach will be tested on-site with photography and videography. This study will provide proof of concept, and subsequent studies will use the technique in "live" on-site conditions.

3.2. Data Processing

Without going into specifics, the model predicts the coordinates of body parts and joints using pre-trained models. Then, using OpenCV, these body parts are redrawn on the original photo. This does not predict the worker's productivity but rather provides visual feedback to ensure that the predicted body parts correspond to the person and that the prediction is correct. Finally, the produced stance is compared to a pre-trained collection of poses in order to determine if the person is working or idling. Retraining MobileNet's last layer accomplished this on a dataset of working and idle workers. Figure 1 shows how the model generates outcomes. These are illustrated using royalty-free photos for demonstration reasons. These figures include subjects that are in the photograph's foreground and take up a large portion of the frame, resulting in better results. It is worth noting that in these illustrations, not all of the body parts need to be seen in order to make a prediction, which is what is likely to happen in real-life settings.

Figure 1: Model prediction showing idle (left) and working (right). (Image source: Freepik.com)

3.3. Experiment Design – Development and Testing of the Model
Model testing in this effort requires video footage. The video is less clear and has a lower resolution than the photographs shown in Figure 1. This is expected to affect prediction accuracy. Furthermore, in their positions, specific productive job tasks resemble standing/idling activities. Perfection is neither required nor expected. The goal is to see if the strategy can help and if some of the accuracy constraints can be addressed when deciding whether to notify decision-makers about non-productive periods.

The model's results must also be statistically significant. This is accomplished by creating an experiment and testing the results within a certain level of confidence. The following methodology was used for this study: Video footage of workers engaged in various activities will be used. A table with the manual observation and model categorization as columns and a 15 to 20 second time interval for the rows, indicated as observation number will be generated. A confusion matrix will be generated for each activity and subsequently for the total observations when the model has been run, and the actual vs. anticipated data have been collected. The confusion matrix is useful because it calculates Type I and Type II errors, which refer to false positives and negatives, respectively. In addition, the model's performance and other relevant features can be estimated. The accuracy, precision, sensitivity, and specificity of the model are typically examined when determining the performance of a classifier. These measures show whether the model is appropriately classifying observations and, if not, what type of error is there.

3.4. Preliminary Model Results and Future Refinements
As discussed, computer vision models have been discussed and evaluated for the purposes of this research effort. The model has been evaluated with a small set of videos culled from instructional videos of construction techniques. These were chosen as they were readily available for use and were royalty-free sources. The videos were evaluated manually to identify observations and compared to the predictive models. Results are shown in Table 1.

Table 1: Correlation of Predicted Results to Observations of Work versus Non-Work States

		Prediction	
		Positive	Negative
Observation	Positive	106	78
	Negative	3	55

Evaluation of the table indicates a small proportion of Type 1 errors with 1.2% false positives. False negatives, reflective of type 2 errors, were quite high at 32%. There are several reasons for false negatives, including similarities of body poses between work and non-work states and a lack of a sufficient population of different work activities for training. This shows the necessity of additional research in this area before reliable results can be implemented on-site.

4. Conclusion

In conclusion, the poor productivity growth recorded by the construction industry over the last several decades was the driving force behind our research. The construction sector is in the midst of a productivity crisis, and this study aims to leverage Construction 4.0 technology to build tools to help enhance labor productivity on job sites. The goal of the study was to combine AI and computer vision models with commonly used field data gathering devices, such as cameras and other data collection methods. The goal was to create a real-time system that would monitor ongoing activities and notify project management people when workers became idle, allowing for possible productivity consequences to be mitigated and work to continue.

The study began with a literature review to better understand current productivity issues and available construction 4.0 technologies for productivity enhancement. Completed research on AI, computer vision, and field data collecting approaches was also highlighted in the literature study. The literature review revealed that there is a wide range of metrics used for assessing productivity. There is no single, unambiguous definition of productivity. Instead, different definitions and equations exist depending on the level of analysis and the purpose of the analysis. In general, productivity is defined as the ratio of input to output. This is measured differently depending on the perspective being evaluated. Productivity evaluation at the macro-level differs from that at the micro- or even nano-level. After further investigation, it was determined that a nano-level analysis is required for the purposes of this research, in which individual productivity is measured rather than evaluating productivity for large components or project phases. Alternative methods of assessing an individual's productivity were investigated. The factors influencing productivity were investigated and documented. The following section of the literature review focused on Construction 4.0 and determining which technologies are required for this research.

The research methodology was developed after gaining a better understanding of the problem through research and literature review. Several neural network-based AI models, such as YOLO, DeepSearch, and TF-Pose-Estimation, were investigated. Each of these was tested, and TF-Pose-Estimation was found to be the most applicable model for the research objectives. The model provided a representation of a worker's pose and joint positions, which is helpful in developing a real-time prediction of whether the person is working or idling. Furthermore, the model provides pose (skeletal) coordinates that can be used in more in-depth analysis. As a result, while the model provides some raw data, the amount of data captured can be valuable if used for the proper analysis.

Preliminary results of the model show that Type 1 errors are minimal, but there are significant numbers of false negatives. This is problematic and may result in alerts being issued when there are no issues with productivity. Further research is required, and there are numerous additional opportunities to expand the research effort. This includes additional refinement and testing, as well as the use of different data sources and site-specific implementations. Continuing efforts show significant promise in developing tools capable of having a significant impact on the industry and bringing about a small amount of positive change. The models can be improved to provide more specific information about the type of activity being carried out. Furthermore, because there is little to no data on nano-level productivity analysis, this method can be used to create a database and baseline productivity for various activities and variables.

Ideally, the research will provide input to a real-time alert system that can distribute information to management staff where extended times of non-productive work periods are observed. This permits active intervention and correction of the causes of the non-productive states at times where the intervention will be effective. It is the hope and goal that such a system, if implemented on a small/nanoscale in the field, would ultimately have a broader impact on increasing the productivity on-site and the profitability of the individual projects.

References

[1] McKinsey&Company 2017 *Reinventing construction: A Route To Higher Productivity.* https://www.mckinsey.com/~/media/McKinsey/Industries/Capital%20Projects%20and%20Infrastructure/Our%20Insights/Reinventing%20construction%20through%20a%20productivity%20revolution/MGI-Reinventing-Construction-Executive-summary.pdf

[2] Dozzi S P and AbouRizk S M 1993 *Productivity in Construction.* Ottawa, Ontario, Canada: National Research Council Canada.

[3] Calvetti D, Mêda P, Gonçalves M C and Sousa H 2020 Worker 4.0: The Future of Sensored Construction Sites. *Buildings,* **10**(169), 1-22.

[4] Ghoddousi P and Hosseini M R 2012 A survey of the factors affecting the productivity of construction projects in Iran. *Technological and Economic Development of Economy,* **18**(1), 99-116.

[5] Hofmann E and Rüsch M 2017 Industry 4.0 and the current status as well as future prospects on logistics. *Computers in Industry,* **89**, 23-34.

[6] Xu, S., Wang, J., Shou, W., Ngo, T., Sadick, A.-M., & Wang, X. (2020, October 19). Computer Vision Techniques in Construction: A Critical Review. *Archives of Computational Methods in Engineering*, 1-15.

[7] Darko A, Chan A P, Adabre M A, Edwards D J, Hosseini M R and Ameyaw E E 2020 Artificial intelligence in the AEC industry: Scientometric analysis and visualization of research activities. *Automation in Construction,* **112**, 1-19.

[8] Cao Z, Hidalgo G, Simon T, Wei S E and Sheikh Y 2018 OpenPose: realtime multi-person 2D pose estimation using part affinity fields. *Computer Research Repository (CoRR),* **80**(8), 1-14.

Using new tools for building construction teaching during the sanitary emergency

X Brioso*, C Calderon-Hernandez and D Fuentes-Hurtado

[1] GETEC Research Group, Department of Engineering, Pontifical Catholic University of Peru, Av. Universitaria 1801, Lima 32, Lima, Peru

* Corresponding author's e-mail address: xbrioso@pucp.edu.pe

Abstract. Covid-19 pandemic has originated that a great number of countries' governments demand universities to teach online classes, presenting a challenge to redesign the courses that had construction site visits included in the curricula. The purpose of this work is to present the successful strategies employed on the teaching of building construction, applied in a third-year civil engineering program of the Pontifical Catholic University of Peru. The methodology was designed so that one can adapt the pandemic and post-pandemic stages proposing a balanced use of virtual platforms and software for video conferences and laboratories, incorporating tools like Building Information Modeling, Lean construction, among other systems. Given the sanitary emergency, the construction site visits were replaced by virtual visits to construction projects generated with scanner laser technology, photogrammetry, and BIM, generating digital twins that were stored in online platforms. Students toured the construction project virtually and in an interactive manner, and along with the description of the professionals in charge of the project they were able to reach a good level of comprehension of the organization and the construction processes. In addition, students developed the layout plans of the temporary works with BIM models. On the other hand, the course taught to generate and analyze the unit prices and quantity surveying of activities using the BIM methodology with 5D software, developing the project budget. The success of this methodology is shown through the notable quality of papers and evaluations taken, through the results of surveys taken on students --which had an effectiveness rate of 96.4 % --, and on the improvement opportunities determined. This paper presents tools relevant that may be adapted by the academia and practitioners related to building constructions, for virtual or face-to-face learning.

Keywords: Teaching, New tools in education, Construction Management, Building Information Modeling, Lean Construction.

1. Introduction

Covid-19 pandemic has arisen in a great number of countries governments, Peruvian among them, to demand universities to teach online classes [1] presenting a challenge to redesign the courses that had construction site visits included in the curricula. In fact, there is a need to implement an integrated vision of virtual education that combines the technology currently available to support teaching and learning [2]. On the other hand, in the construction sector, there is a challenge with using collaborative platforms, having remote access to computer laboratories and that students use the software needed for the course in their personal computers [3]. Moreover, there are previous experiences of solving the challenges of this complicated context with the use of BIM technology based on real projects [4]. However, it still lacks the integration with the many management systems tools in the construction phase.

The purpose of this work is to present the strategies successfully applied to the teaching of the course Building Construction-- now called "Economic Evaluation of Building Projects" -- applied during the pandemic in a third-year civil engineering program at the Pontifical Catholic University of Peru.

2. Course Description

Before the pandemic, the teaching methodology was face-to-face and it consisted of the development of classes, laboratories, and construction site visits, where the construction processes were observed and discussed as well as the unit costs. On the other hand, since 2019 research results of the GETEC research group have been incorporated into the course syllabus, especially the ones that were generated by tools and technologies applied in construction projects. However, once the pandemic started, according to the norm it was mandatory that classes, laboratories, and evaluations were virtual [5]; thus, there was a need to design a methodology based in tools and technologies that allowed the development of competencies and soft skills in students, that were also valued by the industry increasing their employability. For this reason, a flexible integration of construction management tools was proposed, which were combined allowing a work team to operate efficiently, according to the Lean systems [6, 7], Project Management Institute [8], Building Information Modeling (BIM) [9, 10], Prince2 [11], among others.

In addition, an assessment of the technologies currently being applied in the industry was performed, to determine what would be compatible with the course and most beneficial for students.

3. Technologies

3.1. Building Information Modeling (BIM)

BIM is a methodology that integrates the 3D model of a construction Project with geometrical and/or parametrical information and is defined as the shared digital representation of the physical and functional characteristics of any object [9]. BIM describes the tools, processes and technologies that generate the digital documentation for rates, planning, construction and functioning of a building argumentation, idea discussion, decision making, among other communication factors [12]. The BIM focus is based on collaborative planning, argumentation, idea discussion, decision making, among other communication factors, which develops soft skills in participants, and thus, increases the student's employability.

3.2. Virtual Reality (VR)

Virtual reality (VR) is a visualization technique that may be immersive or non-immersive, and it offers a series of applications, from entertainment to marketing, and in this case, building construction [13]. The use of VR and the creation of digital prototypes (digital mock up) for design review are being used more each time during the design and construction phases [13]. Moreover, every day the number of virtual reality software that is compatible with BIM technology of common use is increasing [7,14].

3.3. Laser Scanning

Laser scanning is a method usually used to determine the position of an object, and data surveying which can be then exported as a point cloud [15]. The terrestrial laser scanners allow the surveying of the 3D position of an object's surface and is a usual technology for studying 3D information of construction projects [15]. The laser scanning may later be complemented by photogrammetry of the surfaces where the scanner does not provide enough detail. Both laser scanning and photogrammetry are widely used in projects and aerial photogrammetry through unmanned aerial vehicles (also known as drones) is often used to obtain the orthophotos of surfaces [16].

The information generated by BIM, virtual reality and/or photogrammetry may be converted into the same format, with several technological applications, which make this conversion more automated each time [17].

The technology described was used by our research group in several projects [14, 17, 18, 19]. Case studies were shared with students for a more in-depth vision of the application of the tools and technologies imparted. This was the case of the Architecture Faculty project, which consisted of an extension of an already existing building (see Figure 1). The technology applied for this case study was laser scanning, point cloud importing, 3D, 4D and 5D BIM Modelling.

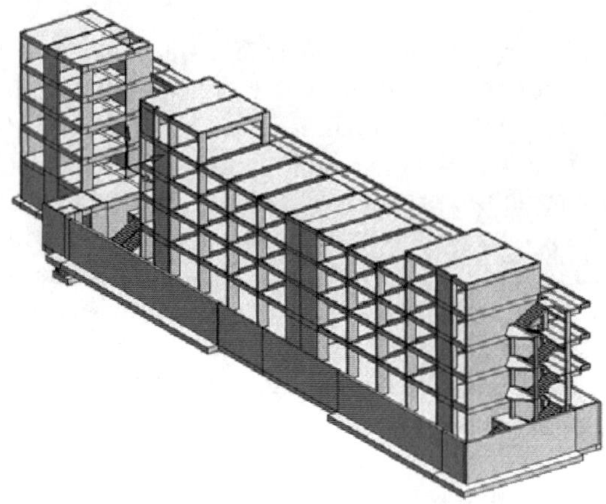

Figure 1. BIM Model in Autodesk Revit of the Architecture Building

4. Methodology

The syllabus needed to be updated for the pandemic times, adapting the integration of the tools proposed in [6, 7, 8, 9,10,11] with the following steps:

4.1. Step 1- Syllabus Revision:

Starting from the needs, principles, and values analysis of the following stakeholders: students, industry, public and private institutions, users, and society as a whole. Moreover, the laws, norms, statutes, regulations, accreditation guidelines, among other guidelines were reviewed and taken into consideration. With this information, the conceptual design alternatives and course preliminary content were generated in an iterative manner.

4.2. Step 2 - Syllabus Re-design:

With the information generated on step 1, the processes are designed and the infrastructure and available technologies of the university, the technological capacity of the student body and the existing logistics are preliminarily evaluated in order to prepare the investment and agreements with the providers according to the budget institution. Along with this evaluation, the initial proposal of the course modification is validated.

4.3. Step 3 - Employability:

The profiles of graduates demanded by the industry and institutions are taken into account, since they will ultimately be assessing the competencies, soft skills and use of technology according to the market changes. During the professional exercise of graduates, the environment is dynamic thus the content, tools, and technologies used in the course must be flexible enough for them to adapt to any possible scenario.

4.4. Step 4 - Course Logistics:

The results of the previous steps serve as an input. The internal and external available infrastructure is analyzed for both the university and the students. This included hardware, software, agreements with providers, among others, finalizing the logistics management with the design and proposal of laboratories as a fundamental part of the course and complement to classes.

4.5. Step 5 - Laboratories proposal:
During the course, students were expected to virtually assist 5 laboratories. For this, the university's Moodle platform named Paideia, as well as the Zoom Platform were used. In Paideia teachers uploaded all the content of the course, the folders where students turned in their assignments and the links for the zoom meetings. Zoom, allowed students to connect to a session either with their computer or mobile device. Professors and teaching assistants were aware that given the emergency, not all students may be able to connect at the time of the session for classes, so these were recorded and later uploaded. Laboratories are explained further with detail.

4.6. Step 6 – Course Update:
The approved update is executed, with feedback being received in real time according to the available information. The syllabus must be flexible for rapid modifications of tools, technologies, and education methods to fulfill the objective of generating employability.

4.7. Step 7 - Evaluation and Feedback:
During these processes the proposals, suggestions and feedback of students is considered, after verifying that they comply with accreditation requirements and institutional internationalization. Soft skills and technological skills developed are evaluated as an indicator of employability and the relation with the industry and the environment is maintained.

5. Laboratories
This course had bi-weekly 2-hour laboratory sessions. Before the day of the session, a video explaining the laboratory and the software to be used was shared with students so that they could start using the software and could prepare questions regarding the assignment. During the laboratory students would share their screen to show any problem they may encounter while using the software.

5.1. Laboratory 1: Virtual visit to a construction project using digital twins
Given the sanitary emergency, the construction site visits were replaced by virtual visits to construction projects generated with scanner laser technology, photogrammetry, and BIM, generating digital twins that were stored in online platforms. Students toured the construction project virtually and in an interactive manner, and along with the description of the professionals in charge of the project they were able to reach a good level of comprehension of the organization and the construction processes. Figure 2 shows a capture of the session, including the virtual tour of the project. In addition, students developed the layout plans of the temporary works with BIM models. The following figure is a screen capture of the teacher showing the digital twin while sharing his screen. Students were also shared a link to access the model and navigate at their own pace.

Figure 2. Digital Twin of the Architecture Building.

For this laboratory, the assignment consisted of preparing a report, with the main information of the project including stakeholders, subcontractors, construction processes, project organizational chart, project areas, and heavy equipment employed. Finally, they included screenshots of the layout plans for temporary works modeled on Revit.

5.2. Laboratory 2: BIM modelling and quantity surveying

In the second laboratory, students had to model the first level of the architecture faculty Project in Revit. For this, the blueprints were shared with students. The deliverables were: 1. Revit model, 2. material take-off for all structural elements including foundations, beams, columns, etc., and 3. A final report with screen captures of different views of the model. Figure 3 shows a capture of the session where students are able to clear any doubts regarding the modelling process, the blueprints and the criteria being used for the model. For material take-off, students were asked to export the Revit tables to an excel sheet by material for concrete, reinforcement steel and formworks.

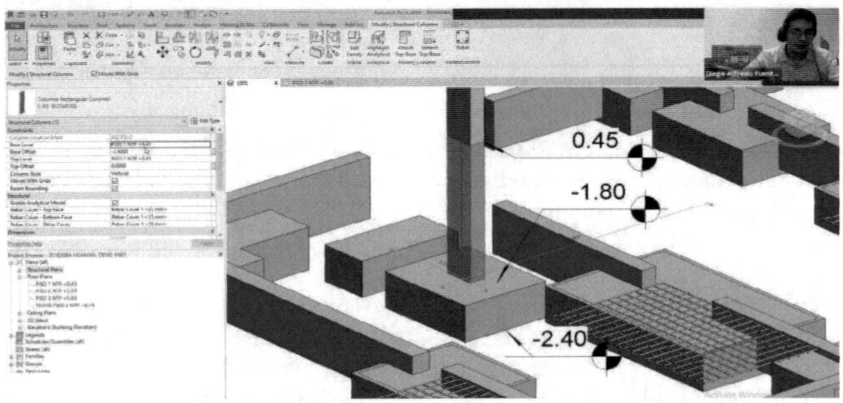

Figure 3. Laboratory Session explaining the modelling process in Revit

It should be mentioned that students also had basic knowledge using the BIM software Revit since it was taught in a previous course, meaning their BIM knowledge was enhanced with the software used.

5.3. Laboratory 3: Scheduling and cost analysis using BIM

For the third laboratory, students had to import the Revit model from laboratory n° 2 to Navisworks, to work on the Project Schedule and construction process simulation. Here, they were able to view the work progress and determine the correct sequence for construction. On figure 4, the construction sequence modeled in Navisworks is observed. As a part of this laboratory, they were also taught how to do Unit Cost Analysis, for the main tasks of a building Project.

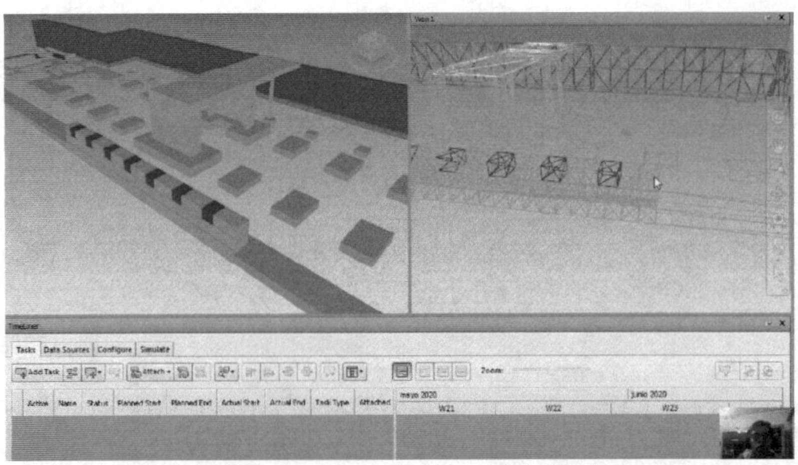

Figure 4. Capture of the video introducing Navisworks

5.4. Laboratory 4: WBS integration and cost analysis using BIM models in IFC format
During this laboratory, students learned how to integrate the work breakdown structure of a Project and the unit cost analysis with the use of the Software Delphin Express, a national software that allows the interaction of BIM models in IFC format. This software is used to create budgets, review unit costs, and link the quantity surveying data directly from the BIM model. With this, the 5D BIM modeling of the Project was completed and students gained a powerful tool for Project control. Figure 5 shows the interface of the software being presented to students, while clearing up any questions regarding the process.

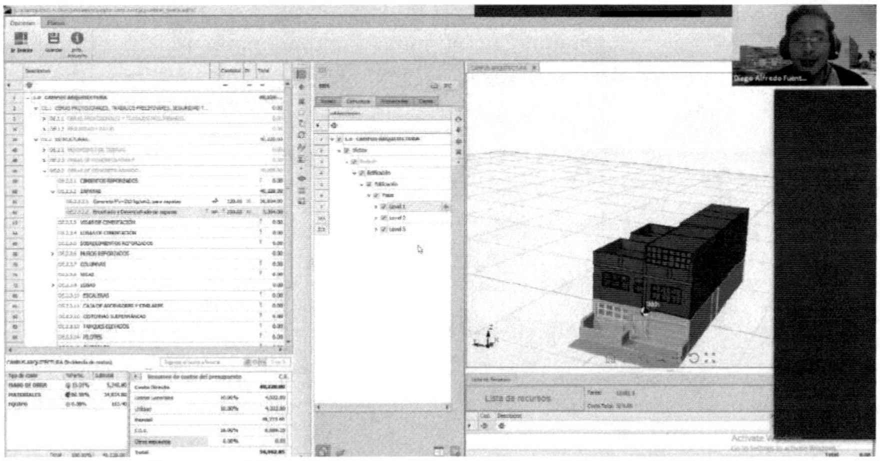

Figure 5. Capture of the session explaining 5D modeling with Delphin Express

5.5. Laboratory 5: Planning, scheduling, and project budget
For this session students were given a small project, an underground buried cistern, to prepare the quantity surveying, unit cost analysis, budget, and planning in an excel sheet. This allowed students to compare the process of working on a project with a more traditional method. Figure 6 shows the real project in execution and the blueprints of the project.

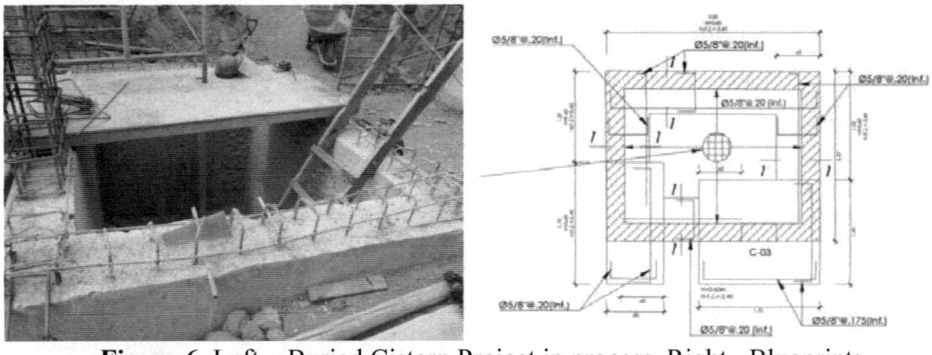

Figure 6. Left – Buried Cistern Project in process, Right - Blueprints

6. Results
During the semester, as students learned more about the tools and technology available, they indicated that the tools used in classes and laboratories were very useful and that they had the perception that it was very similar to a real project design, planning and execution. On laboratories 2 and 3 students learned how to model and plan the construction sequence for a project. Students were able to use these

tools collaboratively, which prepares them for their role in a specific project. The feedback from students was positive and the excellent average of the class confirmed the efficiency of the teaching methodology. On laboratory 5, students were able to compare the process learned with the BIM tools and technology and a more traditional process on an excel sheet. Students agreed that the second one was more difficult in terms of automation, collaborative work, integrated project perception, selecting the best construction sequence and identifying restrictions.

The success of this methodology is shown through the notable quality of papers and evaluations taken, through the results of surveys taken on students --which had an effectiveness rate of 96.4 % -- that exceeded in all categories in comparison to the efficiency of other courses imparted in the engineering department. Table 1 shows the average score obtained for the course after the survey performed at the end of the semester.

Table 1. End of semester survey results.

Survey Section	Average efficiency of courses of the Engineering Department (%)	Course efficiency (%)
A. Learning strategy	87.0	100.0
B. Methodology and resources	84.6	94.8
C. Learning assessment	84.4	93.8
D. Communication and interaction	88.0	98.2
GENERAL QUALIFICATION	**85.8**	**96.4**

7. Conclusions

The use of BIM software in this course gave students a more open vision of construction processes and building construction, regardless of the impediment to visit onsite construction projects, students were able to navigate a Project, from the blueprints to the 3D, 4D, 5D model and digital twin. The methodology was designed so that one can adapt the pandemic and post-pandemic stages proposing a balanced use of virtual platforms and software for video conferences and laboratories, incorporating tools and technologies like Building Information Modeling, Virtual Reality, Lean construction, among other systems. This may be adapted by the academia and professionals related to building constructions, for virtual, and face-to-face learning as well. This will be very useful in countries as Peru, where the Access to construction projects is highly restricted due to the pandemic, since the sanitary norm in the construction industry is strict in terms of forum and maximum allowed number of people in each area of a project [20].

References

[1] Ministerio de Salud 2020 *Resolución Ministerial N° 448-2020/MINSA Lineamientos para la vigilancia, prevención y control de la salud por exposición al SARS-CoV-2.*

[2] Carrillo C and Flores M A 2020 *COVID-19 and teacher education: a literature review of online teaching and learning practices* (European Journal of Teacher Education) 43:4, 466-487.

[3] Boton C 2020 *Remote Teaching of Building Information Modeling During the COVID-19 Pandemic: A Case Study. Sustainability* 12(20):8665.

[4] Leon I Sagarna M Mora F Otaduy J P 2021 *BIM Application for Sustainable Teaching Environment and Solutions in the Context of COVID-19*. Sustainability 2021, 13, 4746.

[5] Ministerio de Educación 2020 *Resolución Viceministerial N° 157-2020/MINEDU Orientaciones para el desarrollo del servicio educativo en los Centros de Educación Técnico-Productiva e Institutos y Escuelas de Educación Superior, durante la Emergencia Sanitaria causada por el COVID-19.*

[6] Forbes L and Ahmed S 2020 *Lean Project Delivery and Integrated Practices in Modern Construction* Routledge, 2nd Ed

[7] Brioso X 2015 *Integrating ISO 21500 Guidance on Project Management, Lean Construction, and PMBOK* Procedia Engineering, 123, pp. 76 – 84.
[8] Project Management Institute 2017 *A Guide to the Project Management Body of Knowledge (PMBOK® Guide)* Project Management Institute, 6th Ed
[9] Sacks R Eastman C Lee G Teicholz P 2018 *BIM Handbook: A Guide to Building Information Modeling for Owners, Designers, Engineers, Contractors, and Facility Managers* Wiley.
[10] Brioso X Murguía D Urbina A 2017 *Comparing three scheduling methods using BIM models in the Last Planner System* Organization, Technology and Management in Construction: an International Journal, 9 (2017), Issue 1, pp. 1604–1614.
[11] Office of Government Commerce 2009 *Managing and Directing Successful Projects with PRINCE2* The Stationery Office, UK.
[12] International Standards Office (2018). *ISO 19650-1:2018. Organization and digitization of information about buildings and civil engineering works, including building information modelling (BIM) — Information management using building information modelling — Part 1: Concepts and principles*. Geneva: ISO.
[13] Paes D Arantes E Irizarry J 2017 *Immersive environment for improving the understanding of architectural 3D models: Comparing user spatial perception between immersive and traditional virtual reality systems* Automation in Construction, 84, 292-303.
[14] Brioso X Calderon-Hernandez C Irizarry J and Paes D 2019 *Using Immersive Virtual Reality to Improve Choosing by Advantages System for the Selection of Fall Protection Measures* Comput. Civ. Eng. Vis. Inf. Model. Simul. - Sel. Pap. from ASCE Int. Conf. Comput. Civ. Eng. 2019, no. June. 146–153.
[15] Wang C and Cho Y 2015 *Smart scanning and near real-time 3D surface modeling of dynamic construction equipment from a point cloud* Automation in Construction vol. 49, pp. 239–249.
[16] Oreni D Brumana R Della Torre S Banfi F Barazzetti L Previtali M 2014 *Survey turned into HBIM: the restoration and the work involved concerning the Basilica di Collemaggio after the earthquake (L'Aquila)* ISPRS Ann Photogramm Remote Sens Spat Inf Sci II-5:267–273
[17] Brioso X. Calderon-Hernandez C Aguilar R and Pando M A 2019 *Preliminary Methodology for the Integration of Lean Construction, BIM and Virtual Reality in the Planning Phase of Structural Intervention in Heritage Structures* RILEM Book Series, Volume 18, 2019, Pages 484-492
[18] Santa-Cruz S Córdova-Arias C Brioso X Vázquez-Rowe I 2021 *Transparency-based protocol for decision-making regarding seismic rehabilitation projects of public buildings*. International Journal of Disaster Risk Reduction 55, 102-116.
[19] Suarez J C Zapata J Brioso X. 2020 *Using 5D Models and Cba for Planning the Foundations and Concrete Structure Stages of a Complex Office Building*. In: Proc. 28th Annual Conference of the International Group for Lean Construction (IGLC). Berkeley, California, USA, 6-10 Jul 2020. pp 913-924.
[20] Ministerio de Vivienda, Construcción y Saneamiento 2020 *Resolución Ministerial Nº 087-2020 Protocolo Sanitario Sectorial*.

IOP Conf. Series: Materials Science and Engineering 1218 (2022) 012043 doi:10.1088/1757-899X/1218/1/012043

Project scheduling algorithms for construction projects with stretchable activities

L Mályusz and Z Vattai

Department of Construction Technology and Management, Budapest University of Technology and Economics

malyusz.levente@epk.bme.hu

Abstract. Project management requires increasingly complex, sophisticated scheduling models, preferably in the simplest possible way, for which the rapid expansion of computing capacity provides an increasing opportunity. In this paper, we examine deterministic models that, as a further development of the traditional CPM and PDM models (with four well-known minimum and maximum priority relationships) allow activity to be stretchable. The aim of the study is a) to implement different algorithms for time analysis on computers, b) to compare their speeds on large-scale real and artificial projects; c) to prepare proposals for selecting the best fitting algorithm(s) to the specific model(s). The comparison was based on real network. The results show that depending on the models, different algorithms perform well, so we recommend that different algorithms be implemented in the scheduling tools and let the tool decide which algorithms will be optimal for the computation time

1. Introduction

Critical path calculation is an essential part of project scheduling problem. Since the beginning of the history of critical path more and more relevant mathematical models were developed for engineering project scheduling problem. In this study we use PDM schedule network with the four traditional precedence relationships (SS,SF,FS,FF) with an improvement to the technique, namely stretchable activities are allowed. The stretchable activities have a minimum and a maximum possible duration and their activity time can be between two these values. This study's general aim is to make an efficient algorithm for this model. This study is a branch of a bigger research project was started in [1].

The complete time analysis that defines the project duration and the early/late start/finish dates for all activities for CPM network schedules was developed in 1957 and introduced to the scientific community in [2,3]. The more flexible PDM results from [4,5,6], among others.

Many algorithms transform traditional precedence relationships into SS relationships [7,8] before time analysis. Maximal SS relationship in its minimal equivalent form results from Roy [4].

Algorithms #1-#3 are iterative algorithms used when maximal relationships are in the network. Algorithm #1, developed by [9] based on the work of the shortest path algorithm for non-directed networks from [10,11]. The modified FIFO algorithm (Algorithm #2), which is explained later, is based on the idea presented by [10,12] for finding the shortest path on non-directed networks. The DEQUEUE implementation used in Algorithm #3 uses the idea of [13], who developed their algorithm for finding the shortest path in non-directed networks. Algorithm #2 and #3 have the same theoretical complexity as Algorithm #1 but are thought to be faster in linear-construction-like networks.

Algorithms #4/a-b are designed explicitly for Monte Carlo analysis based on algorithms #1-#3. In the first instance, they use Algorithm #1-#3, and they save the relationships' calculation order. During the second phase, which involves all the remaining instances, the calculation uses the calculation order defined during the first instance, making the time for calculating a new instance much faster.
Several articles have compared different shortest path algorithms for different applications [14,15,16]; however, none of them investigated project schedules. This paper aims to study the characteristics of Algorithms #1-#3 and their modification developed for Monte Carlo simulation (Algorithm #4 a-b) on construction-like networks.

1.1. Algorithms

In this paper the following algorithms were used to solve the project scheduling problem:
1. algorithm generic label-correcting (critical path algorithm for cyclic network),
2. algorithm modified label-correcting with FIFO implementation,
3. algorithm modified label-correcting with Dequeue implementation,
4. a. two phased version of algorithm of 2,
4. b. two phased version of algorithm of 3.

Algorithms 4.a and 4.b are designed explicitly for Monte Carlo simulation. Algorithm 4.a includes Algorithm 2. for the first instance, with the only difference that the calculation sequence of arcs is stored in an array. This calculation order is used for the rest of the instances. A more detailed description of the algorithms is in [1,17].
This research aims to decide which algorithms perform better for time analysis in a presence of stretchable activities. The following two research objectives have been set to satisfy the goal:
Comparing the speed of the algorithms using the example project
Evaluating the results and making recommendations.
In this study algorithms were investigated through a PDM network of a construction project. The original construction network was modified in accordance with the objectives of the research, namely we created new PDM networks (NET1, NET2,...) increasing the number of the stretchable activities based on the original PDM text.

1.2. Case Study

Our selected project was the architectural reconstruction of the Memorial Building of Nobel-laureate writer Imre Kertész. The project included removal of non-authentic contemporary features of the existing building, partial reconstruction of the load-bearing structures (floors, walls), reproducing historical features such as roofing, cladding, doors and windows, internal wooden wall covers, stairs etc., all kinds of finishes and building mechanics and electric systems and - of course - installing modern multimedia and exhibition facilities. Though the quantities were less than that of a larger family house the quality expectations and complexity of works necessitated a network time model to establish a logical guideline for managing the on-site works. Since the works were financed by public funds strict expectations against the time model were also set by the employer who managed the tendering and the public procurement procedure.
The winner contractor developed a network time model of a hundred-some technological processes (activities) and about 150 technological dependencies (predecessor links) among them together with some calendar restrictions ("employer's schedule", milestones) and some resource-related expectations. Since our examinations were focused on the lean inner logic of the "engineers' network" (engineers' logic) we prepared a kind of CPM network (containing start-to-start relations only) off the original MPM network model, free of any superposed non-network originated features such as calendar constraints, and cost- or resource limitations. Using the purified model, we systematically added the "newer" network elements one-by-one (such as stretchable activities, upper-bound typed limitations, non-positive loops) and studied their effect on the running time of calculations (forward pass and backward pass) by the upper-mentioned algorithms.

To characterize the topology of the network we studied the inner sub-chains and junction extremities of its graphical representation (the "Graph").
We found that the contractors "original" time model involved 99 activities (including 20 milestones), 141 links, one source (originating node), one sink (terminating node), 5 positive sources (with no entering arrow of positive weight), 6 positive sinks (with no leaving arrow of positive weight), 39 FFmin links (27.66%), 92 FSmin links (65.25%), 8 SSmin links (5.67%). Orderliness (links in the upper triangle) was 85.82%, maximum number of links per task was 10, maximum number of predecessors per task was 7 and maximum number of successors per task was 6.
Considering the weights and the time potentials (time results) we found that the overall execution time was 270 t.u. (time units, here workdays), maximum duration 255 t.u. (on-site management), mean duration 26 t.u., maximum lag 260 (would be avoided in general), mean lag 4 t.u., maximum free slack 50 t.u., mean free slack 6 t.u., maximum total slack 50 t.u., mean total slack 15 t.u.
There was one single critical path in the network, involving 76.77% of the activities ("over-tied" or "rigid" network).
Keeping the core of original network our modifications (extra links, maximum-typed limitations) were superposed on that step-by-step.

Table 1. Variants of the example and their parameters

	NET1	NET2	NET3	NET4	NET5	NET6
Number of stretchable activities	1	2	3	5	9	17
number of activities (after transformation)	99	100	101	103	107	115
number of relationships (after transformation)	140	142	144	148	156	172

1.3. Results

Main results are collected in Table 1 and Fig.1. Among the algorithms Dequeue is the slowest but two phased Dequeue has the shortest running time. The advantage of 4.b. is bigger if the calculated number of instances is bigger.

Table 2. Average completion (running) time in nanoseconds for one instance, number of instances is 100000

Number of stretchable activities	**1**	**2**	**3**	**5**	**9**	**17**
1. BF	3005	3033	3147	3284	3450	4937
2. modBF	2389	2406	2437	2492	2582	3116
3. Dequeue	181182	180094	180920	182846	183767	185821
4a. 2phase mBF	1827	1850	1928	1905	1945	2228
4.b.2phase Deq	1711	1729	1729	1758	1827	1987

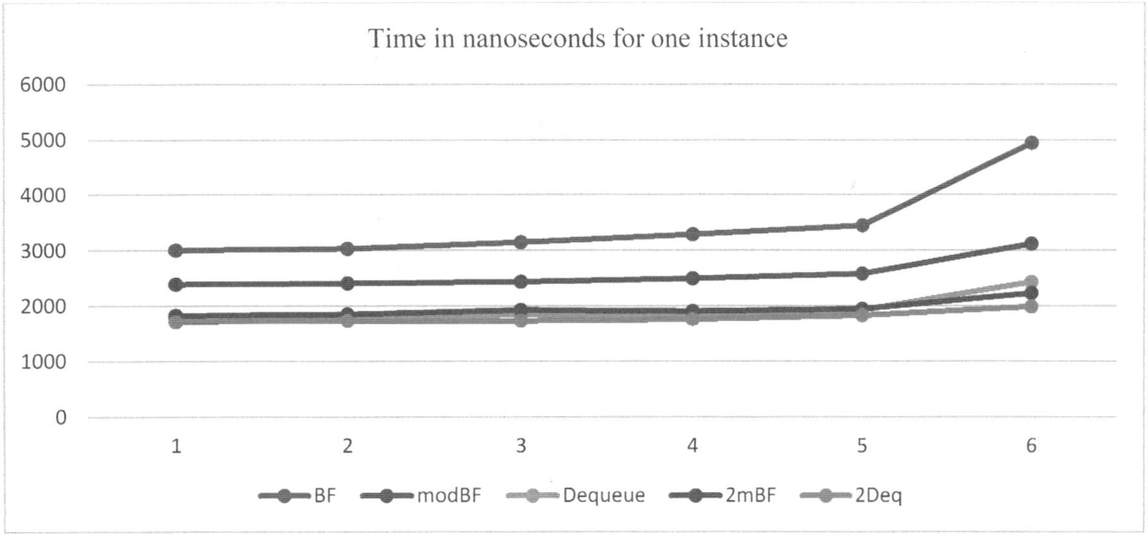

Fig. 1. Time in nanoseconds for one instance

Results (Table1 and Fig.1) show that for Monte Carlo analysis, Algorithm #4/b (based on DEQUEUE implementation) performs better than Algorithm #4/a. The most important practical consequence of this study is that Monte Carlo simulation for large construction-like networks containing stretchable activities can be performed within a couple of seconds.

2. Conclusion and further research

In our example 2phase Dequeue algorithm performed the best. We would like to improve these algorithms' speed further and extend the capabilities of the models, namely improve the flexibility in order of activities.

The speed of different time analysis algorithms was studied in this paper to find the most suitable for analyzing construction-like schedule networks containing stretchable activities. Three algorithms (Algorithms #2-4) have, and their modifications for MC simulation (Algorithms #5/a-c) have been implemented for computers. Algorithms were investigated based on a real life construction project. We found that the two phase DEQUEUE algorithm performed better than the others. The algorithms were implemented in C++ (Visual Studio).

References

[1] Malyusz,L. & Hajdu, M. & Vattai, Z. Comparison of different algorithms for time analysis for CPM schedule networks; vol 127, July 2021, Automation in Construction, 2021, Available at: https://doi.org/10.1016/j.autcon.2021.103697

[2] Kelley, J.E. & Walker, M.R., 1959. Critical-path planning and scheduling. Papers presented at the December 1-3, 1959, eastern joint IRE-AIEE-ACM computer conference on - IRE-AIEE-ACM '59 (Eastern). Available at: http://dx.doi.org/10.1145/1460299.1460318.

[3] Kelley, J.E., 1961. Critical-path planning and scheduling: Mathematical basis. Operations Research, 9(3), pp. 296–320. Available at: http://dx.doi.org/10.1287/opre.9.3.296.

[4] Roy, G.B., 1962 Graphes et ordonnancement (Graphs and scheduling), Revue Française de Recherche Opérationnelle n° 25, 4e trimestre pp323-33

[5] Fondahl, J. W., 1962. A Non-Computer Approach to the Critical Path Method for the Construction Industry, Technical Report #9. 2nd edition. Department of Civil Engineering, Stanford

University, Stanford, CA, 1962, p. 155 Available at: http://hdl.handle.net/2027/mdp.39015000453970 Accessed at: 03/05/2020
[6] IBM.1964, Users' Manual for IBM 1440 Project Control System (PCS)
[7] Kerbosch, JAGM & Schell, HJ., 1975, Network planning by the extended metra potential method (EMPM) TH Eindhoven. ORS, Vakgr. operationele research report, vol. KS-1.1, Technische Hogeschool Eindhoven. https://pure.tue.nl/ws/portalfiles/portal/4368315/104181.pdf.
[8] Elmaghraby, S.E. & Kamburowski, J., 1992. The Analysis of Activity Networks Under Generalized Precedence Relations (GPRs). Management Science, 38(9), pp. 1245–1263. Available at: http://dx.doi.org/10.1287/mnsc.38.9.1245.
[9] Hajdu, M., 1997. Network Scheduling Techniques for Construction Project Management. Nonconvex Optimization and Its Applications, p. 352 Kluwer Academic Publishers. Available at: http://dx.doi.org/10.1007/978-1-4757-5951-8.
[10] Bellman, R., 1958. On a routing problem. Quarterly of Applied Mathematics, 16(1), pp. 87–90. Available at: http://dx.doi.org/10.1090/qam/102435.
[11] Ford, L. R. Jr., 1956. Network Flow Theory. Paper P-923. Santa Monica, California: RAND Corporation. Accessed at: https://apps.dtic.mil/dtic/tr/fulltext/u2/422842.pdf on May 3, 2020.
[12] Moore, EF., 1959. The shortest path through a maze. Proceedings of an International Symposium. on the Theory of Switching 2-5 April, 1957, Part II. Cambridge, Massachusetts: Harvard University Press. 1959, pp. 285–292 MR 114710.
[13] Pape, U., 1974. Implementation and efficiency of Moore-algorithms for the shortest route problem. Mathematical Programming, 7(1), pp.212–222. Available at: http://dx.doi.org/10.1007/bf01585517.
[14] Ajwani, D. et al., 2018. An Empirical Comparison of k-Shortest Simple Path Algorithms on Multicores. Proceedings of the 47th International Conference on Parallel Processing, 2018, pp-1-12 Available at: http://dx.doi.org/10.1145/3225058.3225075.
[15] Xie, D. et al., 2016. Practical private shortest path computation based on Oblivious Storage. 2016 IEEE 32nd International Conference on Data Engineering (ICDE). Available at: http://dx.doi.org/10.1109/icde.2016.7498254.
[16] Wang, X.Z., 2018. The Comparison of Three Algorithms in Shortest Path Issue. Journal of Physics: Conference Series, 1087 (2018), pp. 022011. Available at: http://dx.doi.org/10.1088/1742-6596/1087/2/022011.
[17] Ahuja, R. K., Magnanti T.L., Orlin J.B., 1993. Network flows, Theory, Algorithms and Applications. Prentice Hall, New Jersey. [ISBN-13: 978-0136175490]

Acknowledgements
The research reported in this paper has been supported by the National Research, Development, and Innovation Fund (TUDFO/51757/2019-ITM, Thematic Excellence Program

IOP Conf. Series: Materials Science and Engineering 1218 (2022) 012044 doi:10.1088/1757-899X/1218/1/012044

The Bromilow Time-Cost model: a case study for the infrastructure projects in Romania

A Purnus[1]**, C N Bodea**[2] **and M Stoian**[1]

[1] Technical University of Civil Engineering, 124 Lacul Tei Blvd, 020396, Bucharest, Romania,
[2] Bucharest University of Economic Studies, 6 Piata Romana, Bucharest, Romania

E-mail: augustin.purnus@utcb.ro

Abstract. The Bromilow Time-Cost model is widely applied for estimating the duration of the construction works. The application of this model requires the estimation of different model's parameters, which are capturing the economic, social and technological differences in projects' implementation in different regions/ countries. The paper aims to develop a mathematical Time-Cost model to be applied in Romania for the road infrastructure projects, grouped into four categories: highways, road rehabilitation, road modernization and by-pass. The research carried out by the authors has as objective to confirm the existence of the correlation between the project duration and costs, and to propose a procedure to be applied by the public decision authorities in order to estimate the duration of road infrastructure contracts. Tthe estimation of the implementation time was done through three scenarios: optimistic, most likely and pessimistic. The parameters of the Bromilow Time-Cost model and the regression models were identified

1. Introduction

Road infrastructure projects are characterized by high durations, high costs and significant impact on the economic development. They are tight associated with increased safety and quality of life. Both project owners and contractors need to consider an adequate project duration during bidding.

In Romania, there are no studies on the correlation between the duration of contracts for road infrastructure projects and their cost, which would improve the implementation of project tenders and make the contracted duration of works more predictable. The Romanian National Road Infrastructure Management Company (C.N.A.I.R.), the most important beneficiary of road infrastructure projects, has no procedures or at least public guidelines for estimating the duration of contracts. For this reason, the research presented in this paper, is intended on the one hand to highlight the correlation between the duration of projects and their cost and on the other hand, to provide decision-makers with a method for estimating the duration of road infrastructure contracts taking into account the specificities of our country.

The research conducted by the authors uses data from 89 road infrastructure projects executed between 2001 and 2018, of which 32 highway projects, 19 road rehabilitation projects, 12 road modernization projects and 27 bypass projects. The cumulative value of these projects is 4.577 billion Euros and the total length of the targeted roads is 2,026 km. The purpose of the regression analysis is to develop a mathematical model that allows the estimation of the execution duration of these categories of projects, taking into account their cost, from the stage of feasibility study and organization of tenders.

2. The Bromilow time-cost model for estimating the duration of road infrastructure projects

The duration of execution of construction works is directly proportional to their costs. This observation was analysed, detailed and translated into a mathematical forecasting model for the first time by Frank Bromilow [1] in a paper published in 1969 in the Building Forum, where he established a relationship between the size of a project and the time required to complete the construction, providing a quick and realistic estimate of the duration of construction without applying detailed planning techniques. The formula developed by Bromilow for time as a cost function has the form: T=KC^B, where: T is the actual duration of the construction in working days from the actual start date of the works until their actual completion, C is the final cost of construction (of the beneficiary) in millions of dollars adjusted to the prices of labor and materials, K is a constant feature of execution performance for a $ 1 million project; the shorter K, the shorter it will last (it is more efficient) and B is a constant indicator of the sensitivity of performance over time to cost. One of the characteristics of the Time-Cost Bromilow model is the limitation of its application only to the research area, due to the differences of economic, social, administrative and technological nature between the countries.

Road infrastructure construction projects are of great importance for the development of a country's economy and as a result, researchers have sought to identify through various mathematical models the link between lead time and project cost. Czarnigowska and others [3] analysed 100 road infrastructure projects completed between 2003 and 2008 in Poland, of which 39 were road upgrades, 34 were upgrades and repairs and 27 were new road constructions. Data taken from the official publications of the Directorate-General for National Roads and Motorways were processed using regression analysis and the Bromilow model.

Similar studies were carried out on 41 state and local roads, 27 motorways, 25 roads (roads, tunnels, bridges, etc.) carried out between 1992 and 2003 in Croatia [2]. The sample for large road infrastructure projects comprises 27 road sections built between 2000 and 2010. In order to obtain feasible estimates, the authors propose introducing in the Bromilow model a series of parameters that take into account the magnitude of the orders of variation during the projects, their management capacity and the existence of financial problems faced by contractors in project implementation.

Choudhury [4] has studied the extent to which the Bromilow model can be applied in road infrastructure projects in India and to verify whether such a relationship is good for this sector of the Indian construction industry. For this, 50 infrastructure projects were identified for which the regression analysis based on the Bromilow model was applied. Tawiah [5] studied a set of 70 completed road projects completed between 2011 and 2012 in Ghana. The results of the analysis showed that the Bromilow Time-Cost model can be applied to these categories of projects. Kaka and Price [6] conducted similar research on projects in the UK, analysing a set of 140 road projects with a total value of over £ 120 million. For highway projects in Florida, Shr and Chen [7] investigated possible functional relationships between execution time and cost. Analysing 21 projects carried out by the US Department of Transportation (FDOT) in the United States between 1996 and 1999, Shr and Chen used nine different forms of regression models.

3. Applying Bromilow's time-cost model in road infrastructure projects in Romania

3.1. Research methodology

In order to apply the Time-Cost model developed by Bromilow to the specific conditions of implementation of road infrastructure projects in Romania and to identify its characteristic parameters, 89 projects completed in the period 2001 - 2018 were selected, with a cumulative value of 4.577 billion Euro and totals 2,026 km, starting from data published by C.N.A.I.R. [8].

The following data was considered: year when the project works started, the contractual duration, the duration of the completion of the works, the initial costs of the works and the final costs. In order for the analysis to allow a consistent comparison of the data, the initial and final costs were expressed in Euro and adjusted for the inflation index.

3.2. Data collection

Data was grouped into the following categories: highways, road rehabilitation, road modernizations and bypasses, for the projects carried out between 2001 and 2018. From 31 highway projects, 25 projects were selected for the model development and 6 projects were kept for the model testing., 16 road rehabilitation projects were selected for building the model and al other projects being kept for the model testing. In the case of road modernization, 11 projects were selected for the model development, and one project was used for the model testing. For the bypass variants projects, 19 projects were selected for the model development and 7 projects were used for the model testing. Considering the specific features of each category of works, the regression analysis was performed separately for each project type.

3.3. Data analysis

Data was processed using statistical regression analysis, seeking as far as possible to obtain the highest possible degree of correlation between the duration of the contracts and their cost.

Regression analysis and identification of Bromilow model parameters was performed using the JASP program [9] of the University of Amsterdam

4. Research results

4.1. The results for the highway projects

The shape of the data cloud in the Cost - Duration diagram was drawn. As can be seen in figure 1, he dispersion of actual data is higher than for the contracted data. In addition, most contracted data have the same duration, even if the value of the contracts is significantly different.

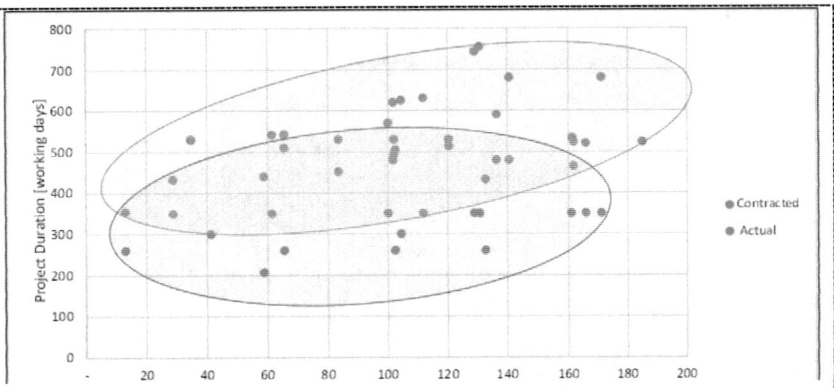

Figure 1. Data cloud distribution for contracted and actual cost and duration in the case of highways.

Unlike the contracted indicators, the actual indicators highlight the correlation of the value of the works with the duration of execution. This fact is also confirmed by the correlation coefficient which for the contracted data is $R^2 = 0.0687$, while for the actual data, it is $R^2 = 0.2276$ (see figure 2).

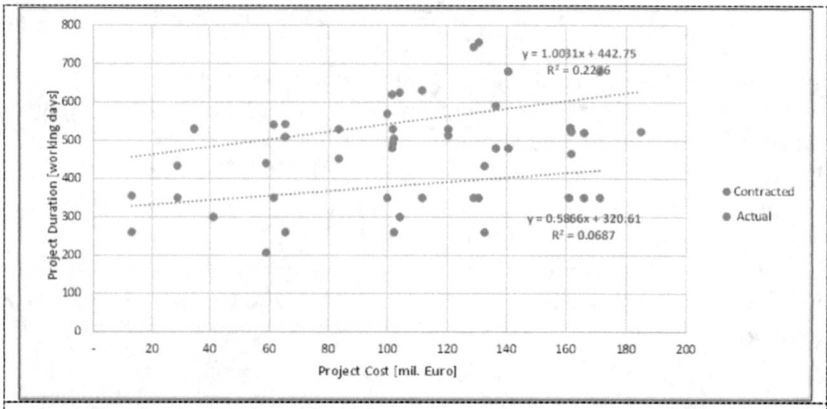

Figure 2. The correlation R-squared for contracted and actual cost and duration in the case of highways.

In order to set the potential values of the parameters in the Bromilow model, the initial and actual durations and costs were transformed by logarithm (see figure 3).

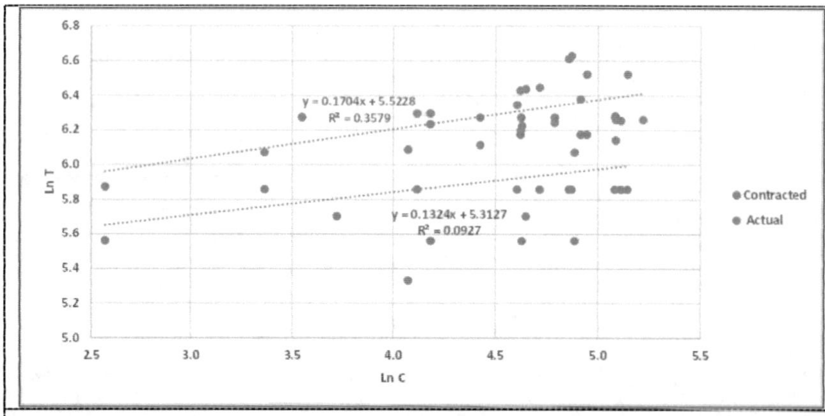

Figure 3. Initial and actual data correlation using the logarithmic transformation.

The logarithm of the initial and actual data results in an improved correlation coefficient for the actual values at $R^2 = 0.3579$, while for the initial data the same non-correlation state is found. By interpreting these results, we can conclude that the contracted durations are not at all correlated with the volume of work, while the factual aspects contradict this state. Most contracts for the execution of motorway lots last for 24 calendar months

In order to correlate the initial data, it was proposed to adjust the duration by defining not a single value, but three values that take into account the potential risks that may occur during execution. In this sense, the average initial and actual duration necessary to build a kilometre of highway was calculated, considering that in this time frame must be included the three durations that will correct the initial duration.

For the analysed data set, the average contracted duration is 20 working days per kilometre, while for the actual data, the average duration is 35 working days / kilometre. Under these conditions, for the optimistic scenario whose probability is usually zero, the duration of 12 working days was chosen, for the most probable scenario whose probability is greater than 50% the duration of 18 working days was chosen, and for the pessimistic scenario, duration of 25 days, corresponding to a probability of 100%. It was found that in this situation, the correlation coefficient R^2 varies between 0.51 for the optimistic scenario, 0.47 for the most likely scenario and 0.56 for the pessimistic scenario (see figure 4). Under these conditions, if we applied the Bromilow model, it could be validated for over 47% of contracts. To determine the model parameters, the values thus obtained are transformed by logarithm.

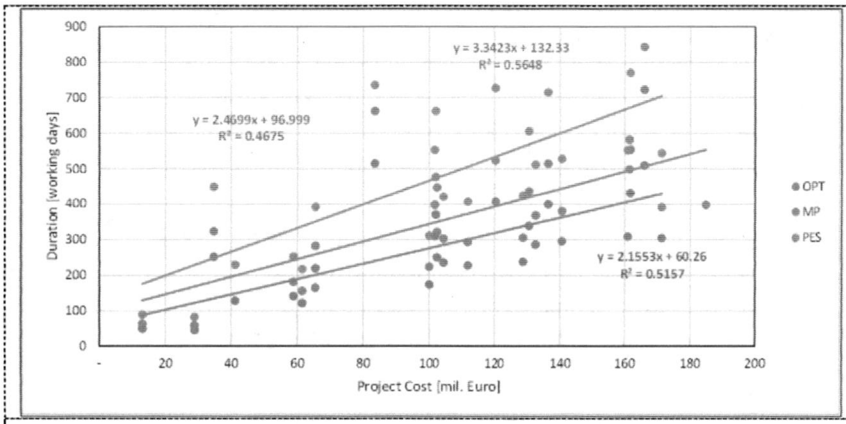

Figure 4. Correlation for adjusted data for highways projects

In this situation, the correlation coefficient R^2 acquires much more acceptable values, ranging between 0.68 for the optimistic scenario, 0.68 for the most probable scenario, up to 0.70 for the pessimistic scenario (see figure 5).

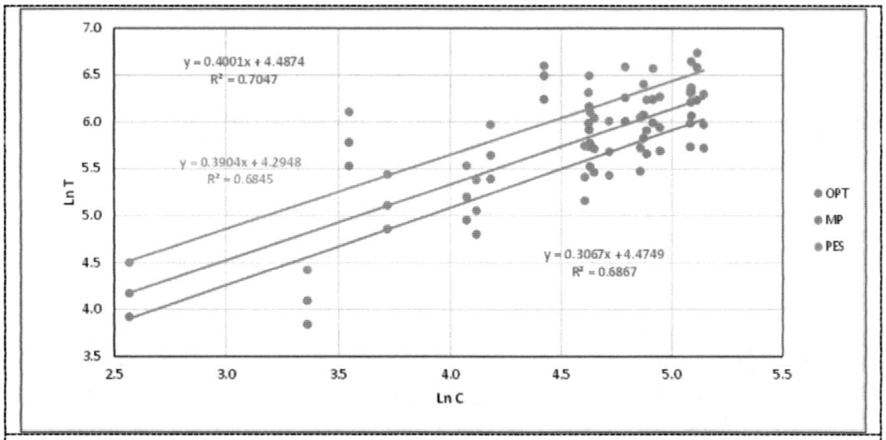

Figure 5. Data correlation using the logarithmic transformation.

The parameters of the Bromilow model are presented in table 1.

Table 1. Bromilow Time-Cost model parameters for the highway projects.

Scenario	Regression equation	R^2	K	B
Optimistic	y = 0.3067x + 4.4749	0.6867	87.7858	0.3067
Most likely	y = 0.3904x + 4.2948	0.6845	73.3175	0.3904
Pessimistic	y = 0.4001x + 4.4874	0.7047	88.8900	0.4001

By comparing the range of variation of the Bromilow model for the three scenarios, we found that it largely covers the variation resulting from contracted and actual data, allowing us to say that this model can be applied in estimating the execution times of highway projects (see figure 6).

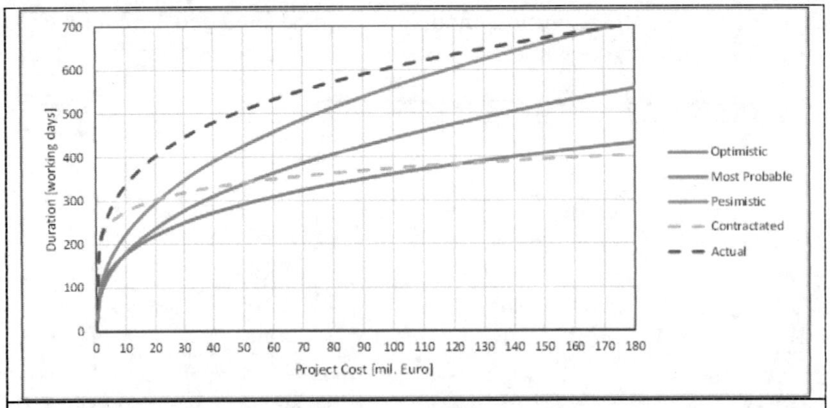

Figure 6. The comparative variation of the duration with the project cost for highways.

Under these conditions, the Bromilow Time-Cost model will have not one, but three relations for estimating the duration of a highway project:

- For the optimistic scenario: $T = 88C^{0.307}$
- For the most likely scenario: $T = 73C^{0.390}$
- For the pessimistic scenario: $T = 89C^{0.400}$

in which the cost is expressed in millions of Euros

4.2. The results for the road rehabilitation projects

From the analysis of the contracted data, it results that the average duration in working days for the accomplishment of a rehabilitation kilometre is 11 days, while the average duration corresponding to the actual data is 20 working days. The three scenarios lead to a significant correlation between the corrected realization time and the cost of the works. The correlation coefficient R^2 varies between 0.76 for the optimistic scenario, 0.75 for the most probable scenario and 0.77 for the pessimistic scenario Using the logarithm transformation, we obtained the regression equations corresponding for the three scenarios.

4.3. The results for the road modernization

From the verification of the degree of correlation between the pairs of values corresponding to the contracted and actual data, it is found that the contracted values - duration, cost - have an extremely low correlation coefficient ($R^2 = 0.1142$), while the correlation coefficient for the actual data is a satisfactory one, it having a value of $R^2 = 0.5487$. However, the analyzed data show a high degree of heterogeneity in terms of the length of the modernized road sectors and the works of art that are included and that affect the accuracy of the Bromilow model. For this reason, from the available data series, 7 projects were selected that recorded similar lengths and costs. However, unlike the previous situations, there is a positive correlation with a correlation coefficient $R^2 = 0.32$ for the contracted data, while for the actual data, it is extremely low, practically non-existent, the correlation coefficient having the value $R^2 = 0.004$. The parameters of the Bromilow model were obtained for which it is found that they have a significant correlation coefficient, equal to $R^2 = 0.77$.

4.4. The results for the by-pass variant projects

Taking into account the high degree of heterogeneity of the data, we selected 15 contracts that have an average length of road sections of 8 kilometres and whose average contracted value is 15 million Euros. For the selected data set, the degree of correlation between the contracted and actual value pairs was analysed, finding that although the correlation is positive, it is insignificant. For the contracted data, three scenarios are proposed, optimistic, most likely and pessimistic, so as to obtain a range of values that allows us to forecast. By transforming the values by logarithm, the parameters of the Bromilow model were obtained for which it is found that they have a correlation coefficient that varies between $R^2 = 0.45$ for the optimistic scenario, $R^2 = 0.42$ for the most probable scenario and $R^2 = 0.38$ for the pessimistic scenario.

5. Conclusions

The paper proposes a mathematical time-cost model which can be applied to the concrete conditions in Romania on road infrastructure projects. Considering the multiple risk events that accompany the complex road infrastructure projects, in the development of the Bromilow Time-Cost model, the estimation of the implementation time was done through three calculation options: optimistic, most likely and pessimistic, to cover extremely varied of the analyzed data.

The general conclusion of the research is that the existing durations of the contracts for the execution of road infrastructure projects must be reviewed by the C.N.A.I.R. to meet the specific conditions of Romania. The accuracy of the developed predictive model is higher the more accurate the processed data. Researching the correlation between execution time and cost requires detailed analysis of projects in the project portfolio of the C.N.A.I.R. for the application of a regression analysis whose results are closer to reality.

However, improving the forecast requires other steps that we recommend for further research:

- Detailed analysis of projects in the C.N.A.I.R. project portfolio for the application of a regression analysis whose results are closer to reality
- Analysis of the productivity of construction processes in the field of road infrastructure projects
- Development of a database with productivity corresponding to the capacity of construction companies operating in Romania and to be used in the realistic estimation of the duration of contracts
- Development of a guide for estimating the duration of contracts to be applied by C.N.A.I.R. in road infrastructure projects
- Using the Critical Path Analysis in the stage of preparation of the tender specifications to allow a realistic estimation of the duration of the contracts

If there are works of art on the road sectors, they will have to be analyzed separately, due to the particularities they involve.

6. References

[1] Bromilow F.J. 1969 Contract time performance expectations and the reality Building Forum 1(3) 70-80.
[2] Kenley R 2003 Financing Construction - Cash Flows and Cash Farming Spon Press, Taylor and Francis e-Library, ISBN 0–415–23207–4
[3] Czarnigowska A, Sobotka A 2013 Time–cost relationship for predicting construction duration - archives of civil and mechanical engineering 13 518 – 526
[4] Choudhury I 2009 Time-Cost Relationship for Infrastructure Projects in India - Texas A&M University, American Society for Engineering Education
[5] Moses T 2014 Using Bromilow's Model and other Regression Model to Predict the Duration of Feeder Road Projects in Ghana, Dissertation Submitted to the Department of Building Technology, Kwame Nkrumah University of Science and Technology Kumasi
[6] Kaka A and Price A 1991 Relationship between value and duration for construction projects, Construction Management and Economics, 9 (4), pp. 383-400
[7] Shr J-F Chen W-T 2006 Functional Model of Cost and Time for Highway Construction Projects, Journal of Marine Science and Technology Vol. 14 No. 3 pp. 127-138
[8] *** C.N.A.I.R.: http://www.cnadnr.ro/, accessed 19.02.2019
[9] *** https://jasp-stats.org/, accessed 19.02.2019

IOP Conf. Series: Materials Science and Engineering 1218 (2022) 012045 doi:10.1088/1757-899X/1218/1/012045

Analysis of Events and Exposures Leading to Construction Injuries in Developing Countries: The Case of Lebanon

M B Hatoum[1], H Nassereddine[1], F Hamzeh[2] and H Khoury[3]

[1] University of Kentucky, Lexington, USA
[2] University of Alberta, Edmonton, Canada
[3] American University of Beirut, Beirut, Lebanon
E-mail: mbh.93@uky.edu

Abstract. There is unanimous agreement that the complex and dynamic nature of construction jobsites is coupled with inherent risks. Global occupational and health agencies revealed that the construction industry leads with the number of fatal injuries incurred every year. Lebanon, a developing country in the Middle East, is no exception. Studies on construction safety in the country showed that the industry has failed to implement safety regulations, while also witnessing a lack of safety incentives, training, and education. The situation worsens with the absence of a national experience multiplier that could reflect on contractors' safety statuses and hold them accountable for their safety performance. This study addresses the problem by proposing a cost model that evaluates construction injuries. An analysis was performed on more than 3,000 accident claims collected from insurance companies to understand the events and exposures that construction labour face. Results showed that there are 17 different types of events and exposures which can be grouped into three clusters depending on their frequency and average cost. This paper presents and expands on the three clusters and identifies the events that fall under each of them. The findings will serve as the base of the cost model in a future study.

1. Introduction and Background

The construction industry is a hazardous industry that is responsible for thousands of injuries annually. Recent statistics show that construction incurred 1 in every 5 deaths in the United States of America (USA) and the European Union (EU) [1,2]. The safety situation was further disrupted with the coronavirus pandemic, which led to unprecedented challenges and concerns raised by construction workers [3]. Lebanon, a developing country in the Middle East, is no exception. The last study that analyzed occupational injuries found that, among all industries in Lebanon, the construction industry incurred the highest number of fatal and non-fatal injuries, with 1 in every 3 deaths attributed to construction [4]. The alarming rate of deaths triggered several construction safety studies that highlighted major safety issues including the absence of government enforcements [5], lack of training and safety certifications [6], overburdened employees and absence of employee engagement in safety decisions [7], absence of visual management practices [8], and lack of awareness and perception of hazardous activities [9]. In addition to the lack of government enforcements, the health and safety laws released by the Lebanese Ministry of Labor are outdated [5]. For example, a survey of contractors found that the most common reason for the lack of adoption of safety systems in Lebanon is that such safety practices are not required by law, in spite of the existence of *Decree No. 11958 of 2004* concerning regulations of safety and labour protection in the construction industry [10].

Moreover, the difference in safety perspectives between contractors and insurance companies causes major challenges. A survey conducted of both parties showed significant differences between the safety practices that contractors prioritize on site and the safety practices that insurance companies consider for controlling premiums [11]. Additionally, insurance companies do not investigate the previous history of any new contractor that they sign in because of inadequate record-keeping, the lack of a database that documents and tracks injuries, and the absence of a national experience multiplier to assess the safety performance of contractors such as the Experience Modification Rate (EMR) in USA [12].

Absent of government enforcements, safety education, and safety communication, and amid the neglection of safety and health laws, insurance companies could become the key player is setting and enforcing safety practices through their premiums. Money can act as a motivator for contractors to improve their safety performance. For instance, higher insurance premiums should be levied on contractors that incur a high number of injuries, forcing these contractors to enhance their safety regulations. However, putting such model into effect requires a better understanding and categorization of the types of construction injuries, especially in the light of the absence of a standardized and unified system among insurance companies that can investigate and evaluate construction injuries. Thus, this study contributes to the existing body of knowledge by filling a gap in the extant literature by proposing a new cost-based safety model that establishes a new basis for evaluating the safety performance of contractors in Lebanon. The rationale behind the proposed model is to analyse the events and exposures that lead to construction injuries. Understanding these causes and their consequences will benefit contractors in developing more robust safety systems and help insurance companies in properly measuring the risks involved with each type of work and assigning the appropriate premiums.

2. Methodology

Since Lebanon does not have a health and safety entity like the Occupational Safety and Health Administration (OSHA) that keeps track of occupational injuries, an archival research approach was adopted in this study to gather and analyze compensation and accident insurance claims from insurance companies [13]. As a result, a total of 3,883 construction-related compensation claims were collected from insurance companies, incurred between 2016 and 2019. Each data point provided information on the injury (event and nature) the injured worker (age and the affected body parts), and the total direct cost paid for the injury.

The development of the cost-based safety model consists of the following tasks: (1) classify the injuries into categories based on the Occupational Injury and Illness Classification (OIICS) Manual released by the United States Department of Labour [14]; (2) provide a descriptive analysis of the categories; (3) cluster the events leading to construction injuries based on their frequency of occurrence and cost; (4) perform bivariate analysis between the types of events and the remaining injury characteristics for every cluster; and (5) develop a cost-based model for insurance companies to evaluate the safety performance of contractors in Lebanon. This paper focuses on classifying the events and exposures of injuries into their OIICS categories (task 1), providing a descriptive analysis of the events and exposures (task 2), and clustering the events into groups based on their frequency and cost (task 3).

3. Data Analysis

3.1. Categorizing and describing the events and exposures leading to construction injuries

The OIIC manual is used by OSHA to classify the characteristics of occupational injuries, illnesses, and fatalities and divides the events into nine categories: (1) Violence and Other Injuries by Persons or Animals, (2) Transportation Incidents, (3) Fires and Explosions, (4) Falls, Slips, Trips, (5) Exposure to Harmful Substances or Environments, (6) Contact with Objects and Equipment, (7) Overexertion and Bodily Reaction, and (8) Non-classifiable. The distribution of the 17 types of events and exposures across the five categories of the OIICS manual are shown in Figure 1. None of the collected events were associated with categories two or three.

Category 6 (C6) had the highest total number of events. Within this category, most injuries were caused by being struck by falling objects or equipment (n=691), injured with handheld objects or equipment (n=625), and struck by discharged or dislodged flying objects (n=584). Other events that occurred less frequently include struck against stationary objects or equipment (n=141), stepped on objects (n=114), and struck against moving objects or equipment (n=16).

Category 4 (C4) had the second highest number of events. Within this category, most injuries were caused by falling to lower levels from ladders, stairs or scaffolds (n=798) and falling on the same level due to tripping (n=668). Other less frequent events include slipping (n=28) or opening (n=21).

Category 5 (C5) had the third highest number of events. Within this category, most injuries were attributed to inhalation of harmful substances (n=12), followed by direct exposure to electricity (n=10), indirect exposure to electricity (n=7), and exposure to harmful substances (n=4).

Other events were distributed between category 1 (C1) (bites and stings from insects with n=2) and category 7 (C7) (overexertion in lifting or lowering with n=4).

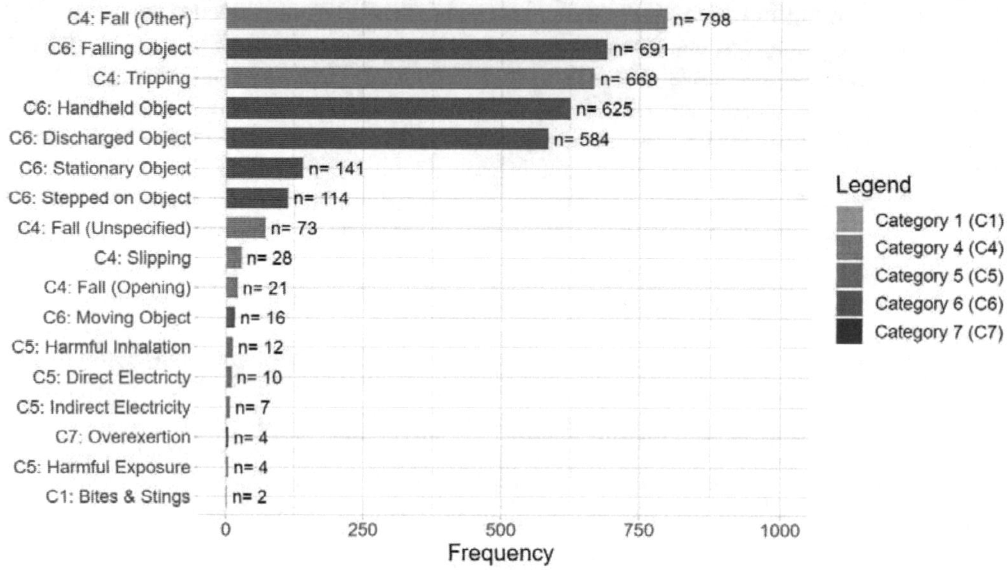

Figure 1. Distribution of events within categories

In addition to obtaining the frequency of occurrence of each of the 17 events discussed above, cost data was also collected. Table 1 shows the total cost and the average cost for every event. Average cost of events varied between US $66.43 (Bites and stings) and US $9,109.93 (Falling to lower levels due to openings).

Table 1. Total cost and average cost of every event (in US Dollars)

Category of Events	Event Description	Total (US Dollars)	Average (US Dollars)
1	Bites & Stings	132.85	66.43
4	Falls on Same Level due to Tripping	188,762.92	282.58
	Fall on Same Level due to Slipping	12,840.99	458.61
	Fall to Lower Level due to Openings [a]	191,308.43	9,109.93
	Fall to Lower Level due to Other Reasons [b]	990,194.13	1,240.84
	Fall to Lower Level (Unspecified)	102,692.43	1,406.75
5	Direct Exposure to Electricity	31,012.50	3,101.25
	Indirect Exposure to Electricity	22,601.03	3,228.72
	Inhalation of Harmful Substances	1,525.43	127.12
	Exposure to Harmful Substances	19,526.82	4,881.71
6	Injured by Handheld Object or Equipment	245,178.39	392.39
	Struck by Falling Object or Equipment	360,246.31	521.34

	Struck by Discharged or dislodged Flying Object	69,382.24	118.81
	Struck Against Moving Object or Equipment	10,207.82	637.99
	Struck Against Stationary Object or Equipment	35,760.28	253.62
	Stepped on Object	16,934.73	148.55
7	Overexertion in Lifting or Lowering	0.00	0.00
8	Unspecified	154,089.74	1,812.82

[a] Openings include elevator shafts, opening in slabs, etc.
[b] Other reasons include falling to lower levels from ladders, stairs, scaffolding, etc.

3.2. Clustering events and exposures based on frequency and average cost
After performing descriptive analysis on the types of events and exposures, k-means clustering was employed to group events based on their frequency of occurrence and average cost. Results of the analysis are shown in Figures 2 and 3. Events in categories C7 (overexertion) and C8 (unspecified) were excluded from the analysis for the lack of information on the event type and cost.

To determine the number of clusters, the elbow method was used (Figure 2). The plot of the variation of within sum of square as a function of clusters showed that the biggest drop happens at k equals to three, indicating that the data can be clustered into three groups. The number of clusters was then verified by running the *NbClust* package in RStudio®, a package that determines the relevant number of clusters for a dataset based on 30 different indices. A majority of 10 of the 30 indices proposed 3 clusters [15]. Thus, the k-means cluster analysis was performed to group the events and exposures into three clusters. The results of the clustering are shown in Figure 3.

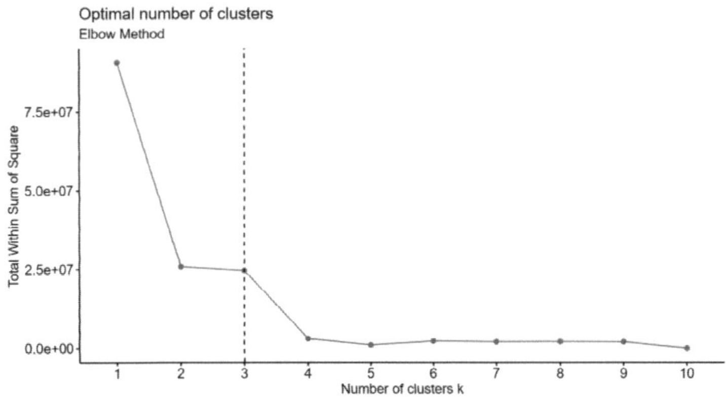

Figure 2. Scree plot for the k-means cluster analysis of the events and exposures. As the number of clusters increases, the variance within-group sum of squares decreases. Based on the elbow method, the elbow at three cluster represents the optimal balance between minimizing the number of clusters and the variance within each cluster.

3.2.1. Cluster 1: Low Frequency, High Cost
Cluster 1 is located at the bottom of Figure 3 and represents events that are the least frequent to occur on construction projects but incur a high cost when they do happen. Even though these events are uncommon, the impact of their occurrence can be extremely dangerous on construction workers because of the significant cost insurance companies pay to cover the claim. Events included in this cluster are direct electricity exposure (5A – where 5 is the category of the event and A is the sequence in Table 1 within the category), indirect electricity exposure (5B), exposure to harmful substances (5D), and falls to lower levels due to openings such as elevator shafts and openings in slabs (4C).

3.2.2. Cluster 2: Medium Frequency, Low Cost
Cluster 2 is located at the bottom-left corner of *Figure 3* and represents events that have a medium frequency of occurrence and that incur a low cost when they happen on construction projects. While the events of this cluster occur more frequently than the events of Cluster 1, their risk on construction workers is not as dangerous because the cost of coverage paid by the insurance companies is low. Events included in this cluster are struck against stationary objects or equipment (6E), struck against moving

objects or equipment (6D), step on objects (6F), slips (4B), inhalation of harmful substances (5C), and bites or stings (1A).

3.2.3. Cluster 3: High Frequency, Low Cost

Cluster 3 is located at the top-left corner of Figure 3 and represents events that have a high frequency of occurring and that tend to incur a low cost when they happen on construction projects. Events of this cluster are very common and frequently happen on construction sites. Despite the low average cost of each event included in this cluster, the high frequency of occurrence of these events could create a financial burden on insurance companies to cover resulting claims. Events included in this cluster are struck by falling objects (6B), trips (4A), handheld objects or equipment (6A), discharged objects or equipment (6C), and falls to lower levels from ladders, stairs, and scaffolds (4D).

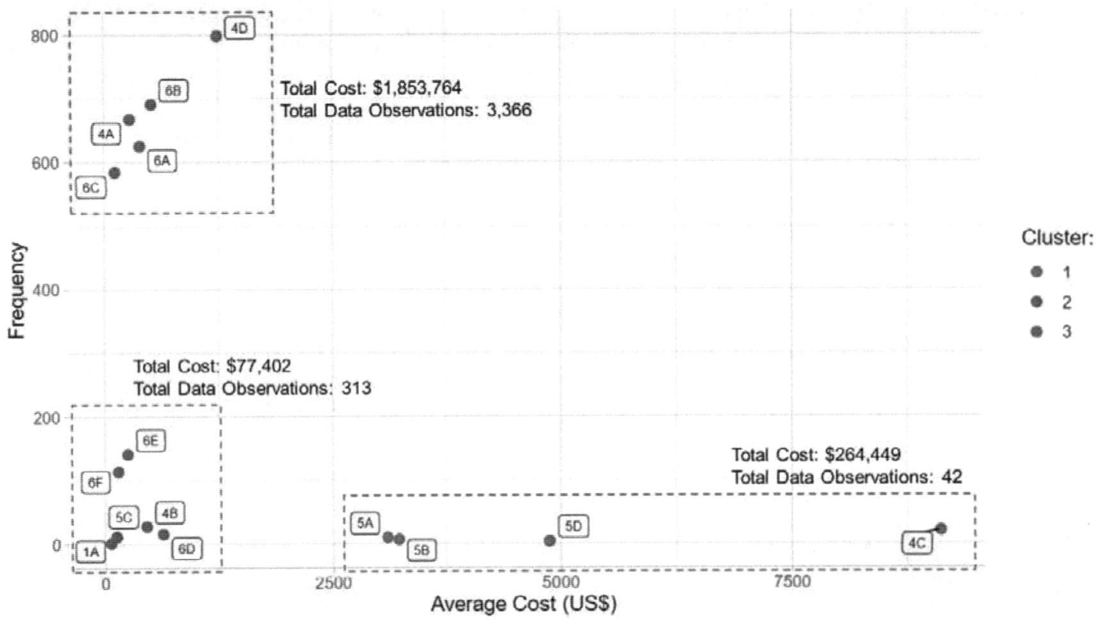

Figure 3. Clustering of Events (acronyms explained in the cluster description)

The results of the cluster analysis show an interesting dynamic between the frequency of occurrence of events and their associated average cost. Understanding the relationship between frequency of occurrence and cost of insurance is a first step to providing a new perspective on how to categorize events in the construction industry and address the absence of a unified system to assign premiums accordingly (discussed in the background section). Correlating insurance premiums to events' frequency of occurrence and costs will provide insurance companies with a strategy to better allocate premiums, forcing contractors to rethink safety on their job sites now that the premiums are driven by the results of their safety practices.

4. Conclusion and Future Studies

Recognizing the need for a new strategy to benchmark and evaluate construction safety practices in Lebanon, this paper presented the preliminary results of an ongoing effort to develop a new model for insurance companies to assess the safety performance of contractors. The scope of this paper focused on analysing 3,883 insurance claims to cluster events and exposures leading to construction injuries based on their frequency and average cost. First, a descriptive analysis using OSHA's OIICS manual was performed to categorize the type of events and exposures. Results showed that the events fell into categories C1, C4, C5, C6, C7, and C8, with most events belonging to C4 (falls, slips, and trips) and C6 (contact with objects and equipment). Next, k-means was employed to cluster events according to their frequency of occurrence and average cost. The analysis resulted in three clusters: a first cluster for low-frequency and high-cost events, a second cluster for medium-frequency and low-cost events, and a third cluster for high-frequency and low-cost events.

The next step of the study will expand on every cluster through performing a bivariate analysis on the events with respect to the age of the injured workers, nature of the injury, part of body affected, and source of injury. The detailed analysis of every cluster will help develop a cost-based model that can assist Lebanese insurance companies in controlling their premiums and enable contractors to re-examine and enhance their safety performance.

References

[1] OSHA 2019 Commonly Used Statistics
[2] Eurostat 2020 Accidents at work statistics
[3] Bou Hatoum M, Faisal A, Nassereddine H and Sarvari H 2021 Analysis of COVID-19 Concerns Raised by the Construction Workforce and Development of Mitigation Practices *Frontiers in Built Environment* **7** 66
[4] Fayad R, Nuwayhid I, Tamim H, Kassak K and Khogali M 2003 Cost of work-related injuries in insured workplaces in Lebanon *Bulletin of the World Health Organization* **81** 509–16
[5] Obeid A R 2015 *Health and Safety in Construction Industry of Lebanon* PhD Thesis (Gazimağusa, North Cyprus: Eastern Mediterranean University)
[6] Awada M A, Lakkis B S, Doughan A R and Hamzeh F R 2016 Influence of lean concepts on safety in the Lebanese construction industry *24th Annual Conference of the International Group for Lean Construction* International Group for Lean Construction (Boston, Massachusetts, USA) pp 63–72
[7] Damaj O, Fakhreddine M, Lahoud M and Hamzeh F 2016 Implementing ergonomics in construction to improve work performance *24th Annual Conference of the International Group for Lean Construction* International Group for Lean Construction (Boston, Massachusetts, USA) pp 53–62
[8] Abdelkhalek E S, Elsibai M D, Ghosson G K and Hamzeh F R 2019 Analysis of Visual Management Practices for Construction Safety *Proc. 27th Annual Conference of the International Group for Lean Construction (IGLC)* International Group for Lean Construction (Dublin, Ireland) pp 1069–80
[9] Abbas M, Mneymneh B E and Khoury H 2018 Assessing on-site construction personnel hazard perception in a Middle Eastern developing country: An interactive graphical approach *Safety Science* **103** 183–96
[10] Awwad R, Jabbour M and El Souki O 2014 Safety practices in the Lebanese construction market: contractors' perspective *ISARC. Proceedings of the International Symposium on Automation and Robotics in Construction* vol 31 (IAARC Publications) pp 354–60
[11] Hatoum M B, Hamzeh F and Khoury H Perspectives of Contractors and Insurance Companies on Construction Safety Practices: Case of a Middle Eastern Developing Country *Construction Research Congress 2020* pp 366–74
[12] Bou Hatoum M N 2019 *Developing an index to enhance construction safety in the global South region.* (Beirut, Lebanon: American University of Beirut)
[13] Das R, Jain K K and Mishra S K 2018 Archival Research: A Neglected Method in Organization Studies *Benchmarking: An International Journal* **25** 138–55
[14] U.S. Department of Labor 2012 *Occupational Injury and Illness Classification Manual* (USA: Bureau of Labor Statistics)
[15] Charrad M, Ghazzali N, Boiteau V and Niknafs A 2014 NbClust: An R Package for Determining the Relevant Number of Clusters in a Data Set *Journal of Statistical Software* **61** 1–36

IOP Conf. Series: Materials Science and Engineering 1218 (2022) 012046 doi:10.1088/1757-899X/1218/1/012046

Construction and Climate Change; Challenges and Opportunities: A Case Study of the Northeast U.S.

Afshin Pourmokhtarian, Payam Bakhshi, Zachary Bannon and Benjamin Everett

School of Management, Wentworth Institute of Technology, Boston, MA 02115, USA

Corresponding Author's Email Address: pourmokhtariana@wit.edu

Abstract. The Northeast megalopolis of the United States, which covers a high-density corridor from Washington, D.C., north to Boston, is one of the most developed environments in the world. It contains a gigantic, complicated, and intertwined network of supporting infrastructure with average score of D+, hence requiring substantial rehabilitation and renewal. The 2020 North American Construction report forecasted a CAGR of 8.4% by 2024 despite the pandemic. The pace of growth will recover from 2021 onwards as the ongoing infrastructure investments and smart city projects will add momentum for the region's construction industry. On the other hand, climate change impacts are underway across the globe and the Northeast of the U.S. is not an exception. Populations and aging infrastructures that they depend on, are highly vulnerable to climate hazards including heat waves, as well as flooding due to a combination of sea level rise, storm surge, and extreme precipitation events.

In this study, future projections of climate change in the Northeast of the U.S. were used to explore their potential impacts on construction industry including but not limited to safety of workforce, selection of building materials and their lifecycle, logistics, scheduling, costs, and insurance. The challenges and opportunities that construction industry faces under climate change were also covered. Later, the feedback loop between the construction and climate change were discussed as well as how sustainable construction practices could mitigate climate change impacts while providing safety for construction workforce. Finally, this paper focuses on resilience, buildings carbon footprint, green infrastructure, sustainability, and a prototype decision-support tool for construction projects to better manage weather risk from contract to project completion, known as Climate-i Construction, and how construction industry can benefit from them under 21st century changing climate.

1. Background

Climate change is a global phenomenon which disturbs all aspects of human lives. These range from rising temperatures, higher sea levels, and more frequent extreme weather events. The result of these changes is rippled throughout all aspects of the environment. The durations and start of seasons are shifting with greater periods on change introducing less stable environmental conditions (NASA, n.d.). The largest changes to seasons are with winters becoming shorter, and spring is becoming longer. As the overall temperatures rise, so are the ocean's temperatures (Dupigny-Giroux, 2018). This warmer water allows for more intense coastal storms with greater frequency. An additional result of a warmer climate is that it produces storms with less predictability and with incidence and intensity, impacting all aspects of daily activities (USGCRP, 2018). Climate change impacts reach all regions of the Earth, with the influences within each region unique to that region-specific characteristics.

Climate change has global implications which when looked at regionally, provide more insight into the effects. The Northeastern Region of the United States is facing specific threats because of climate change. The rising temperatures are occurring at a faster rate than other regions in the world (Dupigny-Giroux, 2018). The area is experiencing warmer winters and hotter summers. The warmer winters are resulting in shifting snow rain lines further north. The hotter summers are particularly concerning as this highly populated region is experiencing large scale urban heat island effects within the previously mild summer cities (Dupigny-Giroux, 2018). This historical trend of mild summer leaves the impacted areas with limited existing ability to lessen the heat. The major urban areas within this region are coastal and built on filled in wetlands. As a result of being near sea, these cities are directly affected by raising sea levels (City of Boston, 2018). The rising sea levels are compounded by an increase in precipitation of 70% from 1985 to 2012. As the rain has increased, flooding has been one of the prominently felt consequences of climate change (US EPA, 2016).

2. Introduction

2.1. Areas being Affected

Climate change has a tangible impact on the physical environment which in turn put the construction sites and the individuals working there at elevated risks for mishaps and incidents. Within the construction industry, climate change has had a direct effect on the increases in overall ambient temperature of jobsites around the country, the duration and intensity of the seasons, the predictability and ferocity of storms. As many construction professionals spend large portions of their shift in either a combination of indoors and outdoors space or entirely outdoors, a rise in ambient air temperature illustrates a significant risk for these workers who often perform activities with a heavy level of exertion (Schulte et al., 2016). Natural disasters can have devastating effects on jobsites and the personnel who are there. Financial losses from natural disasters in the United States have risen from $528 billion from the years of 1981-1990, to an astounding $1,230 billion between 2001-2010 (Kunreuther et al., 2008). This 230 percent increase in losses due to natural catastrophes over the period of this study does not account for the loss of life that has occurred. From 1992-2006 there were (62) floods, (72) hurricanes, and (80) wildfires that resulted in fatalities. While these fatalities were among multiple industries and not solely related to the construction industry, construction did make up a significant amount of these fatalities with restoration activities related to post hurricane reconstruction accounting for (72) deaths alone (Schulte et al., 2016). Climate change brings with it dangerous and destructive forces that continue to elevate the danger level on jobsites and to the employees working there.

Inclement weather and even rare disastrous storms are not uniquely new to the construction industry as every project is subject to the climate and geographical risk of the site's location. Where once there was insurance to stave off significant losses due to weather related setbacks, climate change is also altering the very fabric of certain financial institutions which used to insulate the construction industry from these losses. Insurance companies have always faced a guessing game, but they thrived on the ability to predict when and where disaster would strike. Climate change is altering the game as the frequency and severity of these storms are on the rise. Climate change is on a trajectory to cause a collapse across multiple financial institution, which will in turn void the way in which insurers are able to calculate the risks associated with these projects (Sagatom Saha & Brody Viney, 2020). If insurers are no longer willing to take on the risks associated with climate change, then there will be deep financial consequences as all players in the industry absorb this immense cost.

2.2. Material Availability/New Building Materials

As the construction industry reacts to climate change, there is the expectation that materials will change to reduce construction industry's contribution to the GHGs (Greenhouse Gas) emissions. Some of the ways are through utilizing green materials, low carbon concrete, and sourcing locally. These categories fall within a broader net called embodied carbon. Embodied carbon is the sum of carbon emissions from the extraction, fabrication, transportation, and erection of construction materials on a structure (Kotch, 2020). This is similar in concept to operational carbon output, except this includes the carbon from the start of a materials life to the time it is placed in the structure. Measuring the carbon output from the manufacturing, transportation, and construction phase of a material will help in the selection of materials by selecting truly beneficial products (Building Transparency, n.d.). This practice allows for the entire process to be looked at for an overall reduction in emissions.

Concrete is a large contributor to the carbon emission in construction with cement production being the main culprit contributing 8% of the carbon dioxide (CO_2) emissions globally (Rodgers, 2018). Low carbon concrete is an emerging solution to reducing concrete's emissions. This is done in a variety of methods of including other products to reduce the required amounts of cement in the concrete mix and in some places, using more efficient methods to mix concrete. Some of the materials used are waste by-products from other manufacturing activities such as fly ash or steel slag (Emerson, 2020). Another methodology is to sequester CO_2 and insert it into concrete. Blue Plant uses this idea to create aggregate from CO_2 emissions as well as recycling concrete to use as aggregate (Blue Plant, n.d.). As the focus turns to monitoring and reducing emissions, using products which incorporate recycled and lower emissions will become paramount.

2.3. Impacts on Cost and Schedule

When the effects of climate change culminate into large destructive natural disasters, the construction industry is always at the forefront of institutions most directly affected. Natural disasters effects on the construction industry are twofold as not only do these storms have direct effects by damaging projects that are in progress, but they create damage to already existing roads, utilities, and other infrastructure which in turn need to be repaired (Apurva et al., 2020). These disasters also have an indirect effect on the industry as they create issues with the industries' supply chain that can impact anticipated delivery dates of the project which in turn can lead to cost overruns. Post natural disaster, the combination of several contributing factors such as disruption to basic utility services, stark increases in demand of skilled labour, and massive material price escalation all compounded to increase the duration and cost of the project (Apurva et al., 2020).

2.4. Type of Design and Projects

Climate change is affecting both new construction as well as renovations. One aspect that is most affected is the sustainability design principles being incorporated into new designs. These sustainability features include water conservation, high-efficiency mechanical units, using environmentally sustainable products, optimizing operation and maintenance practices, choosing appropriate building site, and enhancing the indoor air quality (GSA, 2021). New construction incorporates resiliency features in the design which should be implemented in planned renovations as they are a crucial opportunity to upgrade the existing structures to cope with the predicted climatic conditions of the future. These resiliency additions range from planning to incorporate flood barriers to increased ventilation to deal with the predicted extended periods of high temperatures (Button, 2018). Buildings are not alone in the effort to incorporate resiliency and sustainability features into the structure, infrastructure is also an active participant in combating the impacts of climate change (MassDOT, 2021). Infrastructure is particularly unique as the project can transverse multiple regions such as coastal and inland regions. The areas each have specific concerns such as flooding or fire. To better understand where the risks lie with climate change, vulnerability studies need to be conducted to determine the associated risks with their location.

2.5. Current Critical Infrastructure

Vulnerability studies are extremely critical in the determination of where limited money and time are spent by state transportation departments to protect critical infrastructure assets. These studies identify the key assets for moving people and commerce around the northeast region of the United States which are most at risk from the effects of climate change. The Massachusetts Department of

Transportation (MassDOT) has been active in assessing the infrastructure and has been taking action to upgrade the resilience of the state's transportation assets (MassDOT, 2021). With many major facilities within the Boston area being on the water such as Logan International Airport, major port operations for petroleum, and entire sections of the city are at major risk of becoming submerged within the end of the century without action (City of Boston Environmental Department, n.d.; MassDOT, 2021). An example of this would be the subway stop at the Aquarium in Boston. This location is having permanent flood gate systems erected around four access points to prevent flooding as flooding has occurred multiple times in the past and is part of the larger Blue Line Harbor Tunnel Infrastructure Improvements project (MBTA, n.d.). Boston is not alone in having to implement climate change responsive projects, other major cities within the northeast have also followed suit.

New York City (NYC) is currently facing the same concerns over a changing climate as Boston. New York City and Boston are or have looked at the creation of large-scale flood barriers to head off storm surges from storms such as Hurricane Sandy. These structures would seal off the harbours during extreme weather events (Hilburg, 2018; Sustainable Solutions Lab, 2018). While these ambitious projects may not be feasible or may take decades to become reality, but smaller projects are still occurring. NYC is currently working to ensure their stormwater management system can function in future conditions. This includes looking at current and future hightide, sea levels, and precipitation to ensure their systems can still effectively drain. Previously, only in the worst conditions failures were occurring but with increased rain events along with higher tides and sea level, the current systems cannot drain the water due to undersized pipes and the outlets below high tides (Mayor's Office of Resiliency, 2021). In addition to upgrading the city's ability to drain runoff, other projects are looking at protecting low lying areas of the city. One such project is named The Big U, which is located on the Lower East Side of NYC (Rebuild By Design, n.d.). The large undertaking is proposed to provide flood protection barriers, improve stormwater drainage, and elevated open space along the waterfront (East Side Coastal Resiliency Staff, n.d.). This project is like many other projects in other major coastal cities within the Northeast region providing upgrades to existing shorelines to protect coastal populations.

In addition to rising sea levels, coastal and all urban areas are facing a specific threat from the increasing temperatures. With an ever-increasing number of days above 90 degrees Fahrenheit, the urban heat island effect is a major threat to vulnerable communities (City of Boston, 2018). This heat effect occurs in areas with more concrete, asphalt, and steel, which absorb heat and are areas that tend to have less natural vegetation which help dissipate the heat (BWSC Staff, n.d.). The simple and logical fix would be to create more green space within urban areas. While this is feasible in areas of new development, in areas already heavily developed the feasibility decreases drastically. Possible ways to reduce urban heat islands is to incorporate green design into infrastructure projects, increase urban vegetation, green roofs, and most importantly conduct studies to determine where this work needs to occur within a city (EPA, 2020). Boston uses the term urban forest when referring to the city's trees, which are maintained in part to help cool the city (BWSC Staff, n.d.). In some cities within the Northeast, tax credits are being offered for the construction of green roofs. Philadelphia and NYC are among the cities that are offering these incentives to build green roofs to help cool and green the cities (Department of Revenue, 2019; Mayor's Office of Resiliency, 2021). As cities explore the threat of urban heat islands, the risks to human health are present and the strain of the mass use of air conditioning on the electrical grid is apparent (City of Boston, 2018).

3. Current Solutions

3.1. Legislation Changes and New Building Codes

The construction industry is not known to be the most reactive industry to implement change. With the introduction of stricter building codes by municipalities, which better protect structures from new 100-year storm data, change within the industry is being enacted. Research looking into 100-year storm events shows and predicts the higher frequency of large-scale storm events. This includes storms with increased precipitation in shorter periods of time (Massachusetts Climate Adaptation Partnership,

2017). The studies which present this type of information feeds into policy throughout the Northeast region. NYC has enacted local law 97, which is one of the most ambitious laws of its type (Urban Green, 2021). This law establishes restrictions of GHG emissions for buildings over 25,000 square feet (Sajip, 2021). If these constraints are not met, the owners of the buildings are imposed fines. These fines are motivating renovation projects to meet the emission standards. Boston has enacted zoning code article 37 which incorporates green building and climate resiliency into the building code. This requires major buildings to reach a certifiable level with Leadership in Energy and Environmental Design (LEED). This code offers guidance and requests information on the buildings to determine what risks there are and what measures are in the design to protect the buildings from the impacts of climate change (BPDA, n.d.). Building departments are not alone in introducing or providing guidelines to produce resilient buildings within the construction industry.

Nongovernmental organizations have also played a role in combating climate change. LEED part of the U.S. Green Building Council (USGBC) has different levels of ratings which corelate with how "green" the building is. LEED has numerous different building categories, which range from new construction, renovations, operations, and maintenance (USGBC, n.d.). These are rank at different levels such as certified, silver, gold, platinum, and zero (USGBC, n.d.). LEED Zero is similar to that of Passive House Institute US (PHIUS), where the mission is to certify and provide goals for buildings to meet. The goals are to create homes and buildings which are net zero (USGBC, n.d.; PHIUS, n.d.). Net zero refers to a building that uses, over a course of a year, a net zero electricity and water. LEED and PHIUS both have different standards and levels with PHIUS originating with homes growing into commercial buildings with LEED focusing on commercial and growing in residential (USGBC, n.d.; PHIUS, n.d.). The organizations provide a symbol for the owners to express their role in protecting the environment and following the social responsibility trend of being environmentally aware.

3.2. What Companies Are Doing to Overcome New Legislation (Private Sector)

With new laws and regulations being introduced to the construction industry as well as new third-party authorities on the subject matter such as USGBC, firms operating in the private sector are being held to the highest standards of sustainability and resiliency when designing new projects. The construction industry represents a huge player in the United States' contribution to climate change as buildings are responsible for 39% of CO_2 emissions, 41% of all U.S. energy usage, and 73% electricity usage (Ahn et al., 2016). In 2007, Boston updated Article 37 of its zoning code to ensure there is a minimum green building standards that all new projects must reach as well as requiring developers to integrate consideration for current and future climate conditions into a formal resiliency checklist (BPDA, n.d.). Designers, developers, and contractors have all become acutely aware of the policies created by the United States Green Building Council. While the USGBC is the official third-party organization that oversees and regulates the standards for Leadership in Energy and Environmental Design (LEED), there are certain alternatives to LEED such as the Green Globes Levels that are overseen by the Green Building Initiation. These types of certifications which have become greatly sought after as they represent an understanding of the responsibility developers have to the community as well as attractive tax benefits. Being able to classify a building as a LEED certified building, provides advantages to the marketability of the product and brand, can create better work and living environments, and reduce negative ecological impacts as well as reducing carbon emissions (DeLisle et al., 2013). To deliver the highest quality and least environmentally impactful product, projects are heavily integrating all members of the construction team to design and deliver the foremost environmentally conscious building. Developers are implementing a wide array of green systems into the design wherever they can to achieve the highest level of certification. These systems include reuse of existing structure, high efficiency envelop systems, smart energy management systems, renewable energy (solar, geothermal HC/energy pile, fuel cell), rainwater utilization systems, natural ventilation and daylighting, and roof & wall greening (Ahn et al., 2016).

3.3. New Technologies

The construction industry use of heavy equipment is a symbolic picture of the industry. The use of fossil fuelled equipment adds to the emissions that the industry release annually. To counter these emissions, electrical heavy equipment is being experimented with. Caterpillar has created an all-electric 26-ton excavator, which is the first one of its size (Lambert, 2019). Caterpillar is not alone in the electrified equipment push, Volvo as well as other large manufacturers are working on releasing all electric equipment as well (Volvo, n.d.; Raczon, 2020). In addition to making electric power equipment, many companies are now offering remote controlled equipment as well (Caterpillar, n.d.). These new technologies help keep workers safe and fight climate change. The widespread use of all electric equipment is lagging, which is preventing the large-scale manufacturing of these machines.

Technology is not limited to upgrading currently used equipment but is also on pace to provide a replacement for human labour. Robotics has advanced and continues to do so at a rapid pace. While increasingly higher temperatures affect human labour, robots can operate without any effect. While climate change may not be a predominate force spurring the creation of these robots, it does provide another benefit for them. For instance, Advance Construction Robotics manufactures automation tools designed for tying and placing rebar on horizontal planes (Advance Construction Robotics, n.d.). Drones are another tool being used to increase productivity in all aspects of construction, ranging from inspections, surveying, mapping, progress walk throughs, and trials are being done to use drones to complete tasks such as painting (Knight, 2020). As technology advances so will the use of robotics, as the benefits of increased productivity, ability to work in extreme conditions, and cost reduction their application will become desirable.

4. Moving Forward

Not only are the current effects of climate change here to stay, but as the construction industry looks to the future the evidence suggests that there will be future climate related issues that will compound with the current problems that will need to be identified, analysed, and accounted for in future design. Ensuring that the projects that are being constructed, are on the forefront in terms of resiliency are critical factors for both designers and developers. As greener technologies become available and green building become the industry standard, the starting point for many of these projects begins adapting the project delivery method. While traditional design-bid-build (DBB) method was the staple process by which projects use to be completed, changing technologies, and ever evolving legislation that defines green construction call for all team members to communicate and be more involved. The first step in creating a resilient building is implementing an integrated construction process (ICP) and using integrated project delivery (IPD) method. IPD ensures that all major team members in the construction process are collaborating and supporting each other to successfully integrate the necessary technologies and meet the criteria of the regulatory administrations to be considered green and resilient projects (Ahn et al., 2016). Ensuring that all project team members are capable of delivering on the desired product is essential to projects success.

Changes in municipal regulation and third-party certification requirements can have adverse impacts on a project. Throughout the Northeast many cities and towns have organized committees and regulating authorities to investigate, report, and regulate policies in regard to climate change. In Boston, Article 37 of the zoning code also created the Interagency Green Building Committee (IGBC) which is responsible for ensuring that projects comply with the cities policies towards resiliency and green buildings. Currently the City of Boston requires that projects consider the potential impacts of the results of climate change such as extreme heat, extreme precipitation, rising sea levels, etc. and provide a framework for how these buildings will combat these issues (BPDA, n.d.). New York city has also taken action to bring resiliency regarding their stormwater and sewer systems to the forefront of construction issues. Local Law 172 required the city identify which areas of the city are most susceptible to flooding and the NYC Green Infrastructure Program is offering grants for new or redevelopment properties that comply with the stormwater retention program and that incorporate green infrastructure into their projects (NYC Stormwater Resiliency Plan, 2021). In addition to the governing laws and legislation, projects must also consider the requirements set forth by the USGBC to attain the agreed upon LEED certification. Projects must be innovative and utilize green technologies in their design. Project teams must collaborate to deliver the most desirable green product. Incorporating innovations such as enhanced energy efficient HVAC and water usage, use of green material and resources, improved indoor environment, and capitalizing on the existing site and its proximity to transportation all need to be taken into consideration (USGBC, n.d.). Project teams need to continue pushing the envelope of green building innovation to ensure that they continually stay ahead of the curve and to ensure that their projects are compliant with the local regulations.

Understanding the changes that are occurring with the climate and being able to better predict the location and degree to which severe storms will affect an area is also an important aspect of the future of resiliency. Looking ahead to only the rising sea level, it is estimated that by 2070 as many as 150 million people worldwide could be adversely affected by rising sea levels, and the Northeast with its many highly populated coastal cities will be at elevated risk (Yang et al., 2020). Early warning systems that can better predict and notify individuals in the affected area of a severe impending storm would be critical to ensure adequate preparations can be made.

In addition to new technologies that incorporate early warning systems, incorporating new strategic building designs can play a large factor in preventing damage to critical building systems. Designing modern buildings with electrical, HVAC, and mechanical equipment on upper floors as opposed to their traditional location in basements and other below grade areas helps protect against critical equipment failure in the event of disasters striking (USGBC, n.d.). In situations where flooding either already happens consistently or is expected to occur regularly with the rise in frequency of storms, some owners will have to consider abandoning the lowest floor entirely. This option comes with a heavy loss of opportunity as abandoning an entire floor limits the projects usable space. However, abandoning the lowest floor does have a large window of effectiveness as this method can be utilized with little to no additional maintenance cost for up to 50 years, and the method has been proven to be highly effective at mitigating damage to the rest of the property. While not as drastic as abandoning the entire first floor of the project, elevating the buildings first floor can be an appropriate option in certain situations. By elevating the lowest floor, teams can mitigate damage to building elements and still retain use of the first floor. This option also is not without trade off as it limits the square footage of the first floor and there is potential that storms could still damage portions of the first floor if it is a severe storm (FEMA, 2015).

5. Conclusion

Climate change is an imminent multi-faceted concern that encompasses a large array of industries and will continue to evolve as the extent of this force is continuously explored and researched. In its current capacity climate change has already had a direct impact on the construction industry. Climate change has impacted all aspects of a construction projects from the type and sources of building materials to the equipment and methods used in the physical construction, to the planning and designing of future projects with the intent of minimizing climate changes effects. Like the construction industry, climate change is a numbers game and being at the forefront of climate change, innovation and resiliency projects can effectively strategizes and minimizes its adverse effects while simultaneously delivering a superior and greener product.

6. References

[1] Advance Construction Robotics. (n.d.). *TyBot*. Retrieved July 6, 2021, from Advance Construction Robotics: https://www.constructionrobots.com/

[2] Blue Plant. (n.d.). *Products*. Retrieved March 24, 2021, from Blue Plant: http://www.blueplanet-ltd.com/#products

[3] BPDA. (n.d.). *Article 37 Green Building and Climate Resiliency Guidelines*. Retrieved June 20, 2021, from Boston Planning & Development Agency: http://www.bostonplans.org/planning/planning-initiatives/article-37-green-building-guidelines

[4] Building Transparency. (n.d.). *EC3 FAQ*. Retrieved July 5, 2021, from Building Transparency: https://www.buildingtransparency.org/ec3-resources/ec3-faq/

[5] Button, R. (2018, March 6). *Resilience Planning in Design and Construction.* Retrieved from Constructable Trimble: https://constructible.trimble.com/construction-industry/resilience-planning-in-design-and-construction

[6] BWSC Staff. (n.d.). *Climate Ready Boston*. Retrieved June 11, 2021, from Boston Water and Sewer Commission: https://www.bwsc.org/environment-education/climate-resiliency

[7] Caterpillar. (n.d.). *CAT® COMMAND*. Retrieved June 6, 2021, from Caterpillar: https://www.cat.com/en_US/by-industry/construction-industry-resources/technology/command.html

[8] City of Boston. (2018, October 17). *CLIMATE READY SOUTH BOSTON.* Retrieved from Boston.gov: https://www.boston.gov/departments/environment/climate-ready-south-boston

[9] City of Boston Environmental Department. (n.d.). *Coastal Resiliance Project Tracker*. Retrieved Febuary 14, 2021, from Climate Ready Boston: https://boston.maps.arcgis.com/apps/Cascade/index.html?appid=c438a8a66e9d470b8b305a973fc192f7

[10] Department of Revenue. (2019, July 22). *Green Roof Tax Credit*. Retrieved from City of Philadelphia: https://www.phila.gov/services/payments-assistance-taxes/tax-credits/green-roof-tax-credit/

[11] Dupigny-Giroux, L. E.-S. (2018). *Impacts, Risks, and Adaptation in the United States: Fourth National Climate Assessment, Volume II* . Washington, DC: u.s. Global Change Research Program.

[12] East Side Coastal Resiliency Staff. (n.d.). *Construction Progress Construction Project Areas for East Side Coastal Resiliency*. Retrieved May 26, 2021, from New York City Website: https://www1.nyc.gov/site/escr/project-updates/construction-progress.page

[13] Emerson, J. (2020, November 9). *Low Carbon Concrete – Starting from the Ground Up.* Retrieved from Zero Energy Project: https://zeroenergyproject.org/2020/11/09/low-carbon-concrete-starting-from-the-ground-up/

[14] EPA. (2020, November 2). *Reduce Urban Heat Island Effect*. Retrieved from Environmental Protection Agency : https://www.epa.gov/green-infrastructure/reduce-urban-heat-island-effect

[15] GSA. (2021, March 07). *Sustainable Design*. Retrieved from U.S. General Services Administration : https://www.gsa.gov/real-estate/design-construction/design-excellence/sustainability/sustainable-design

[16] Hilburg, J. (2018, July 10). *Army Corps of Engineers proposes swinging sea gates for New York Harbor.* Retrieved from The Architect's Newspaper: https://www.archpaper.com/2018/07/army-corps-of-engineers-closing-new-york-harbor-swinging-sea-gate/

[17] Knight, W. (2020, November 20). *Robots Invade the Construction Site.* Retrieved from Wired: https://www.wired.com/story/robots-invade-construction-site/

[18] Kotch, M. (2020, January 02). *The climate case for construction*. Retrieved from Greenbiz: https://www.greenbiz.com/article/climate-case-construction

[19] Lambert, F. (2019, January 29). *Caterpillar unveils an all-electric 26-ton excavator with a giant 300 kWh battery pack.* Retrieved from Electrek: https://electrek.co/2019/01/29/caterpillar-electric-excavator-giant-battery-pack/

[20] Massachusetts Climate Adaptation Partnership. (2017). *Storms and Floods.* Retrieved from Climate Action Tool: https://climateactiontool.org/content/storms-and-floods

[21] MassDOT. (2021). *MassDOT Statewide Climate Change Adaptation Plan Objectives*. Retrieved from Mass: https://www.mass.gov/service-details/massdot-statewide-climate-change-adaptation-plan-objectives

[22] Mayor's Office of Resiliency . (2021, May). *New York City Stormwater Resiliency Plamn.* Retrieved from New York City: https://www1.nyc.gov/site/orr/index.page

[23] MBTA. (n.d.). *Aquarium Station Floodproofing*. Retrieved from Massachusetts Bay Transportation Authority: https://www.mbta.com/projects/aquarium-station-floodproofing

[24] NASA. (n.d.). *Climate Change: Vital Signs of the Planet*. Retrieved June 02, 2021, from The Effects of Climate Change: https://climate.nasa.gov/effects

[25] PHIUS. (n.d.). *Passive House Principles*. Retrieved June 30, 2021, from Passive House Insitute US: https://www.phius.org/home-page

[26] Raczon, F. (2020, July 21). *Current State of Electric Equipment.* Retrieved from Construction Equipment,com : https://www.constructionequipment.com/current

[27] Rebuild By Design. (n.d.). *The Big U*. Retrieved May 25, 2021, from Rebuild By Design: http://www.rebuildbydesign.org/our-work/all-proposals/winning-projects/big-u

[28] Rodgers, L. (2018, December 17). *Climate change: The massive CO2 emitter you may not know about.* Retrieved from BBC News: https://www.bbc.com/news/science-environment-46455844

[29] Sajip, J. (2021, May 12). *Local Law 97 of 2019: Understanding the NYC Building Emission Limits.* Retrieved from New York Engineers: https://www.ny-engineers.com/blog/local-law-97-of-2019

[30] Sustainable Solutions Lab . (2018, May 29). *UMass Boston Report: Boston Harbor Barrier Costs Would Outweigh Benefits.* Retrieved from UMAss Boston: https://www.umb.edu/news/detail/umass_boston_report_boston_harbor_barrier_costs_would_outweigh_benefits

[31] Urban Green. (2021, May). *ALL ABOUT LOCAL LAW 97*. Retrieved from Urban Green Council : https://www.urbangreencouncil.org/content/projects/all-about-local-law-97

[32] US EPA. (2016, Decemeber 22). *Climate Impacts in the Northeast*. Retrieved from Climate Change Impacts: https://19january2017snapshot.epa.gov/climate-impacts/climate-impacts-northeast_.html

[33] USGBC. (n.d.). *LEED v4.1*. Retrieved July 1, 2021, from U.S. Green Building Council: https://www.usgbc.org/leed/v41#bdc

[34] USGBC. (n.d.). *LEED Zero*. Retrieved from U.S. Green Building Council: https://www.usgbc.org/programs/leed-zero

[35] Volvo. (n.d.). *Emobility*. Retrieved June 8, 2021, from Volvo Construction Equipment: https://www.volvoce.com/global/en/our-offer/emobility/

Greenhouse Gas Emissions of Global Construction Industries

Robert H. Crawford

The University of Melbourne, Parkville, Australia

rhcr@unimelb.edu.au

Abstract. The global construction industry is responsible for considerable effects on the environment and society. The construction and use of our built environment accounts for 39% of global greenhouse gas emissions. Population growth and rising standards of living are further exacerbating these problems. Understanding the contribution of construction industries and those activities responsible for the greatest share of greenhouse gas emissions is crucial in order to identify opportunities for emissions mitigation. This study uses multi-regional input-output analysis to analyse the embodied greenhouse gas emissions of global construction industries. A structural path analysis (SPA) was conducted for the construction sectors of 44 countries using top-down economic input-output data from EXIOBASE3. The findings were analysed to compare the embodied greenhouse gas emissions of each construction sector to ascertain their emissions intensity. The structural path analysis was used to disaggregate each sector into unique nodes or pathways, with each representing a single or group of activities and their associated emissions. This was used to identify critical hot spots of emissions within the various construction supply chains. The results show that there is wide discrepancy between the emissions intensity of the construction industry for the 44 countries, ranging from 0.165 $kgCO_2e$/Euro to 2.05 $kgCO_2e$/Euro. In addition, the most significant contributors to emissions for the most emissions intensive countries are the production of concrete and steel. This in-depth analysis of global construction industries using multi-regional input-output data provides critical information needed to identify opportunities for reducing global construction-related greenhouse gas emissions. This will help prioritise future emissions reduction efforts within the construction industry and target specific solutions to achieve the greatest improvements to the overall environmental performance of global construction industries.

Keywords: Construction sector, greenhouse gas emissions, multi-regional input-output analysis.

1. Introduction

The construction and use of our buildings and infrastructure contribute more than any other single sector to global energy use (36%) and greenhouse gas (GHG) emissions (39%) [1]. These have continued to increase, and in 2018, they reached record highs, up 7% from 2010 levels, despite global efforts to reduce them. Improvements to building envelopes and building energy systems have offset some of the growth in building related GHG emissions in recent years. However, GHG emissions have continued to rise due to increasing demand for electricity. Global population growth and rising living standards will put further pressure on these GHG emissions, fueled largely by an expected threefold increase in air conditioning use by 2050 [2] and increasing demand for materials. Buildings and infrastructure thus represent one of the most important and cost-effective options for mitigating global GHG emissions [3]. A critical part of this will be addressing embodied GHG emissions - those emanating from the manufacture of building materials and the construction process – as they account for around a third of all building-related emissions [4]. As our buildings head towards net-zero operational GHG emissions, these embodied GHG emissions will become even more significant. In response to this, the World Green Building Council [4] has called for a 40% reduction in construction-related embodied GHG emissions by 2030 and net-zero embodied GHG emissions by 2050.

The construction industries of the world have a critical role to play in reducing the embodied GHG emissions of our buildings and infrastructure projects as they are ultimately responsible for their construction, including the specification of materials and components. To achieve these emissions reductions, an understanding of the source of emissions is critical. Construction companies have most control over activities involved in the construction process, which are mostly project site based. However, the majority of emissions occur upstream of these main site-based construction activities [5]. The role of designers, engineers and other consultants becomes even more important in this context due to their influence over the materials and construction approach used. A general understanding of the need to reduce the use of cement-based products, increase the use of sustainably sourced renewable materials and reuse existing materials is now well known. Despite this, a holistic understanding of the broad areas for emissions reduction across the entire construction supply chain is lacking. A key reason for this stems from the inability of commonly used emissions assessment techniques to capture supply chain wide processes and associated emissions.

In an attempt to address this issue, environmentally extended input-output models have been used. When used to disaggregate the supply chain of an entire industry, the most significant emissions sources can be easily identified, allowing them to be prioritised for future mitigation. These models have been used to analyse various environmental flows for the construction industry of several individual countries, such as China [6], UK [7], Ireland [8] and Australia [9]. In many cases, these analyses are limited to a single region, or connections to the global supply chain are aggregated at the highest possible level. As construction supply chains are increasingly globally connected, it is just as important to be able to interrogate imported sources of emissions as it is for those of locally sourced goods and services. This is made possible with a multi-regional input-output model, such as EXIOBASE [10], WIOD [11] or EORA [12].

The aim of this study was to compare the GHG emissions of selected global construction industries and identify critical hot spots for emissions reduction using a comprehensive MRIO analysis.

2. Research method

This section describes the method used to quantify the GHG emissions associated with the construction industry for 44 countries, including the identification of key emissions hotspots for the top five most emissions intensive construction industries.

2.1. Quantification of greenhouse gas emissions

This study uses an environmentally extended multi-regional input-output (MRIO) analysis to quantify greenhouse gas emissions. This section describes the MRIO model used, including the process for determining the GHG emissions for each country's construction industry. The process of conducting a structural path analysis is then described in order to identify the key emissions sources within each construction industry. It should be noted that the GHG emission values calculated are intensities, rather than absolute emissions. This enables a comparison between countries regardless of construction activity. In reality, absolute emissions will vary based on actual construction activity which will be non-identical for each country. However, as the value of construction activity per annum will likely change, sometimes significantly, an average intensity can be a much more useful indicator of environmental efficiency of construction activity in a particular country.

2.1.1. MRIO model

The MRIO model used for this analysis is based on the input-output tables from EXIOBASE3 [10]. This version includes input-output tables for 44 countries as well as five rest of world (ROW) regions, disaggregated into 163 industries per country/region (a total of 7,987 industries). These input-output tables provide information on the relationship within and between individual industries and products, such as the construction industry. The input-output tables are processed as per [13], with the direct and total GHG emissions associated with each industry determined.

2.1.2. Structural path analysis

In order to better understand the source of GHG emissions associated with each country's construction industry, a structural path analysis (SPA) was used to disaggregate their supply chains into individual nodes or processes. There are a potential maximum 32 quintillion nodes for each construction sector at just five stages upstream in the supply chain (7,987 industries at stage 1, $7{,}987^2$ industries at stage 2, $7{,}987^3$ industries at stage 3 etc.). The SPA was conducted as per [13] using the same user interface (GUI) and python package [14]. The input-output database is loaded into the GUI. The SPA is conducted for each construction sector, one by one. To meet current computing processing capabilities, nodes representing less than 0.0001% of the total emissions intensity for each construction sector were excluded and the analysis was performed to a maximum of five stages upstream. The output of the SPA is a series of mutually exclusive nodes, each representing a good or service provided from one stage to another within the construction supply chain analysed. A series of nodes, corresponding to a chain of transactions leading to the construction sector being assessed, is referred to as a pathway – for instance, iron ore purchased by the reinforcing steel bar sector, in turn purchased by the construction sector for an apartment building. Each node contains data on its associated GHG emissions intensity (in $kgCO_2e$/Euro). The model analyses multiple flows at the same time and automatically ranks nodes in order of importance. This helps identify emissions hotspots at various stages of the supply chain.

3. Results and discussion

This section presents the total GHG emissions intensity associated with the construction industry of 44 countries, based in EXIOBASE3 data (Figure 1). This shows that the total GHG emissions intensity for these countries varies from 0.165 $kgCO_2e$/Euro to 2.05 $kgCO_2e$/Euro with the top five most GHG emissions intensive construction industries being in China, India, South Africa, Taiwan and South Korea. The least emission intensive construction industry was found to be that of the United Kingdom.

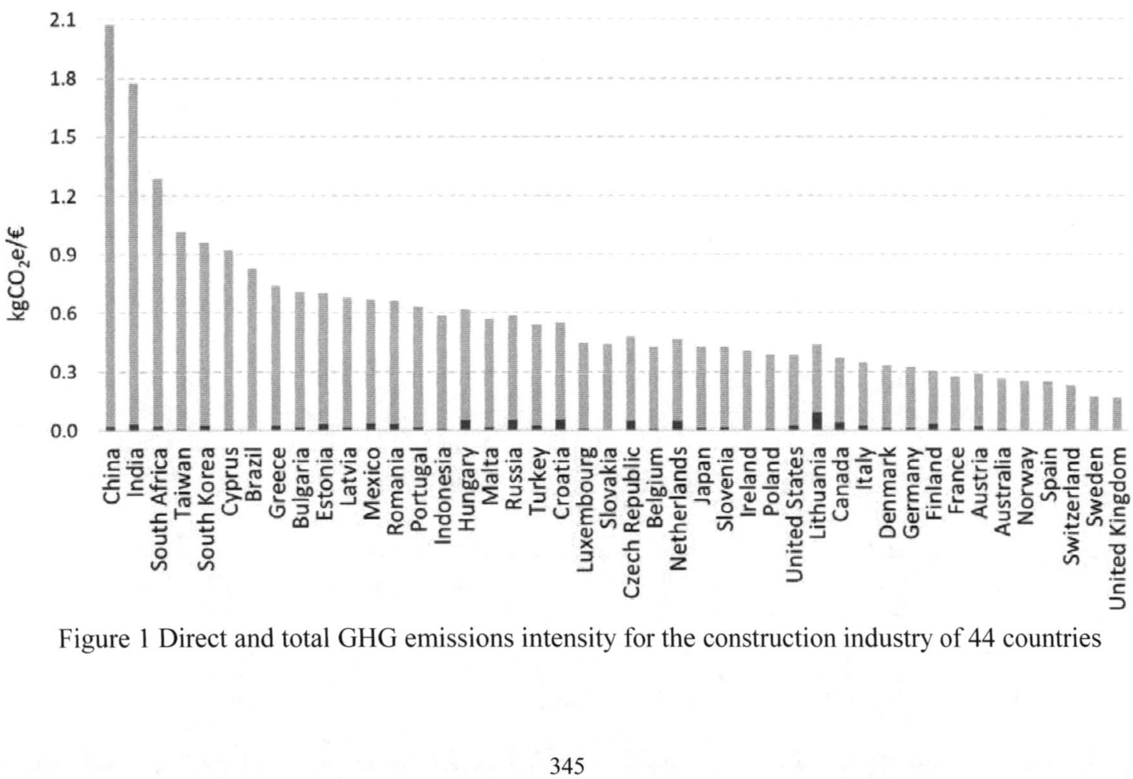

Figure 1 Direct and total GHG emissions intensity for the construction industry of 44 countries

Figure 1 also shows the direct emissions intensity for the construction industry of each country. Direct emissions, also known as scope 1 emissions, account for the emissions resulting from mainly site-based activities, such as the use of machinery and equipment that requires the burning of fuel. These direct emissions generally represent a small proportion of the total emissions of the construction industry due to the emissions intensive processes that occur further upstream in the construction supply chain (from the production of cement, steel and electricity, for example). The direct emissions were found to represent between 0.33%, or 0.0027 $kgCO_2e$/Euro (Brazil) and 26.6%, or 0.092 $kgCO_2e$/Euro (Lithuania) of the total emissions across the 44 country's construction industries (average of 5% or 0.023 $kgCO_2e$/Euro). A higher direct emissions intensity may mean a country relies on more emissions intensive construction processes, simpler construction with less materials, or uses materials with lower GHG emissions intensity.

There may be multiple reasons for a high indirect emissions intensity, such as that seen in China and India. This could include an emissions intensive local fuel mix, large reliance on products or services with emissions intensive manufacturing processes (e.g. transport and concrete), or large reliance on goods and services imported from countries with an emissions intensive fuel mix.

Table 1 provides the results of the structural path analysis of the top five most GHG emissions intensive construction industries, based on Figure 1. While a very large number of nodes were identified by the SPA, only the top five processes/inputs are shown, with the remainder representing less than 2.8% of the total emissions intensity for each country's construction sector, each. This shows that the manufacture of 'cement, lime and plaster' and steel products are responsible for the majority of GHG emissions for these most emissions intensive construction industries.

Table 1 Source and proportion of total emissions for top five most emissions intensive construction industries

	Stage 1	Stage 2
China		
29.1%	Manufacture of cement, lime and plaster	
4.8%	Re-processing of ash into clinker	
3.6%	Manufacture of basic iron and steel and of ferro-alloys and first products thereof	
2.5%	Manufacture of other non-metallic mineral products nec	
2.2%	Manufacture of cement, lime and plaster	Mining of coal and lignite; extraction of peat
India		
17.9%	Manufacture of cement, lime and plaster	
9.2%	Manufacture of basic iron and steel and of ferro-alloys and first products thereof	
6.8%	Manufacture of basic iron and steel and of ferro-alloys and first products thereof	Mining of coal and lignite; extraction of peat
3.7%	Re-processing of secondary steel into new steel	
3.0%	Construction	Manufacture of cement, lime and plaster

	Stage 1	Stage 2
South Africa		
10.8%	Manufacture of other non-metallic mineral products nec	
5.6%	Manufacture of cement, lime and plaster	
4.8%	Re-processing of ash into clinker	
3.5%	Manufacture of rubber and plastic products	Production of electricity by coal
3.2%	Manufacture of other non-metallic mineral products nec	Manufacture of cement, lime and plaster
Taiwan		
8.4%	Manufacture of other non-metallic mineral products nec	Manufacture of cement, lime and plaster
7.8%	Quarrying of sand and clay	
5.0%	WA-Manufacture of other non-metallic mineral products nec	
4.6%	Manufacture of other non-metallic mineral products nec	Quarrying of sand and clay
4.6%	Manufacture of other non-metallic mineral products nec	Manufacture of cement, lime and plaster
South Korea		
16.7%	Manufacture of cement, lime and plaster	
3.8%	Re-processing of ash into clinker	
2.9%	DIRECT Stage 0	
2.6%	Manufacture of other non-metallic mineral products nec	
2.0%	Manufacture of cement, lime and plaster	Manufacture of cement, lime and plaster

While this study provides useful and detailed insight into the potential environmental effects of some of the world's major construction industries, it should be noted that relying on input-output data alone has major limitations. This data enables a more holistic, whole-of-economy perspective of the transactions between industries and their associated environmental flows. However, as it is based on financial, rather than physical quantities, and a number of assumptions are used in the compilation and use of this data, the reliability and relevance when applied to specific goods or services is limited. Further in-depth analysis of the SPA results is also needed to: identify trends across a larger number of countries; target more detailed analysis of construction supply chains using bottom-up data; and recommend appropriate GHG emissions reduction strategies.

4. Conclusion

The aim of this study was to compare the greenhouse gas emissions intensity of 44 countries and explore key areas for emissions mitigation. An environmentally extended multi-regional input-output analysis was conducted to identify the emissions intensity of each country. This found China and India to have the most emissions intensive construction industries, at 2.05 $kgCO_2e$/Euro and 1.74 $kgCO_2e$/Euro, respectively. The least emissions intensive construction industry was that of the United Kingdom at just 0.165 $kgCO_2e$/Euro. With direct emissions accounting for on average 5% of total emissions associated with construction activity across all countries, this indicates that the benefit in addressing these emissions is likely to be very minimal. Thus, the major emissions mitigation efforts should be focused on emission sources further upstream in the supply chain.

Previous studies and data on the GHG emissions of global construction industries generally provide emissions information on site-based activities and key related industries within the construction supply chain (e.g. concrete and steel production). However, this study is unique as it provides a more holistic analysis by including the entire supply chain, including emissions associated with both goods (e.g. materials) and supporting services. It also provides greater insight into the key areas for emissions mitigation, using a SPA to disaggregate the construction industry supply chains into discrete pathways and nodes for each of the top five countries in order to identify the upstream source of emissions for each country's construction industry. This showed that future emissions reduction efforts should focus on 'cement, lime and plaster' and steel products in order to achieve the greatest emissions reductions. Significant inroads have already made into reducing the GHG emissions associated with the production of cement, concrete and steel, with SCM as a substitute for cement and low emissions steel being increasingly used around the world. As the emissions intensity of these materials declines across the world, the MRIO and SPA techniques used in this study will be unvaluable for identifying the next GHG emissions hotspots and targeting future emissions reduction efforts.

References

[1] IEA, *Global status report for building and construction 2019*. 2019, Paris: IEA.
[2] IEA, *The future of cooling*. 2018, Paris: IEA.
[3] IPCC, 2014 *Climate change 2014: Mitigation of climate change. Contribution of working group III to the fifth assessment report of the intergovernmental panel on climate change*, Cambridge University Press: Cambridge.
[4] Adams M, Burrows V and Richardson S, *Bringing embodied carbon upfront*. 2019, London: World Green Building Council.
[5] Crawford R H 2013 Improving the environmental performance of the construction supply chain *CIB World Building Congress 2013* Brisbane
[6] Hong J, Shen Q and Xue F 2016 A multi-regional structural path analysis of the energy supply chain in china's construction industry *Energy Policy* **92** 56-68
[7] Pomponi F and D'Amico B 2018 Carbon mitigation in the built environment: An input-output analysis of building materials and components in the UK *Procedia CIRP* **69** 189-193
[8] Acquaye A A and Duffy A P 2010 Input–output analysis of irish construction sector greenhouse gas emissions *Building and Environment* **45**(3) 784-791
[9] Yu M et al. 2017 The carbon footprint of Australia's construction sector *Procedia Engineering* **180** 211-220
[10] Stadler K et al. 2018 Exiobase 3: Developing a time series of detailed environmentally extended multi-regional input-output tables *Journal of Industrial Ecology* **22**(3) 502-515
[11] Timmer M P et al. 2015 An illustrated user guide to the world input–output database: The case of global automotive production *Review of International Economics* **23**(3) 575-605
[12] Lenzen M et al. 2013 Building EORA: A global multi-region input-output database at high country and sector resolution *Economic Systems Research* **25**(1) 20-49
[13] Stephan A, Crawford R H and Bontinck P-A 2018 A model for streamlining and automating path exchange hybrid life cycle assessment *The International Journal of Life Cycle Assessment* **24**(2) 237-252
[14] Stephan A and Bontinck P-A, 2019 *Pyspa - an object-oriented python package for structural path analysis (version 1.4)*, Github.

The Automated Farmlands of Tomorrow: An IoT Integration with Farmlands

Priyanka Singh1, Saurav Dixit2*, Debaroti Sammanit1, and Paritosh Krishnan 3

1Civil Engineering Department, Amity University Uttar Pradesh, Noida, 201313, India

2Associate, Dean Research & Director International Collaboration, K. R. Mangalam University, Gurgaon, 122103, India.

3Mechanical Engineering Department, Amity University Uttar Pradesh, Noida, 201313, India.

*Corresponding author: Saurav.dixit@krmangalam.edu.in; sauravarambol@gmail.com

Abstract. The rapid urbanization and industrialization of the population have resulted in a critical shortage of agricultural land and manpower to meet the growing population's needs. The future of tomorrow requires impetus through concepts such as IoT Automation in farmlands to meet the increased demand for labour and inspection while maintaining economic efficiency without depleting the environment. Monitoring becomes difficult with a reduced manpower base, which may result in crop failures. The purpose of this paper is to propose how IoT, in conjunction with basic service automation, can be used to influence climate control. Thus, a working model is created using sensors that control the humidity and required temperature conditions, thereby maintaining a suitable growth environment while monitoring and analyzing in real time.

Keywords: Automation; Climate control; Humidity; IoT; Temperature; Remote monitoring; wireless data transmission.

1. Introduction

Natural irrigation is now under pressure due to increasing water shortages, primarily due to population increase and climate change. It is therefore imperative to track water sources to optimise the allocation of retained water. Climate change has been shown to have a significant impact on agricultural crop production over the past two decades, especially in the Indian sub-continent. However, the prediction of good harvests before harvesting allows both farmers and government officials to take adequate steps to market and store the crops. Few approaches for predicting and modelling crop yields have been developed but do not consider climate characteristics and are observational [1-3].

Agriculture is known as one of the significant sources of a country's GDP. Many developing and underdeveloped countries are dependent on agriculture to increase their economic wealth. In this modern technological age, technology will play a significant role in the agricultural field. Advanced technology can automate several agricultural phases, including drainage, fertilization, harvesting and

more. We are using smart sensors to detect water levels to make the cultivation phases smarter, photo sensors to ensure ample sunlight for plant growth, sensors to detect nitrogen content, and warn the farmer to take steps towards proper fertilization [4-5]. Several experiments are being conducted in this area and even more, are now advancing in the laboratories. We analysed the various standard IoT techniques used in the agriculture sector based on hardware and software. We thus extracted the current challenges to make farming much smarter and more efficient.
The Indian economy is solely dependent on agriculture, where the agriculture industry contributes 17% of the total GDP and a source of employment to over 60% of the population [25–31]. Thus, one of the sole motives is the need to meet food demands for the growing population. With the intervention of new technologies like the Internet of Things, smart farms' tomorrow is more of a reality now. The implementation of new technologies affects a positive yield. Creating IoT advancements can assist with gathering a lot of environmental and harvest presentation information. "IoT includes numerous new clever ideas for utilizing right away, for example, brilliant home, savvy city, shrewd transportation, and keen cultivation." The excellent use of sensors for monitoring. The real-time data analysis helps in predicting growth and crop lifecycle [6-9].
This paper presents a water system observing and controlling the water system framework utilizing IoT. To store data in the cloud server via an application that will send notifications on the farmer's mobile.

2. Literature review

However, in an age of advanced technology and intelligent systems, the Internet of Things is extremely popular. The Internet of Things is a rapidly growing field that has the potential to revolutionize nearly every industry and discipline, including agriculture, health, home robotics, aviation and transportation, defence and military applications, and much more. The Internet of Things (IoT) is the networking of physical machines, cars, homes, and other objects via hardware, applications, sensors, actuators, and network communication [10-11].
It enables the collection and exchange of data between these objects; the Internet of Things (abbreviated IoT) is a systematic networking mechanism. The Internet of Things can be used to make objects or products smarter by sensing and controlling them automatically. The Internet of Things is referred to by a variety of names and is closely related to the Internet [12-13]. The Internet of Things (IoT) is a term that refers to a device that consists of multiple physical objects and sensors that are connected to or hooked up to the Internet via wireless or wired Internet access. The Internet of Things (IoT) is critical in a variety of industries, including agriculture, aviation, tourism, and health care. Our research developed and refined the IoT-based framework for agriculture and smart agriculture.
According to the plant nutrition handbook, a plant requires 17 essential nutrients, including carbon, hydrogen, and oxygen. Several accessories are obtained through air and water, while others are obtained through dirt and supplement arrangements. Each plant grows optimally under a variety of supplement conditions, which are converted to a fluid solution for transportation convenience. Different plants have multiple seasons, distinct sexes, and levels of profundity. The current framework is designed with a General Jamming Gripper that enables a mass of granular material encased in a flexible layer to wrap a material that enables the system to grasp and deliver a wide variety of objects. A comparative instrument exists that makes use of all family unit materials available on a Do-It-Yourself website [14-19].

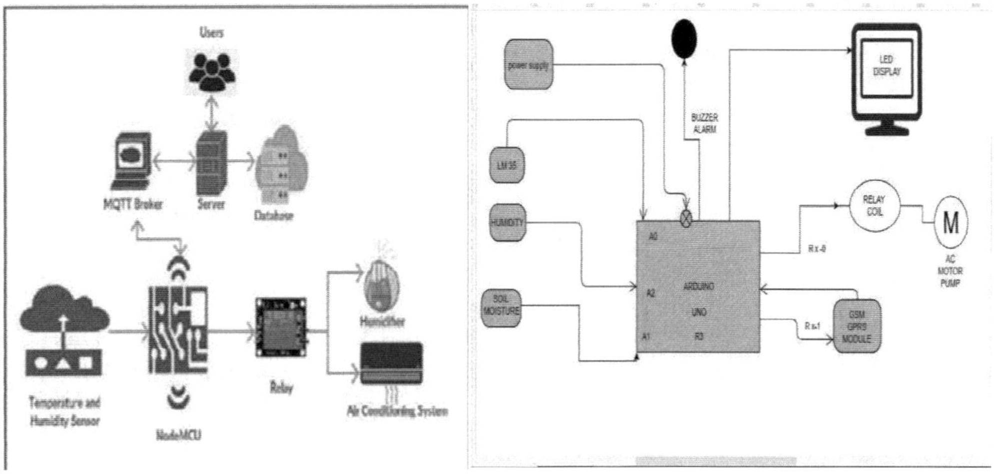

Figure 1.a & b *Design of a completely mechanized atmosphere controlled aqua-farming ranch.*

3. Proposed Architecture

With the in-depth study of making various models, the work is based on the area of focus of low economy and more viable accessibility. proposed methodology encourages clients to improve the quality and proportion of their ranch yield by distinguishing incorporating temperature, and persistence needs to be concerned with their soil sponginess worth and water level of the tank from the field with no human intervening. The proposed system overviews contain the pictorial representation of remote sensor units denoted as node1 and node2 as demonstrated up in fig.1 put in the farmland to get the continuous updates. Admission community highlight gets and sends captured and analyzed data to the control locale and a control piece that controls the streams for the watering subsystem. The water framework structure is devised to give water frame appropriateness, which will allow saving water and improve the yield quality. A wise water structure system can revive crop quality and yield to perceive limits like soil sogginess, air temperature, lightness, and water level of the tank. It uses up of the water model with its proficiency in computing the control variables and asks the not available sensor in the local area directly, whether at the start or terminating the watering process. The structure screens the tank's water level using a water level sensor so that if the water level is underneath the base, then the water framework will not be started.

3.1. Mobile System

When transferred to the mobile system for easy tracking and staying updated at the minimum, attention gives way to the cloud infrastructure. The android portable application is utilized to provide clients a literary and graphical portrayal of the data obtained by the remote sensor organization, which is put away in the cloud stage.

3.2. Additional data requirements

a. Data about plants - The observing framework gives the most recent data about the plant, where this data will be a spot to add client knowledge about the plant, unique bean stew plants.

b. Table of information examination -This element is the client can get the information from each sensor in detail. The data is spoken to as a table, which contains the sensor's estimations and furnished with the sensor information recovery time so ranchers can handle the entire information sensors. This table structure is a delegate for ranchers to see whether the plants are in acceptable condition.

c. Graphs of investigator information- The framework will screen explicitly; we configuration checking framework in investigator information graph. This framework will give data as information sensors are spoken to in graphical structure. It plans to give itemized data on their stew plants, where clients can see individually finished data, for example, pH, EC (Electrical Conductivity), Soil Dampness, Temperature, and Humidity. With this component, ranchers will get a diagram of data on their stew plants; an outline is a piece of important information.

d. Smart Sprinklers - The excellent state of a nursery can make the plants green and thrive with the short water supply and right supplements obtained from the dirt ripeness with the self-ruling nursery arrangement guide; some have electronic clocks. The brilliant sprinkler can be related to substantial usefulness, such as a coordinated sprinkler framework that makes some planned memories for sprinkling water on the plants that can be distantly controlled on your gadget in your non-appearance. The sprinkler regulator can hold 8000 gallons of water over the year. The implicit advantage is the wise sprinkler regulator's interoperability with different frameworks. For example, smoke cautions and cameras that use climate forecasts and nursery-determined data create no pretty much water needed by the plant at any second.

e. Climate Stations - The climate stations, in some cases called sensors, give more knowledge into the nursery. The accommodating input is inferred by checking examination; for example, temperature, water, and wind guarantee that the plants fill through the greenhouse.

f. Smart Soil Sensors -Water is key to any plant's appropriate development as the correct soil is fundamental. To learn the right level for both the dampness and manure, it is past the simple visual perception that will require sensitive sensors to investigate information across the measurements that identify when plants need watering and confirm whether the temperature causes hindered advancement.

g. Mechanical Lawn Mowers - In our nursery, exercises that can be time-devouring require energy and can be generally monotonous; adding a mechanical touch to the lawnmowers can be a gigantic forward leap the hands-off experience as an automated vacuum.

4. Conclusion

The central idea of fusing independent mechanical technology into farming is decreasing dependence on difficult work while expanding effectiveness, item yield, and quality. The improvement has killed the significant work factor, making the clients/ranchers invest more energy at work, investigating robot coding, breaking down information, and fixing apparatus. The difficulties presented by the conventional technique are the absence of oversight, exercise in futility, supporting as per the arrangement of water and legitimate the executives because of the dull day by day timetables of clients/ranchers which hampers the fast development of plants which bugs and weeds might take over. This investigation uncovers the different Smart Garden mechanization with an administration framework that can be actualized and effectively joined with ease. The Smart Farm has a firm spine of sensors and IoT, which depends on the capacity to speak with one another and with the client/rancher as it works self-governing. The Smart Garden computerization framework brings the client closer to their ideal plants. It brings recommendations and improvement with techniques being executed and guidelines are given to the computerization machine distantly or modified per task.

5. Acknowledgement

The authors would like to acknowledge the support and facility provided by the *K. R. Mangalam University Gurgaon* and *Amity University Noida* for casting the samples and proving the testing equipment's.

References

[1] T. Qiu, J. Liu, W. Si, M. Han, H. Ning and M. Atiquzzaman, "A Data-Driven Robustness Algorithm for the Internet of Things in Smart Cities," in IEEE Communications Magazine, vol:55, no:12, p. 18-23, DECEMBER 2017.
[2] A. Bahga and V. MAdisetti, Internet of Things: A hands on approach, Universities Press, First edition, ISBN: 97888173719547.
[3] Y. Erlich, "A vision for ubiquitous sequencing," Genome Research, ISSN 1088-9051, 2015, p. 1411–1416.
[4] L. Dan, C. Xin, H. Chongwei and J. Liangliang, "Intelligent Agriculture Greenhouse Environment Monitoring System Based on IOT Technology," International Conference on Intelligent Transportation, Big Data and Smart City, Halong Bay, 2015.
[5] "Zigbee," Digi, Available at: https://www.digi.com/resources/standards-and-technologies/rfmodems/zigbee-wireless-standard. [Accessed 2017 March 20].
[6] V. V. h. Ram, H. Vishal, S. Dhanalakshmi and P. M. Vidya, "Regulation of water in agriculture field using Internet of Things," IEEE Technological Innovation in ICT for Agriculture and Rural Development (TIAR), Chennai, 2015.
[7] J. Shenoy and Y. Pingle, "IOT in agriculture," 3rd International Conference on Computing for Sustainable Global Development (INDIACom), New Delhi, 2016.
[8] "Polyhouse farming advantages and benefits", Available at: http://www.agrifarming.in/polyhouse-farming-profits/. [Accessed 3 March 2017].
[9] "Soil pH," Wikipedia, Available at: https://en.wikipedia.org/wiki/Soil_pH.
[10] A. Abdullah, S. A. Enazi and I. Damaj, "AgriSys: A smart and ubiquitous controlled-environment agriculture system," 3rd MEC International Conference on Big Data and Smart City (ICBDSC), Muscat, 2016.
[11] "Multiple Input Multiple Output (MIMO)," Radio Electronics.com, [Online]. Available: http://www.radio-electronics.com/info/antennas/mimo/multiple-input-multiple-output-technology-tutorial.php. [Accessed 26 March 2017].
[12] N. Gondchawar and R. S. Kawitkar, "Smart Agriculture Using IoT and WSN based Modern technologies," International Journal of Innovative Research in Computer and Communication Engineering, vol:4, no:6, 2016.
[13] "Raspberry PI", Available at: https://www.raspberrypi.org/.
[14] S. K. Nagothu, "Weather based smart watering system using soil sensor and GSM," 2016 World Conference on Futuristic Trends in Research and Innovation for Social Welfare (Startup Conclave), Coimbatore, 2016.
[15] N. Agrawal and S. Singhal, "Smart drip irrigation system using raspberry pi and arduino," International Conference on Computing, Communication & Automation, Noida, 2015.
[16] "Arduino," Arduino, Available at: https://www.arduino.cc/. [Accessed 24 March 2017].
[17] N. Putjaika, S. Phusae, A. Chen-Im, P. Phunchongharn and K. Akkarajitsakul, "A control system in an intelligent farming by using arduino technology," Fifth ICT International Student Project Conference (ICT-ISPC), Nakhon Pathom, 2016.
[18] S. Khashirunnisa, B. K. Chand and B. L. Kumari, "Performance analysis of Kalman filter, fuzzy Kalman filter and wind driven optimized Kalman filter for tracking applications," 2nd International Conference on Communication Control and Intelligent Systems, Mathura India, 2016 .
[19] A. Kapoor, S. I. Bhat, S. Shidnal and A. Mehra, "Implementation of IoT (Internet of Things) and Image processing in smart agriculture," International Conference on Computation System and Information Technology for Sustainable Solutions (CSITSS), Bangalore, 2016.
[20] C. A. et. al, "Wireless sensing and control for precision Green house management," Sixth International Conference on Sensing Technology (ICST), Kolkata, 2012.
[21] P. M. Jacob, Muhammed Ilyas H, J. Jose and J. Jose, "An Analytical approach on DFD to UML model transformation techniques," 2016 International Conference on Information Science (ICIS), Kochi, 2016, pp. 12-17.
[22] Mary Shaw, David Garlan, 'Software Architecture: Perspectives on an Emerging Discipline', Prentice Hall, 1996.

[23] Y. Ducq and D. Chen, 'How to measure interoperability: Concept and approach,' 2008 IEEE International Technology Management Conference (ICE), Lisbon, 2008, pp. 1-8.

[24] P. M. Jacob and M. Prasanna, "A Comparative analysis on Black box testing strategies," 2016 International Conference on Information Science (ICIS), Kochi, 2016, pp. 1-6.

[25] Dixit S, Sharma K and Singh S 2020 Identifying and Analysing Key Factors Associated with Risks in Construction Projects Emerging Trends in Civil Engineering, https://doi.org/10.1007/978-981-15-1404-3_3 vol 61 (Springer, Singapore) pp 25–32

[26] Dixit S, Stefańska A and Musiuk A 2021 Architectural form finding in arboreal supporting structure optimisation Ain Shams Eng. J. 12 2321–9

[27] Shah M N, Dixit S, Kumar R, Jain R and Anand K 2021 Causes of delays in slum reconstruction projects in India Int. J. Constr. Manag. 21 452–67

[28] Dixit S, Stefańska A, Musiuk A and Singh P 2021 Study of enabling factors affecting the adoption of ICT in the Indian built environment sector Ain Shams Eng. J. 12 2313–9

[29] Dixit S 2021 Impact of management practices on construction productivity in Indian building construction projects: an empirical study Organ. Technol. Manag. Constr. 13 2383–90

[30] Dixit S 2020 Study of factors affecting the performance of construction projects in AEC industry Organ. Technol. Manag. Constr. an Int. J. 12 2275–82

[31] Dixit S and Saurabh K 2019 Impact of Construction Productivity Attributes Over Construction Project Performance in Indian Construction Projects Period. Polytech. Archit. 50 89–96

Response strategies used to mitigate the effect of extreme weather on rural and remote housing in Australia

Christopher A. Jensen[a] **Jessamie Yule**[a]

University of Melbourne, Parkville, Victoria, Australia

cjensen@unimelb.edu.au

Abstract. In the extreme climates of Australia, rural and remote locations have special site and dwelling requirements and therefore special design solutions. This can include construction constraints, bushfire risk, resource efficiency, comfort and respect for the surrounding environment and heritage. Existing regulations targeted at extreme weather impacts such as bushfire, flood, cyclones, and heatwaves often impact the design and material choices for rural and remote housing due to their exposed locations, including housing proposed for rebuild following damage resulting from an extreme weather event. These regulations also lead to increased costs and uncertainty about the suitability of rural land for construction, making it challenging for rebuilding communities in affected areas. Such well-intentioned regulations also create limitations on innovation through experimental / bespoke building design, such as; 1) novel solutions to extreme weather mitigation and resistance 2) material choices for construction; and 3) achievement of operational and embodied energy reductions.

Experimental buildings provide significant innovation benefits to industry as is often seen in the development of sustainable and high-performance buildings. This research examines the academic and industry knowledge of current design regulations for extreme weather events and implications for experimental and innovative design, highlighting the challenges for buildings to achieve increasing standards of environmental performance whilst ensuring resilience in the face of increasing extreme weather events, with a specific focus on rural communities. Case study analysis is used to provide an understanding of the focus and strategies used by different groups in different locations to address the impacts of extreme weather.

Keywords: Construction innovation; resilience; innovation; sustainability; emissions.

1. Introduction

Extreme weather events are part of a long-term trend of worsening extreme weather since the 1980s, both globally and in Australia. The last decade has presented the very worst of extremes as a result of human-induced climate change. Over the past several decades, heatwaves in Australia have increased in duration, frequency and intensity in many parts of the country; southern Australia has experienced a rainfall decline in the cooler months, heavy rainfall events have accounted for an increased proportion of total annual rainfall, and extreme fire weather days have increased at 24 out of 38 sites [1].

There is no simple definition of an extreme weather event, noting that some weather events can be classified separately as severe events, rare events, extreme events, or high impact events [2]. For the purposes of this research, we adopt the IPCC (2012) definition that extreme weather is "an occurrence of a value of a weather or climate variable beyond a threshold that lies near the end of the range of observations for the variable" [3]. Examples include very high / low temperatures, uncharacteristically heavy precipitation, and very high wind speeds. Extreme events usually occur only rarely, and they are noticeable because they are so different from the usual weather and climate, and because they are associated with adverse impacts on humans, infrastructure and ecosystems [4]. The spatio-temporal scale is a primary determinant of extreme weather events, defined by geographic location, and local geography. Seasonal variation amongst other factors [5] is an important determinant of extreme weather events. Despite geographic isolation, Australia is made up of many climates, but in consideration of extreme weather can be broadly separated into the tropical climate characterised by cyclone and flood, and the sub-tropical or temperate zones characterised by bushfire and heatwaves. Cyclone and bushfire are the deadliest in terms of lives lost, and along with flood represent the most significant economic impact. Heatwaves are not usually associated with the word 'disaster' but have a dramatic effect on health and wellbeing.

In Australia, regulations targeted at extreme weather impacts such as bushfire, flood, cyclones, and heatwaves have predominantly impacted the design (including structural integrity) and material choices for rural and remote housing due to their exposed locations, including new build or rebuild following damage resulting from an extreme weather event. However, such well-intentioned regulations create limitations on; 1) site-based design response and amenity, 2) material choices for construction; and 3) embodied energy. These regulations also lead to increased costs and uncertainty about the suitability of rural land for construction, making it challenging for rebuilding communities in affected areas. Rural and remote housing have considerable additional demands over urban residential projects as they are characterised by challenging sites, difficulty with delivery due to location, additional costs (such as transport and onsite labour) and a small population with limited trade workers. put greater demand on air conditioning and occupant comfort.

Despite extensive work on preparedness for extreme weather by governments, solutions by industry and research on impacts ex-post vary greatly according to the spatio-temporal variances by a number of different groups at different levels. there is no research that assembles the understanding of the different strategies in use by designers to mitigate against such disasters. Experimental buildings are often ahead of the regulations in the building industry, and this research captures this knowledge for the benefit of future building designs

2. Background and Literature Review

The literature extensively covers the issues related to the impact of climate change on buildings [6] often considered inline with IPCC predictions for temperature and sea level rise. Climatic drivers of extreme weather are less well understood in terms of predictability [7] but it is generally accepted that extreme weather events will continue to increase. Macro climate change research is concerned with the appropriateness of the weather files and accuracy of building simulation, overheating risk and adaptation strategies and resilience.

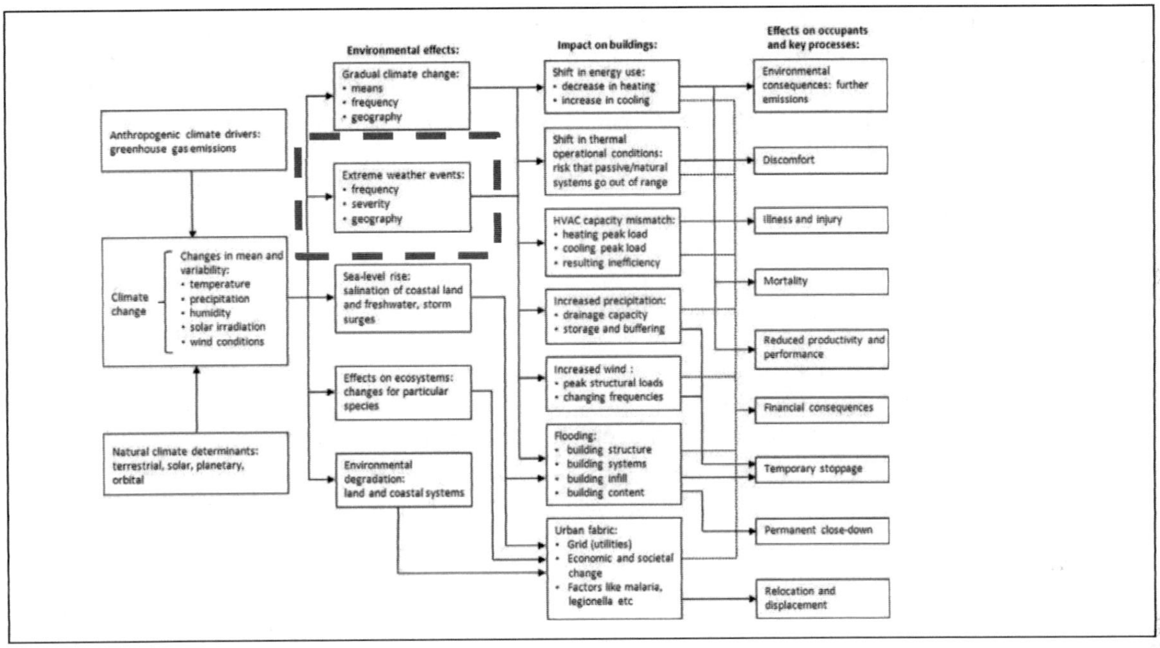

Figure 1: Schematic overview of climate change impact on buildings showing impacts of extreme weather events (highlighted) [8].

Australian research on extreme weather impact on buildings and appropriate solutions are understandably location specific, and is focused on event specific conditions and solutions such as:

- cyclones, rebuilding, resilience
- flood: planning, structures and economic impacts
- bushfire: land use and assessing risks
- overheating: design considerations, health and wellbeing

Below is a summary of the extreme weather events in each type of impact:

2.1. Cyclone

In 1974 Cyclone Tracy caused 65 deaths and damaged 70 percent of Darwin homes. More recently, cyclones Vance (1999), Larry (2006) and Yasi (2011) showed that updated regulations and standards have resulted in much less building damage and consequent loss of life [9]. During Cyclone Yasi, for example, 12 per cent of older homes suffered severe roof damage, but only three per cent of newer homes [10]. Cyclones are categorised on a scale of 1 - 5, represented by wind speeds of 125km/h up to 280 km/h.

2.2. Flood

Often experienced in tandem with cyclones, both inland and coastal flooding present their own unique challenges to the built environment. Inland flooding is associated with heavy rainfall and vulnerability to the failure of infrastructure such as dams and local catchments [4]. Coastal flooding, as well as contributing to inland 'storm surge' flooding, presents a sustained challenge for those living within low-lying coastal areas in proximity to rising sea levels. In a built environment designed and built for the climate of the 20th century, Australian housing is "vulnerable and unprepared for the changes in climate that are likely to increase the intensity of coastal storms" [12]. There are many coastal flooding forecasting models being used to predict the range of impact on housing and infrastructure [13], and all forecast that in the lower range, tens of thousands of homes on Queensland's shoreline are at high risk of coastal flooding from sea level rise. At the higher end of this forecasting, the impact scales to hundreds of thousands of residential buildings exposed to the combined impact of inundation and shoreline recession [14]. This is of particular concern considering "most insurance policies do not cover actions of the sea. Insurers do not cover the future impact of climate change" [15].

2.3. Bushfire

Summer in Australia is synonymous with devastating bushfires, with the 2019 / 2020 summer being one of the worst seasons on record. By March 2020 Black Summer fires burnt almost 19 million hectares, destroyed over 3,000 houses, and killed 33 people [16]. The future looks set for more dramatic events, as stated by the CSIRO and BOM after the devastating 19/20 fires, "There has been an increase in extreme fire weather, and in the length of the fire season, across large parts of Australia since the 1950s, especially in southern Australia and it is predicted that both the south and east of Australia will experience a longer fire season and increase in the number of dangerous fire weather days." [1].

2.4. Heatwave

Despite being a continent of extreme heat waves, there is currently no comprehensive requirement for overheating in Australian buildings, despite the extremes of the continent such as on 4 January 2020, where Penrith (New South Wales) was officially the hottest place on Earth at 48.9°C. [17]. In 2009 Victorian heatwave that began on the 27 January with daytime temperatures topping 43°C across 3 days, with night-time minimums of above 25°C [18]. Understandably the 2009 heatwaves coincided with catastrophic bushfires in Victoria. Although changes are progressively improving the National Construction Code (NCC) for energy efficiency, heat stress and thermal comfort is not explicitly included as a metric in the assessment of proposed designs, and the increase of building performance regulations has led to increased overheating, particularly in existing apartment buildings. This regulatory limitation is partially addressed by NCC 2019, which introduced a 'cap' on heating and cooling load per dwelling when calculating the energy performance. The Council enforced 'Better Apartment Design Standards' (BADS) introduced even more stringent cooling load limits, reflecting concern for overheating risks in apartments. Preceding the latest revision of NCC 2022, following a consultation process of 121 submissions on to updates of residential codes, an outcomes report was released by the ABCB identifying "An urgent need to respond to climate change, including protection from extreme heatwaves" [19] which demonstrates an understanding of the importance of the issue.

3. Case Study & Discussion

Case studies are appropriate to be used to test or generate theory and are particularly appropriate for areas where the research is still in its infancy, formative stages where there are no solid theoretical foundations. It is preferred when "how" or "why" questions are being posed [20]. According to [21], the case study research method represents an inductive approach that allows the in-depth study of individual cases (or innovations), which will provide a detailed understanding that would not be possible using more cross-sectional methods. It can include data from direct observation and systematic interviewing as well as from public and private archives. Any fact relevant to the stream of events describing the phenomenon is a potential datum in a case study, since context is important [22]. Of importance to this study, a critical case is used in order to generalise about innovations. The case study data is collected via content analysis [23].

This research analyses three groups responding to the issue to understand their strategies to mitigate the effect of extreme weather in Australian house design. This includes:

Innovators - designing solutions to achieve best practice. Innovators engage in rigorous testing, prototyping and research and development. They raise the benchmark for what is possible, creating the case for standards to move beyond minimum standards in an open conversation that engages the wider community. These innovative projects often connect to regulators through intermediaries such as insurers and government bodies.

Insurers - addressing risks related to housing damage and claims from extreme weather. Insurers are core to disaster recovery response and hold invaluable data and potential insight into both risk and recovery. They conduct their own their own research and development to advise and work directly with both regulatory bodies and innovators, directly contributing to the coordination of funding and mitigation efforts.

Regulators - addressing the appropriate minimum standards for extreme weather resilience. The regulatory system is a complex of policy makers and government bodies who are informed by a hierarchy of independent boards and advisory groups made up of state and local government representatives, technical experts, industry representatives, businesses, academics and community experts who are informed by research of innovators in these areas.

3.1. Case Study 1: Joost Bakker Fire Proof house, CSIRO

The Fireproof house was built by Joost Homes and was able to passed the worst-case CSIRO fire simulation of a major fire front. The unique design could withstand external temperatures of 1000°C, and maintain internally the air temperature only reach 40degC and internal surface temps reaching 55°C [24]. The design is also an exemplar of both innovative resilient combined with the rarer elements of sustainable design (green roof), non-toxic and recyclable materials, and challenges traditional construction methods and materials, using straw bale insulation set into a recyclable steel frame. Specifically for the project, Joost initiated the development and production of a fire-resistant, formaldehyde free plywood and charcoal embedded Magnesium Oxide (MgO) board that has 1/10 the carbon footprint of fibre cement sheet [25]. Embracing biomimicry principles allowed the plywood from sustainably harvested forests to use soy-based adhesives replacing formaldehyde as the glue that bonds the layers together. Dr. Li, a professor at Oregon State University's College of Forestry, found that mussels secrete proteins known as byssal threads provide superior strength and extraordinary flexibility and discovered that soy proteins can be modified to perform similarly to byssal threads. And not only did they deliver phenomenal adhesion, they also offered exceptional water resistance. This breakthrough led to 'PureBond', proving that enhanced environmental quality and increased product performance can go hand in hand" [26]. The Fireproof house demonstrated that an innovation need created by extreme weather could lead to innovative solutions.

3.2. Case Study 2: ONE House to Save Many -, CSIRO, Suncorp, James Cook University, Room 11 Architects

'One House' is a home specifically designed to protect against floods, fires, and cyclones in order to compare innovative strategies against a standard family home [27]. The experimental test facility allowed rigorous testing, including:

- Wind-driven debris and a roof section pressure demonstration simulating the effects of cyclonic wind
- A water ingress test to understand the vulnerable aspects of standard windows and doors in the event of wind driven rain and inundation
- A bushfire resilience test on a One House prototype measuring the impacts at different burnover levels using specialised simulated bushfire flame fronts.

The project collaboration responded to research that showed that homeowners prioritise aesthetics over the strength and protection of the home, and developed a number of key recommendations including, high performance mesh screens, dual water tanks for firefighting, electrical safety solutions for flooding, as well as design recommendations to provide strength and minimize risk. The project demonstrated best practice strategies, noting that "there's an assumption that current building codes will fully protect you, and whilst they do protect life they don't necessarily support building resilience into a building, and depending on when you built your home and how the code responded at that point in time." [28].

3.3. Case Study 3: Insurance industry

Insurance companies have a responsibility to minimise harm to their clients and to provide financial returns. This has led to the identification of the need to assist homeowners before extreme weather impacts occur to minimise the payout, noting that in 2019/2020 the combination of all extreme weather related disaster cost $8.8 billion to the Australian GDP, and noted that without insurance some communities will never recover from disaster - This is especially the case in regional Australia, where there is increased risk of disasters and communities have a narrower economic base [29] (SGS Economics and Planning Pty Ltd 2020). As illustrated in the "One House" case study, insurance companies are already preemptively funding innovative designs to mitigate against extreme weather impacts that cause damage to buildings and properties.

Australia's national insurance body, the Insurance council of Australia (ICA) has identified that while "at present no region in Australia is uninsurable… it is possible some regions may become difficult to insure in the future unless governments invest in appropriate physical mitigation and adaptation strategies. Implementation of stronger building codes, improved land-use planning and permanent physical mitigation measures, where necessary, will be key to ensuring an 'Insurable Australia'." [15]. Tasked to raise the standards, the ICA is teaming with both innovators and regulators to move ahead of the next revision period for Australia's building code revision, NCC 2022, on projects which will raise the bar for new developments and promoting more stringent standards for resilient design.

Teaming with an independent authority of sustainable buildings, the Insurance Council of Australia (ICA) and the Green Building Council of Australia (GBCA), created a new rating system for residential homes which goes beyond minimum standards to withstand natural disasters such as bushfires, flooding, and heat stress [30]. The worry of rising insurance premiums has also prompted the development of climate-ready home specification a locality of South Australia. The 'Where we

Build, What we Build' research project found that retrofitting housing stock across six regional local government areas in South Australia would translate into a net present value of $72 million. This research was conducted through heat and fire hazard mapping, comparing insurance industry data with Council data to assess the resilience of homes to natural hazards and risk exposure of specific housing archetypes. The resulting hone specifications include detail materials, construction, planning climate hazard mapping information sheets that residents, organisations and local government can used for improving the climate resilience of new build and retrofit homes [31].

3.4. Case Study 4: Regulator perspective

Australian building standards have seen a now decades-long discussion amongst regulatory groups, discussing the pathway to introducing climate resilience [32, 33, 34, 35, 36]. The National Construction Code (NCC, or the Code) is a model code directed by two entities, the Building Ministers Forum and the Council of Australian Governments (COAG). The NCC is represented by the Building Code of Australia (BCA) – NCC Volumes 1 and 2; and Plumbing Code of Australia (PCA) – NCC Volume 3. These provide standards for extreme weather impacts related to settlements and infrastructure, emergency planning and response, insurance, and human health. Residential buildings and structures codes come under the domain of an Australian Building Codes Board (ABCB) which oversees the standards administration and development.

In 2019, as part of The Global Resiliency Dialogue, the ABCB participated in a global survey with government and non-governmental stakeholders from eight countries to assess how climate-based risks are considered within their national building codes and standards. The surveys concluded that none of the participating countries, including Australia, addressed future risk [37] So, while the Australian Standards are adequate to preserve life, they do not address resilience. Australia's building standards relating to effects of extreme weather are produced by an independent, non-governmental organisation, Standards Australia, which utilises committees of technical experts, businesses, academics, government bodies and representatives and community experts to develop, assess and review changes.

The NCC provisions (2019) has numerous provisions related to the design for extreme weather impacts such as Specification B1.2, 'Design of buildings in cyclonic areas' which contains requirements for the design of buildings in cyclonic areas in addition to the requirements of Australian Standard AS/NZS 1170.2. This chain of command illustrates how the NCC interprets standards as regulation for the building industry, each of which is independently adopted by States and Territories. Since 2014 the ABCB has worked towards addressing a number of impacts of extreme weather events including to:

- Comprehensively review and consider the impacts of extreme weather events in relation to all relevant new regulatory initiatives;
- Investigate climate related natural hazard events as they occur to determine whether the NCC scope and provisions are adequate;
- Liaise with the planning, building, industry and insurance sectors, to improve the relationship between building and planning and to explore ways of better defining responsibilities, especially in high hazard areas.

The case study analysis demonstrates considerable efforts by all groups but with different strategies to mitigate impacts form extreme weather for rural housing, with a focused effort by the innovators on practical solutions, a community level effort by insurers who have a vested interest in reducing the scale of damage to houses from extreme weather events, and by the regulators attempting to define the requirements of new or rebuild construction projects to save lives and minimise damage. It is reassuring to see that in most cases, even if at the high level, the groups with mitigation strategies are communicating in order to achieve the common goal.

4. Conclusion

New housing projects always exist somewhere between the minimum standards of regulation and the best practice innovation of designers, suppliers and builders. At present the regulatory aspects of managing extreme weather appear to be lagging due to previously identified challenges, the need for robust consistent data, government/community support, and affordability. Much like sustainability initiatives that are a response to climate change, extreme weather responses are required to achieve a balancing act that is part way between the efforts of the innovators, insurers and regulations.

5. References

[1] BoM and CSIRO (2020) State of the Climate 2020.
[2] Stephenson, D. B., Diaz, H., Murnane, R. J. C. e., & society. (2008). Definition, diagnosis, and origin of extreme weather and climate events. *340*, 11-23.
[3] Field, C. B., Barros, V., Stocker, T. F., & Dahe, Q. (Eds.). (2012). Managing the risks of extreme events and disasters to advance climate change adaptation: special report of the intergovernmental panel on climate change. Cambridge University Press.
[4] Steffen, W., Hughes, L. and Karoly, D. (2013) The Critical Decade: Extreme Weather.
[5] Cattiaux, J., & Ribes, A. J. B. o. t. A. M. S. (2018). Defining single extreme weather events in a climate perspective. *99*(8), 1557-1568.
[6] De Wilde, P. (2014) The gap between predicted and measured energy performance of buildings: A framework for investigation. In: 41. 2014 40–49.
[7] Kornhuber, K., Osprey, S., Coumou, D., Petri, S., Petoukhov, V., Rahmstorf, S. & Gray, L. (2019). Extreme weather events in early summer 2018 connected by a recurrent hemispheric wave-7 pattern. Proceedings of Environmental Research Letters, 2019.
[8] De Wilde, P., & Coley, D. (2012). The implications of a changing climate for buildings.
[9] ACCC (2018) Northern Australia Insurance Inquiry First Interim Report. Australian Competition and Consumer Commission.
[10] CSIRO (2020) Climate and Disaster: Technical Reports Available from https://www.csiro.au/en/research/natural-disasters/bushfires/report-climate-disaster-resilience [accessed 25 April 2021].
[11] Steffen, W. and Alexander, D. (2016) Super-Charged Storms in Australia: The Influence of Climate Change. Climate Council of Australia Ltd.
[12] Steffen, W., Hunter, J. and Hughes, L. (2014) Counting the Costs: Climate Change and Coastal Flooding. Climate Council of Australia Ltd.
[13] Department of Climate Change and Energy Efficiency (2011b) Climate Change Risks to Coastal Buildings and Infrastructure. In: 2011 Commonwealth of Australia.
[14] Insurance Council of Australia (2021). Flood insurance explained. Insurance Council of Australia. https://insurancecouncil.com.au/resource/flood-insurance-explained/ [2021-06-01]
[15] Filkov, A.I., Ngo, T., Matthews, S., Telfer, S. and Penman, T.D. (2020) Impact of Australia's catastrophic 2019/20 bushfire season on communities and environment. Retrospective analysis and current trends. In: 1. 2020 44–56.
[16] BoM (2020) Special Climate Statement 73—extreme heat and fire weather in December 2019 and January 2020. Commonwealth of Australia.
[17] Jensen (2020). Impact of curtainwall facades on apartment overheating, The 54th International Conference of the Architectural Science Association (ANZAScA) 2020
[18] Australian Building Codes Board (2019) Energy efficiency: NCC 20022 and Beyond - Outcomes Report. Commonwealth of Australia.
[19] Yin, R. K. (1994). Discovering the future of the case study. Method in evaluation research. Evaluation practice, 15(3), 283-290.
[20] Eisenhardt, K. M. (1989). Building theories from case study research. Academy of Management Review, 14(4), 532-550.
[21] Leonard-Barton, D. (1990). A dual methodology for case studies: Synergistic use of a longitudinal single site with replicated multiple sites. Organization science, 1(3), 248-266.

[22] Bell, E., & Bryman, A. (2007). The ethics of management research: an exploratory content analysis. British Journal of Management, 18(1), 63-77.
[23] Leonard, J., Webb, A., Macindoes, L. and Brown, S. (2012) Bushfire Burnover Assessment of the By Joost Greenhouse : February 2012. CSIRO Materials Science and Engineering.
[24] Built by Joost Brochure (n.d.).
[25] Edgar, R. (2014) Joost Bakker house designed to resist bushfires Available from https://www.smh.com.au/entertainment/joost-bakker-house-designed-to-resist-bushfires-20141215-127j7q.html [accessed 25 April 2021].
[26] One House to Save Many | Suncorp Insurance (n.d.) Available from https://onehouse.suncorp.com.au/ [accessed 1 June 2021].
[27] The Glue Society (n.d.) One House to Save Many Documentary.
[28] SGS Economics and Planning Pty Ltd (2020). Economic recovery after disaster strikes - volume two: When communities face flood, fire and hail
[29] Green Building Council of Australia (2020). Green Star for Homes: A strategy for the future. Green Building Council of Australia.
[30] Siebentritt, M. and Auld, M. (2020) What We Build, Where We Build Project Report. Edge Environment.
[31] Australian Building Codes Board (2010) An Investigation of Possible Building Code of Australia (BCA) Adaptation Measures for Climate Change. Canberra: .
[32] Department of Climate Change and Energy Efficiency (2011a) Barriers to Effective Climate Change Adaptation: A Submission to the Productivity Commission.
[33] Productivity Commission (2012) Barriers to Effective Climate Change Adaptation: Productivity Commission Inquiry Report. Canberra: Commonwealth of Australia.
[34] Commonwealth of Australia (2015). National Climate Resilience and Adaptation Strategy. Commonwealth of Australia.
[35] Department of Climate Change and Energy Efficiency (2013) Australian Government response to the Productivity Commission report: Barriers to Effective Climate Change Adaptation. Commonwealth of Australia. Australian Building Codes Board (2014) Resilience of buildings to extreme weather events.
[36] Australian Building Codes Board (2021) The Use of Climate Data and Assessment of Extreme Weather Event Risks in Building Codes Around the World: Survey Findings from the Global Resiliency Dialogue. International Code Council.

SUBCONTRACTORS' HEALTH AND SAFETY COMPLIANCE ON A WATER BOARD'S PROJECTS

L Moshanyana and J Smallwood

Nelson Mandela University, Port Elizabeth, South Africa, lmoshaya@randwater.co.za john.smallwood@mandela.ac.za

Abstract. Compared to other industrial sectors, the construction industry is responsible for a disproportionate number of fatalities and injuries. Subcontracting constitutes a significant portion of all construction projects, and it is noted for its adverse H&S influence in the industry. Health and safety (H&S) risks related to subcontracting pose a significant business risk to principal contractors (PCs) who employ subcontractors (SCs). Some PCs report that their sites have been shut down because of poor H&S compliance by the SCs. This situation has a negative impact on time schedules, and the overall profitability of projects.

The study seeks to examine why construction SCs undertaking projects for a Water Board located in Johannesburg do not comply with H&S regulations. Quantitative methods were employed for the study. Literature was reviewed, followed by a questionnaire survey to obtain primary data. A total of 174 construction managers, supervisors, and H&S officers in the employ of PCs and SCs participated in the survey to gather data relative to SCs' degree of compliance with H&S legislation, H&S regulations, and the Water Board's H&S specifications.

The literature suggests that there is general H&S non-compliance on construction projects by SCs. The salient findings include: SCs on the Water Board's projects do not comply with H&S legislation and regulations, and H&S specifications; SCs' management is not trained in terms of H&S; PCs do not issue work stoppage orders relative to SCs' unsafe acts / behaviour; SCs further subcontract; the contractual relationships between PCs and SCs are not clear due to unauthorised subcontracting, and H&S enforcement by PCs and the Department of Employment and Labour (DEL) is inadequate.

Recommendations include: PCs must ensure that H&S practices are implemented; PCs and the DEL must enforce H&S requirements, and a framework must be evolved for ensuring SCs' H&S compliance on construction.must be evolved for ensuring SCs' H&S compliance in construction.

Keywords: compliance, health and safety, practices, principal contractor, subcontractor.

1. Introduction

The construction sector is notorious for being responsible for numerous occupational deaths, injuries, and illnesses in many countries [1]. Mbachu highlights that subcontracting constitutes a significant portion of all construction projects in the building industry [2]. Earlier research estimates that 80% of construction work undertaken by United Kingdom (UK) PCs is subcontracted [3]. Furthermore, subcontracting is prevalent in the South African construction industry, with up to 70% of the value of building and 30% of the value of civil construction projects subcontracted [4]. This shows the importance of subcontractors (SCs) in the construction industry.

Outsourcing has become an important tool by which large organisations seek to increase competitiveness, reduce costs, bypass regulatory controls, and secure more flexible employment arrangements [5]. However, despite its economic benefits, subcontracting has adverse H&S

consequences in construction and other industries, which is a situation that has long persisted. Some of the problems associated with subcontracting have been identified as communication problems, competence issues, especially in the case of multi-layer subcontracting, the interest of SCs in making their profit with less commitment to H&S, and the unfamiliarity of subcontractors with the H&S practices of PCs. It is anticipated that, given the increasing specialisation in the construction industry, the adverse impact of subcontracting could worsen with its increase [6].

On the other hand, according to the Preferential Procurement Policy Framework Act 2000 (PPPFA), Preferential Regulations, 2017 Clause 9(1) 'subcontracting as a condition of the tender' requires that if it is feasible to subcontract for a contract above R30 million, an organ of state must apply to subcontract to advance designated groups. Based on the information above, the practice of subcontracting suggests its inevitability. The emphasis is clearly on the need for concentrated efforts to address the adverse H&S effects of subcontracting to avoid not only losses, but also the liability that the PC maybe exposed to by SCs non-compliance.

Given the adverse H&S effects of subcontracting in the South African construction industry; the requirements of H&S legislation, and the PPPFA, a study was conducted amongst construction managers, supervisors, and H&S officers in the employ of PCs and SCs involved in the projects of a Water Board, the objectives being to determine the frequency at which:

- PC organisations undertake seven H&S practices;
- SC organisations undertake eight H&S practices;
- PC organisations undertake six contractual practices;
- PC organisations undertake five evaluation practices, and
- SC organisations undertake two actions relative to employment practices.

2. Review of the literature

2.1. Introduction

The literature review attempts to highlight the possible causes of SC non-compliance with H&S legislation and regulations on the Water Board's projects, and the H&S practices implemented by SCs nationwide.

2.2. Management Practices

Management practices are the backbone of any successful business [7]. According to Agumba et al. [8], H&S management practices constitute a vehicle to improve H&S performance outcomes. H&S practices have been developed and adopted by various countries in managing the adverse H&S effects of subcontracting. However, the extent of the adoption of such measures would need weighing against factors, such as the cost and time implications, the human resources required for their operation, and the characteristics of the projects for which the measures are intended [6]. The study conducted by the Construction Industry Development Board (cidb) reports that many SCs, mainly smaller and trade SCs' H&S management practices are poor [4].

2.3. H&S legislation

The Construction Design and Management (CDM) 2007 Regulations have been noted as providing mitigation against the issues responsible for the adverse H&S effects of subcontracting [6]. The authors examined specific provisions and features of the CDM 2007 critically (i.e., competence assessment; training and induction; coordination and co-operation, and clear delineation of duties for PCs), which might address causative factors of the adverse H&S effects of subcontracting. Agumba et al. [9] report that poor H&S performance in the South African construction industry drove H&S stakeholders to take H&S seriously, as legislation alone was inadequate to address the problems of managing workplace H&S. PCs have the responsibility to go beyond compliance with regulatory requirements to proactively develop and implement measures that will prevent adverse H&S outcomes derived from subcontracting [6].

Manu et al. [6] extrapolate that in the spirit of striving for H&S excellence, the identification of other measures or practices, which may complement the mitigation offered by the CDM 2007 provisions, is thus warranted. Furthermore, the identification of such measures or practices provides useful organisational learning opportunities to PCs who often sublet work packages.

Likewise, Horbury and Hope [10] suggest a generic framework for employers to manage PCs or SCs. Manu et al. [6], however, claim that the framework recommends measures and practices that are similar to the regulatory requirements of the CDM 2007. These include the need for training SC workers, work method statements, and consultation with SCs.

2.4. Six interconnected elements of Pérezgonzález's model

According to Manu et al. [6], as a means of addressing the issues presented by SCs, PCs implemented H&S measures as part of their overall H&S management systems.

The authors use the following six interconnected elements of Pérezgonzález's model [11], widely used as the basis for identifying management practices:

- Policy: a general statement and overall guiding principle regarding the H&S of an organisation;
- Organisation: this includes the roles, responsibilities, and provision of resources to control H&S issues effectively;
- Planning and implementing: this includes goal setting, and operating the system;
- Measuring performance: this includes monitoring implementation to ensure that the set targets are achieved;
- Review performance: this includes procedures to ensure that the organisations learn from experiences at the measuring stage to improve performance, and
- Auditing: this covers monitoring of the overall system to ensure that it is achieving continuous improvement.

2.5. The performance improvement report

The performance improvement report encourages established enterprises to introduce best practice systems for H&S, quality management, and environmental management to improve their performance [6]. In terms of the cidb Best Practice Contractor Recognition System, the Board has identified the following as best practices for PC enterprises in Grades 5 to 9: a) certification by SANS ISO 9001 Quality Management System, SANS ISO 14001 Environmental Management Systems and SANS OHSAS 18001 H&S Management System, or b) certification by SANS 1393 Construction Management System [12]. Thus, there are H&S practices that aim to improve the status of H&S in terms of subcontracting. The study evaluates their effectiveness.

2.6. PC measures stemming from UK regulatory requirements

Practices based on legislation and regulations in the UK include the following:

- employing competent SCs;
- the risk assessment conducted by SCs and submitted for approval;
- PCs providing SCs with induction and training, and
- PCs arranging meetings with SCs to discuss H&S issues [6].

2.7. PC in-house measures

Some PCs have in-house H&S measures that address specific needs relating to their environments and areas addressed. These include the following:

- restricting layers(tiers) of SCs;
- using regular SCs;
- H&S reward scheme for SCs, and
- competency assessment being done before SCs commence work [6].

Agumba, Pretorius and Haupt suggest that continuous monitoring and reviewing of H&S management practices are required to improve poor H&S performance, and enhance overall legal compliance [9].

3. Research

3.1. Research method and sample stratum

The researcher selected a quantitative research method for the research study. A questionnaire was developed and distributed to contractors who were sampled using probability sampling. The researcher randomly selected one hundred and seventy-five (175) PCs, and two hundred and three (203) SCs for the whole exercise. Ninety-four (94) responses were received from PCs and eighty (80) from SCs, which equates to response rates of 53.7% and 39.4%, respectively, and an overall response rate of 46.0%. One hundred and seventy-four (174) responses were included in the data analysis. The target group selected to complete the surveys were construction managers, construction supervisors, and H&S officers or representatives. A computerised statistical programme 'Microsoft Excel 2016' was used to process and analyse the primary data collected through questionnaires. Closed-ended questions with five-point Likert scales, which included an 'Unsure' response option, were used. The analysis of the data entailed the computation of descriptive statistics in the form of frequencies, and a measure of central tendency in the form of a mean score (MS).

3.2. Research findings

65.9% of the PC respondents had worked for their current employer for > 1 year, 44.6% for > 5 years, and 24.4% for > 10 years. The mean length of time respondents had worked for their current employer was 6.4 years. Therefore, the respondents can be deemed to have had exposure to their organisations, which contributes to the reliability of the findings.

76.7% of the SC respondents had worked for their current employer > 1 year, 49.2% for > 5 years, and 22.9% > 10 years. The mean length of time respondents had worked for their current employer was 6.8 years. Therefore, the respondents can be deemed to have had exposure to their organisations, which contributes to the reliability of the findings.

88.3% of the PC respondents had worked in construction > 1 year, 74.5% for > 5 years and 44.7% > 10 years. The mean length of time respondents had worked in construction was 14.1 years. This implies that the respondents can be deemed experienced. This contributes to the reliability of the findings.

75% of the SC respondents had worked in construction > 1 year, 43.7% for > 5 years, and 27.5% > 10 years. The mean length of time the respondents had worked in construction was 9.5 years. This implies that the respondents can be deemed experienced, which contributes to the reliability of the findings.

50% of the PC respondents were ≤ 40 years of age, and 50% were > 40 years. Per definition relative to workers, people > 40 years of age are 'older workers'. The mean age was 38.8 years. 18% of the PCs respondents were female, and 76% were male. In terms of qualifications, 38.2% of the PC respondents had qualifications higher than a National Diploma. Therefore, the respondents can be deemed experienced and educated, which contributes to the reliability of the findings.

51.3% of the SC respondents were ≤ 40 years of age, and 48.7% were > 40 years i.e. 'older workers. The mean age was 37.8 years. 17.7% of the SCs respondents were female, and 82.5% were male. In terms of qualifications, 27.2% of SC respondents had qualifications higher than a National Diploma. Therefore, the respondents can be deemed as experienced, but not as well educated as the PC respondents, which to a degree, contributes to the reliability of the findings.

Table 1 indicates the frequency at which the PC respondents' organisations undertake seven H&S practices in terms of percentage responses to a scale of never to always, and a MS ranging from 1.00 to 5.00.

It is notable that all the MSs are above the midpoint score of 3.00, indicating that in general, the PC organisations can be deemed to undertake the H&S practices frequently, as opposed to infrequently. However, given that the MSs are all > 4.20 ≤ 5.00, the PC organisations can be deemed to undertake the H&S practices between often to always / always. It is notable that 'All SC employees receive H&S induction training before commencing employment by the PC' was ranked first, followed by 'SCs submit plans for all projects', and 'There is a budget for H&S at organisational level' is ranked last. It is notable that although the latter was ranked last, the MS is 4.51.

Table 1. The frequency at which the PC organisations undertake seven H&S practices.

H&S practice	Response (%) Unsure	Never	Rarely	Some times	Often	Always	MS	Rank
All SC employees receive H&S induction before commencing employment by us	1.9	0.0	0.0	0.0	5.8	92.3	4.94	1
SCs submit H&S plans for all projects	3.8	0.0	0.0	3.8	3.8	88.5	4.88	2
H&S performance is discussed with the SC employees by us	3.8	0.0	1.9	0.0	11.5	82.7	4.82	3
Our H&S policy is discussed with all SC employees	1.9	0.0	3.8	1.9	9.6	82.0	4.75	4
H&S is priced in the tender documentation during procurement by us	5.8	0.0	3.8	3.8	9.6	76.9	4.69	5
There is a budget for H&S at project level	7.7	0.0	0.0	5.8	21.2	65.4	4.65	6
There is a budget for H&S at the organisation level	5.8	0.0	3.8	7.7	19.2	63.5	4.51	7

Table 2 indicates the frequency at which the SC respondents' organisations undertake seven H&S practices in terms of percentage responses to a scale of never to always, and a MS ranging from 1.00 to 5.00.

It is notable that all the MSs are above the midpoint score of 3.00, indicating that in general, the SC organisations can be deemed to undertake the H&S practices frequently, as opposed to infrequently. However, given that the MSs are all $> 4.20 \leq 5.00$, the SC organisations can be deemed to undertake the H&S practices between often to always / always. Furthermore, it is notable that 'All SC's employees receive H&S induction training before commencing employment' was ranked first, followed by 'Daily H&S task instructions are communicated to all employees before they start with work', and 'The H&S policy is discussed with all employees' was ranked last. It is notable that although the latter is ranked last, the MS is 4.58.

Table 2. The frequency at which the SC organisations undertake eight H&S practices.

Factor H&S practice	Response (%) Unsure	Never	Rarely	Some times	Often	Always	MS	Rank
All employees receive H&S induction before commencing employment	1.9	0.0	0.0	0.0	0.0	99.1	5.00	1
Daily H&S task instructions are communicated to all employees before they start with work	0.0	0.0	0.0	0.0	3.8	96.2	4.96	2
An H&S plan is prepared for all projects	0.0	0.0	1.9	1.9	7.7	90.4	4.88	3
H&S is priced in the tender documentation during procurement	30.8	0.0	0.0	7.7	1.9	59.6	4.75	4
There is a budget for H&S at organisation level	28.8	0.0	0.0	5.8	11.5	53.8	4.68	5
There is a budget for H&S at project level	26.9	0.0	0.0	7.7	9.6	55.8	4.66	6
H&S performance is discussed with the employees	3.9	0.0	0.0	13.7	5.9	76.5	4.65	7
The H&S policy is discussed with all employees	0.0	0.0	0.0	11.5	19.2	69.2	4.58	8

Table 3 indicates the frequency at which the PC respondents' organisations undertake six contractual practices in terms of percentage responses to a scale of never to always and a MS ranging from 1.00 to 5.00.

It is notable that all the MSs are above the midpoint score of 3.00, indicating that in general, the PC organisations can be deemed to undertake the employment practices frequently, as opposed to infrequently. However, given that the MSs are all > 4.20 ≤ 5.00, the PC organisations can be deemed to undertake employment practices between often to always / always. Furthermore, it is notable that 'We verify if SC letters of good standing are valid' is ranked 1st, followed by 'We assess SCs H&S competencies before awarding contracts' 2nd. Although 'The SCs are appointed at the appropriate cidb grade level' is ranked last, it is notable that the MS is 4.48.

Table 3. The frequency at which the PC organisations undertake six contractual practices.

Contractual practice	Response (%)						MS	Rank
	Unsure	Never	Rarely	Some times	Often	Always		
We verify if SCs' letters of good standing are valid	0.0	0.0	1.9	0.0	7.7	90.4	4.87	1
We assess SCs' H&S competencies before awarding contracts	1.9	0.0	0.0	7.7	3.8	86.5	4.80	2
We undertake an H&S due diligence check on SCs before awarding a contract	5.9	3.9	3.9	3.9	3.9	82.4	4.75	3
There is a contract between the SCs and us	3.8	0.0	3.8	5.8	11.5	75.0	4.64	4
We appoint SCs registered with the cidb	11.5	0.0	0.0	13.5	15.4	59.6	4.52	5
The SCs are appointed at the appropriate cidb grade level	11.5	1.9	3.8	9.6	7.7	65.4	4.48	6

Table 4 indicates the frequency at which the PC respondents' organisations undertake five evaluation practices in terms of percentage responses to a scale of never to always, and a MS ranging from 1.00 to 5.00.

It is notable that all the MSs are above the midpoint score of 3.00, indicating that in general, the PC organisations can be deemed to undertake the evaluation practices frequently, as opposed to infrequently. However, given that the MSs are all > 4.20 ≤ 5.00, the PC respondents can be deemed to undertake the practices between often to always / always. Furthermore, it is notable that 'We assess SCs' H&S competencies before awarding contracts' is ranked 1st, followed by 'There is a contract between the SCs and ourselves' 2nd, and 'We appoint SCs registered with the cidb' is ranked last. It is notable that although the latter is ranked last, the MS is 4.27.

Table 4. The frequency at which the PC organisations undertake five evaluation practices.

Practice	Response (%)						MS	Rank
	Unsure	Never	Rarely	Some times	Often	Always		
We assess SCs' H&S competencies before awarding contracts	0.0	0.0	0.0	2.0	3.9	94.1	4.92	1
There is a contract between the SCs and us	0.0	0.0	0.0	1.9	5.8	92.3	4.90	2
We verify if SCs' letters of good standing are valid	15.4	1.9	3.8	1.9	5.8	71.2	4.66	3
The SCs are appointed at the appropriate cidb grade level	19.2	0.0	7.7	5.8	1.9	65.4	4.55	4
We appoint SCs registered with the cidb	13.5	0.0	7.7	13.5	13.5	51.9	4.27	5

Table 5 indicates the frequency at which the SC respondents' organisations experience or undertake two practices in terms of percentage responses to a scale of never to always, and a MS ranging from 1.00 to 5.00.

It is notable that both the MSs are above the midpoint score of 3.00, which indicates that in general, the SC organisations can be deemed to experience or undertake the practices frequently, as opposed to infrequently. However, 1 / 2 (50.0%) of the MSs is > 4.20 ≤ 5.00, indicating the SC organisations can be deemed to experience the practice between often to always / always. Given that the MS for 'We sub-subcontract' is > 3.40 ≤ 4.20, the practice can be deemed to occur sometimes to often / often.

Table 5. The frequency at which the SC organisations experience or undertake two practices relative to concluding contracts.

Practice	Response (%)						MS	Rank
	Unsure	Never	Rarely	Some times	Often	Always		
PCs assess our H&S competencies before concluding contracts	2.1	0.0	2.1	23.4	4.3	68.1	4.41	1
We sub-subcontract	2.1	0.0	4.2	39.6	10.4	43.8	3.96	2

4. Conclusions

The reliability of the results considered the mean length of time that PC and SC respondents had worked for their current employers, their mean experience in construction, and level of education. Based on the MSs obtained, both PC and SC respondents are deemed to have had exposure to their organisations, including being experienced. However, although 38.2% of the PC respondents had qualifications higher than a National Diploma, only 27.2% of SC respondents had the same.

Given that PC respondents' organisations undertake seven H&S practices, namely 'All SC employees receive H&S induction before commencing employment by us', 'SCs submit H&S plans for all projects', 'H&S performance is discussed with the SC employees by us', 'Our H&S policy is discussed with all SC employees', 'H&S is priced in the tender documentation during procurement by us', and 'There is a budget for H&S at project level' frequently, as opposed to infrequently, it can be concluded that the PCs manage their SCs' H&S effectively.

Given that SC respondents' organisations undertake seven H&S practices, namely 'All employees receive H&S induction before commencing employment', 'Daily H&S task instructions are communicated to all employees before they start with work', 'An H&S plan is prepared for all projects', 'H&S is priced in the tender documentation during procurement', 'There is a budget for H&S at organisation level', 'There is a budget for H&S at project level', 'H&S performance is discussed with the employees', and 'The H&S policy is discussed with all employees' frequently, as opposed to infrequently, it can be concluded that PCs influence SCs' H&S management commitment and practices, and that SCs are committed to H&S.

Given that PC respondents' organisations undertake six H&S contractual practices, namely 'We verify if SCs' letters of good standing are valid', 'We assess SCs' H&S competencies before awarding contracts', 'We undertake an H&S due diligence check on SCs before awarding a contract', 'The SCs are appointed at the appropriate cidb grade level', 'We appoint SCs registered with the cidb', and 'There is a contract between the SCs and us' frequently, as opposed to infrequently, it can be concluded that PCs integrate H&S into their procurement systems, and that H&S is a criterion for awarding contracts to SCs.

Given that PC respondents' organisations undertake five H&S evaluation practices namely 'We assess SCs' H&S competencies before awarding contracts', 'There is a contract between the SCs and us', 'We verify if SCs' letters of good standing are valid', 'The SCs are appointed at the appropriate cidb grade level', and 'We appoint SCs registered with the cidb' frequently, as opposed to infrequently, it can be concluded that PCs integrate H&S into their procurement systems, and that H&S is a criterion for awarding contracts to SCs.

Given that SC respondents' organisations experience / undertake two practices, namely 'PCs assess our H&S competencies before concluding contracts', and 'We sub-subcontract' frequently, as opposed to infrequently, it can be concluded that PCs integrate H&S into their procurement systems, that H&S is a criterion for awarding contracts to SCs, and multi-level subcontracting occurs.

5. Recommendations

Although the study revealed that poor H&S culture does not result in SCs not complying with H&S legislation and regulations, it is recommended that PCs continually implement H&S practices to improve H&S culture on projects. Furthermore, increased inspections by the DEL can assist with improving H&S compliance. PCs should implement only authorised subcontracting by adopting H&S contractual and evaluation practices when selecting SCs. It is recommended that further studies be conducted, so that a framework can be evolved to assist SCs with compliance with H&S legislation and regulations.

References

[1] Manu P, Mahamadu A-M, Phung V M, Nguyen T T, Ath C, Heng A Y and Kit S C 2018 Health and safety management practices of contractors in South East Asia: A multi country study of Cambodia, Vietnam, and Malaysia *Safety Science* **107** 188–201

[2] Mbachu J 2008 Conceptual framework for the assessment of subcontractors' eligibility and performance in the construction industry *Construction Management and Economics* **26** 471–84

[3] Manu P, Ankrah N, Proverbs D and Suresh S 2011 Managing the adverse health and safety influence of subcontracting: findings of a qualitative inquiry. In: Laryea S, Leiringer R and Hughes W (Eds.) *Proceedings of the S West Africa Built Environment Research (WABER) Conference* Accra Ghana 19-21 July 2011 735-744

[4] Construction Industry Development Board (cidb) 2013 *Subcontracting in the South African Construction Industry; Opportunities for Development* cidb Pretoria

[5] Johnstone R, Mayhew C and Quinlan 2003 *Outsourcing Risk? The Regulation of Occupational Health and Safety Where Subcontractors are employed* University of Illinois College of Law Chicago 351

[6] Manu P, Ankrah N, Proverbs D and Suresh S 2013 Mitigating the health and safety influence of subcontracting in construction: The approach of main contractors *International Journal of Project Management* **31** 1017–26

[7] Construction Industry Development Board (cidb) 2009 *Construction Health & Safety in South Africa Status & Recommendations* cidb Pretoria

[8] Agumba J, Pretorius J M and Haupt T 2013 Employee involvement and empowerment in health and safety: A perception of small and medium contractors in South Africa In: *Proceedings*

[9] *Safety and Health in Construction Selected papers presented at the CIB World Building Congres Construction and Society*, Brisbane, Australia, 5-9 May 2013 85-96

[10] Agumba J, Pretorius, J M and Haupt T 2013 *Health and safety management practices in small and medium enterprises in the South African construction* Acta *Structilia* **20**(1). 66-88

[11] Horbury C and Hope C 1999 *The impact of procurement and contracting practices on health and safety - A review of literature* HSL Buxton

[12] Pérezgonzález J 2005 *Construction Safety Management, A Systems Approach (Knowledge Management Edition) Paperback* (USA: Lulu) 23-24

IOP Conf. Series: Materials Science and Engineering 1218 (2022) 012051 doi:10.1088/1757-899X/1218/1/012051

The reuse of discarded plastic water bottles as an insulating material for external walls in buildings

E Mousakhani[1], A Simons[2], D Harris, M Yavarkhani, and A Anvari

[1]MBC Graduate, BSCI, Auburn University, US
[2]Assistant Professor, BSCI, Auburn University, US

Email: ellisap@auburn.edu

Abstract. The thermal conductivity of exterior wall insulation affects the overall energy consumption of a given building. Fiberglass insulation is often used for this application. This research investigated the feasibility of replacing typical fiberglass insulation with used empty Polyethylene Terephthalate (PET) bottles. In addition to potential cost savings, replacing fiberglass insulation with PET bottles on a large scale could help alleviate the environmental impact of sending these bottles to landfills. A small 8-inch by 8-inch wall section constructed with sheet metal and drywall was used for testing, maintaining a 3-inch airspace for proposed insulation. Thermal resistance was measured with the airspace packed with empty water bottles and comparisons were made to the same wall sections void of any water bottles. Testing was conducted both with and without a ½-inch foam board present for comparison. Kapton heaters and temperature sensors were used for the simulation. Preliminary testing showed promise for the proposed application. The effectiveness of the plastic water bottles at providing insulation properties increased with decreasing voltage in both test setups (both with and without the foam board) and was most effective when the foam board was present. Further research is needed to determine how the use of PET bottles for exterior wall insulation can be implemented on a larger scale and how R-values can be improved.

1. Introduction and background

Large amounts of waste are generated throughout the world on a daily basis with the amount of disposable products rising rapidly each year. In particular, there has been a fast growth in production of plastic materials in recent years. Plastic production has increased from 1.5 million metric tons in 1950 to 368 million metric tons in 2019 [1]. In the case of Polyethylene Terephthalate (PET) bottles specifically, global production reached 485 billion PET bottles in 2016, forecasting to touch 583.3 billion bottles in 2021 [2]. Although recycling rates of PET bottles have increased in many countries [3], plastic production has greatly outpaced recycling efforts. It is presumed that many unrecycled PET bottles will be damped at the environment, which will consequently lead environmental and health issues.

Several studies have examined using waste materials in new building construction. One study demonstrated that environmental concerns prompt many of us to seek environmentally friendly alternatives, as we explore green alternatives, and seek to respect our ecology by utilizing plastic bottles as a building material [4]. Many times, further processing is needed for waste material to be properly recycled. The research lists three types of recycling technology, including reuse production, mechanical recycling, and chemical recycling [5]. While recycling technology is necessary for many applications, some studies propose that recycling technology may not always be economically valuable [6].

Much research has been devoted to the use of recycled materials in various construction applications, both with and without the use of recycling technology. One study tested the usage of lower grade recycled PET bottles, using a mechanical recycling method, to produce a so-called PET brick [7]. A product called Heatlok Soya serves as spray foam insulation is created from plastic bottles and soy beans by first converting the bottles into liquid and the soy beans into oil [8]. Other studies have evaluated the use of PET bottles to replace concrete block. One study found that concrete block containing PET bottles could structurally and thermally be utilized as a building unit, replacing traditional concrete blocks [6]. Another study found that concrete block packed with empty water bottles showed a 57% increase in compressive strength compared to hollow block of the same size [9]. These studies combine to show promising and practical usefulness of an increasingly abundant waste product.

Most buildings require insulation materials be used for all external walls. The usual approach is to use batt, foam, or other insulation in the air space between the inner wall material (typically drywall) and exterior wall sheathing (typically plywood or foam board). These two surface materials are attached to the internal studs (made of lumber or steel). The spacing between the studs create air pockets that span across the studs at the thickness of the studs (i.e. 2 to 4 inches) and extend the entire floor to ceiling height. The air space this geometry forms is a significant portion of the overall surface area between the inner surface and the external wall surface. This air space offers very little resistance to heat flow and is typically insulated using a fiberglass insulation. The study presented herein sought to evaluate the usefulness of discarded water bottles (PET bottles) as an insulation material for exterior wall applications.

2. Research objective and methodology

The objective of this research was to gauge the effectiveness of using empty plastic water bottles as insulation material for exterior building walls. This objective was accomplished by comparing the increased thermal resistance of a test wall packed with empty water bottles in relation to an empty air space under the same conditions. A simple composite wall section for testing was fabricated whereby a 3-inch wide inner air space was sandwiched between two 3/8-inch drywall panels. The testing section also allowed for a ½ inch section of foam board to be placed between the airspace and the drywall panel facing the laboratory ambient air. The section was 8-inches by 8-inches square and was constructed using a L-channels fabricated from sheet metal and fastened by rivets. The composite wall test section is shown in Figure 1 below.

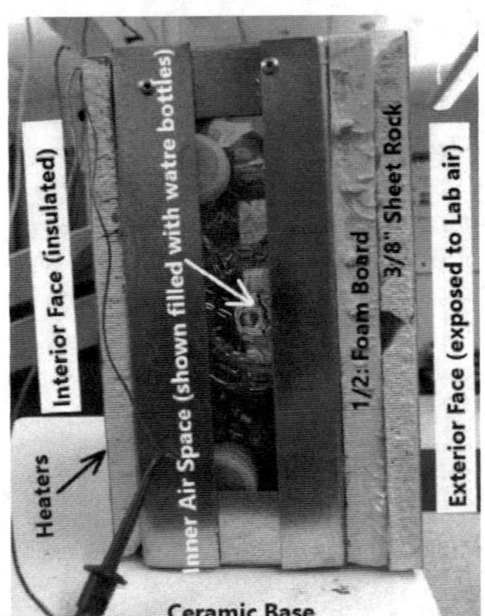

Figure 1: Composite wall test section *with* foam board

One of the drywall panels was exposed to the laboratory air while the other drywall panel was insulated and had thin film Kapton heaters attached on its external face. The entire test section was then instrumented with RTD temperature sensors on the two external drywall faces. The entire test section was then insulated from the external laboratory environment except for the exterior face opposite of the heaters. The testing approach was to apply heat to the inner insulated face thereby forcing all the heat to have to travel through the stack, including the airspace, and onto the external side of the exposed drywall face. The heat would be removed by natural convection off of the exposed face and into the laboratory air. Once the test section was thermally settled (usually 24+ hours), the temperatures of the inner and outer drywall faces were recorded versus the amount of heat applied by the Kapton heaters. The ratio of the temperature difference over that of the heat being applied provided a direct measurement of thermal resistance. Several power levels were tested for both conditions of an empty inner airspace and an airspace packed with empty water bottles. This comparison was done for both arrangements of having the foam board in place and without the foam board present. Therefore, there are four arrangements that were tested as follows:

- Arrangement A – Packed airspace with the foam board present
- Arrangement B – Empty airspace with the foam board present
- Arrangement C – Packed airspace with no foam board
- Arrangement D – Empty airspace with no foam board

Figure 2 below shows the testing setup with insulation, power source, and sensors.

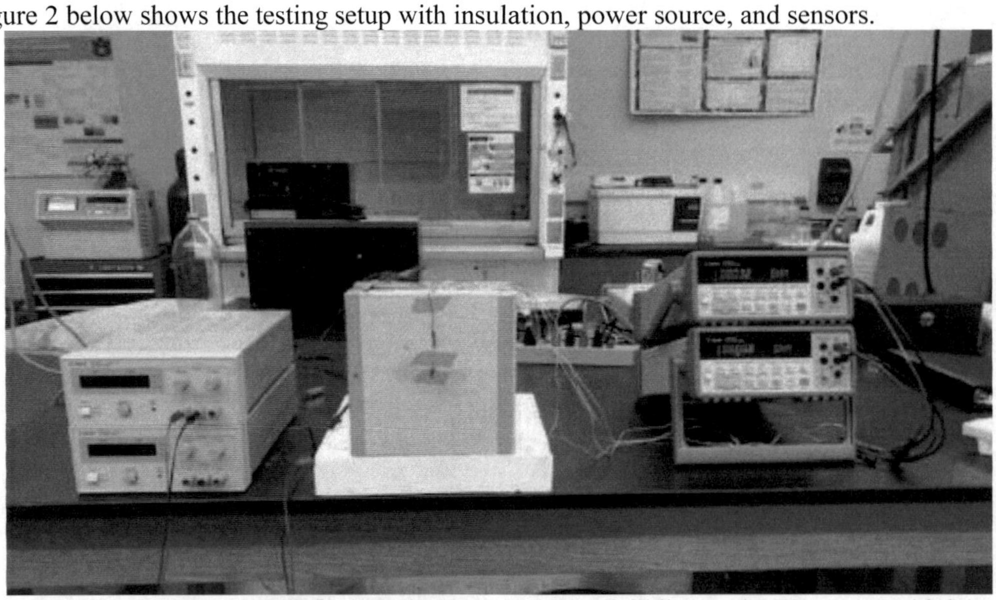

Figure 2: Power source (left), sample, test specimen (middle), and RTD sensors (right)

3. Results and discussion

The four arrangements (A-D) described above were tested over three separate power levels. Two arrangements (A&B) included foam board in the wall section stack up while two other arrangements (C&D) did not include the foam board. Measurements and comparisons were made between these arrangement pairs when the air space was filled (A&C) and un-filled (B&D) with the empty water bottles. These arrangements and the results are shown in Table 1 below.

Table 1: Test results

	Foam	Bottles	POWER Volt	Amps	Power	Flux	TEMPERATURE RTD1		RTD2		ΔT [°C]	U [W/°C]	U" [W/m^2·°C]	R [°C/W]	R" [°C·m^2/W]
Arrangement A			[V]	[I]	[W]	[W/m^2]	[Ω]	[°C]	[Ω]	[°C]					
Test 1	☑	☑	**9.69**	**0.006**	0.058	1.408	**110.23**	26.6	**108.22**	21.4	5.22	0.011	0.27	89.80	3.71
Test 2	☑	☑	**13.00**	**0.009**	0.117	2.833	**111.79**	30.6	**108.35**	21.7	8.94	0.013	0.32	76.37	3.15
Test 3	☑	☑	**20.00**	**0.140**	2.800	67.80	**116.45**	42.7	**108.79**	22.8	19.90	0.141	3.41	7.11	0.29
Arrangement B															
Test 4	☑	☒	**9.69**	**0.006**	0.058	1.408	**110.05**	26.1	**108.33**	21.6	4.47	0.013	0.32	76.84	3.17
Test 5	☑	☒	**13.00**	**0.009**	0.117	2.833	**111.40**	29.6	**108.40**	21.8	7.79	0.015	0.36	66.60	2.75
Test 6	☑	☒	**20.00**	**0.140**	2.800	67.80	**115.71**	40.8	**108.83**	22.9	17.87	0.157	3.79	6.38	0.26
Arrangement C															
Test 7	☒	☑	**9.70**	**0.006**	0.058	1.409	**109.90**	25.7	**108.43**	21.9	3.82	0.015	0.37	65.60	2.71
Test 8	☒	☑	**13.00**	**0.009**	0.117	2.833	**111.24**	29.2	**108.87**	23.0	6.16	0.019	0.46	52.61	2.17
Test 9	☒	☑	**20.00**	**0.14**	2.800	67.80	**114.53**	37.7	**109.59**	24.9	12.83	0.218	5.28	4.58	0.19
Arrangement D															
Test 10	☒	☒	**9.70**	**0.006**	0.058	1.409	**109.66**	25.1	**108.40**	21.8	3.27	0.018	0.43	56.23	2.32

Table 1 provides the three different Voltages used and associated Ampere in each stage. From the values given here, the amount of provided Power (W) (Power = Volt x Amps) and Heat Flux (W/m^2) (Heat Flux = Power / Area) have been calculated based on the known sample area (8^{in} x 8^{in} or 0.2023^{m} x 0.2023^{m}). The next column states Resistance of RTD (ohm or Ω) and Temperature (°C) over the inner and outer drywall faces. The right side of Table 1 shows the Temperature Difference over inner vs outer face (ΔT). Thermal Conductivity (U) is the inverse of R and indicates how well the element conducts heat or the rate of transfer of heat (in watts) through one square meter of a sample. Finally, R-value (or Thermal Resistance) is calculated by Temperature Difference times Area per Heat Transfer flow rate (or Heat Flux).

During testing, an IR Camera was used to capture a thermal infra-red image of the face exposed to the laboratory air. These images were taken at the completion of each thermal soak, lasting 24 – 48 hours, and included all 12 tests (3 tests for each of the 4 testing arrangements). Figure 3 (a)-(l) show these IR images at the end of thermal soak (thermal equilibrium). The thermal images shown in Figure 3 help to visually illustrate the numerical findings included in Table 1.

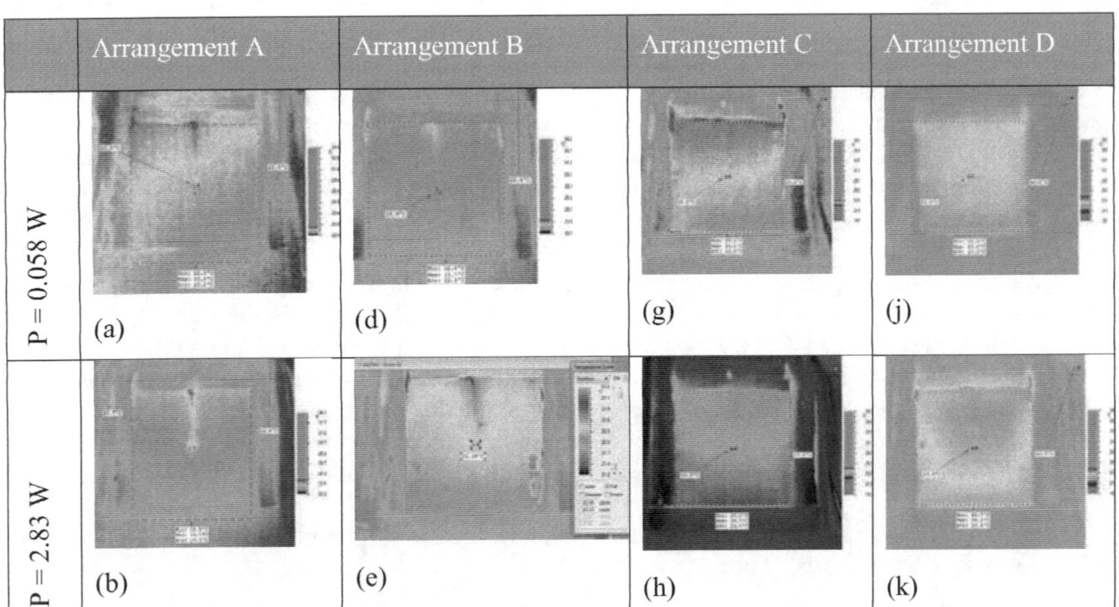

P = 2.80 W

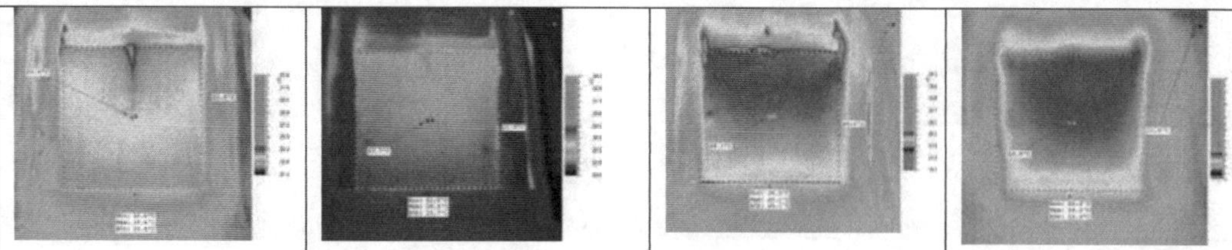

(c) (f) (i) (l)

Figure 3 (a)-(l): IR images taken at thermal equilibrium for all 12 testing cases

While the experimental results, including calculated values, are included in Table 1 above, the authors sought to determine the insulation value gained from the addition of water bottles on a numerical and a percentage basis. Table 2 below provides the percent increase in R-value for each test setup (with and without foam board) at each of the 3 experimental voltages.

Table 2: Percentage increase in R-value for each test setup and voltage

Voltage	ΔR (C-D)	% Increase in R (with Foam Board)	ΔR (A-B)	% Increase in R (without Foam Board)
9.70	0.54	17.0%	0.39	16.8%
13.00	0.40	14.5%	0.21	10.7%
20.00	0.03	11.5%	0.01	5.6%

As shown in Table 2, the plastic water bottles provided greater insulation value (R-value) at lower voltages (i.e. lower temperatures). Additionally, the increase in R-value with the addition of plastic water bottles was greater when foam board was present in the testing setup. Generally, the effectiveness of the plastic water bottles at providing insulation properties decreased with increasing voltage in both setups, but even more so when the foam board was not present. The greatest increase in R-value (0.54 or 17%) was found with the foam board present at 9.7 Volts while the smallest increase in R-value (0.01 or 5.6%) was found at 20 Volts in the absence of foam board. It is evident that the foam board and water bottles worked together to provide greater and more sustained insulation properties.

4. Conclusions and recommendations

Results of this study indicate that discarded plastic water bottles can contribute to insulation value for exterior building walls, thus providing construction cost savings and reducing environmental waste. Each test setup (with and without ½ inch foam board insulation) exhibited an increase in R-value with the addition of empty plastic water bottles at each test voltage (9.7 Volts, 13 Volts, and 20 Volts). The effectiveness of the plastic water bottles at providing insulation properties increased with decreasing voltage in both test setups. Additionally, the water bottles were more effective at providing insulation properties (i.e. greater R-value) when the ½ inch foam board was present in the sample. The greatest increase in R-value (0.54 or 17%) was found with the foam board present at 9.7 Volts while the smallest increase in R-value (0.01 or 5.6%) was found at 20 Volts when the foam board was not present.

This study is limited to the test setup utilized, including size and makeup of the test specimen and specific voltages utilized. Future research should examine larger air spaces (allowing room for a larger amount of water bottles) and alternative insulating materials which may be used in conjunction with discarded PET bottles to reduce cost while contributing to sustainability. Further research should also determine how the use of PET bottles for exterior wall insulation can be implemented on a larger scale and other ways in which R-values can be improved.

5. References

[1] Statista 2021 Production of plastics worldwide from 1950 to 2019 (in million metric tons) https://www.statista.com/statistics/282732/global-production-of-plastics-since-1950/

[2] Tiso I 2021 Global PET bottle production 2004-2021 https://www.statista.com/statistics/723191/production-of-polyethylene-terephthalate-bottles-worldwide/#statisticContainer

[3] Welle F 2011 Twenty years of PET bottle to bottle recycling-An overview Resources, Conservation and Recycling 55 865-875 https://www-sciencedirect-com.spot.lib.auburn.edu/science/article/pii/S0921344911000656?via%3Dihub

[4] Writer C 2011 Best of eco-friendly homes built with recycled material https://greendiary.com/2011-eco-friendly-homes-built-recycled-material.html

[5] Hamad K, Kaseem M, and Deri F 2013 Recycling of waste from polymer materials: An overview of the recent works. Polymer Degradation and Stability 98 2801–2812 https://doi.org/10.1016/j.polymdegradstab.2013.09.025

[6] Mansour A and Ali S 2015 Reusing waste plastic bottles as an alternative sustainable building material. Energy for Sustainable Development 24 79–85 https://doi.org/10.1016/j.esd.2014.11.001

[7] Nováková K, Šeps K, and Achten H 2017 Experimental development of a plastic bottle usable as a construction building block created out of polyethylene terephthalate: Testing PET(b)rick 1.0 Journal of Building Engineering 12 239–247 https://doi.org/10.1016/j.jobe.2017.05.015

[8] Greer S 2007 New priorities prompt innovation in materials Toronto Star https://eds-b-ebscohost-com.spot.lib.auburn.edu/eds/detail/detail?vid=9&sid=b5190063-30ca-4d70-85c2-abced7bf6dc8%40sessionmgr102&bdata=JnNpdGU9ZWRzLWxpdmUmc2NvcGU9c2l0ZQ%3d%3d#AN=6FP3166209003&db=pwh

[9] Safina S and Alkalbani A 2016 Use of recycled plastic water bottles in concrete blocks. Creative ConstructionConferencehttps://www-sciencedirect-com.spot.lib.auburn.edu/science/article/pii/S1877705816339534?via%3Dihub

IOP Conf. Series: Materials Science and Engineering 1218 (2022) 012052 doi:10.1088/1757-899X/1218/1/012052

BIM and Lean for value generation in the built asset industry: an information management perspective

D Mehran, E A Poirier and D Forgues

École de technologie supérieure, Montréal, Canada

Donya.Mehran.1@ens.etsmtl.ca

Abstract. The built asset industry's notorious productivity gap is being targeted through digitalization, operationalized through Building Information Modelling (BIM), and the application of Lean philosophy to the planning, design, delivery, maintenance, and management of the built environment. While both approaches grow in popularity, their development has remained largely on parallel tracks. Both approaches have existed, for the most part, as two independent initiatives aimed at improving performance and productivity. An increasing amount of work however is pointing to the significant potential that can be achieved through the integration of both approaches. This paper investigates the mutual relations and synergies between BIM and Lean from an information management perspective. The paper also presents the key similarities, differences, synergies, and interactions between these two main drivers of built asset industry reform, while providing an investigation into the principles guiding the application of both within the construction industry.

1. Introduction

The built asset industry plays an integral part in the global economy through its role in urbanization and enhancing quality of living. As Lee [1] states "*projects in the construction industry are becoming increasingly large and complex, with construction technologies, methods, and the like developing rapidly*". Construction problems are well known and their impact on productivity and quality have been documented in the past [2]. Several solutions have been provided to relieve the chronic difficulties in construction. Two of the most recent and notable efforts to resolve these problems are considered in this paper. First is Lean Construction (LC) which involves the use of a conceptual approach to construction management borrowed from the manufacturing sector. Second are the transformative advancements in digitalization, operationalized through Building Information Modelling (BIM). Both are seen as a mechanism to increase efficiency, productivity, reduce project duration and lead times, enhance quality, and improve sustainability [3] The concepts of BIM and Lean have largely existed as two separate initiatives in the past. Research in the field is increasingly pointing towards the commonalities and synergies between both approaches. Yet, their implementation in tandem has been seen to cause confusion as explained by Sacks [5] "*Their parallel adoption in state-of-the-art construction practice is a potential source of confusion when assessing their impacts and effectiveness*". While both Lean construction (LC) and BIM are bringing important changes to the industry in their own way, the integration of both principles promises to provide important benefits to the Architectural Engineering and Construction (AEC) sector. However, there is still significant work to be done to fully reap these benefits. Among this work, setting the proper theoretical foundations can help to better explain and predict future outcomes as well as serve to better frame and understand how to effectively implement both approaches in tandem. Indeed, many researchers point to the lack of theoretical foundation in construction as an overall hinderance to progress [9].

The purpose of this paper is to investigate the synergies of BIM and Lean from an information management perspective. More specifically, it explores this question from a theoretical point of view through a review of past work in the field. The paper also provides a review of the background literature related to BIM and Lean principles to explore their likenesses and variances as well as potential integration and highlighting the existence of literature limitation and gap.

2. Literature Review

2.1. Theory of production as foundational theory of construction

The AEC industry suffers from poor quality and productivity due to problems related to a number of root causes which impact client decision making, design and construction management, among many others. Significant parts of these issues are initiated by limited view on production [6].One of the main barriers to construction improvement is known as the lack of theoretical foundation as pointed out by several scholars (e.g., Koskela [7], [6] Ballard [11], Demirdöğen[13], Tauriainen [[14]). Adopting the theory of production as a foundational theory of construction, can help in better framing pathways for improvement [7]. For instance, Sacks [8] propose "*In the current context of lean construction and BIM, we propose that for comprehensive realization of benefits, not only should changes in information and material processes be coherently based on these two (IT benefits and understanding construction peculiarities), but that all three – process changes, BIM tools themselves, and of course lean construction principles – should be rooted in conceptual understanding of the theory of production in construction*" which is explained in Figure [8].

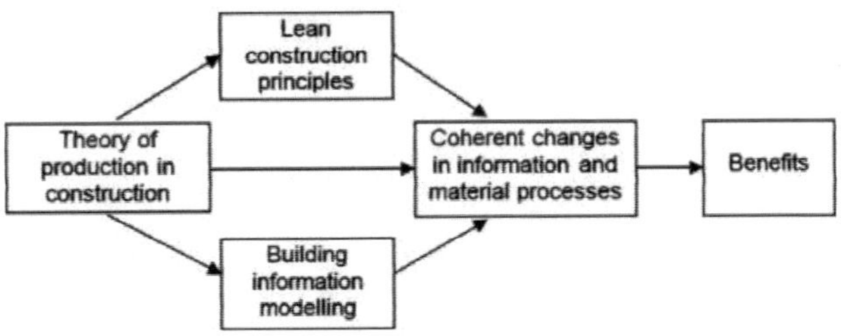

Figure 1 The dependence of benefits realization through process changes in construction on lean construction principles, BIM, and a theoretical understanding of production in construction [8].

Three different concepts of production have been developed. These exist as Transformation, Flow, and Value generation. None of these have been accepted as a generalized theory of production from either an industrial or a scientific point of view. Indeed:

- The Transformation (T) or conversion concept was dominant as a foundational theory of production, and it was formed based on Economic science. Two fundamental principles of the transformation model which view production as a conversion of inputs into outputs and dividing the total transformation into smaller, more manageable conversions known as analytical reductionism. Nevertheless, as transformation view is not particularly helpful in identifying how to avoid using unnecessary resources or how to certify that customer requests are met in the best approach, it tends to become ineffective and inefficient [9].
- The Flow (F) concept has its roots in Japan in the 1940s and challenged the transformation model, as the "Time" factor in transformation was neglected and it is identifying waste and non-value adding activities through the process [10]. The Lean production system was developed based on this theory. As such, the flow view is represented in lean production and lean construction. In the flow view, the essential thrust is to eliminate non-value adding or waste

from processes. Consequently, such principles as lead time, variability reduction and simplification are endorsed.

- The Value generation (V) initiated in 1930's where researchers debated that the value of a product can be determined only in reference to the customer, and the total goal of production is satisfying customer needs [11]. In the value generation view, the basic tenant is to achieve the best possible value from the point of view of the customer. The quality movement has attempted to translate this view into useful practices for the industry.

The fundamental goal guiding the development of a "theory of production" was to form a unified conceptualization of production, in lieu of three incomplete conceptualizations. The TFV theory of production developed by Koskela [9] proposes an integrated view of all three views of production. He suggested that the three views are all necessary for production management and they should be used simultaneously, in an integrated and balanced manner. The Integrated TFV view on production is shown in Table 1 [9]. The TFV theory of production enhances construction by improving performance through elimination of waste and increase value where possible by reduction of variability [2]. The TFV theory has demonstrated its potential to enhance construction performance, however it requires discipline integration that focus on design, control, and enhancement of production system.

Table 1 TFV Theory of Production [9]

	Transformation view	Flow view	Value generation view
Conceptualization of production	As a transformation of inputs into outputs	As a flow of material, composed of transformation, inspection, moving and waiting	As a process where value for the customer is created through fulfilment of his/her requirements
Main principle	Getting production realized efficiently	Elimination of waste (non-value-adding activities)	Elimination of value loss (achieved value in relation to best possible value)
Methods and practices	Work breakdown structure, MRP, organizational responsibility chart	Continuous flow, pull production control, continuous improvement	Methods for requirement capture, quality function deployment
Practical contribution	Taking care of what has to be done	Making sure that unnecessary things are done as little as possible	Taking care that customer requirements are met in the best possible manner
Suggested name of practical application of the view	Task management	Flow management	Value management

2.2. Conventional vs. Novel Project Management

Project management (PM) is a crucial factor in achieving the desired outcomes in the construction industry. Yet, conventional project management and its associated delivery systems follow a 'task'-based rationale to project delivery. They largely ignore both value maximization and waste reduction. This approach confuses the 'task' view with managing the project. The new theoretical foundation of project management (PM) contains extended theory of project and theory of management. The theory of project consist of transformation, flow and value generation considers the factor of time, variability, and customer satisfaction into the transformation model. The theory of management on the other hand is comprised of three main constructs: planning, execution, and control. The theory of production (TFV) is considered a valuable basis for novel project management Koskela claim [9] construction should be considered as from a system of production perspective.

These novel approaches to project delivery and management are instantiated through mechanisms and tools such as Last Planner (LP) or Last Planner System (LPS) developed by Ballard [11] as a response to failure of conventional PM. The Lean Project Delivery System (LPDS) specifies what must be done and who is the responsible party for the task at the very beginning of the project [12].The traditional delivery versus LPDS is mainly different in terms of project phases and participants in each phase, value, and control concepts. Traditionally, project design and definition are performed by the architect in isolation or with the client, however the LPDS concerns overlapping iterations over five main project phases as shown in **Hiba! A hivatkozási forrás nem található.** [12]. The LPDS enables teams involved in every stage to be involved from project start to completion. The conventional design management is frozen in the design phase and any change causes rework and extra cost. However, the LPDS design decisions are systematic and enable more time for exploration of design alternatives which enhance value generation as per client requirement. From a control point of view, traditional project control refers to monitoring the actual versus planned performance and identifying possible variance. In contrast, the LPDS control is based on systematic process for making tasks to be formed along with explicit organization liability at production level to what need to be released to client in the next plan phase. Moreover, in traditional project delivery, the information about operation, maintenance and decommissioning is not often considered as part of the process which leads to customer dissatisfaction and neglecting value generation. In contrast, the LPDS consider end-user value generation as the information about operation and maintenance are considered from beginning of project. Therefore, For the LPDS successful implementation, early-involvement, aligned incentives besides early-involvement of stakeholders are required [2],[11] ,[12].

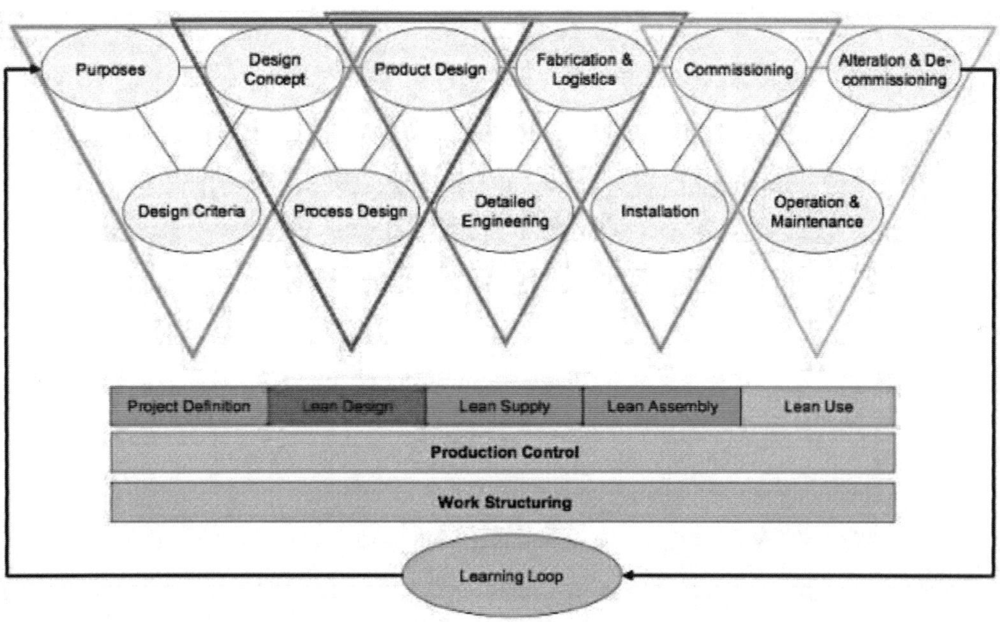

Figure 2 Lean Project Delivery System [12]

2.3. key Similarities and Differences between Lean Construction and BIM

Lean and BIM notions have largely been developed as two distinct initiatives to enhance the performance of AEC industry, each possessing different communities and interests [4]. Lean is a conceptual approach to project construction management and BIM is considered as Transformative information technology which cause fundamental change in the AEC industry [8].The main similarities and differences between BIM and Lean are presented from various perspectives as backgrounds, application, design management, lifecycle, and supply chain information management.

- The background and generation of Lean and BIM concept are different. Since mid1990s Lean construction (LC) has appeared as a novel theory in the construction domain adapted from manufacturing [5], [7] It is an adaptation of core concepts and principles of the Toyota Production systems (TPS) [10]. In mid 1970s, Eastman [3] introduced the basic concept of BIM as a computable interpretation of a building object, transforming the traditional approach based on descriptive geometry used to signify the objects to be constructed. However, the idea of BIM implementation could not be applied until 1990s, as computers technologies improved and enabled practical implementation of the BIM idea [3], [8], [13].

- The application and adoption of both (Lean and BIM) are similar as both concepts require organizational cultural change which is challenging. For instance, Koskela [2] explain: "*Changing procedures, techniques and corporate systems is the easy part; changing minds is the real challenge.*"
- Design management follows different approaches to the application of Lean and BIM. Project design management by using Lean tools, enhance operations, minimize waste, improve value generation and client satisfaction [14]. During this stage, design process is understood as a flow and value creation that can be improved by using Lean tools. Using lean tools facilitate coordination and collaboration among project team regardless of their physical geographic location and facilitate effective information sharing between project team. Therefore, overall design duration reduces, and client satisfaction is ensured. The BIM process during design stage includes several design disciplines that need to work effectively to achieve client requirements to prevent conflicts. However, as usually there is no standard for collaboration among design teams and responsibilities are not well defined, the whole process face challenges [[13].
- Lifecycle information management and supply chain management is similarly challenging for both Lean and BIM as the nature of projects in the AEC industry is fragmented [15]. The management of supply chain across construction industry is a challenge as due to the temporary nature of construction as well as unwillingness of project team to cooperatively share information due to existing contracts contents and clauses. Therefore, this led to challenges in trust and cooperation [16].

2.4. Application of both (BIM & Lean) from information management perspective

Over the last decade, BIM and Lean concepts have begun to diffuse into advanced practice. As Sacks [8] claims, "*BIM technology could also be used as an enabler or catalyst for lean transformation*". Today, the emerging methods of BIM and Lean are reshaping the worldwide business environment of the AEC industry [15],[17]. The insight about both initiative's synergies and benefits is being increasingly explored by academia and leading practitioners as illustrated in figure 3 [4]. As Koskela [8] state: "*lean construction arguably facilitates the implementation for BIM, especially when this is based on robust and reliable technology*."

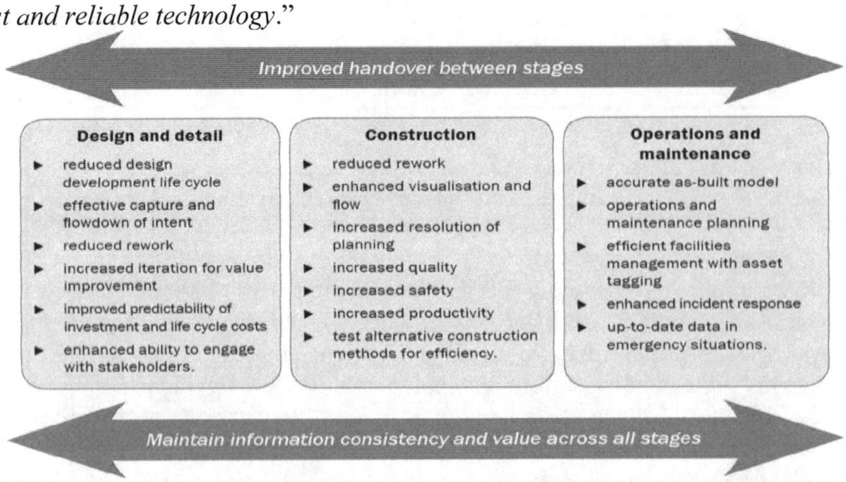

Figure 3 Benefits of using Lean and BIM [4]

One way to bridge the gap and enable a Lean approach to project delivery supported through BIM, is to look at the question through the lens of information management (IM). Several standards and bodies of knowledge are beginning to be developed in this regard. For instance, the International Standard

Organization has published the ISO 19650 series to provide direction for information management, collaborative working, information delivery plan creation, software's application, roles, and responsibilities of different stakeholders across the built environment. On one hand, creation of standards and protocols cause positive effect by enabling information exchange consistency across project team. On the other hand, it causes negative effect on the innovation process as it promotes robust and solid factors for innovative technologies, mainly because they are developed with modest or without theoretical foundation. Accordingly, Succar [18] claim: *"Foundational questions remain as to what should or should not be standardized in the fast-evolving domain of construction technologies and, more consequentially, questions relating to the appropriateness of these standards for information integration between overlapping domains."*

In the BIM and Lean integrated approach, the balance between people, process and technology is fundamental. As Dave [4] state "*A Lean and BIM project is people-and process-oriented, where BIM as a technological platform acts as an enabling tool*". An example of information exchange and information flow among supply chain to support key appointing party decisions is illustrated in figure 4 [19].The information from project teams flows and exchanges across project phases which require validation and verification at every project stage. To outline, manage, and integrate lifecycle information exchange and transformation, physically and digitally, among project team, there is requirement for foundation of information management (IM) [18].

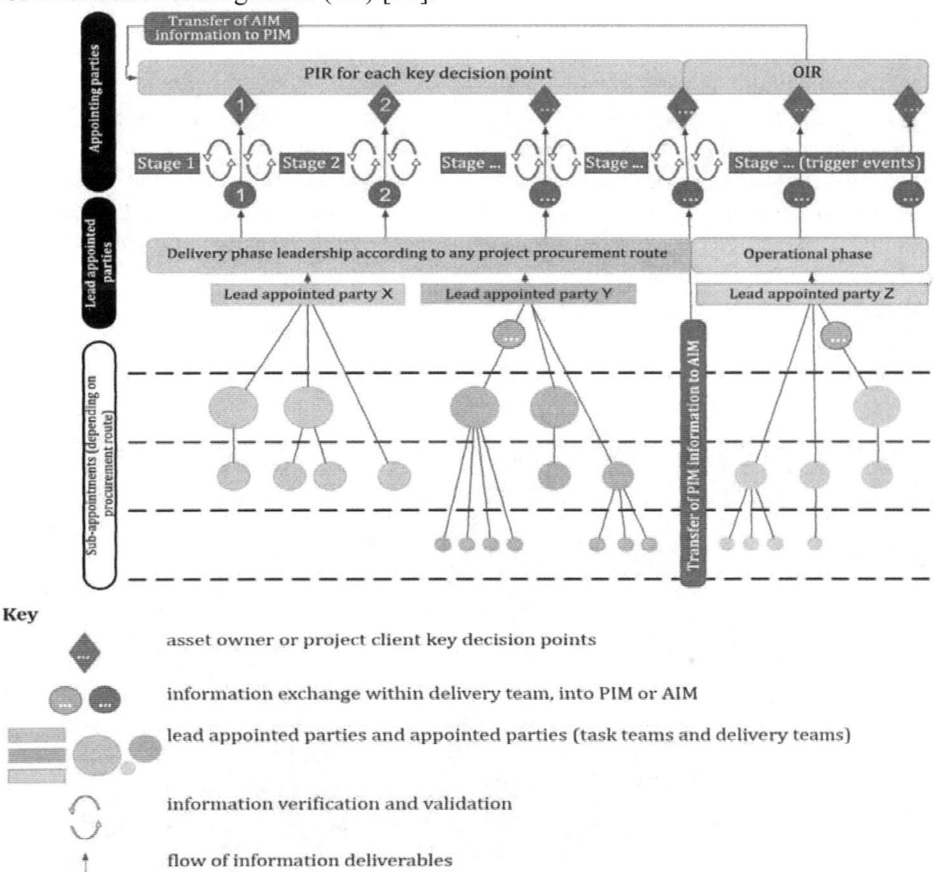

Figure 4 Example of Information delivery through information exchange to support key appointing party decisions [19]

The question of lean and effective IM practices enabled through BIM is also garnering considerable attention in academia. As such, researchers have investigated ways to better define, manage, and integrate asset and project information across their lifecycles. For instance, the Lifecycle Information

and Exchange (LITE) framework [18] (figure 5) provides a structured approach to understanding how information flows and transforms across an asset's lifecycle . The LITE framework acts as a visual rendition summary of many conceptual components and their relationships in one Information Cycle. It is composed of several conceptual contents that require independent and collective recognition. The ten components of this Information Cycle are described by Succar and Poirier [18]. They include the status of information, which defines the continuous information transformation through lifecycle defined as two statuses of Targeted Status and Actual Status. The States refer to the information transformation from initial intend to deliver new asset upon its delivery which are represented as three states: Purposes, Deliverables, and Resources and Methods. The Sets refer to collection of information covering all targeted and existing information as the information transforms and exchanges between Statuses and States. There are eight Milestones as manifest track of asset's lifecycle information where each milestone is a clear transformation from being an intent of an asset, to a digital representation and to become Physical Asset itself. The Flows refer to the movement of information within and in between the Milestones which can move forward, reverse, inward and outward directions and the Gates refer to the mechanism of controlling these movements. The Routes refer to the information once they enter into the Loops which are three categories as Assisted Flow, Automated and Automatic Flows, and Autonomous Flow. The Loops consist of open loops where information flows once along a route, whereas close loops information flows more than once along a route during asset lifecycle. The Actions are collections of granular information activities, sub-Activities, Tasks and sub-Tasks shows the connections between information and milestones. The Shortcuts refer to the information flows that avoid milestone(s) with a valid reason of potential risk or a divergence from routes. The Tiers refer to organizing and sharing of information through a Common Information Environment to minimize duplication of contents and avoid reliance on disconnected documents.
The LITE framework puts emphasis on transformation of information flow from defining the purpose to deliver new asset to recognizing the actual Digital/Physical Assets to managing during entire lifecycle. The LITE framework, similarly to Lean Construction, provides information flow primacy to pull activities. However, Succar and Poirier [18] state that *"the Lean Construction focus on the product and production system, it focuses on the information flow itself across Milestones and transforms through Statuses and States". F*urther development and investigation toward similarities and differences between LITE framework, and other emerging process-oriented approaches are valuable to engage other communities [18].

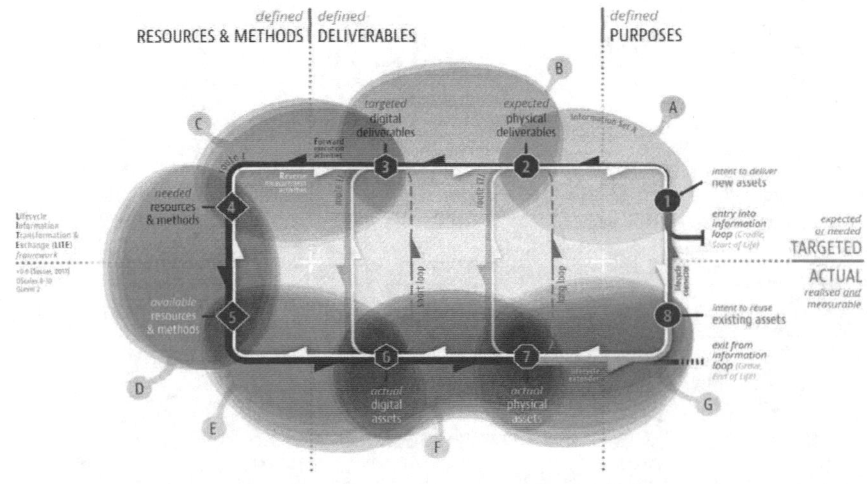

Figure 5 The LITE framework [18]

IOP Conf. Series: Materials Science and Engineering 1218 (2022) 012052 doi:10.1088/1757-899X/1218/1/012052

3. Key limitations and gaps in current literature

Digital transformation in construction implies a departure from conventional practice, demands the implementation of advanced technological solutions across a project's lifecycle. There are two gap and limitation respectively identified.

First, there is a gap in integration of BIM and Lean in practices. It is understood from the existing literature review that many interesting connections have been highlighted in theory from academia about the integrated BIM/Lean. However, there is lack of BIM/Lean application in practice [8]. As Dave [4] mention, currently organizations practice either BIM or Lean. Therefore, further efforts from both academia and practice are needed to bridge the gap in knowledge and to enhance productivity and efficiency of the construction industry.

Second, to fully benefit from potential advantages in the innovative technological resolutions, parallel innovation in policies, processes and practices are needed. As a result, many countries around the world formalized standards, protocols, specifications nationally and internationally for information management across the built environment [18]. The LITE framework concepts identified requires further testing, validation and development. The BIM and Lean integration through the LITE framework require clear definition of milestones, value mapping and value generation. Additionally, identification of adequate resources and suitable methods that bridge both approaches are the gaps to be addressed. Furthermore, the Lean coupling of assets, resources, deliverables, etc. requires continuous synchronization of actual Digital Asset with their Physical counterparts [18] is another field of study in line with the emergence of Digital Twin as a concept which is beyond scope of this paper.

4. Conclusion

The construction industry has suffered from poor productivity and performance for many years. The individual areas of BIM and Lean have emerged as two separate initiatives within the AEC, and both have been explored extensively in latest years. Nevertheless, there seems to be a deficiency of research to explore their synergies and integration of both concepts. The TFV theory of production as foundational theory for construction was discussed to identify the potential benefits of BIM and Lean. The two concepts of BIM and Lean were explored from different perspectives such as background, application, design management, lifecycle information management, tools, methods, and enablers, etc. to identify similarities and differences. The application of both concepts were investigated from an IM perspective by identifying their relationships, benefits, tools, and techniques for their integrated implementation. The paper highlights a gap in integration of BIM and Lean in practices as there are limited studies on BIM and Lean projects delivery both in theory and practice and further efforts are needed to bridge this gap. Future work will investigate the synergies in a structured manner from an IM perspective and pull from existing frameworks, namely ISO 19650 and the LITE framework. The concepts and their relation to BIM and Lean integrated approach require further investigation and development. A follow-up study will be conducted to address these gaps.

5. References

[1] Lee DG, Park JY and Song SH 2018 BIM-based construction information management framework for site information management. Jr. Advances in Civil Engineering 2018. https://doi.org/10.1155/2018/5249548

[2] Koskela L, Howell G, Ballard G and Tommelein I 2002 The foundations of lean construction (Design and construction: Building in value 291) pp 211-26. https://www.academia.edu/39126373/The_foundations_of_lean_construction?from=cover_page

[3] Sacks R, Eastman C, Lee G and Teicholz P 2018 BIM handbook: A Guide to Building Information Modeling for Owners, Designers, Engineers, Contractors, and Facility Managers. ed John Wiley and Sons.

[4] Dave B, Koskela L, Kiviniemi A, Tzortzopoulos P and Owen R 2013 Implementing lean in construction:LeanconstructionandBIM[UK:CIRIAGuideC725],

https://eprints.qut.edu.au/71034/1/71034.pdf

[5] Sacks R, Dave B, Koskela L and Owen R 2009 Analysis framework for the interaction between lean construction and building information modelling (Taiwan: 17th Annual Con. of the Int. Group for Lean Construction IGLC 17) http://eprints.hud.ac.uk/

[6] Koskela L 1996 Towards the theory of (lean) construction. (Finland: In Proc. of the 4th Annual Con. of the Int. Group for Lean Construction IGLC) pp 26-27, https://www.iglc.net/Papers/Details/8

[7] Koskela, Lauri et Gregory Howell. 2002. « The theory of project management: explanation to novel methods, https://www.researchgate.net/publication/228918258_The_theory_of_project_management_explanation_to_novel_methods

[8] Sacks R, Koskela L, Dave BA and Owen R 2010 Interaction of lean and building information modeling in construction *J. of construction engineering and management.* **136** pp 968-80, https://doi.org/10.1061/(ASCE)CO.1943-7862.0000203

[9] Koskela L 2000 *An exploration towards a production theory and its application to construction* (Finland: VTT Technical Research Centre of Finland), https://aaltodoc.aalto.fi:443/handle/123456789/2150

[10] Womack JP, Jones DT and Roos D 2007 The machine that changed the world: The story of lean production--Toyota's secret weapon in the global car wars that is now revolutionizing world industry. Simon and Schuster

[11] Ballard HG. 2000 *The last planner system of production control* (Doctoral dissertation, University of Birmingham) http://etheses.bham.ac.uk/id/eprint/4789

[12] Ballard G 2008 The Lean Project Delivery System: An Update. *Lean Construction*Pp 1-19 https://www.researchgate.net/publication/228362769_The_lean_project_delivery_system_An_update

[13] Demirdöğen G, Işık Z and Arayici Y 2020 Lean Management Framework for Healthcare Facilities Integrating BIM, BEPS and Big Data Analytics *J. Sustainability.* **12** p 7061. https://doi.org/10.3390/su12177061

[14] Tauriainen M, Marttinen P, Dave B and Koskela L 2016 The effects of BIM and lean construction on design management practices *Procedia Engineering* **164** pp 567-74 https://doi.org/10.1016/j.proeng.2016.11.659

[15] Hoeber H and Alsem D 2016 Life-cycle information management using open-standard BIM *J. Engineering, Construction and Architectural Management* **23** pp 696-708. https://doi.org/10.1108/ECAM-01-2016-0023

[16] Zhang J, Li H, Golizadeh H, Zhao C, Lyu S and Jin R 2020 Reliability evaluation index for the integrated supply chain utilising BIM and lean approaches *Engineering, Construction and Architectural Management* **27** pp 997-1038 https://doi.org/10.1108/ECAM-12-2018-0542

[17] Koskela L, Owen RB and Dave B 2010 *Lean construction, building information modelling and sustainability.* (Sweden: In Proc. of the 2010 ERACOBUILD Workshop on BIM and Lean) pp 1-8 https://eprints.qut.edu.au/71044/

[18] Succar B and Poirier E 2020 Lifecycle information transformation and exchange for delivering and managing digital and physical assets *J. Automation in Construction.* **112** p103090 https://doi.org/10.1016/j.autcon.2020.103090

[19] ISO 19650-1:2018 Organization and digitization of information about buildings and civil engineering works, including building information modelling (BIM) — Information management using building information modelling — Part 1: Concepts and principles ISO https://www.iso.org/cms/render/live/en/sites/isoorg/contents/data/standard/06/80/68078.html

Building Information Modelling Pedagogy: A Step in the Right Direction for the Construction Industry

K Dithebe, W D Thwala, C Aigbavboa and M Ramaotswa

SARChl in Sustainable Construction Management and Leadership in the Built Environment, Faculty of Engineering and the Built Environment, University of Johannesburg, Johannesburg, South Africa

khdithebe@gmail.com

Abstract. The aim of this paper is to assess the incorporation of BIM content in construction education for the benefit of productive and competitive construction firms. The study adopted a quantitative approach, where 44 out of 50 questionnaires were returned, representing 88% response rate. The respondents were inclusive of head of departments, lecturers, assistant lecturers, and doctoral candidates in different departments of architecture, construction management and quantity surveying, civil engineering and town and regional planning. Using both percentile and descriptive analysis, results revealed that collaborative learning ensured fast and effective compilation and interpretation of data, better communication, improved teamwork. From the findings it is evident that there was limited incorporation of BIM in the construction education, 45% of the respondents agreed that there was no incorporation of BIM in the current curriculum. The findings further revealed that teaching BIM at a higher level produces competent graduates with enough technological advanced knowledge to participate in the industry, and they stand a better chance of employment than graduates that are not familiar and well-equipped with BIM knowledge. With the collaboration of key stakeholders, the initiative of intensifying the incorporation of BIM in construction education can improve communication and teamwork within the construction firms, and the industry at large. The findings of the study allow institutions of higher learning to improve their programmes on construction education to include BIM content. Additionally, the construction firms must play a more meaningful role towards contributing to programmes of BIM in the construction education. An understanding between the BIM programmes of institutions of higher learning and the use of BIM in the construction industry is also a requirement to produce firms that are more productive and technologically competitive.

Keywords: Building Information Modelling, BIM pedagogy, construction education, collaborative learning, authentic learning.

1. Introduction

The emergence and incorporation of Building Information Modelling (BIM) in the construction firms has recorded multiple success stories, including the ability to effectively remove human errors during the compilation and interpretation of quantity take-offs and cost estimates [1] using automation. BIM provides accurate compilation of cost estimates, and quick extraction of quantities in a BIM model [2].

The importance and use of BIM in the construction firms has improved over the years, Smith and Tardif [3] and Olatunji [1] concur that many firms are adapting into the era of digitalised planning, designing, constructing and managing infrastructure. Clevenger et al. [4] adds that half of the construction firms in the United States of America use BIM.

However, the construction industry is and has been known for many years for its resistance to adapt to new and improved Information Communication Technology (ICT) systems, such is the introduction of BIM [5; 6]. BIM is a series of information model-based interconnected applications that efficiently plan, design, construct and manage infrastructure using a variety of collaborators [7]. For its successful application in the construction firms there needs to be employees in the form of construction graduates that are knowledgeable in the use of BIM. Construction graduates that are knowledgeable in BIM can only be attained if the construction curriculums in institutions of higher learning are committed to incorporating BIM content.

Like any other emerging concepts and content in institutions of higher learning BIM has also faced its own difficulties. Panuwatwannich et al. [8] and Abbas et al. [9] add that the type of course content to include in the curriculum, lack of resources to develop a new curriculum, expensive software packages, lack of qualified personnel, limited space in the existing curriculum to add BIM content and methods of teaching BIM act as barriers to the incorporation of BIM in construction education.

Given a wide range of existing information on BIM this study bestows its focus on the benefits of leading the construction industry in the right direction by embracing the methods of teaching BIM in construction education. More so, the study discusses different methods of inducing BIM content in the construction education with priority given not only to student commitment but to teachers of BIM too.

2. Benefits and learning approaches of BIM in construction education

The need for knowledgeable and skilled graduates in BIM in the construction industry has become an important aspect for construction firms. Daily operations in the construction firms of taking-off quantities and estimating costs have shifted from conventional ways to digital methods of quantification. The following studies have dealt in detail with the incorporation of BIM in architectural, engineering and construction (AEC) programmes in institutions of higher learning; Clevenger et al. [4], Panuwatwanich et al. [8] and Sacks and Pikas [10]. Thus, as a response to the construction industry's concerns of absorbing graduates that are unaccustomed with BIM the study discusses different methods of inducing and improving BIM education in institutions of higher learning.

BIM is packaged in different forms for different projects, Šadauskienė and Pupeikis [11] differentiates between Revit, AutoCAD, Navisworks, Civil 3D and MagiCAD. The use of different forms of BIM packages has created a niche in the AEC industries that needs to be explored by newly participants in the job market. Naviswork Revit amongst other packages is isolated as a key BIM software in information modelling of construction processes [7]. The software designs and assesses 3D structural building models and incorporates appropriate construction methods [7]. The incorporation of a variety of BIM software in construction education is needed for the development of the graduate's skills and competency level. BIM academic forum [12] asserts that the construction industry is in demand for BIM-ready graduates to take up space. It is thus clear from the review that graduates with BIM knowledge stand a good chance of being employed compared to graduates without the knowledge of BIM and how it works. Forsythe *et al.* [13] and Jin *et al.* [14] concur that graduates with BIM knowledge understand the significance of decision-making processes, collaboration (between clients, designers, contractors and occupants), coordination, communication, scheduling, visualisation and information gathering and sharing skills better than graduates without the knowledge. Eastman *et al.* [15] and Allen Consulting Group [16] share that the incorporation of BIM in the construction industry has ensured that the quality of the end-product is improved, there is greater transparency and accountability in the project's decision-making, minimised cost and time overruns and improved information sharing between stakeholders.

Ghosh et al. [17] cited by Abbas et al. [9] shares that BIM can be incorporated in four ways in the curriculums; revamping the current curriculum to include BIM content, incorporating BIM fundamentals into the existing construction curriculum [10], holding workshops and training sessions on BIM fundamentals and its technicalities, lastly BIM can be induced in higher education level as a new degree program. Sacks and Pikas [10] further shares that despite various methods of incorporating BIM into the construction curriculum institutions still find it difficult to successfully integrate teachers into BIM education, especially with the obsolete knowledge and skills they currently hold. The use of OpeBIM known as TeacherBIM is one of the pivotal courses needed to provide theoretical and practical knowledge to educators regarding the fundamentals of BIM [18].

A perceptible benefit of enriching students with BIM knowledge and skills in their undergraduate to postgraduate level is that it favours AEC firms that are desperately in need of competent graduates that are knowledgeable in BIM and can adequately apply digital information modelling to various occupational challenges of the projects [19] with improved efficiency. To realise the benefits of incorporating BIM in construction education it is important to encourage the reformation of the existing curriculum to allow the entrance of new and improved content on information modelling.

The realisation of BIM benefits in the construction firms is possible when literacy on BIM fundamentals is prioritised through detailed BIM software trainings. BIM training and workshops gives students the ability to apply appropriate parametric models in the field of construction technology using Revit BIM software that focuses on information modelling of parametrization [7]. Revit is a well-known BIM software that develops a model using 3D drawing. Graduates that are knowledgeable in 3D drawing are equipped with the design skills that allow them to apply knowledge to practical BIM issues relating to construction products using parametric modelling.

Olowa and Lill [20] asserts that digital modelling in the construction industry is improving as a result graduates in institutions of higher learning to improve their employability status are expected to comprehend and apply digital software like BIM in order. The incorporation of BIM content in construction education has become pivotal in the success of producing BIM-ready graduates [21]. Different methods of teaching BIM and which one is more beneficial are discussed in this current study. Olowa and Lill [21] makes emphasis that currently there has not been a single straightforward pedagogy of teaching BIM in construction. However, Mills and Treagust [22], Barison and Santos [23] and Jin *et al.* [14] differentiates between project, problem and case-based learning. While Olatunji [1] differentiates between action learning, constructivist learning, reflective learning, authentic learning and cooperative learning. Revans [24] indicates that action learning relates to learning based on real life problems, whereby learners learn from their own actions or work in teams. Action learning is classified into three approaches [25]: (i) scientific approach (learners use their perspectives in understanding and issue in relation to external factors); (ii) experiential approach (leaners collaborate and work in teams to understand an issue); (iii) critical reflection approach (based on learner's individual effort in understanding an issue) [24].

In terms of constructivist learning, "leaners build knowledge by developing their own subjective representations of an objective reality" [1], instead of depending on personal interpretations [26]. Pawson *et al.* [27] opines that constructivist learning unlike other approaches enable learners to focus on specific content of the subject at a time instead of exploring the entire aspects of the subject. On the other hand, there is reflective learning, which advocates learning through hands-on experience and theory [1]. Reflective learning advances the approach of learners seeing what they need to see by themselves. It is important that learners can share knowledge and help each other understand concepts, this type of approach is known as authentic learning. In authentic learning learners are less likely to compete against each other but interact and collaborate with each other to understand concepts such as BIM [28; 29]. The choice on which learning approach to employ depends on the level of adoption of BIM content in the curriculum. A cooperative learning approach is suitable in a case where learners are required to work independently at the same time have the skills and intelligence to work in a group [1]. Lee and Hollar [30] and Zhao *et al.*, [31] concur that collaborative learning and BIM-integrated learning are best suited for teaching BIM and beneficial to the construction industry.

3. Methodology

The study conferred its focus on BIM incorporation in the construction education for the benefit of productive and competitive construction firms. The study adopted a quantitative survey approach where a structured questionnaire was distributed to the respondents. A five-point Likert scale was adopted in the questionnaire to determine the level of agreement whilst mean item scores of the questionnaires were used to rank potential BIM benefits to the construction firms. The structured questionnaire was distributed to head of departments, lecturers (part-time and full-time) and assistant lecturers who are at the fore front of construction education at the University of Johannesburg, South Africa and doctoral students whose research area is on BIM in the built environment. The affiliation of the researcher with University of Johannesburg, having access to the construction departments at the university and the project being funded and supervised by the institution was the reason for selecting it over the 25 existing public universities in South Africa. The research instrument had two sections, the first section was on the background information of the respondents and the second on 'incorporating BIM in the construction education for the benefit of construction firms'. The study nominated respondents from two faculties at the University of Johannesburg, these are key faculties teaching construction education from project inception to its close-out phase. Within the Faculty of Engineering and Built Environment three departments were identified; Department of Civil Engineering Technology, Department of Construction Management and Quantity Surveying and the Department of Town and Regional Planning, whilst under the Faculty of Art, Design and Architecture the Department of Architecture was selected. Programmes on construction education and knowledge on Building Information Modelling were the main criteria for the selection of the departments, more importantly the departments produce graduates that are informed on items of each stage of the construction project which is the main attribute of Building Information Modelling – sharing and linking information between the construction professionals at the early stages and throughout the project.

Purposive sampling was adopted, the non-random technique requires the researcher to develop a criterion to select the most suitable respondents, unlike in a case of random sampling technique where potential respondents are selected without a specific criterion. Given the limited number of head of departments, lecturers, assistant lectures and doctoral candidates only 50 questionnaires were distributed. To eliminate responses dominated by a single department, each department was served with 12 questionnaires, the Department of Construction Management and Quantity Surveying was served with 2 extra questionnaires for having more PhD candidates whom their research is based on BIM in the built environment to make it 50. The returned questionnaires from the distributed 50 was 44, representing a response rate of 88%. According to Moser and Kalton [32] a response rate over 20-30% represent results that are significant and impartial. The high response rate can be attributed to the researcher having access to the respondents in different departments. Data collection for the study took three weeks. Face validity in the form of a pilot study was undertaken to check the validity of the research instrument, professors and doctors in the Construction Management and Quantity Surveying Department were used for this purpose. According to Sushil and Verma [33] face validity involves experts reviewing the contents of the research instrument for appropriateness. More so, the reliability of the study was undertaken. The reliability tests revealed a cronbach's alpha value of 0.73, showing internal consistency. Moser and Kalton [32] attest that when the value of the cronbach's alpha tends towards 1 the research instrument is reliable.

Subsequently, the analysis on the background of the respondents using frequency and percentile was undertaken, as well as analysis on the benefits of BIM in the construction firms was carried out using mean-item scores, standard deviation and ranking.

4. Findings and discussion

4.1. Background Information of Respondents

From the results obtained it was revealed that 64% of the respondents were female whilst 36% were male. Additionally, background results of the respondents revealed that 25% had doctoral degrees, 61% had master's degree, 10% of the respondents had honours degree and 4% had diploma certificate. From the University of Johannesburg 23 (48%) respondents came from the architectural department, 12 (25%) from the civil engineering department, 7 (14%) from the construction management and quantity surveying department and 6 (13%) from Town and Regional Planning Department. Higher return from the Department of Architecture can be attributed to the fact that at the time of distribution other Departments were preoccupied with assessments, only a few could respond to the questionnaires. Lastly, 45% of the respondents said there was no incorporation of BIM in the construction education, as shown in figure 4.1, 17% said yes whilst 38% were not sure. From the findings it is evident that a total of 83% of the students indicated that there was no or not sure of any BIM content in the existing curriculum. The remaining 17% that said yes could be referring to the BIM seminars and training held at the University from time to time.

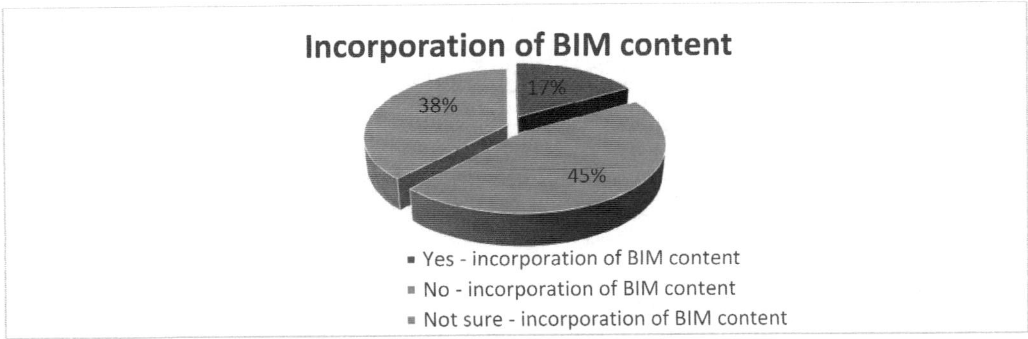

Figure 4.1 Incorporation of BIM content in the construction education

4.2 BIM pedagogy in the construction education for the benefit of productive and competitive construction firms

The results of incorporating BIM content in the construction education for the benefit of productive and competitive construction firms were captured and analysed and compared to the reviewed literature. From Table 1 it is evident collaborative learning was an important method of learning BIM and it was ranked first with a mean value of 4.64 and a standard deviation of 0.65. Given the five-point Likert scale employed, Badu et al. [34] supports that a factor with a mean value of 3.5 and above is deemed important. It is for this reason that all the methods of teaching BIM in Table 1 are deemed important. The results of the study that collaborative learning enables learners to work independently and in groups will ensure fast and effective compilation and interpretation of data, and this concurred with Babatunde et al. [6], that fast and efficient quantity-take offs and cost estimations was amongst the top-rated benefits of BIM.

Collaboration further ensures effective communication, and vice versa. Azhar et al. [35] and Šadauskienė and Pupeikis [11] that effective communication encourages collaborative partnerships where the accessibility of accurate information from one application to the next is made available. Jäväjä et al. [36] further cites that a collaborative partnership and teamwork ensures the completion of projects before or on time, within/ using lesser budget. With new and digitalised forms of communication between construction professionals, construction firms are required to abandon the old methods and techniques and invest in new emerging forms of communicating for the benefit of improved collaborations. Brown *et al.* [28] and Collins *et al.* [29] further posits that even authentic learning (MIS=4.43; SD=0.80) which is ranked 2nd in the study is rated amongst the prominent methods of teaching BIM in construction education, and it involves learners collaborating in understanding concepts.

The use of problem-based learning (MIS=4.43; SD=0.87) which is ranked third in Table 1 is used to improve the understanding and knowledge of leaners regarding construction management and building materials in the daily operations of construction firms. The ability to undertake constructability, review the processes of construction management from inception to the final stage of the project using BIM elements such as visualisation puts the construction industry in a better position to compete and deliver expected outcomes timeously and within the agreed budget requires a project-based learning approach learning (MIS=4.41; SD=0.84).

The study further indicates that though BIM usage is crucial for the construction firms stepping in the right direction of productivity and competitiveness so is the detailed training on action learning approaches (MIS=4.39; SD=0.84). Given the benefits of the BIM software packages there is also a need for training on the fundamentals and usage of BIM in the construction industry. Clevenger et al. [4] indicates that limited expertise is required from graduates that are only beginning to use BIM [36], it is through training and experience that the expertise of BIM users will develop. More so, AUTODESK [7] shares that BIM training and workshops gives users the ability to virtually apply parametric models in the field of construction technology using Revit BIM software and enables them to adapt to international standards [1]. The results of the study attests that the employability status of most competent and knowledgeable graduates in the BIM discipline can be improved with constructivist learning (MIS=3.93; SD=1.02) from primarily being able to apply 3D drawing concept in practical situations The results of the study also concurs with Olatunji [1] and Olowa and Lill [20] that that being BIM competent opens extended opportunities to construction graduates and gives them an edge to compete for opportunities locally and abroad than graduates without BIM knowledge. However, institutions of higher learning are still faced with the problem of determining an effective learning approach that must be adopted in teaching BIM, to this end an approach that prioritises cooperativeness and collaborations is most beneficial.

Table 1. BIM pedagogy in the construction education

BIM pedagogy	Σ	$\bar{x}$	**R**
Collaborative (cooperative) learning	0.650	4.64	1
BIM-integrated learning	0.789	4.43	2
Problem-based learning	0.873	4.43	3
Project-based learning	0.844	4.41	4
Action learning	0.841	4.39	5
Case-based learning	1.037	4.25	6
Authentic learning	0.912	4.23	7
Constructivist learning	1.021	3.93	8
Reflective learning	1.187	3.82	9

5. Conclusion and recommendations

The objective of the study was to assess the methods of teaching BIM and the benefits of incorporating BIM in the construction education for the benefit of producing construction graduates that are productive and technologically competitive, with the goal of leading the construction industry in the right direction. The right direction referred to in the study refers to the envisaged complete digitalised era of the construction industry. Fast and effective compilation and interpretation of data was the most prominent benefit to stimulate digitalisation in the construction firms through collaborative and cooperative learning. Collaborative learning ensured better communication for improved teamwork, including improved understanding of visualisation and enhanced literacy on BIM fundamentals which was made possible through detailed BIM software training. Problem-based learning approach improved understanding and knowledge of building materials and construction management. The study recommends urgent steps to address lack of resources to introduce BIM education, limited space in the existing curriculum to add BIM content and lack of methods to teach BIM. Additionally, a collaborative effort between relevant stakeholders to develop an educational BIM programme that is well-structured and cost effective is required. For improved understanding of design fundamentals BIM teachings must also be considerate and inclusive of software application, collaborative modelling and BIM-based scheduling as well as advanced visualisation and communication techniques. This is all to drive the agenda of providing construction firms with well informed and competent BIM graduates. More so, rather than the initiative being about students only (which is key) focus and resources should also be placed in stimulating and equipping teachers of BIM with new and improved methods of teaching BIM for the benefit of productive and technologically competitive construction firms. Construction firms are to benefit and lead the construction industry in the right direction through competent construction graduates that are knowledgeable in BIM. It is also significant to note that due to the use of a single institution, attempts to generalise the results of the study across the 24 other universities in South Africa was not possible.

6. Implications

The objective of the study was to assess Building Information Modelling (BIM) as a practice used in the construction firms to lead the construction industry in the right direction. The objective of the study was achieved by assessing and ranking the benefits of BIM in the construction education for the benefit of productive and technologically competitive construction firms. The results of the study have an impact on the existing construction curriculum used in institutions of higher learning. The curriculum must be amended to include more BIM content. It was critical for the study to probe the existing construction curriculum to determine the knowledge construction graduates exit with from institutions of higher learning. From the findings there is limited BIM content incorporated in the curriculum, an area institution needs to investigate. A gap was also evident between what is taught in institutions of higher learning and what is expected by the construction firms. It is therefore clear that amongst many other things that the industry can only be heading in the right direction when the construction graduates are well equipped with BIM knowledge to practically apply in the construction firms. Having construction firms with employees in the form of construction graduates that are sufficiently equipped with BIM allows firms to be productive and technologically competitive. Further research will have to be undertaken to compare existing BIM content in the construction curriculums of all the existing public institutions of higher learning before any attempts can be done to compare or generalise the findings of this study to other countries.

References

[1] Olatunji O A 2019 Promoting Student Commitment to BIM in Construction Education *Engr. Const. & Archit. Mgmt.* **36** 1240-1260
[2] Boon J 2009 *Preparing for a BIM Revolution* 13th Pacific Association of Quantity Surveyors Congress (PAQS)
[3] Smith D K and Tardif M 2009 *Building Information Modelling: A Strategic Implementation Guide for Architects, Engineers, Constructors, and Real Estate Asset Managers* John Wiley & Sons Hoboken NJ
[4] Clevenger C Glick S and del Puerto C L 2012 Interoperable learning leveraging Building Information Modelling (BIM) in construction education *Int. J. of Const. Edu. and Res.* **8** 101-118
[5] Kim J L 2012 Use of BIM for effective visualization teaching approach in construction education *J. of Pro Issues in Engr Edu and Practice* **138** 214-223
[6] Babatunde S O *et al* 2018 Analysis of the drivers and benefits of BIM incorporation into quantity surveying profession *J. of Engr. Des. & Tech* **16** 750-766
[7] AUTODESK 2019 *What are the Benefits of BIM?* [ONLINE] Available at: https://www.autodesk.com/solutions/bim/benefits-of-bim. [Accessed 2 September 2020]
[8] Panuwatwanich K *et al* 2013 Integrating building information modelling (BIM) into Engineering education: an exploratory study of industry perceptions using social network data *In C. Lemckert, G. Jenkins & S. Lemckert (Eds.), AAEE2013: Aust. Assoc. for Engr. Edu. Conf. 1-9*

[9] Abbas A *et al* 2016 Integration of BIM in construction management education: an overview of Parkistan Engineering universities *Int. Conf. on Sust. Des. Engr. and Const.* **145** 151-157
[10] Sacks R and Pikas E 2013 Building information modelling education for construction engineering and management. I: Industry requirements, state of the art, and gap analysis *J. of Const. Engr. & Mgmt.* **139** 05013002, 2013
[11] Šadauskienė J and Pupeikis D 2018 Review of BIM implementation in Higher Education. *J. of Sust. Archi. & Civil Engr* **1** 99-109
[12] BIM academic forum 2013 *Embedding Building Information Modelling (BIM) within the taught curriculum* BIM Academic Forum (BAF)
[13] Forsythe P *et al* 2013 Building Information Modelling in Tertiary Construction Project Management Education: A Programme-wide Implementation Strategy *J. for Edu in the Built Env* **8** 16-34
[14] Jin R *et al* 2018 Project-based pedagogy in interdisciplinary building design adopting BIM *Engr. Const. & Archi. Mgmt.* **25** 1376-1397
[15] Eastman C *et al* 2008 *BIM handbook: a guide to building information modelling for owners, managers, designers, engineers, and contractors* Hoboken, NJ, USA: John Wiley & Sons, Inc
[16] Allen Consulting Group 2010 *Productivity in the buildings network: assessing the impacts of Building Information Models* Report to the Built Environment Innovation and Industry Council Report Sydney
[17] Ghosh A *et al* 2013 From BIM to collaboration: A proposed integrated construction curriculum *American Society for Engr. Edu. (ASEE) Annual Conf.* Atlanta, Georgia
[18] Suwal S *et al 2014* BUM Education: Implementing and Reviewing "OpeBIM" – BIM for Teachers *Computing in Civil and Bldg. Engr.* ASCE. Helsinki: Metropolia
[19] Masood R *et al 2014 Is BIM* Adoption Advantageous for Construction Industry of Pakistan? *Procedia Engr* **77** 229-238
[20] Olowa T *et al* 2020 Conceptualising building information modelling for construction education *J. of Civil Engr & Mgmt* **26** 551-563
[21] Tang L *et al* 2015 Launching the innovative BIM module for the architecture and built environment programme in China *Building Information Modelling (BIM) in Design, Construction and Operations* **149** WIT Transactions on The Built Environment, WIT Press 145-156
[22] Mills J E and Treagust D 2003 Engineering education, is problem-based or project-based learning the answer *Aust. J. of Engr. Edu.* **3** 2–16
[23] Barison M B and Santos E T 2018 Advances in BIM Education. In I. Mutis, R. Fruchter, & C. C. Menassa (Eds.) *Transforming engr. Edu.* 45–122 American Society of Civil Engineers, Reston, VA
[24] Revans R 2011 *ABC of Action Learning* Gower Publishing, Farngam
[25] Marsick VJ and O'Neil J 1999 The many faces of action learning *Mgmt Learning* **30** 159-176
[26]Munoz L 2011 *Constructivist learning theory* available fro,: https://lynnmunoz.wordpress.com/learning-theories/constructivist-learning-theory/ [Accessed 11 December 2020].
[27] Pawson E *et al* 2006 Problem-based learning in geography: towards a critical assessment of its purposes, benefits and risks *J. of Geography in Higher Edu* **30** 103-116
[28] Brown J S *et al* 1988 Situated cognition and the culture of learning (Report No. IRL 88-0008) *Institute for Research on Learning*, Palo Alto, CA.2009
[29] Collins A *et al* 1989 Cognitive apprenticeship: teaching the craft of reading, writing, and mathematics, in Resnick, L.B. (Ed.), Knowing, Learning, and Instruction: Essays in Honor of Robert Glaser, Lawrence Erlbaum Associates, Hillsdale, NJ, 453-493
[30] Lee N and Hollar D A 2013 Probing BIM education in construction engineering and management programs using industry perceptions *49th ASC Annual Int. Conf. Proceedings*, San Luis Obispo, CA

[31] Zhao D *et al* 2015 Building collaborative construction skills through BIM-integrated learning environment *Int. J. of Const. Edu. & Res.* **11** 97-120

[32] Moser C A and Kalton G 1999 *Survey Methods in Social Investigation* 2nd Edition. Gower Publishing Company Ltd, Aldershot

[33] Sushil S and Verma N 2010 Questionnaire Validation Made Easy *EU. J of Scientific Res* **46** 172 – 178

[34] Badu *et al* 2012 Barriers to the implementation of innovative financing (IF) of infrastructure *J. of Financ. Mgmt. of Prop. & Const.* **17** 253-273

[35] Azhar S *et al 2012* Building information modelling (BIM): now and beyond *Const Econ. & Bldng* **12** 15-28

[36] Jäväjä P *et al* 2013 Enhancing customer orientation in construction industry by means of new technology. *In 7th Nordic Conf. on Const. Econ. and Org.* **2013**

IOP Conf. Series: Materials Science and Engineering 1218 (2022) 012054 doi:10.1088/1757-899X/1218/1/012054

Comparing 360° Virtual Reality Learning Configurations for Construction Education

J Kim

Auburn University, Auburn, Alabama, United States

Abstract. Virtual reality is rapidly becoming an effective tool for training in construction. Hazardous environments can be simulated and used for training purposes without endangering students. Furthermore, these virtual environments can be saved and re-used, on-demand, allowing students to observe temporal changes to an environment that would normally require multiple visits to a real project site to achieve. While the virtual construction site visit should not be considered a replacement for the real thing, some site visits get passed over because they are costly and logistically difficult to coordinate. Having options would be desirable. Virtual reality construction site visits using 360° photographs are a novel approach toward substituting for the real thing – yet they can be ineffective. Therefore, considering this media as a replacement for the real experience is not advisable – without some additional upgrades. This research documents a between-groups experiment that compared student's self-reported learning performance when using 360° photographs that were annotated versus 360° photographs that were not annotated. The annotations included pop-ups, visual cues, audio clips, video clips, and quizzes to direct the learning when viewing the 360° photographs. Results indicated increased performance with the students that used the annotated 360° photographs. This study anticipated a difference in performance between the two groups and therefore used eye tracking technology with the students that used the non-annotated 360° photographs to obtain a better understanding of why their achievement differed.

1. Introduction

Instructors of construction management programs recognize the uniqueness and importance of the construction field trip that is used to augment the passive teaching that occurs in the classroom [1]. The construction management student gains valuable experience from a field trip in much the same way as a medical student that visit a hospital. However, there are many challenges in managing a successful field trip [2] and these challenges often become excuses for not incorporating the field trip in the curriculum. Therefore, it is imperative that we find ways that broaden the student's experience base and makes the task of incorporating these experiences less burdensome for the instructors. With this need in mind, this paper focuses on the betterment of the virtual construction site field trip.

2. Background and Research Rationale

There is an ever-growing collection of research studies that examine the use of virtual reality (VR) for training and general education. To begin this examination, there needs to be a distinction made about the various realities that are available and the content used to create them. Furthermore, there needs to be an understanding how the cognitive ability of humans are affected when using technology tools to increase learning performance.

2.1. A Perspective of Virtual Reality

It is not uncommon nowadays to confuse the various types of virtual and augmented realities that are available. In their seminal study on classifying the "continuum" of "reality-to-virtuality", Milgram et al. [3] put forth a concept that is used in nearly all VR research to define how reality is experienced when using different technologies. Their spectrum defines everything from absolute reality to fully simulated reality. Considering the technologies involved in making reality virtual, it is generally understood that VR devices tend toward fully simulated reality – the computer mimics almost all of what the user experiences. Albeit there is also a spectrum of how much reality is simulated in a virtual environment and Kim et al. [2] examined the immersion of VR devices. Figure. 1 illustrates the most common types of modalities of VR used and aligns them based on their level of immersion [4], presence [5], and telepresence [6].

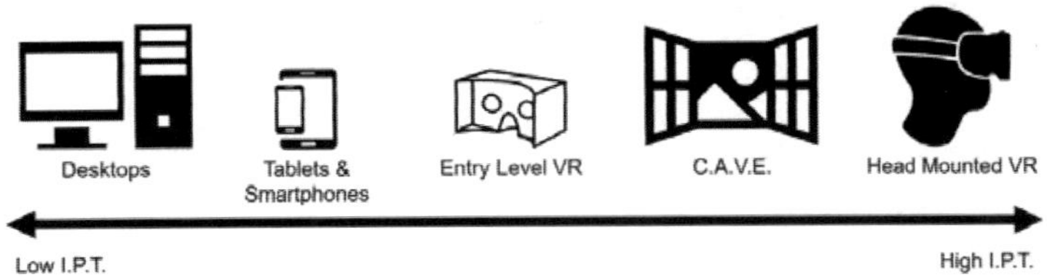

Figure 1. Comparison of Immersion, Presence, and Telepresence (I.P.T.) in various modalities with today's VR technology [2]

From their figure, the reality is more engrossing when devices on the right side of the spectrum are used. Therefore, devices that fully replace the user's view, such as the head mounted VR device, will render the most immersive experience that also gives a sense of truly being in the simulated environment, thus increasing the realism of the experience.

2.2. An Educational Challenge for Virtual Reality

VR can be an effective educational tool as has been indicated in the research [6]-[9], yet there are others that have suggested that for the VR to properly enhance learning, there are some key relationships between people and their environment (i.e. the difference between the real environment and the virtual environment) that may lead to "counter-intuitive" findings for supporting VR as and effective training tool [10]. Li et al. conducted a systematic literature review concerning mixed realities used to train construction safety and identified the following challenges faced when using VR:

- Hardware and software limitations
- Insufficient connectivity
- Complications due to hybridized training (some virtual and some non-virtual)
- Symptomatic effects (motion sickness)

These are significant factors for a technology that is viewed as technical and difficult to implement [2]. Therefore, there remains much work that can be done to research where VR can be used most effectively in the training and educational process.

2.3. Technology and Cognitive Load

The factors enumerated in the previous section are significant, yet one must also consider that people can become cognitively "overloaded" or distracted [11] as they incorporate the use of technology in their learning experience. The cognitive load can be so overwhelming that it prohibits the student from storing newly obtained information from their learning experience [11]. Despite this undesirable outcome when using technology in the classroom, if applied properly, it has the potential to augment the learning experience. As Kim et al. [12] found in their research using augmented reality, they observed that while students exhibited some learning improvement when using extended reality (XR) tools to assist in their understanding of spatial skills they also discovered that the student's perceived a favorable outcome in their learning performance. In their study they indirectly measured cognitive overload by using tools that allowed students to self-report perceived effort when using the technology tools during the experiment. This method represents a novel way to measure cognitive overload and could be used in preliminary studies that seek to address the conundrum of cognitive overload.

2.4. Virtual Reality Content

The virtual environment is mostly digital, in the sense that it replaces the viewer's perception of the world around them. Many sensory replacements can be mimicked, such as sounds, touch, motion, temperature, and smell. Everything within the virtual environment is said to be symbolic of something within reality. VR is intended to act on the viewer's perceptions in such a way as to cause experiences and potentially feelings to encourage behavior [13]. Despite the literature that supports, or casts doubt on VR, there is an underlying notion that the content is what truly differentiates the VR experience [9], [5, (p.1249)], and has effect on the learning experience. Instead of merely creating a virtual environment and testing if the outcome had a benefit or not, there needs to be a more thoughtful approach toward measuring the content that is represented and its impact on the VR learning experience, currently the research is lacking in this attribute of the technology's use.

2.5. Research Rationale and Objective

VR, among other XR technologies, is finding new opportunities as a useful tool for enhancing the learning process. Using VR in the classroom encourages active learning pedagogies that liven the learning experience for students [14]. There are tangible incentives proven from past research that encourages more research in the use of VR in the classroom [7]–[9]. Therefore, aligning a need to increase experiential learning with a method that makes it more available for the student would be prudent.

The objective for this research study was to ascertain the effectiveness of using 360° photographs as a supplement to the construction site field trip. Because content has been demonstrated to be an important element in the success of VR in the classroom, this study will undertake a between groups examination of content that contains annotated features (pop-ups, visual cues, audio clips, video clips, and quizzes) versus content that is not annotated. The researcher anticipates that the annotated content will result in higher self-perceived performance and so a further examination the students using the non-annotated 360° photographs would elicit a better understanding of why their achievement may differ.

3. Methodology

3.1. Demographics

This research included, for convenience, students from a 4-year construction management program in the Southeastern United States. Students that graduate from this program are sought for employment in the commercial construction sector of the US. Table 1 below summarizes the characteristics of the population for this study.

Table 1. Student Demographics

Characteristics	TEST GROUP VR Experience Included Annotations (n=64)	CONTROL GROUP VR Experience DID NOT Include Annotations (n=17)
Age Group		
Younger than 19	1 (1.6%)	-

20 – 25 years 26 – 30 years	62 (96.9%) 1 (1.6%)	16 (94.1%) 1 (5.9%)
Gender Female Male	 4 (6.3%) 60 (93.8%)	 2 (11.8%) 15 (88.2%)
Academic Year (Postsecondary) Junior (3rd year) Senior (4th year)	 49 (76.6%) 15 (23.4%)	 14 (82.3%) 3 (17.6%)
Experience Working in the Construction Industry Yes No	 44 (68.8%) 20 (31.3%)	 9 (52.9%) 8 (47.1%)
Have You Ever User VR in the Classroom? Yes No	 44 (68.8%) 20 (31.3 %)	 11 (64.7%) 6 (35.3%)

3.2. Controlled Experiment

A controlled between-groups experiment was arranged so that the students from each group were divided and separated from one another to prevent either group from knowing what type of VR experience they would encounter. The test group (TEST) viewed a 360° photograph that included annotations (annotations are described in the next section of this paper). The control group (CONTROL) viewed a non-annotated 360° photograph. Both groups were given a head mounted VR headset (Oculus Quest with hand controllers) and asked to view the 360° photograph (see Figure. 3a). The students were allowed up to ten minutes viewing time and upon completing their viewing were asked to self-report their performance in understanding what they observed. Figure. 2 illustrates the experiment workflow.

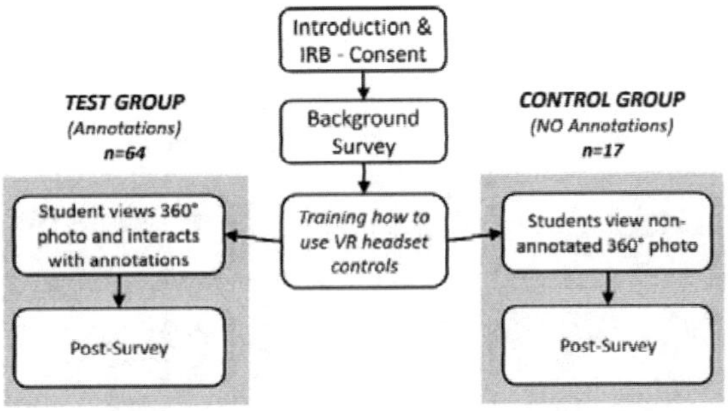

Figure 2. Experiment Workflow

3.3. 360° Photographs and VR Annotations

This section will describe the elements included in the TEST environment and constitutes the independent variable of the study. The elements described here were viewed by the students in the TEST group by using the head mounted VR device mentioned in the previous section of this paper.

3.3.1. The 360° Photograph

The same 360° photograph was used for both the TEST and CONTROL groups (see Fig. 3a). The subject of the photograph was a music auditorium project that was currently under construction. A student could change their viewing perspective by rotating their head. The VR headset would respond by displaying a different perspective of the 360° photograph. The effect of the simulation mimics a person viewing their surroundings as if they were at the actual construction project site. For the purposes of this study, the photograph was a static image that did not contain any motion or audio, aside from the video annotation (see section 3.3.2. Annotations) included in the TEST group's viewing. Fig. 3 illustrates the 360° photograph and how it was viewed by the students.

(a) (b)

Figure 3. The 360° Photograph (a) and Students Viewing by way of a VR Headset (b)

3.3.2. Annotations

The TEST group's VR experience included various annotations that were part of their VR environment. The annotations appeared as graphical notations in the 360° photograph that the students could interact with using a hand controller. The following annotations were used in this experiment and are depicted in Figure. 4.

- **Textual Annotations** – these annotations were staged as visual pop-up text boxes that students could read and obtain more information about what they were currently viewing.
- **Video Annotation** – these annotations were designed to play a short video of a construction topic that also included an audio track.
- **Quiz Annotation** – this annotation was used to assess a student on some element of the VR experience – they interacted with the quiz with their hand controller and were given immediate scoring feedback from their response.

(a) (b)

(c)

Figure 4. Annotations used with the TEST group
(a) Textual Annotations, (b) Video Annotations, and (c) Quiz Annotations

4. Results

After viewing the 360° photograph in the VR headset, all the students (TEST and CONTROL) were asked to self-report learning performance. Figure.5 summarizes the self-reported learning performance on nine subscales using a 5-point Likert scale.

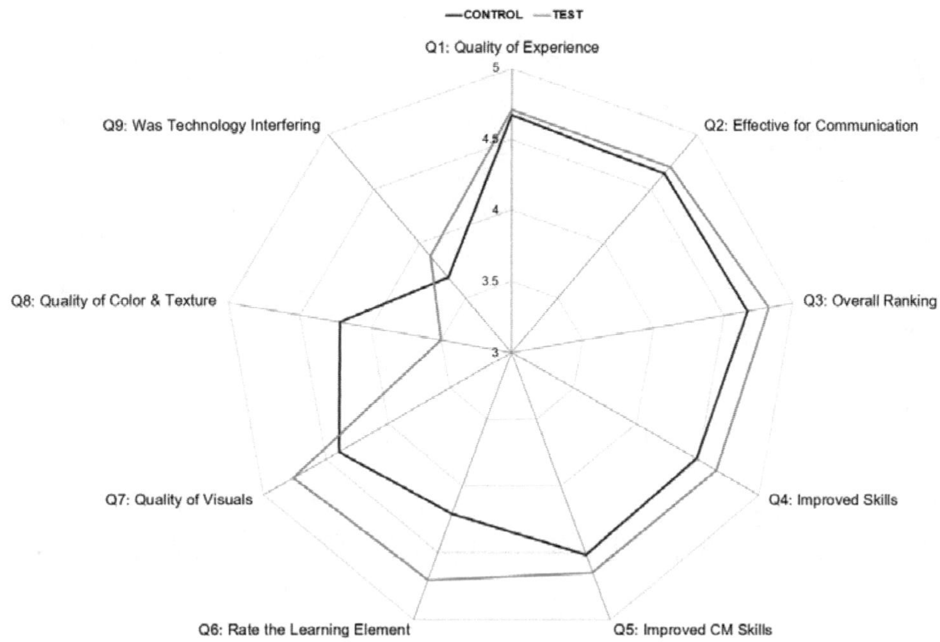

Figure 5. Self-Reported Learning Performance on Nine Subscales

5. Analysis and Discussion

The student's self-reported scores shown in Fig.5 demonstrates a higher perceived learning performance by the TEST students in all but one category. It is presumed that Q8 (Rate the quality of color and texture) was ranked lower because the students may have been rating the annotations and not the 360° photograph. The CONTROL group did not have annotations and their responses would have been directed only at the quality of the 360° photograph. The annotations did appear to use a lower resolution graphics, due in part to a technical issue when converting the annotation images from a 2D layer to a simulated 3D layer of the 360° photograph. While this self-reporting survey was not statistically compared, it does represent a tendency, from the student's experience, that the tool was configured in such a way to enhance their perceived learning experience. As previously mentioned, this was not unexpected. Therefore, the researcher chose to further investigate the CONTROL group's VR experience.

5.1. Eye Tracking Analysis of the CONTROL Group's Experience

Half of the participants from the CONTROL group were asked to participate in a follow up examination using head motion tracking sensors and eye tracking sensors that were attached to a different VR headset (HTC Vive). The same 360° photograph (see Figure. 3a) was presented to the students again and they were asked to freely observe their surroundings – as they did before. The gathered motion and eye tracking data were compiled into heat maps so that the researchers could observe areas of interest to the students. It was anticipated from this follow up examination that the researcher could identify regions of the 360° photograph that the students seemed more interested in and thereby inform where annotation elements could be placed to better suit the student's interests and in-turn their perceived learning performance. An example image included in Figure. 6 illustrates regions of ductwork in the project that students seemed most interested in by indication of their gaze in the eye tracking VR headset. The area of the heat map colored in red indicated the highest viewing intensity and diminished from there to yellow then to green which indicated some viewing, but less than yellow and red. These areas represented focus areas for content authors to add more annotations that could further educate the students about what they are looking at when presented with a 360° photograph.

Figure 6. Areas of Interest Heat Map of a 360° Photograph

Some limitations in this study existed and should be considered if this study were to be expanded.

- This experiment was conducted during the height of the world-wide COVID epidemic (from 2020 March to 2020 May). This required the researcher to enhance measures requiring social distancing and limiting personal contact. In doing so, the population pool was relatively small and severely unbalanced (TEST n=64 and CONTROL n=17). While not tested, this imbalance may yield different results if the population were to be expanded and the groups were more evenly balanced.
- The results from the eye tracking study were not reincorporated into the original study and re-tested to determine if the eye tracking data could be useful in determining where annotations should be placed in the 360° photographs. This is a recommendation for further research and could validate the effectiveness of the workflow presented in this paper.

6. Conclusion

The objective of this research study was to examine a supplement to the construction project field trip that students commonly take to augment their classroom learning. These field trips can be costly, time consuming, and unsafe for students [1] but we should not discontinue them altogether. We need to find solutions that supplement the classroom field trip. This research study examined the use of 360° photographs that were annotated and non-annotated and compared the perceived learning experience from their use. Overall, the annotated 360° photographs provided a better perceived learning experience, and the data suggests that the non-annotated versions were also significant as a learning tool. As this result was somewhat expected the researchers sought to understand if a method could be employed to close the gap between the two different VR experiences (annotated versus non-annotated). It was determined that using head motion and eye tracking software on non-annotated viewing users while they were observing 360° photographs could be a useful workflow to add annotations to a 360° photograph.

Again, the overall intent of this research study is not to advocate for the replacement of the construction site field trips. Those experiences offer intangible learning opportunities despite their difficulties in being coordinated and executed [1]. However, it is because of these shortcomings that alternatives for the construction site field trip must be sought out. Within this study, the researchers considered the content of a VR experience as a significant element of the learning experience and in so doing advocate that the comprehensiveness of the content is also a significant contributor to the learning experience. Furthermore, the researchers advocate a workflow like the one designed in this study as a method for testing the content in a VR experience that can improve a student's perceived learning outcome.

References

[1] R. Eiris Pereira and M. Gheisari, "Site Visit Application in Construction Education: A Descriptive Study of Faculty Members," Int. J. Constr. Educ. Res., vol. 00, no. 00, pp. 1–17, 2017.
[2] J. Kim and T. Leathem, "Virtual reality as a standard in the construction management curriculum," in International Conference on Construction Futures, 2018.
[3] P. Milgram, H. Takemura, A. Utsumi, and F. Kishino, "Mixed Reality (MR) Reality-Virtuality (RV) Continuum," Syst. Res., vol. 2351, no. Telemanipulator and Telepresence Technologies, pp. 282–292, 1994.
[4] L. Freina and M. Ott, "A Literature Review on Immersive Virtual Reality in Education: State of the Art and Perspectives," eLearning Softw. Educ., vol. 1, no. July, pp. 199–205, 2015.
[5] H. Brenton, M. Gillies, D. Ballin, and D. Chatting, "The 'Uncanny Valley': Does it Exist and is it Related to Presence?," in Presence Connect, 2005, pp. 1–8.
[6] J. Seibert and D. M. Shafer, "Control Mapping in Virtual Reality: Effects on Spatial Presence and Controller Naturalness," Virtual Real., vol. 22, no. 1, pp. 79–88, 2018.
[7] A. H. Behzadan and V. R. Kamat, "Enabling Discovery-Based Learning in Construction Using Telepresent Augmented Reality," Autom. Constr., vol. 33, pp. 3–10, 2013.
[8] O. Golovina, C. Kazanci, J. Teizer, and M. König, "Using serious games in virtual reality for automated close call and contact collision analysis in construction safety," in Proceedings of the 36th International Symposium on Automation and Robotics in Construction, ISARC 2019, 2019, pp. 967–974.
[9] M. Wolf, J. Teizer, and J. H. Ruse, "Case Study on Mobile Virtual Reality Construction Training," in Proceedings of the 36th International Symposium on Automation and Robotics in Construction, ISARC 2019, 2019, pp. 1231–1237.
[10] J. Whyte and D. Nikolic, "User experience in VR systems," in Virtual reality and the built environment, 1st ed., New York: Routledge, 2018, pp. 14–42.
[11] J. Sweller, "Cognitive Load Theory," Psychol. Learn. Motiv., vol. 55, pp. 37–76, Jan. 2011.
[12] J. Kim and Javier Irizarry, "Assessing the Effectiveness of Augmented Reality on the Spatial Skills of Postsecondary Construction Management Students in the," 34th Int. Symp. Autom. Robot. Constr. (ISARC 2017), 2017.
[13] J. I. Montana et al., "The Benefits of emotion Regulation Interventions in Virtual Reality for the Improvement of Wellbeing in Adults and Older Adults: A Systematic Review.," J. Clin. Med., vol. 9, no. 2, 2020.
[14] J. Conklin et al., A Taxonomy for Learning, Teaching, and Assessing: A Revision of Bloom's Taxonomy of Educational Objectives Complete Edition. JSTOR, 2005.

IOP Conf. Series: Materials Science and Engineering 1218 (2022) 012055 doi:10.1088/1757-899X/1218/1/012055

Design of multi-layered building structures using BIM method

D Szalai and T Horváth

Széchenyi István University, Egyetem tér 1. Győr, H-9026, Hungary

szalaidory@gmail.com

Abstract. Today BIM is an inevitable concept in construction design and while the workflow of consultant designers is mostly influenced to a small extent, new methods require a new kind of communication between them. This research focuses on the workflows of construction designers because their design works have a big impact on many other disciplines. This paper presents an examination about the processing methods of the building structure designers and the use of BIM with a questionnaire survey. In this research, a new way was also introduced to integrate definitions of multi-layered structures (created by construction designers) into the BIM/CAD model itself. The goal was to allow the data in the layer structure documents to be automatically linked to the model, and make it possible for all collaborators to continue working with the data separately. This paper proofs that this automated workflow greatly increases efficiency, accuracy and reduces human errors. ArchiCAD and Excel were used as our development framework: former one to create CAD/BIM models, latter one to handle the database of multi-layered structures. Individual codes were used to identify each unit of the layer structure and link them to the model itself. This way, it is not necessary to manually add the data to the elements, but the material properties can be read automatically into the model and the quantities can be measured. Similar scale, size, and complexity buildings were selected as a test environment where the ordinary and the new method were used in parallel throughout the design process. This provided a good comparison and demonstrated the effectiveness of database-based design. With the new approach, time required for first-round quantification was reduced by 41% and the time of a repeated quantification due to modifications was reduced by 56%, in average.

1. Introduction

Construction design, although 2D-based design is a long way off, in many places uses outdated technologies. This lag can be felt especially in the workflows of discipline designers. The main tool for digitalization in construction is the BIM workflow. The BIM (Building Information Modelling) is a model-based design method, which basically means 3D elements with extended information content [1]. In the construction industry, the use of BIM is becoming more widespread [2] [3]. At the same time, the spread of BIM processing towards discipline designers has not yet started or has started to just a small extent. The aim of this article is to present a developed workflow which can integrate the method of writing multi-layered structure documentation into model-based design. This database-based way can increase efficiency and extend BIM systems to disciplines.

As a first step, current building construction design methods and work processes should be studied. The Hungarian construction designers' habits were surveyed through a questionnaire. Then based on the results, a solution system was developed to integrate their tasks into the BIM workflow.

IOP Conf. Series: Materials Science and Engineering 1218 (2022) 012055 doi:10.1088/1757-899X/1218/1/012055

2. Working methods of building construction designers

The building construction designer is usually a specialized architect who deals with the design of building structures and / or the preparation of construction plans and has gained extensive experience. It is important to note, that such a specialist building construction designer is not involved in all projects. For smaller ones, this task is usually performed by the architect team. However, for larger projects, the building construction designer may not be a single person, but a team with different competences such as energetics, building insulation, acoustics, fire protection and so on.

Building construction designers' responsibility is to determine the most appropriate structural designs taking into account the complex effects and uses on building structures, and

- in the case of layered structures, they prepare layer documentation with a list of the planned materials, the expected performance characteristics, material thicknesses and construction technology instructions;
- in the case of problematic section or floor plan details, they draw up detailed plans to determine the correct design of the appropriate structural connections;
- and they prepare performance specifications about the properties of materials, structures and about the expected properties (performance characteristics) to determine the range of interchangeable products that can be used for a specific purpose.

3. Survey about building structural design methods

In the form of an online questionnaire [4], a survey was conducted on the working methods of building construction designers in Hungary. The questionnaire was available online for a month, the survey was posted on several professional forums to reach the designers. The aim of the survey was to map the processing strategies used by the construction designers in connection with the performance of each subtask, as well as the impact of BIM systems on their work. The following questions were included:

1. How many years of experience do you have in the field of building structure design?
2. What scale of projects do you usually work on?
3. In which program do you make the construction details?
4. What program do you use to create multi-layered structure documentation?
5. What program do you use to create performance specifications?
6. Does the spread of BIM-based design affect on your working methods at present?
7. Will the spread of BIM-based design have an impact on your working methods in the future?

The questionnaire was completed by a total of 35 designers engaged in building construction design. 60% of the respondents work only in this field, and the rest are architects who do this part of the work themselves within the whole design task, so they have gained a lot of experience in it. 50% of the respondents have more than 10 years of experience and 23% have 3-5 years of experience. Mainly highly experienced designers took part in the survey, so conclusions can certainly be drawn from the collected data on the conditions in Hungary.

The examination of software use shows that digitization has advanced in the field of architectural design. The results of the programs used for the three characteristic tasks of building construction design (detail drawing, layering and performance characterization) are shown in Figures 1-2-3. Detail designs are almost always created with the help of a program capable for vector graphic designs, which also facilitates editing and collaboration with related disciplines, as the software is able to link the drawings to the model elements by sending the given detail drawings back and forth on a .dwg basis. Thus, for example, detail drawings of branch designers can be linked to architect plans, where the location of the recording can also be accurately indicated.

50% of the respondents use Word and 39% use ArchiCAD to create and maintain multi-layered structures. These two programs are fully suitable for flowing text editing, but data can only be further processed manually, making it impossible to connect them to BIM systems. 82% of respondents also use Word software to create performance statements. This data is remarkably high, from which two conclusions can be drawn. On the one hand, experts do not solve administrative tasks in a way that can be further developed and effectively combined with other software, but they concentrate on the given task in the most convenient way. On the other hand, they do not currently have a software solution that can help the task with sufficient efficiency along the lines of BIM.

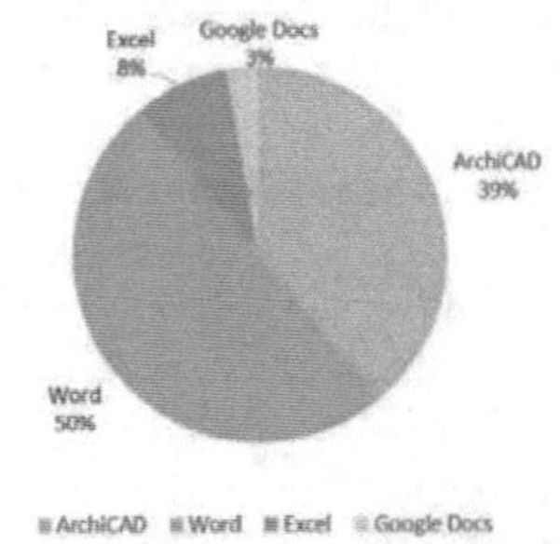

Figure 1 Software used for multi-layered structure documentation

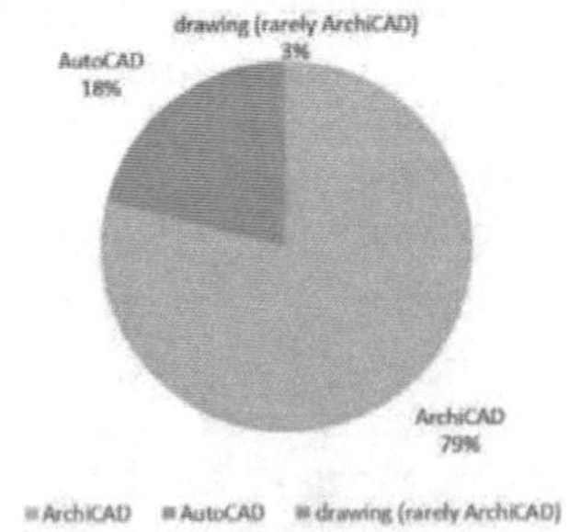

Figure 2 Software used for detail drawing

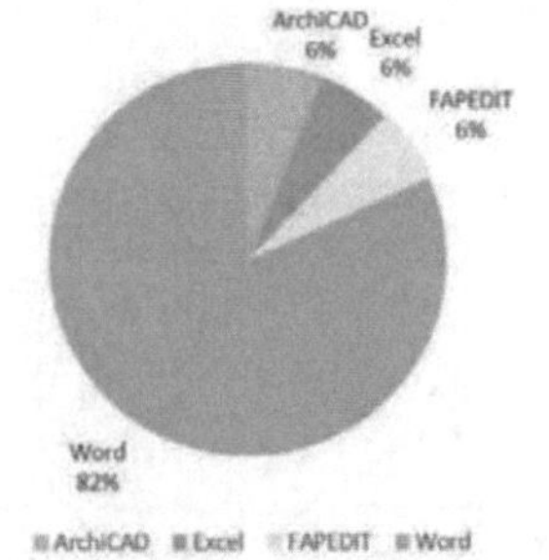

Figure 3 Software used for performance display

Figure 4-5 show the current and future impact of BIM-based architectural design and design processing on the work of building construction designers based on the opinions of the respondents. The difference between the work processes that can be envisioned according to the current and the BIM principles shows well that this area has so far been only marginally affected by the new approach. At the same time, it is also clear that most construction designers expect significant change in this area.

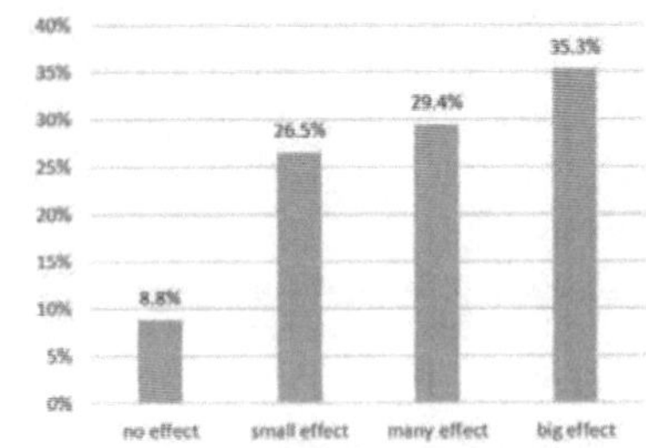

Figure 4 The current impact of BIM on the work of building structure designers

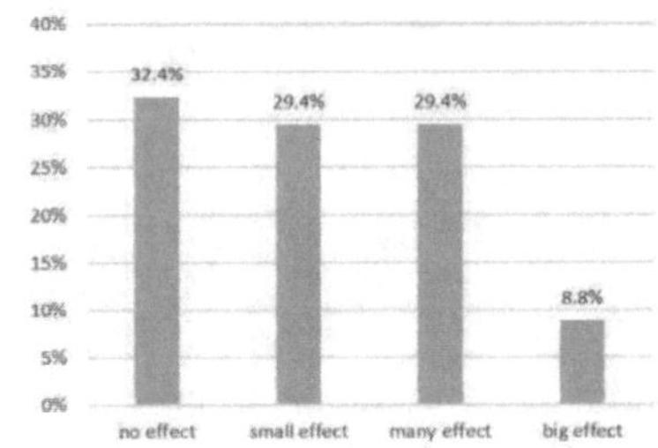

Figure 5 Expected impact of BIM on the work of building structure designers in the future

4. Dynamic Information Raise method for multi-layered structures

The examination of the construction designers' workflows make obvious that flowing text data writing is essential in building design, as many other disciplines use this data, and they have an important role in the final design documentation of the building. The results of the survey reported above also show that there is a need to increase digitization and BIM integration in the field of building construction design [5] [6].

One obvious IT solution for big data management is the application of databases. A database can be almost anything that can manage recorded values in a structured way, organize them into tables, and create relationships between them, and so on. Creating multi-layered structure documentation in a systematic form would provide an opportunity to automate certain work processes. One such process is the quantification of the designed materials, as the BIM model provides the correct geometry, so the quantities, and the exact material definition would be provided by the database-based multi-layered structure documentation [7].

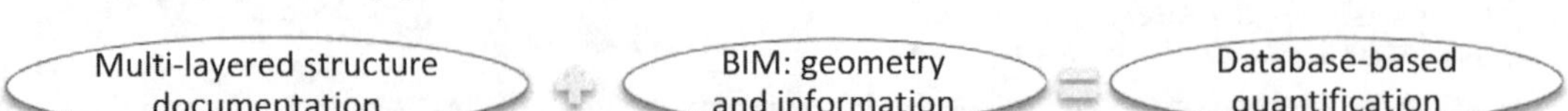

Figure 6 Database based material quantity logic

The concept of Dynamic Information Raise (DIR) has to be introduced, which means this process. Systematic, structured processing of data created by construction designs in a database-like environment and then automated loading into the BIM model and further use by related disciplines.

4.1. New method of making multi-layered structures documentation

A new method has been developed to create multi-layered structure documentation on a database basis, so it is possible to link it to models created in 3D CAD software, which also operate as databases. Interconnection allows automation of quantification and loading of material properties.

Excel was used as the framework for the development because it is available on all colleagues' computers, it is easy to manage, and it is also suitable for programming with the help of the VBA (Visual Basic for Applications) development environment. The goal was to create a solution that is easy to use but stores all the necessary information in a form suitable for further processing. Therefore, a structured system was created, the logic of the system is shown in Figure 7. A hierarchical, extensible system has been defined that more accurately describes the properties of the products designed into each structure.

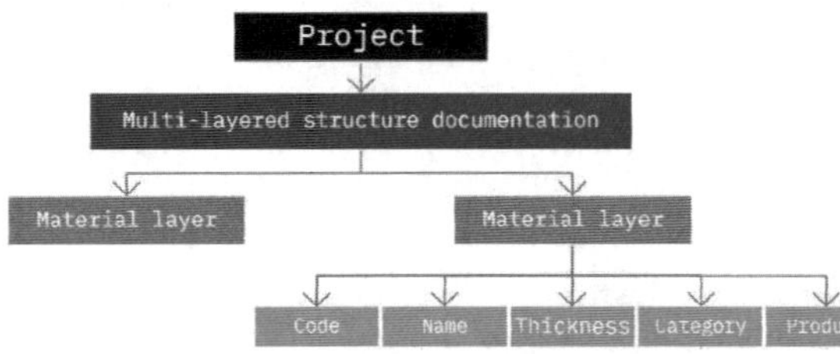

Figure 7 Positioning layer documentation in project hierarchy

The Excel-based module is divided into three main groups, into three worksheet groups: building materials (orange), multi-layered structure documentation interface (blue), assistant sheets (green) (figure 8). The designer can group the construction products thematically and collect all the product properties on the orange worksheets. The assistant sheets contain the background settings necessary for the operation of the program, the designer has nothing to do on these tabs. The main function is available on the blue tab, where construction designers can create the layer sequence of the multi-layered building structures. In the upper right corner are buttons required for operation, which are programmed in VBA, for example a button for the creation of a new multi-layered structure unit.

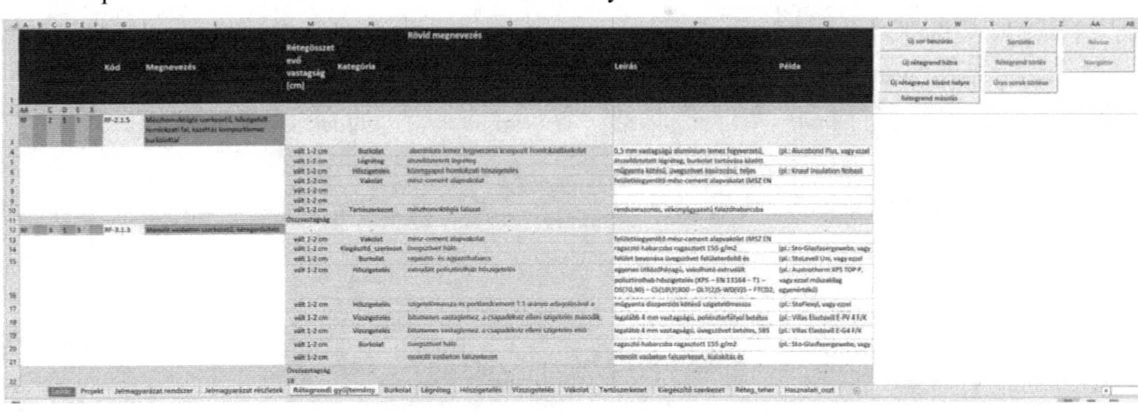

Figure 8 Excel development for multi-layered structure documentation

The operation of these buttons is ensured by VBA programming. The principle is that the blue background cells can be edited freely, while the modification of other cells need an administrator permission. Excel's non-editable cells and code are also password-protected to prevent accidental errors and ensure the proper work of the file.

4.2. Connection with the model file

An important feature of the new method is the ability to connect to a 3D BIM model. The main advantage of its application is that Excel data can be linked to the information of the model elements. The model must be prepared for the integration of the subsystem, since automated operation can only be ensured if the following conditions are met:

- geometrically correct model elements are created;
- layer unit codes are defined, assigned to model elements, either in the name of the composites or fixed as a property value, etc.

An architectural design software usually has a function that can sort its information content into a list according to a filter condition system. This function can be used to extract the required data, so the multi-layered structure layer unit code and the quantity value. A BIM manager is needed to supervise the building of the right model and ensure the right operation.

Excel files that contain the documentation of multi-layered structures must be linked to the BIM files that contain quantities. Along the principle of database-based data linking, with the help of Excel's Power Query add-on, the two amounts of data can be linked using a layer unit code and converted into a large table (figure 9). The completed table is containing all the data, and the required data can be queried and formatted. It is quick to filter and summarize data, but other operations can be also performed on the data. The main advantage of the above method is manifested when the source files are modified, the queries can be updated, thus the documents can be generated quickly.

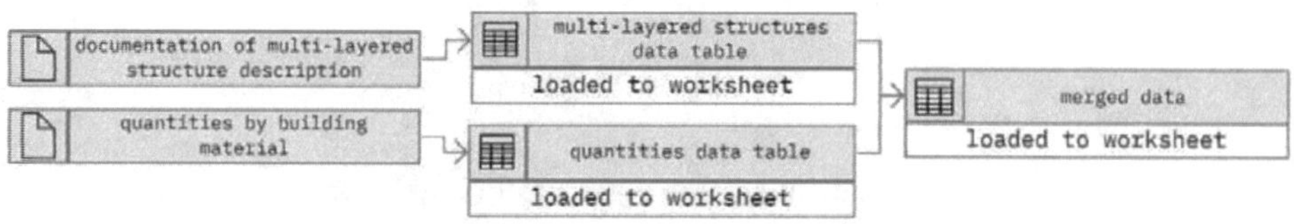

Figure 9 Merge of quantity and material data

4.3. Project testing

The developed method and program have been tested on real projects. Ongoing design tasks provided a good opportunity to compare the ordinary and the developed method. The development was tested on three living design projects, so that the building material quantity take off (QTO) was also completed using the ordinary, manual calculations and the new method. An efficiency comparison was made by measuring the time spent on these processes. The basic data of the test projects and the results of the comparison are summarized in Table 1.

Table1 Comparison of the quantification of constructional structure of three buildings

	Project1	**Project2**	**Project3**
designed function	bookstore	office building	military base
designed floor area (circa)	11 000 m^2	25 000 m^2	40 000 m^2
design task status	new	renovation	new
design task complexity	average	complex	simple
QTO with ordinary method	28,0 h	31,5 h	24,5 h
QTO with new method	15,5 h	13,5 h	19,0 h
QTO due to modifications with ordinary method	3,0 h	4,0 h	3,5 h
QTO due to modifications with new method	1,5 h	1,0 h	2,0 h

Testing three projects with different functions and scales, the comparison of quantification with the ordinary and the new method showed significant saving of working time. Although this test procedure is not suitable for exact comparison, as small number of different projects were observed. The timing of quantification depends on many factors (e. g. building complexity, easy understanding of drawings, colleague practice, etc.). However, the positive effect can still be stated, and it is likely that this method will increase the accuracy and reduce the planning time. The test showed reduction in the quantification time by an average of 41%, and the re-quantification was also achieved in an average of 56% less time due to some modification.

5. Conclusion

The aim of the research was to create a development that can actively channel construction designers into BIM workflows. A survey was used to assess the BIM impact of the area. The results clearly show that this sector of the industry still uses ordinary design methods, using a word processing program (mostly Word) to create written documents. In the digital design environment, the rapid flow of information and its connection to a model is playing an increasingly important role.

A unique programmed solution was created in Excel framework, thus, the multi-layered structure information can be connected to the model and the information can be further processed. The quantity take off method was examined primarily. Using the suggested solution the quantities can be calculated from the building material and the geometric data of the BIM. This developed method is called Dynamic Information Raise.

The developed new solution was tested on three design projects. The results clearly show a significant reduction in time requirements in the case of QTO tasks. In addition to the time saving, the loss of information content can also be minimized, as the document prepared by the building construction designer is connected directly to the BIM.

Multi-layered-structure data generated using the DIR method also allows for additional automated processes, it can provide connected and updatable data for energy calculation about the building structures, or the layered weight data can be used during structural dimensioning. There are many more connection possibilities, which would be worth to be examined in a later research.

6. Acknowledgement

Special thanks to the TSPC Group for the opportunity and the support of the development, for the testing opportunity, and for the comments helping the research. Thanks to Épszerkinfó Kft. for the application of the development, for providing useful ideas and for the support in the implementation.

References

[1] C. Eastman, P. Teicholz, R. Sacks, K. Liston, BIM Handbook: A guide to building information modeling for owners, managers, designers, engineers, contractors. John Wiley, Hoboken, 2008, 490 p.

[2] K. Zima, E. Plebankiewicz, D. Wieczorek, "A SWOT analysis of the use of BIM Technology in the Polish construction industry" Buildings, tom 10, nr 1, https://doi.org/10.3390/buildings10010016

[3] A.J. Garcia, S. Mollaoglu, M. Syal, "Implementation of BIM in Small Home-Building Businesses" Practice Periodical on Structural Design and Construction, tom 23, nr 2, doi: https://doi.org/10.1061/(ASCE)SC.1943-5576.0000362

[4] Y.M. Brenda, S. Arash, Z. Li Hao, W. Mark, C. Yuan, University of Toronto: 2nd Annual BIM Report 2019; Canada

[5] C. Sun, S. Jiang, M.J. Skibniewski, Q. Man, L. Shen, "A literature review of the factors limiting the application of BIM in the construction industry" Technological and development of economy, tom 23, nr 5, pp. 764–779, https://doi.org/10.3846/20294913.2015.1087071

[6] H. Kim, Z. Shen, I. Kim, K. Kim, A. Stumpf, J. Yu, "BIM IFC information mapping to building energy analysis (BEA) model with manually extended material information" Automation in Construction, tom 68, nr 8, pp. 183–193, http://dx.doi.org/10.1016/j.autcon.2016.04.002

[7] M. Cheng, N. Chang, "Dynamic construction material layout planning optimization model by integrating 4D BIM" Engineering with Computers, tom 35, pp. 703–720, doi: https://doi.org/10.1007/s00366-018-0628-0

IOP Conf. Series: Materials Science and Engineering 1218 (2022) 012056 doi:10.1088/1757-899X/1218/1/012056

Redefinition of procurement process with BIM in the Czech Republic

R Bouška, R Schneiderová Heralová, P Matějka and Z Rudovský

Faculty of Civil Engineering, Czech Technical University in Prague, Thákurova 7, 166 29, Prague, Czech Republic

robert.bouska@cvut.cz

Abstract. The paper is describing process of preparation of the evaluation methodology that will help public and private contracting authorities to evaluate an appropriate supplier of the BIM methodology for their projects. Such a methodology will accurately define requirements for BIM implementation for different phases of the project. It will also evaluate companies' fulfilment of these requirements from previous projects or based on the presented technologies of the company. The evaluation methodology will find its application both in the state administration and in the private sector of Czech civil engineering industry. In the current market, there is a need to quantify the level of implementation of work processes and outputs of the BIM methodology. Such a methodology assessment will help improve and make the procurement process more selective based on BIM.

1. Introduction

A number of developed countries, including the Czech Republic, have relied on the BIM methodology as a tool to help prepare the public sector and the entire construction industry for the challenges of the digital world. In the Czech Republic, the Concept for the introduction of the BIM methodology was adopted by the Czech government in the autumn of 2017. The state decided to motivate public and state administration, i.e. public contracting authorities, to work digitally first, and through them also to support the standardisation of digitalisation in the private sector.

The resolution of the Czech government from 1st January 2021 [1] and the approval of the update of the schedule of the BIM Implementation concept will make it possible to reconcile the gradual introduction of the use of building information modelling in construction contracts financed from public sector and the start of digital construction management, as stipulated by the already valid Act No. 47/2020 Sb. This act also makes use of BIM mandatory for above-limit public construction works (in the sense of procurement), financed from public budgets, including the manufacture of their preparatory and project documentation, with taking into account the conclusions of the evaluation of the pilot projects and with taking into account specificities of each type construction. This part of the law is scheduled to go into an action from July 2023. [2]

The data from the building information model can be used as one of the ways to prepare documentation for individual stages of the construction procedure very easily. Information and Digital Modelling Act envisages the creation of information models during the individual phases of the construction life cycle, which are linked to the digital procurement. [3] [4]

2. Main delivery methods used for construction projects in the Czech Republic

In current construction practice, several types of different delivery systems are used to varying degrees, depending on the suitability of their application. [5] The delivery systems applied in different countries vary slightly according to the traditional division of labour, but the main features are common and applicable in any market economy. The Czech Republic is influenced in this respect by the short history of the market economy and although it can be said that in principle all types are used, not all are respected and used to the same extent as in other advanced market economies.

This paper will focus on the following two main delivery systems used in the Czech Republic. [6]

2.1. Design-Bid-Build

The traditional supply chain system, also called Design-Bid-Build is undoubtedly the most used supply system in the Czech Republic and in other countries. The reason for its frequent use is not only its tradition, but above all its transparency in controlling project costs, which is why it is mostly used for construction contracts financed from public budgets, or its use is required by the legislation of a given country.

In the Design-Bid-Build system, the investor concludes two contracts for work. One with the independent designer and the other with the general contractor. The latter can hire subcontractors for selected works. [7]

2.2. Design-Build

In the Design-Build method, the responsibility for the development of the design of the documentation is the responsibility of the contractor. The contracting authority will award the entire project to one contractor who will carry out the project from start to finish. This contractor might not necessarily be selected based on the lowest tender price. The selection process should determine whether a particular contractor can complete the entire project.

One of the main advantages of the DB method is the openness of the procurement proposal, where the project is specified only through the requirements for the purpose and function of the building. The contractor is thus free to choose the technologies and work processes according to his capabilities and experience. Another advantage is the unification of the design and implementation of the building under one contractor, which allows for greater financial and time savings. [1]

3. Public procurement process in the Czech Republic

The procurement process and the tendering procedure vary according to the subject of the procurement and its expected cost. The award of a public contract entails an obligation to publish information in publicly accessible places such as the Public Procurement Bulletin, the contracting authority's profile and, where appropriate, the Official Journal of the EU. However, it can be generalised and divided into several steps. [8] These steps are described in the following subsections.

3.1. Defining the type of procurement

The first step is to choose the type of procurement procedure. The contracting authority may choose from several types of procurement procedures - open procedure, restricted procedure, negotiated procedure with publication, negotiated procedure without publication, competitive dialogue procedure, innovation partnership procedure, concession procedure, simplified procedure and simplified below-limit procedure. The choice of the appropriate form of procurement procedure is influenced primarily by the subject of the public contract, its expected price and the characteristics of the contracting authority (business, municipality, subsidized contracting authority, etc.).

3.2. Call for submission

Once the contracting authority has chosen the type of procurement procedure, the procedure is launched by means of a notice of initiation for publication or a call for tenders (simplified invitation to procurement). Here the contracting authority shall specify in detail the subject of the contract and the conduct of the procurement. Often, the invitation to submit tenders is accompanied by the so-called procurement documentation, which specifies in detail the subject of the public contract within a specific procurement procedure.

3.3. Determination of the method of submission of tenders
Tenders shall be submitted in written or electronic form. The envelope must bear the name of the procurement and the address of the tenderer. In the event of receipt after the deadline for submission of tenders, the envelope containing the tender shall be returned to the tenderer. Tenders received after the deadline for submission shall not be opened. Each tenderer is limited to submitting only one tender. A variant solution shall be considered as one tender.

3.4. Defining the requirements for demonstrating qualifications
Meeting the required qualification conditions is an essential element for proper participation in the procurement procedure. The law recognizes four basic types of qualification requirements. Basic and professional qualifications must always be required by the contracting authority. The economic, financial and technical qualifications are chosen by the contracting authority according to its requirements in relation to the public contract and need not be required at all.[8]

All qualification requirements are defined by the contracting authority in the procurement documentation. Each supplier should be able to determine whether it meets the qualifications from the procurement documentation.

3.5. Determination of the deadline for submission of the tender
The time limit for receipt of the required documents proving compliance with the qualification, as well as the time limit for receipt of requests to participate in a restricted procedure, a negotiated procedure with publication or a competitive dialogue, may not be less than 37 days for public contracts above-limit public contracts. For urgent objective reasons, the time limit may be reduced to 15 days. For below-limit public contracts, the time limit shall not be less than 15 days, or at least 10 days for urgent objective reasons. [8]

The time limit for the submission of tenders may not be less than 52 days for above-limit procurement procedures, less than 40 days for restricted procurement procedures or less than 10 days for restricted procurement procedures for urgent objective reasons. [8]

For below-limit public contracts, the law provides for a minimum time limit of 22 days in an open procedure, 15 days in a restricted procedure and a simplified below-limit procedure, or at least 7 days in a restricted procedure and a simplified sub-limit procedure, unless a standard minimum time limit cannot be set for urgent objective reasons. [8]

The time limits shall begin to run on the day following the day on which the procurement procedure is opened. [8]

3.6. Determination of methods for evaluation of tenders
The contracting authority has the right to choose between two basic evaluation criteria. It may choose between the economic advantage of the tender or the lowest tender price.

3.7. Issuance of the decision of the most suitable tender
Immediately after selecting the winner of the procurement, the contracting authority issues a decision on the selection of the most suitable tender. The contracting authority must send a notification of the selection of the most advantageous tender to all tenderers concerned within 5 working days of the decision. The notification must include the details of the contracting authority and the title of the contract. It shall also identify the winner of the procurement, stating its full name. This shall be followed by a justification of the actual selection, specifying the reasons for the selection in accordance with the applicable conditions set out in the procurement documents.

3.8. Conclusion of the contract

After receiving the notification of the selection of the most advantageous tender, the contracting authority is obliged to allow each tenderer a period of time to lodge objections to the contracting authority's decision to select the most advantageous tender. Each tenderer will normally have a 15-day period to object separately from the date of receipt of the notice of selection of the most advantageous tender. The contracting authority may ask the tenderers concerned to waive their right to object in order to accelerate the time limit. If the contracting authority receives no objections during the objection period, it may proceed to sign the contract with the successful tenderer.

After concluding a contract with the selected tenderer, the contracting authority is obliged to publish a notice of the result of the procurement procedure on the Public Procurement Bulletin portal, the profile of the contracting authority and, in the case of above-limit contracts, on the Official Journal of the European Union.

4. Main challenges of implementing BIM in public procurement process

As was described in the previous part of the paper, public procurement is generally a highly formalised procedure aimed at selecting a supplier of supplies, services or works. The contracting authority must comply with a myriad of rules and obligations when awarding contracts. Where the contracting authority has a 'free hand', it must also ensure that the setting of various requirements (requirements for the subject-matter of the contract, requirements for the qualification of suppliers, etc.) does not violate the principles of public procurement non-discrimination and compliance with equal treatment, transparency and proportionality. The less innovative the approach, the safer the choice of supplier. At the same time, the less innovative the approach taken, the less likely the investor is to select with certainty a high-quality supplier and to realistically check its experience before the actual implementation phase of the public contract, i.e. before the moment of conclusion of the contract and subsequent performance of the contract.

Adoption of BIM in the Czech Republic is governed by Ministry of industry and trade and subsequently by Czech standardization agency. They oversee preparation of BIM standards and guidelines for public investors. Topic of this paper is complementing their proposed legislation, since methodology of evaluating BIM supplier have not been presented yet.

5. Proposed changes to public procurement process

This paper is focused on defining the requirements for demonstrating qualifications which are described in subsection 3.4. of this paper. It is necessary to be able to define and consequently be able to evaluate BIM related requirements. During the procurement process suppliers are asked to provide a proof of qualification in the form of previous reference. Since there have not been finished a lot of projects using BIM in the Czech Republic, it could be discriminatory for most suppliers. This paper is proposing another approach to determine supplier's ability deliver required qualification. For the scope of the paper this requirement will be demonstrated on designers.

From the professional experience of authors of the paper, designers BIM knowledge in the Czech Republic can be divided into several categories:

- Designer team have no basic knowledge of BIM, they do not have BIM software tools and work in 2D
- Designer team have BIM software tools, but utilize it just for 3D visualization and for generating 2D documentation
- Designer team fully utilizes BIM tools, but do not have any data standards for information embedded in model
- Designer team fully utilizes BIM tools and have data standards for information embedded in model

One of the possible ways how to determine their real ability to work with BIM is to give them opportunity to demonstrate their abilities. In order to select a quality BIM contractor, this paper proposes not to take the path of least resistance, but to make the necessary effort to examine the capabilities and capacities of the contractors during the procurement process. This procedure certainly increases the administrative and technical complexity of preparing and implementing the procurement procedure on the part of the contracting authority. It certainly adds to the complexity of the preparation of the tender by the suppliers, but in such a complex area as the use of BIM methodology is, it is worthwhile for the contracting authorities.

5.1. Defining the requirements for demonstrating BIM qualifications

The paper proposes that in order to ensure that tenderers meet investors qualification criteria, they would be asked to prepare small scale BIM model. Detailed instructions for BIM will be provided as a part of procurement documentation, in the form of Employers Information Requirements (EIR) Annex. Tenderers will be also provided with project documentation with defined sample size. Size of the sample should provide contracting authority with enough information about tenderer's BIM skills and should not be too big and complex, so it would discourage tenderers from participating in procurement. Type of the documentation would depend on the project type, delivery method and stage of the project. Then the tenderer will deliver model sample and filled BIM execution plan (BEP) to demonstrate BIM qualifications as a part of the tender.

The key in this section is the correct compilation of the BIM sample, the completion of the BEP and the completion of the supporting documents in order to evaluate tenders against the individual evaluation criteria. Of course, it is necessary to document all other requirements for the preparation of tenders (e.g. demonstration of basic competence, professional competence, technical qualification, etc.). Here, the designer demonstrates advanced knowledge of BIM model creation and the ability to communicate detailed information about it to other project participants, i.e. the construction contractor.

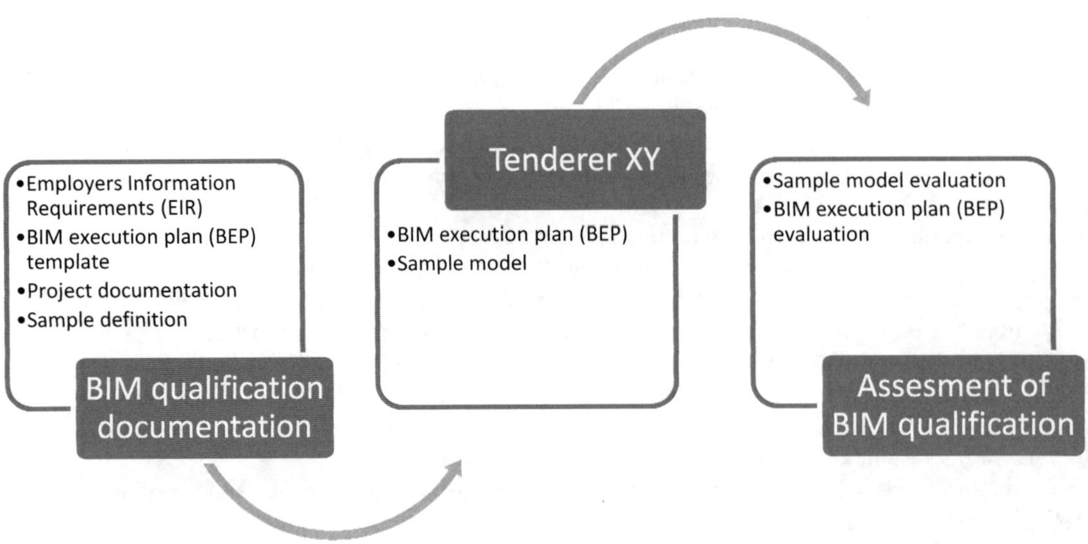

Figure 1. Process of BIM qualification assessment

5.2. Determination of method for evaluation of tenders from the point of BIM

The evaluation of tenders is not based on knowledge of working with BIM (this is assumed, e.g. by preparing a sample and completing the BEP). The tender with the highest score (the most economically advantageous tender) is subject to a closer examination. In addition to the assessment of the general parts of the tender (e.g. documents to demonstrate basic competence), the BEP and the sample submitted are also assessed. This process is shown in Figure 1.

The assessment of the BEP and the sample represents a key (and perhaps the most challenging) point in the selection process. In the Czech Republic, BIM model checking by the investor is virtually unknown. Perhaps because the investor needs to be equipped with advanced knowledge and tools to check the model. This step is crucial to avoiding selecting tenderer without required knowledge of BIM methodology. Only tenderers with completed BEP (that relates to presented sample) and sample (that matches required size and quality) can compete in procurement.

6. Conclusion

The paper focuses on public procurement process, because of its complexity. However, introduced principles are applicable for private sector as well. Despite of continuous effort from all interested parties, implementation of BIM into projects as a standard practice has still few major obstacles to overcome. [9,10] Presented research is focusing on technical aspect of selecting skilled and reliable supplier. Since BIM represents such a major disruption to established environment, the process of change is at some aspects painfully slow. It touches every aspect of building industry. Presented improvement of procurement process is currently being tested on pilot projects and results will be published once enough data will be gathered and processed.

Acknowledgments
This work was supported by the Grant Agency of the Czech Technical University in Prague, grant No. SGS21/100/OHK1/2T/11.

References

[1] Paníček, P., 2019. *Metoda Design-Build a její použití pro zadávání v ČR*. (Brno: Vysoké učení technické v Brně)
[2] Czech goverment. 2020. *47/2020 Sb.*. [online] Available at: <https://www.zakonyprolidi.cz/cs/2020-47/zneni-20200312> [Accessed 02 June 2021].
[3] Ministry of industry and trade, 2021. *Vláda schválila aktualizaci harmonogramu Koncepce zavádění metody BIM v ČR a vzala na vědomí informaci o jejím plnění* [online]. Available at: <https://www.mpo.cz/cz/stavebnictvi-a-suroviny/bim/vlada-schvalila-aktualizaci-harmonogramu-koncepce-zavadeni-metody-bim-v-cr-a-vzala-na-vedomi-informaci-o-jejim-plneni--259142/> [Accessed 02 June 2021].
[4] Aguiar Costa, A. and Grilo, A., 2015. *BIM-Based E-Procurement: An Innovative Approach to Construction E-Procurement.* (The Scientific World Journal) pp.1-15.
[5] Suresh Tiwari, S., Chan, S. and Mubarak, M., 2018. *Critical analysis of procurement techniques in construction management sectors*. IOP Conference Series: Materials Science and Engineering, 342, p.012100.
[6] Zilcher, R., 2018. *Comparison of construction supply delivery methods used in Czech Republic and Argentina.* (Praha: České vysoké učení technické v Praze)
[7] Schaufelberger, J., 2017. *Management of construction projects*. (Taylor & Francis)
[8] Czech goverment. 2016. *134/2016 Sb. Zákon o zadávání veřejných zakázek*. [online] Available at: <https://www.zakonyprolidi.cz/cs/2016-134> [Accessed 02 June 2021].
[9] Bouška, R. and Heralová, R., 2017. *Opportunities for use of Advanced Visualization Techniques for Project Coordination*. (Procedia Engineering), 196, pp.1051-1056.
[10] Vitasek, S., 2019. *Using Building Information Modelling (BIM) in construction budget: benefits and barriers*. (18th International Scientific Conference Engineering for Rural Development) Pages 1699 - 1706.

SCAN TO BIM: a systematic literature review network analysis

S A Adekunle[1]*, C Aigbavboa[1] and O A Ejohwomu[2]

[1] SARChl in Sustainable Construction Management and Leadership in the Built Environment, Faculty of Engineering and the Built Environment, University of Johannesburg, RSA

[2] School of Mechanical, Aerospace and Civil Engineering, The University of Manchester, UK

adekunlesamueladeniyi@gmail.com

Abstract. Building information modelling (BIM) has transformed the AEC industry tremendously, and its adoption has been reported to be high due to the immense latent benefits. One of the inherent benefits of BIM is information management and the ability to integrate with other technologies to achieve this has become the gap being explored by researchers. Laser scanning technology is one of the technologies integrated with BIM. This is done to achieve BIM compliance of pre-BIM edifices. In recent years, substantial research has been conducted on the integration of BIM and laser scanning technologies. This study performed a state of the art review of these research efforts using Systematic literature review network analysis (SNLA) (this is a combination of systematic literature review (SLR) and citation network analysis (CNA). The objective of the study is to identify research trends and gaps for further research. The coverage of this study is from 2007 to 2020. Data was sourced from the Scopus database of published research works and analysed using network analysis. The study identified the publication trends, research trends among others. The study identified quality control and the integration of photogrammetry as the hot topics in this space. This study provides researchers with an insight into the scan to BIM research globally.

1. Introduction

The idea of BIM was first conceptualized as building design system (BDS) by Eastman, (1975). Eastman's work, "The Use of Computers Instead of Drawings in Building Design", described the functionalities of the currently known Building information modelling. Although the transformation of Eastman's concept took 25 years before its actualization, and it is still being developed to achieve his predictions [2]. However, presently BIM has metamorphosised and new horizons are being unravelled to achieve a seamless information management process in the construction industry. BIM is a technological platform that supports the collaboration of professionals to achieve the planning, design and construction throughout the construction project lifecycle. It allows for the management of a building throughout its lifecycle. It promotes productivity, efficiency, precision and accuracy.

Heretofore, the construction industry has experienced the migration from lines and arcs to 2D and 3D but the BIM is an all-encompassing model beyond object dimensions. BIM is beyond 3D but nD; it has added time, cost, resource management, sustainability, among others in the fourth, fifth, sixth, seventh dimensions [3]. The BIM dimension has also been extended to the tenth dimension; this consists of the following dimensions: sixth dimension (sustainability and CO_2 accounting), seventh dimension(operation and maintenance), eighth dimension (Health and Safety), ninth dimension (Lean management) and tenth dimension (digitisation) [4]. Although Koutamanis, (2020) argues that BIM should not go beyond the fourth dimension as the other dimensions do not qualify as dimensions. Generally, BIM dimensions, especially beyond the third dimension, is not standardized and thus are not harmonized [6]. These latter dimensions beyond the fourth do not have a synchronized characterization by researchers.
For this study, the crux of these different BIM dimensions is that BIM is central to many transformations in the construction industry. Also, it possesses potentials that can be unravelled with time, these latent potentials are achievable through its integration with other technologies. Overall, its inherent dimensions are directed towards industrializing and transforming the construction sector [4]. Hence, it is expanding in dimensions in order to achieve its benefits of efficiency, data accuracy, productivity real-time data delivery, among others. However, the most significant purpose is the achievement of clients' objective and not been confined to the dimensional boundaries.
In order to achieve this, BIM has been exploring and conquering more technological boundaries through the integration with other emergent technologies in the construction industry. These emergent technologies, especially in the area of the field to BIM processes, includes but not limited to radio frequency identification (RFID), ultra-wideband (UWB), global navigation satellite system (GNSS), 2D imaging, photogrammetry, 3D terrestrial laser scanning, among others. An important integration of technologies is the application of laser scanning for a field-to-BIM process as opposed to the more popular BIM-to-field processes. The former is employed mostly in the creation of BIM, where it is inexistent. It can be employed in new buildings (without a preexisting BIM model), existing building (with or without a preexisting BIM model) [7,8]. Laser scanning technology is one of the techniques employed; it produces 3D point clouds to generate as-built information for existing buildings through lasers(amplified lights) [9]. This study is aimed at observing the trend in the global scan to BIM research. Various studies have been conducted on the integration and application of these two technologies in the construction industry. This study seeks to explore the trend and status of the integration using a SLNA approach.

2. Methodology
To achieve the objectives, a systematic literature review network analysis (SLNA) was adopted. This methodology is a combination of systematic literature review (SLR) and citation network analysis (CNA) [10]. Other researchers have adopted this method in carrying out reviews [11,12].

2.1 Data collection
The data collection was done through SLR. The SLR includes the selection of keywords and a systematic approach to identifying themes in the field of study. This study adopted an automated search approach to achieve its data collection systematically. The data was collected from Scopus. Scopus is one of the two largest indexing databases; the other one is Web of Science, Scopus, however, have a wider coverage (Mongeon and Paul-Hus, 2016). This position of Scopus having a wider coverage more than other databases and is more informative was alluded to by [14]. Consequently, this study adopted the Scopus database as its automated search source. The search string for the study is given below:

TITLE-ABS-KEY("BIM" OR "Building information modeling" AND "laser Scanning" OR "laser scanner" OR "laser application") AND (EXCLUDE (SUBJAREA,"COMP") OR EXCLUDE (SUBJAREA,"SOCI") OR EXCLUDE (SUBJAREA,"EART") OR EXCLUDE (SUBJAREA,"PHYS") OR EXCLUDE (SUBJAREA,"MATH") OR EXCLUDE (SUBJAREA,"MATE") OR EXCLUDE (SUBJAREA,"ARTS") OR EXCLUDE (SUBJAREA,"BUSI"))

The search focused on the studies that examined the BIM and laser scanning and adopted the use of the Boolean connectors: "AND" and "OR". The study, however, was non-restrictive in terms of the document type. This is in order to achieve a wider coverage of publications. The coverage for this study were all research publications in the study focus area between 2007 and August 2020.

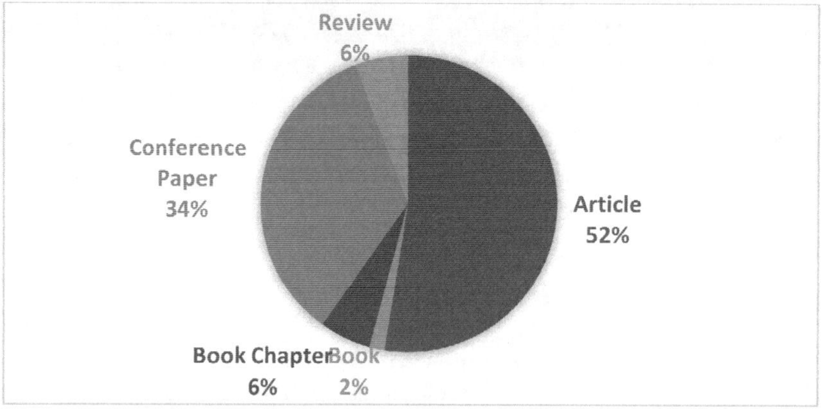

Figure 1: Publication type

The search included all publications without exclusions in order to achieve an all-inclusive and wide coverage of data. Figure 1 gives an overview of document types constituting the data collected for this study.

2.2 Data analysis

To achieve the objective analysis of the collected data through CNA, the VOS viewer software was adopted. It is an open-access software for data visualization, the positions of analysed data on the visualization reflect their relationship [15]. It is also flexible and easy to use with a great graphical presentation compared to other bibliometric visualization tools. It has been adopted in previous bibliometric visualisation studies, for instance [16,17].

3. Discussion

3.1 Annual publication trend

The trend of publication is presented in Figure 2 from 2007 to 2020. It is evident that the first publication in BIM-3D scanning integration was in 2007. The publication has risen steadily from 1 number to 12 numbers as of August 2020 although irregular.

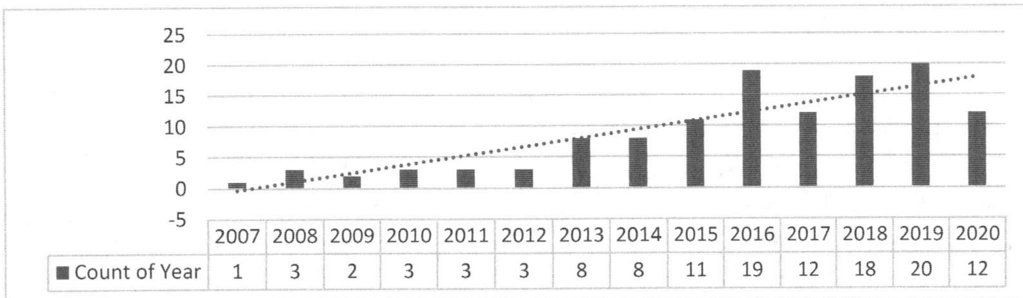

Figure 2: Annual publication trend

The publication trend evidences the growing interest in the application of these two technologies to achieve the field to BIM objectives.

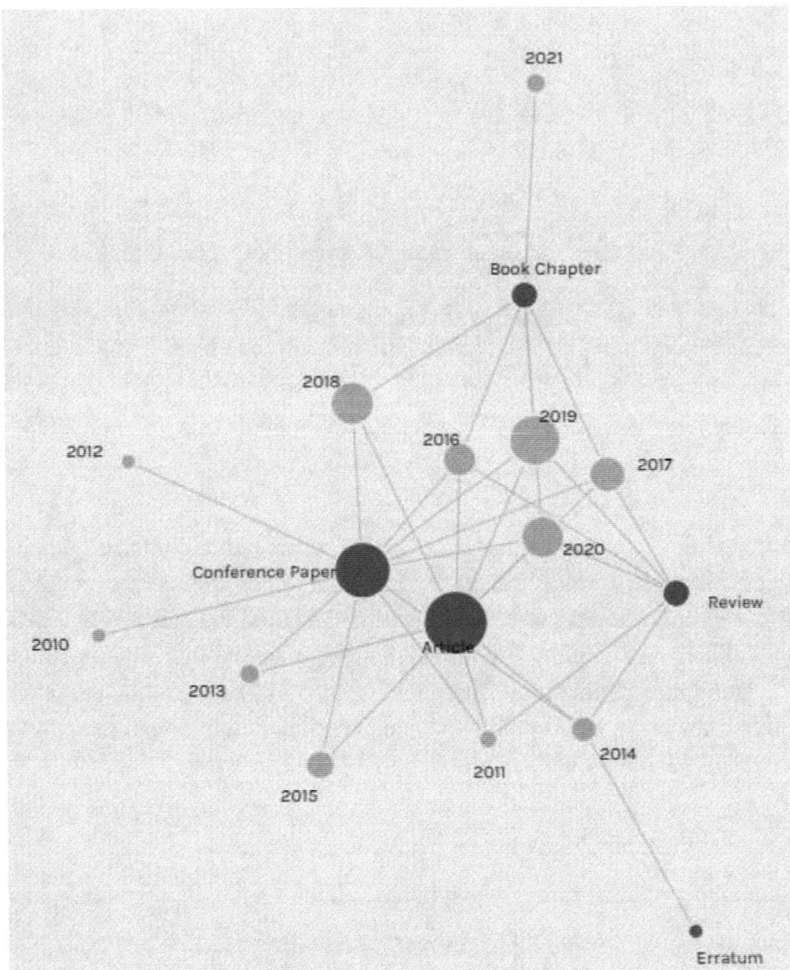

Figure 3: Publication type network

Figure 3 shows, the publication trend based on the publication type. The size of the node indicates the relative magnitude of the variable and the links shows the relationship between the nodes [18]. These links are represented by the lines in the network (Figure 3). It is evident that Articles are the most published, followed by conference papers.

3.2 Country-wise publication trend

Most publications were contributed by the United States of America, the United Kingdom, China, Spain, and Germany. The output in these countries is not unexpected as they are advanced in BIM research, for instance, USA, South Korea, England (UK) [19]. Thus, they are expected to be advanced in the integration with emergent technologies. However, it is encouraging to note that Egypt and some other developing countries are on the list, it shows that BIM research is gaining momentum in the developing countries.

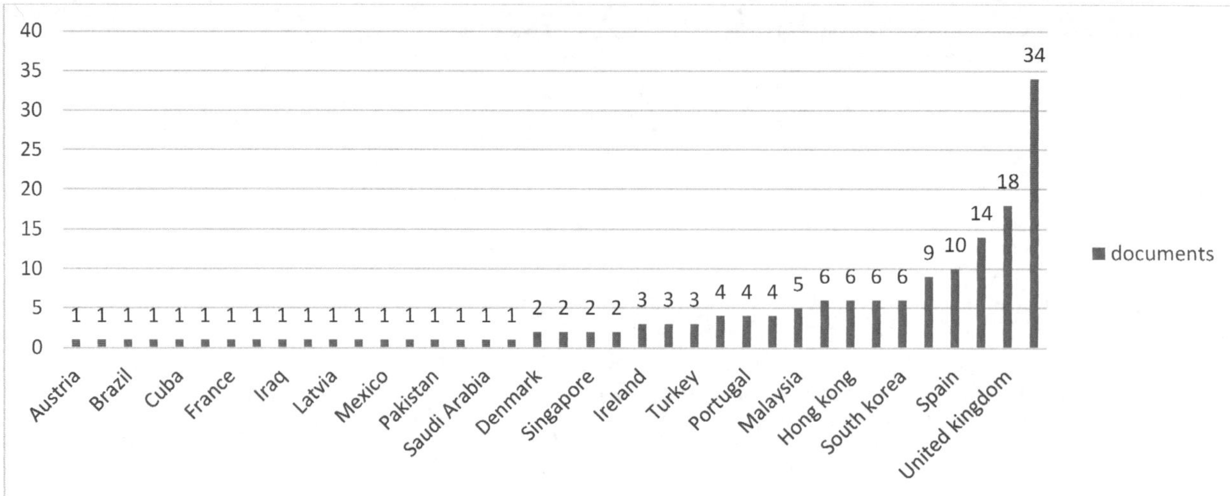

Figure 4: Country-based publication trend

3.3 Publication sources

According to the data collected from the Scopus website, publications in the BIM and scanning research is from different sources. The sources include but not limited to journals and conference proceedings. Table 1 and figure 5 shows the various publication sources. The study identified the top 20 most productive publication sources. From table 1, Automation in construction occupies the first place with 60 documents and 2653 citations. It is good to note that this source is observed as being productive and a leading contributor to BIM research in the construction industry. This is alluded to by the study results by [19–22]; hence it can be said that this journal is very productive in this research area.

Table 1. Publication sources

	Source title	Documents	Citations
1	Automation in Construction	60	2653
2	ISPRS Annals of The Photogrammetry, Remote Sensing and Spatial Information Sciences	29	102
3	Journal of Computing in Civil Engineering	17	321
4	32nd International Symposium on Automation and Robotics in Construction and Mining: Connected to the Future, Proceedings	7	53
5	Applied Sciences (Switzerland)	7	94
6	Congress on Computing in Civil Engineering, Proceedings	6	15

7	Computer-Aided Civil and Infrastructure Engineering	5	85
8	Engineering, Construction and Architectural Management	5	13
9	Journal of Building Engineering	5	38
10	Proceedings of the 36th International Symposium on Automation and Robotics in Construction, ISARC 2019	5	3
11	Sensors (Switzerland)	5	21
12	Computing in Civil Engineering 2019: Visualization, Information Modeling, and Simulation - Selected Papers from the ASCE International Conference on Computing in Civil Engineering 2019	4	6
13	ISARC 2016 - 33rd International Symposium on Automation and Robotics in Construction	4	24
14	Journal of Management in Engineering	4	76
15	Sustainability (Switzerland)	4	5
16	Construction Innovation	3	30
17	Construction Research Congress 2018: Construction Information Technology - Selected Papers from the Construction Research Congress 2018	3	7
18	IOP Conference Series: Materials Science and Engineering	3	4
19	Journal of Cleaner Production	3	104
20	Journal of Construction Engineering and Management	3	28

The density visualization is presented in Figure 5, the cluster around Automation in construction is the most accentuated. This is followed by the ISPRS Annals of The Photogrammetry, Remote Sensing and Spatial Information Sciences, and Journal of Computing in Civil Engineering.

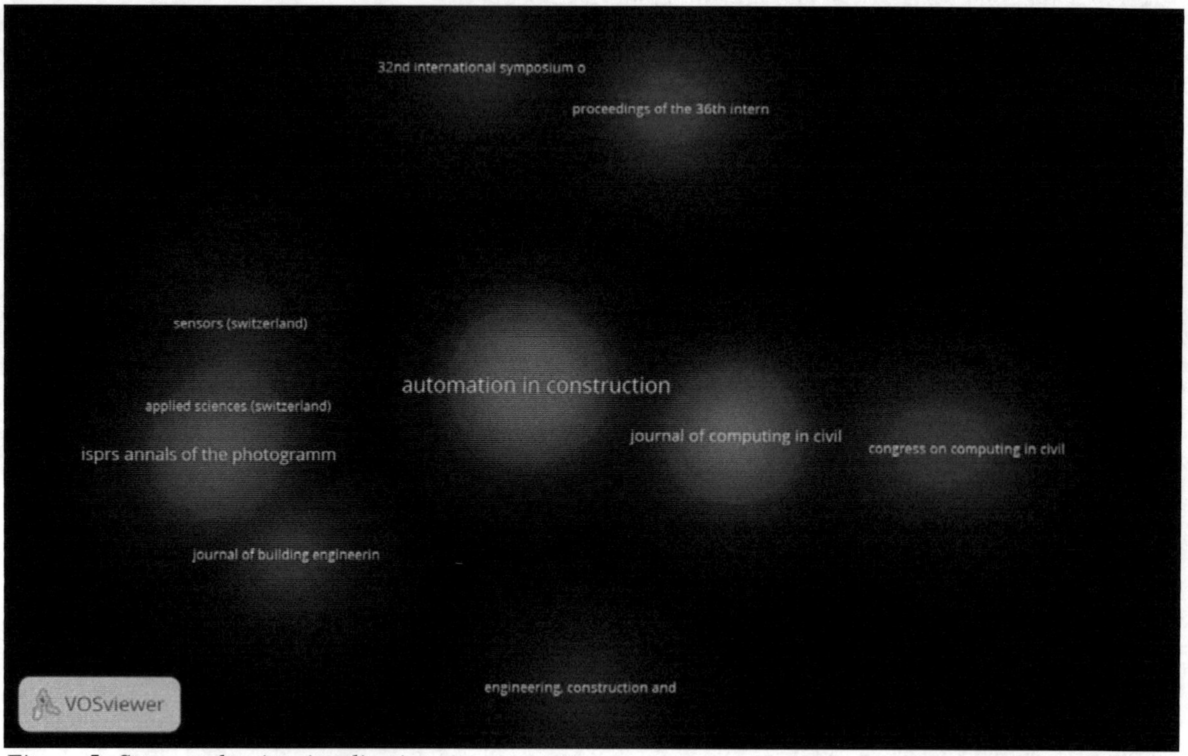

Figure 5: Sources density visualization

3.4 Country wise productivity

From the analyzed data, Table 2 shows that scholars from the United States of America rank first in terms of the number of publications and citations. This is followed by the United Kingdom. However, moving down the countries, it is observed that there is no direct correlation between the number of document and citation garnered. For instance, China has 14 documents and 78 citations whereas countries like Spain, Canada, Hong Kong, Italy, South Korea have 10, 6, 6, 6 and 6 documents respectively but with 479, 208, 180,104 and 399 citations, respectively. It can be inferred that the number of documents does not automatically command the same number of influence in terms of citations. An analysis based on the geographic location of these countries reveals that most of the productive countries are from Europe and Asia. However, there is an African country (Egypt) with two documents, this shows that this research domain is busy with activities from both developing and developed countries.

Figure 6 provides the network visualization of the country productivity. It can be observed that countries like Singapore, Italy and the Netherlands have more recent publications in this research area. This is shown in the time bar at the bottom right corner of the figure. The colour scheme represents the annual publication trend, the colour on the timeline denotes the year against the country node on the figure. Thus, the USA, UK and China cannot be said to have recent publications like the earlier listed countries. Their publications are around 2017.

Table 2. Details of the first 20 countries productivity

S/N	Country	Documents	Citations
1	United states	34	1715

2	United Kingdom	18	546
3	China	14	78
4	Spain	10	479
5	Germany	9	124
6	Canada	6	208
7	Hong Kong	6	180
8	Italy	6	104
9	South Korea	6	399
10	Malaysia	5	21
11	Australia	4	50
12	Portugal	4	58
13	Russian federation	4	1
14	Ireland	3	172
15	Netherlands	3	24
16	Turkey	3	20
17	Denmark	2	0
18	Egypt	2	0
19	Singapore	2	1
20	Sweden	2	1

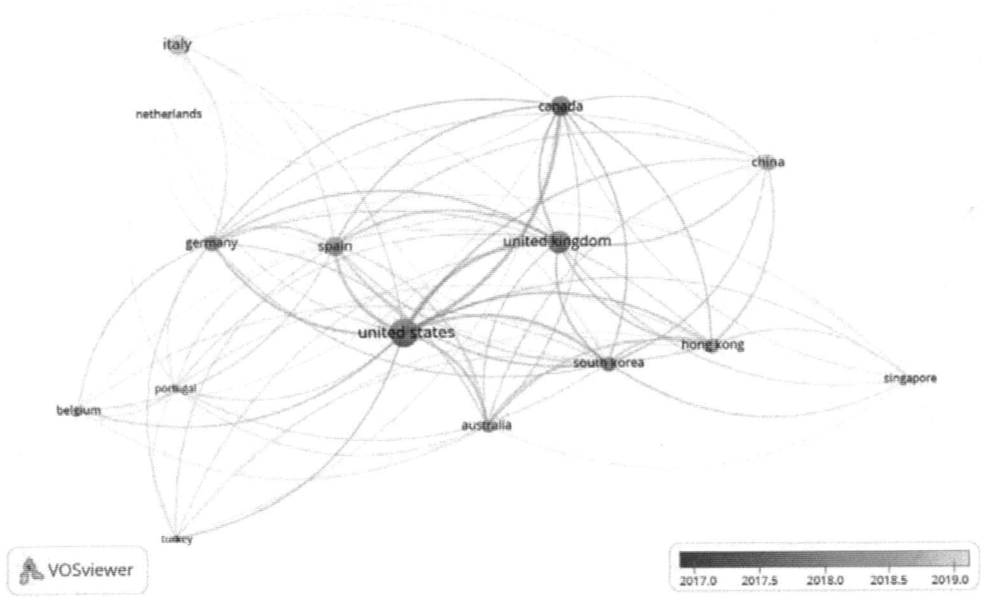

Figure 6: country productivity network

3.5 Research trend

To achieve the visualization of the research trends in the scan to BIM research publications, a co-occurrence map of the bibliographic data using the keywords was generated. The use of keywords to achieve the patterns and research clusters has been adopted in studies such as [16,17,22] The default number of 5 occurrences was adopted for this study. A critical look at the clusters generated shows that there exist 5 clusters representing five research trends. It is observed that the research focus has been on the

IOP Conf. Series: Materials Science and Engineering 1218 (2022) 012057 doi:10.1088/1757-899X/1218/1/012057

application of scan to BIM to facility management, retrofitting and especially for historic preservation. The second focus was majorly on the integration of Augmented reality (AR) and Virtual Reality(VR) in this field for information management in the area of project management, and construction management. also, the research focus has been aimed at the adoption of laser scanning for lifecycle management through reverse engineering and surface analysis. Research focus four and five are aimed at the introduction of quality control and the integration of photogrammetry in this space. Clusters four and five however are emergent and thus are regarded as hot topics.

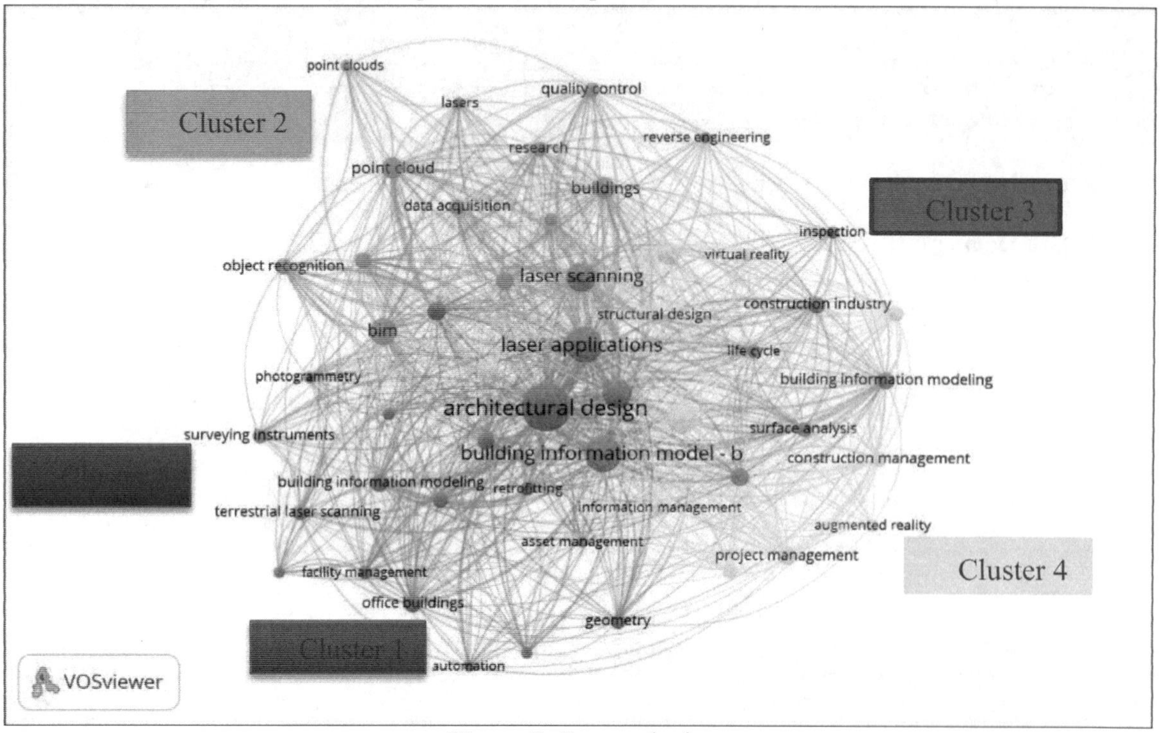

Figure 7: Research clusters

4. Conclusion

This study focused on the state-of-the-art review of the scan to BIM integration focusing mainly on the use of laser scanner technology. The major area of focus for this study are the country trend, annual publication trend and the research trend.

The field of scan to BIM has received attention especially in integrating erstwhile non-BIM buildings into BIM. This study focused on articulating the research efforts in this area and also established the various research trends heretofore. It is worthy to mention that there is a lot to explore in the BIM space in order to achieve its adoption especially while integrating it with other emerging technologies.

However, there exists some limitation from this study as there are other areas not covered like author productivity and influence, citation networks, co-citation networks among others. Also, the data was sourced from one database, Scopus. Thus, further work can still be done in this space.

References

[1] Eastman C 1975 The Use of Computers Instead of Drawings in Building Design. *AIA J.*

[2] Sacks R, Girolami M and Brilakis I 2020 Building Information Modelling, Artificial Intelligence and Construction Tech *Dev. Built Environ.* 100011

[3] Charef R, Alaka H and Emmitt S 2018 Beyond the third dimension of BIM: A systematic review of literature and assessment of professional views *J. Build. Eng.*
[4] Arnal I P 2018 Why don't we start at the beginning? | BIMCommunity
[5] Koutamanis A 2020 Dimensionality in BIM: Why BIM cannot have more than four dimensions? *Autom. Constr.* **114**
[6] Wildenauer A A 2020 Critical assessment of the existing definitions of bim dimensions on the example of Switzerland *Int. J. Civ. Eng. Technol.* **11** 134–51
[7] Sanei Sistani N 2017 *An investigation on Success Factors for Scan to BIM Process* (Berlin: Metropolia UAS and HTW Berlin)
[8] Volk R, Stengel J and Schultmann F 2014 Building Information Modeling (BIM) for existing buildings - Literature review and future needs *Autom. Constr.* **38** 109–27
[9] Hossain M A and Yeoh J K W 2018 BIM for Existing Buildings: Potential Opportunities and Barriers *IOP Conference Series: Materials Science and Engineering* vol 371
[10] Colicchia C and Strozzi F 2012 Supply chain risk management: A new methodology for a systematic literature review *Supply Chain Manag.* **17** 403–18
[11] Baryannis G, Validi S, Dani S and Antoniou G 2019 Supply chain risk management and artificial intelligence: state of the art and future research directions *Int. J. Prod. Res.* **57** 2179–202
[12] Comerio N and Strozzi F 2019 Tourism and its economic impact: A literature review using bibliometric tools *Tour. Econ.* **25** 109–1331
[13] Mongeon P and Paul-Hus A 2016 The journal coverage of Web of Science and Scopus: a comparative analysis *Scientometrics* **106** 213–28
[14] Guz A N and Rushchitsky J J 2009 SCOPUS: A system for the evaluation of scientific journals *Int. Appl. Mech.* **45**
[15] Jan van Eck N and Waltman L 2006 VOS: A New Method for Visualizing Similarities between Objects *ERIM Rep. Ser. Res. Manag.* 11
[16] Aghimien D O, Aigbavboa C O, Oke A E and Thwala W D 2019 Mapping out research focus for robotics and automation research in construction-related studies: A bibliometric approach *J. Eng. Des. Technol.*
[17] Saka A B and Chan D W M 2019 A scientometric review and metasynthesis of building information modelling (BIM) research in Africa *Buildings* **9**
[18] Palladio 2020 Home *Palladio*
[19] Zhao X 2017 A scientometric review of global BIM research: Analysis and visualization *Autom. Constr.* **80** 37–47
[20] Liu Z, Lu Y and Peh L C 2019 A review and scientometric analysis of Global Building Information Modeling (BIM) Research in the Architecture, Engineering and Construction (AEC) industry *Buildings* **9** 1–36
[21] Santos R, Costa A A and Grilo A 2017 Bibliometric analysis and review of Building Information Modelling literature published between 2005 and 2015 *Autom. Constr.* **80** 118–36
[22] Yin X, Liu H, Chen Y and Al-Hussein M 2019 Building information modelling for off-site construction: Review and future directions *Autom. Constr.* **101** 72–91

Influence lines for schedule networks

A Lepel[1] and M Hajdu[1,2]

[1] Budapest University of Science and Technology, Hungary
[2] Chaoyang University of Science and Technology, Taiwan

lepel.adrienn@edu.bme.hu

Abstract. An influence line graphs for project planning show how the changes of a selected activity duration affect the project duration. The name comes from the structural analysis where influence lines show on the ordinates of a graph the variation of the magnitude of, e.g., the bending moment at a specific point as a unit load traverses across the structure. This paper examines the principles of project planning influence lines: it discusses how to depict the six criticality types of activities, and it also shows what kind of transitions are possible among the criticality types. The findings help the greater understanding of the effects of the activity duration changes on the project duration.

Keywords: CPM, Precedence Diagramming Method (PDM), Influence lines

1. Introduction and literature review

One of the most crucial tasks of the planning process is to adjust the project duration to be within the expected project deadline. For speeding up projects, planners usually apply two types of actions: either select some from the critical activities, modify their durations, or change the project's logic, thus creating new and shorter critical paths. The recently developed influence lines can help to speed up this process. The influence line for an activity shows how a selected activity's duration changes affect the project duration. Fig 1 presents an example of the influence line.

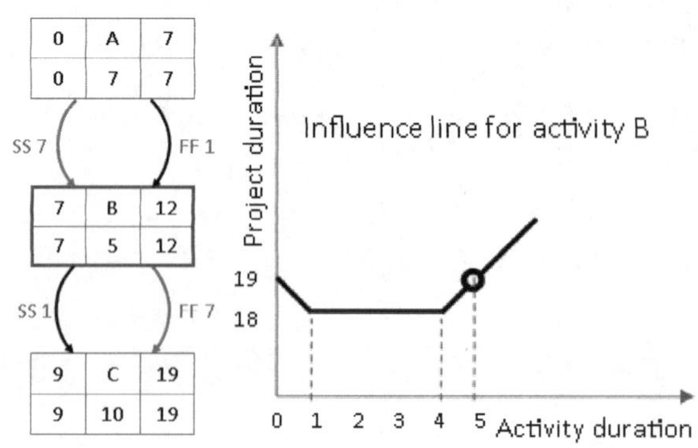

Figure 1. An example for influence line.

Figure 1 shows a network with three activities A, B, and C, and four precedence relationships. The project duration is 19 days. The influence line for activity B shows the project duration for different activity durations.

A one-day decrement (from five to four days) in B's duration accelerates the project; two to four days decrements do not result in any further changes; it leaves the project duration unaltered. However, an additional one-day decrement in the duration, from one day to zero, increases the project duration. Increasing the duration of B with any amount results in the same amount of increment in the project duration.

Influence lines for project planning result from Hajdu [1]. The authors have shown that activity criticality types can change, and they introduced the influence lines as a tool to visualize the effects of changing duration on the project duration. Different types of criticality were first discussed by Weist [2]. He has shown that in Precedence Diagramming Method (PDM) [3], [4], [5], changing activity durations can have a diverse effect on project duration. He described only four criticality types in his paper: the normal critical, the reverse critical, the neutral critical, and the bi-critical (called perverse critical in the article). However, contrary to Weist's results, six criticality types exist in PDM, even in the simplest case [1]. The simplest case assumes that activities are non-stretchable and non-splittable, and only the well-known four precedence relationships, namely Start-to-Start (SS), Finish-to-Start (FS), Finish-to-Finish (FF), and Start-to-Finish (SF), are allowed. The influence lines graphically display the effect of changing activity durations on the project duration; therefore, they point out the dynamically changing nature of criticality. However, Hajdu et al. did not discuss the internal rules that define the shape of influence lines. The goal of this research is to fill this research gap.

2. Research goals and limitations

This research aims to define the rules shaping the form of the influence lines in PDM networks. The simplest version of PDM is the subject of the investigation where

- activities are assumed to be linear, non-splittable, and non-stretchable;
- only the four minimal end-point relationships are allowed;
- calendars are not allowed;
- hard constraints (such as Must Start at, Start Earlier than) are not allowed;
- resources and their limitations do not affect the results of the time analysis.

One can see that the different generalizations of the PDM method are not allowed. Extending the scope of the study for these generalizations can serve as possibilities for future research.

3. Depicting criticality types in influence lines

Criticality types of critical activities indicate how the project duration changes if increasing or decreasing the activity's duration by one day. (For the sake of simplicity, duration and relationship lags are given in days.) Table 1. shows the six criticality types.

Table 1. Criticality types based on the effect of changing the activity duration.

Criticality types		Decreasing activity duration		
		Increasing project duration	**Decreasing project duration**	**No effect on project duration**
Increasing activity duration	**Increasing project duration**	Bi-critical	Normal critical	Increasing normal – decreasing neutral
	Decreasing project duration	Reverse critical		
	No effect on project duration	Increasing neutral – decreasing reverse		Neutral critical

Criticality types of critical activities can be defined based on the incoming and outgoing critical paths. In the case of end-point relationships (SS, FF, FS, SF), critical paths can enter a critical activity at its start, finish, or both points and leave at its start, finish, or both points (figure 2). The nine possible arrangements define six criticality types. Relationships in red stand for the critical relationships, the black ones represent the non-critical relationships.

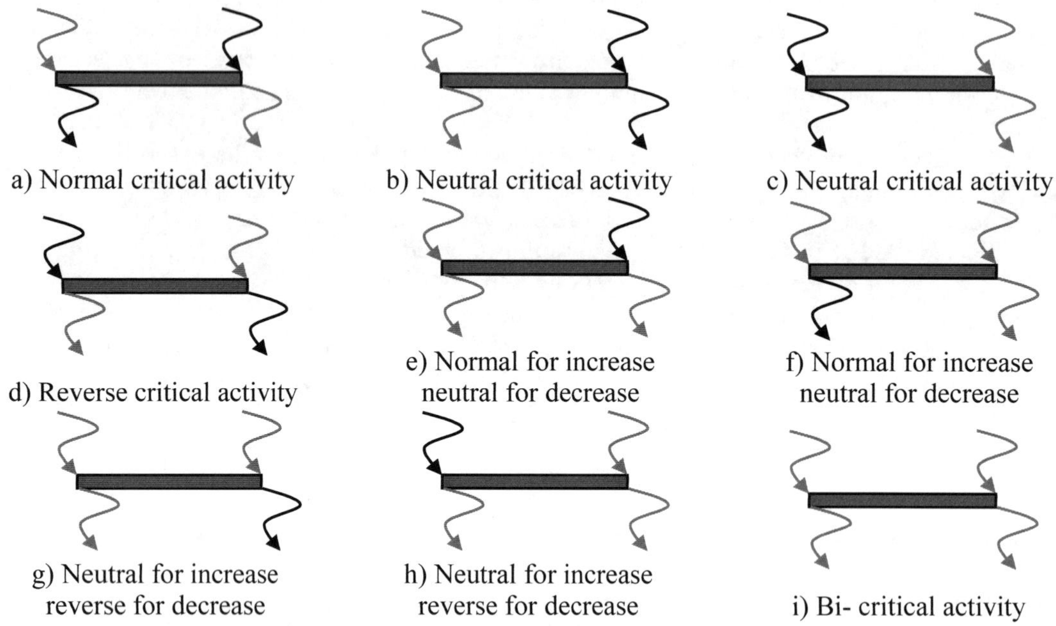

Figure 2. Criticality types based on the incoming and outgoing critical paths.

3.1. Normal critical activities

Activities are normal critical if the critical path enters at the start of the activity and leaves at its finish (Figure 2a). Increasing normal critical activities' duration by one day increases the project duration by one day. In comparison, a one-day decrement in the activities' duration results in decreasing project duration by one day. Figure 3. illustrates this case.

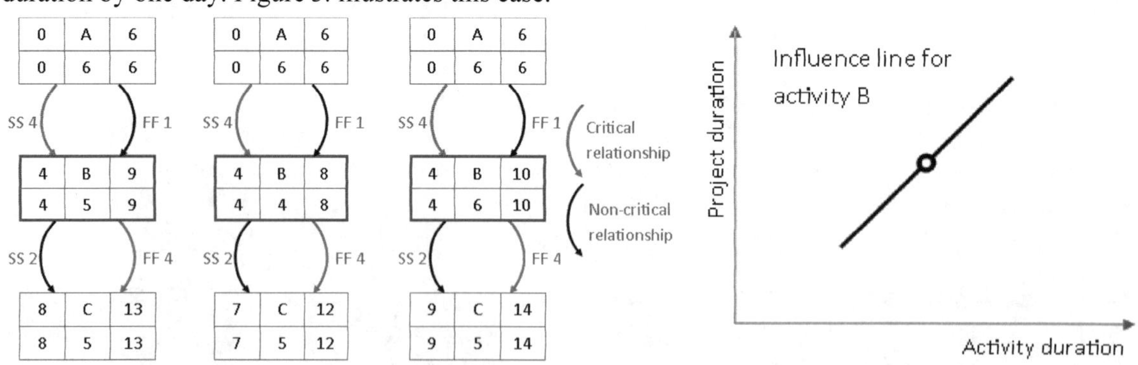

Figure 3. Example for normal critical activity and its influence line.

3.2. Neutral critical activities

An activity is neutral critical if the critical path enters and leaves the activity at the same point, either at the start or at the finish (figure 2.b)-c)). In this case, changing the activity's duration does not affect the project's duration. Figure 4. demonstrates the first case.

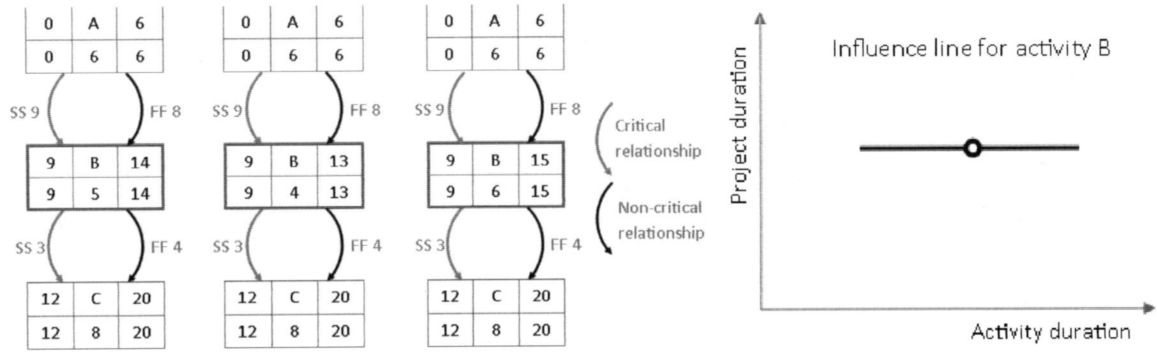

Figure 4. Neutral criticality type based on figure 2.b).

3.3. Reverse critical activities

Reverse critical activities have the critical path of entering at their finish and leaving at their start (figure 2.d)). As figure 5. presents, in this case, the project duration moves against the direction of the activity duration.

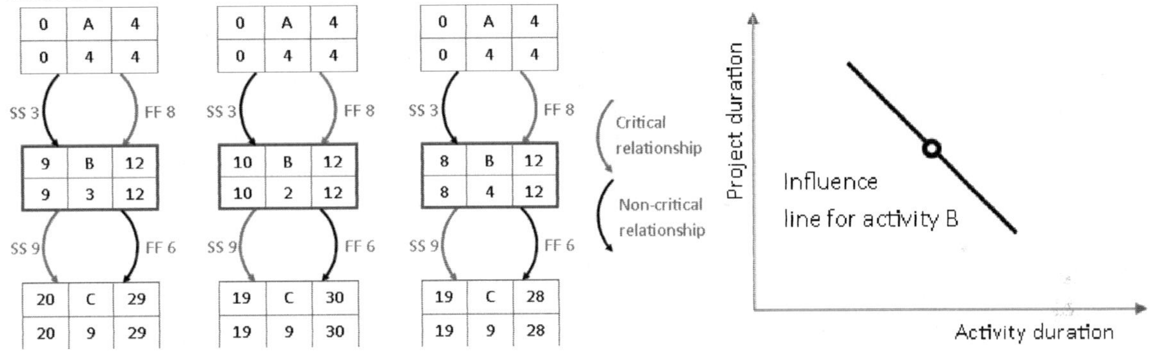

Figure 5. Reverse criticality type.

3.4. Normal for an increase – neutral for a decrease critical activities

This situation emerges as a combination of normal and neutral criticality (figure 2.e)-f)). The first critical path enters at the start of the activity and leaves at the finish, similar to the normal critical activities. The third critical relationship is either entering and the finish or leaving at the start. The effect on the project duration is also the combination of normal and neural criticalities.

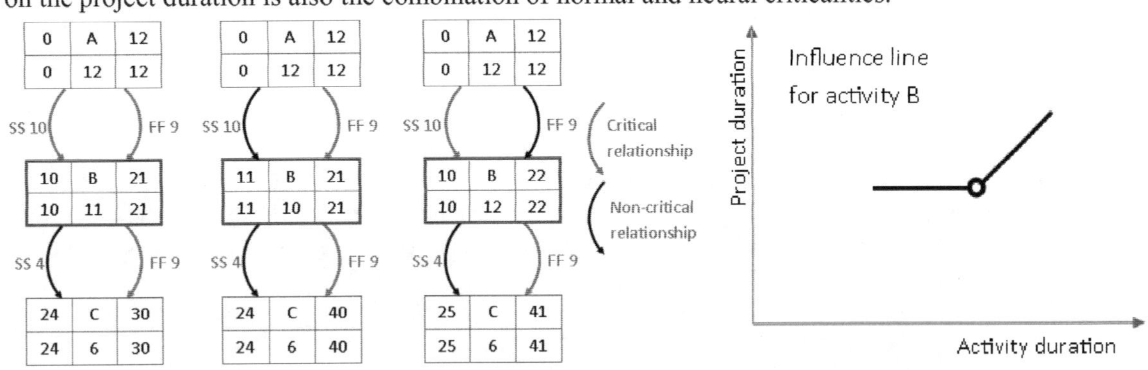

Figure 6. Normal for an increase -neutral for a decrease criticality type based on figure 2.f))

Normal criticality governs the changes of the project duration if the duration of the selected activity is increasing. If activity duration is decreasing, then neutral criticality defines the changes of the project duration. Figure 6. explains the joint effect of these two criticalities on the project duration.

3.5. Increasing neutral – decreasing reverse

Increasing neutral – decreasing reverse activities have the critical path entering at their finish and leaving at their start like reverse critical activities have. The third critical relationship can enter at the start (figure 2.g)) or leave at the finish (figure 2.h)). In this case, increasing the activity duration does not affect the project duration, while decreasing the activity duration results in increasing project duration. This is shown in figure 7.

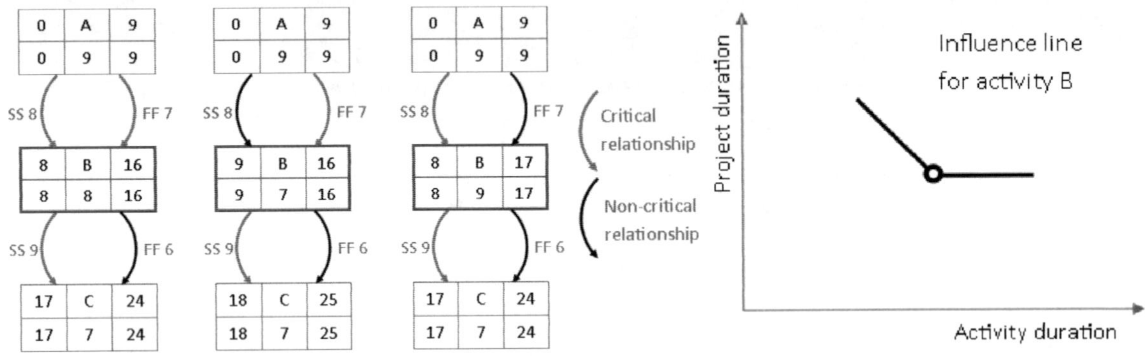

Figure 7. Neutral for an increase -reverse for a decrease criticality type based on figure 2.g).

3.6. Bi-critical

If the critical path enters and leaves at both start and finish, then the activity is bi-critical. It is normal, reverse (and neutral) at the same time. Increasing or decreasing a bi-critical activity's duration both result in increasing project duration. This criticality type is presented in figure 8.

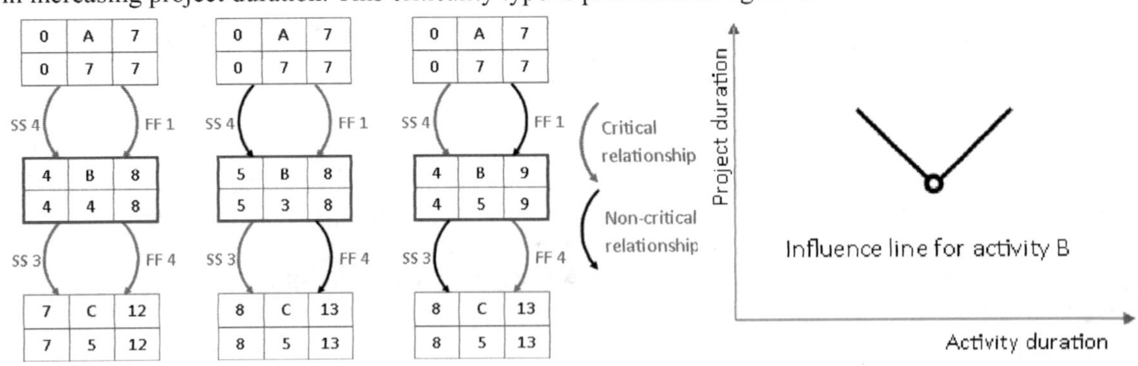

Figure 8. Bi-critical activity type.

4. Rules governing the shape of the influence lines

The transition between criticality types follows specific rules. Increasing and decreasing the duration of the activity can result in changing the criticality type of activities. Table 2. shows the possible transitions from a criticality type to another.

Table 2. Transitions between criticality types.

Criticality type	Increasing activity duration	Decreasing activity duration
Normal critical	*no change*	normal-neutral or bi-critical
Neutral critical	normal-neutral	neutral-reverse
Reverse critical	neutral-reverse or bi-critical	no change
Normal-neutral	normal	neutral-reverse
Neutral-reverse	neutral-normal	reverse
Bi-critical	normal	reverse

4.1. Possible changes for normal critical activities

Increasing the activity duration of normal critical activities increases the project duration, and the criticality type does not change because the critical path still enters at the start of the activity and leaves at its finish. Decrement in the activity duration results in decreasing project duration to a point where the activity becomes normal-neutral or bi-critical: the critical path enters not just at the start but also at the finish of the activity. Figure 9 illustrates the two possible transitions from normal to normal-neutral (figure 9.a)) and bi-directional criticality types (figure 9.b)).

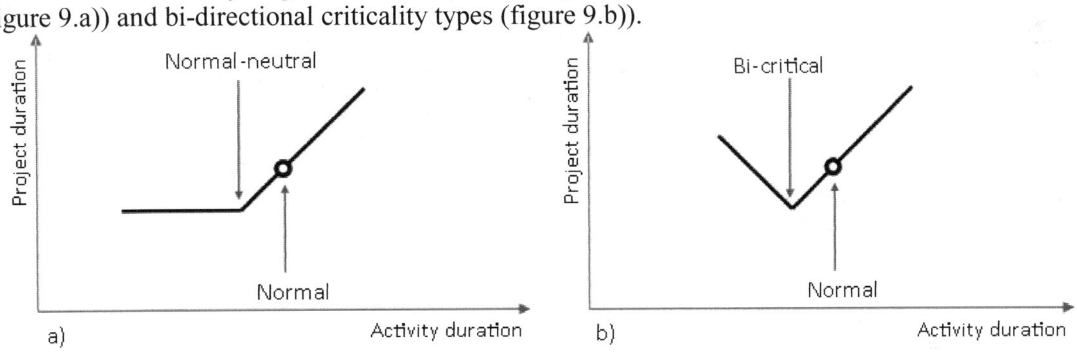

Figure 9. Influence line changes for normal criticality type.

4.2. Neutral critical

The critical path enters and leaves the activity at the same point, so the project duration does not change until the activity duration is increased or decreased to a point where a third relation becomes critical. An increment in the activity duration results in a normal-neutral (further increment changes it to normal critical) activity, while decreasing the activity duration leads to a neutral-reverse (with further decrement reverse critical) activity. The effects of changing the duration are shown in figure 4., the influence line is shown in figure 10.

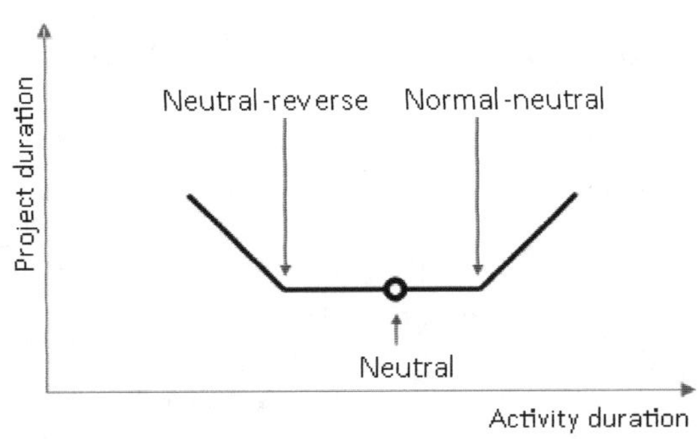

Figure 10. Influence line changes for neutral criticality type.

4.3. Reverse critical

Reverse critical activities act the opposite way as normal criticals. Increasing a reverse critical activity's duration decreases the project duration until the critical path entering at their finish and leaving at their start also enters at their start, where the activity becomes neutral-reverse or bi-critical. Decreasing the activity's duration, on the contrary, increases the project duration. The changing activity durations are shown in figure 5., the influence line changes are presented in figure 11.a) and 11.b).

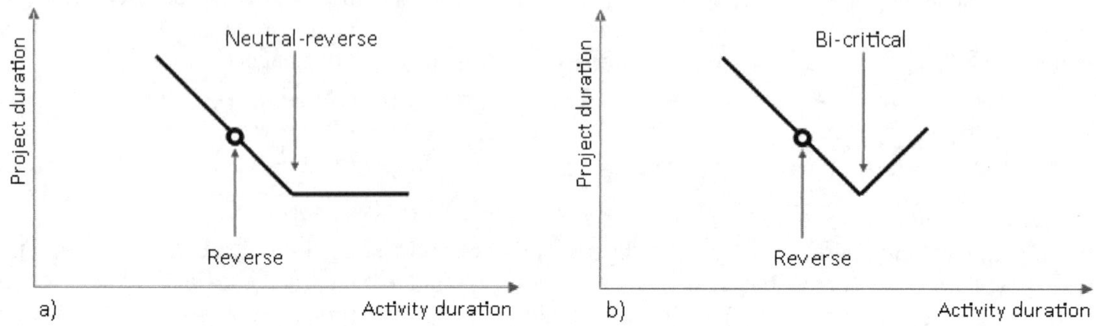

Figure 11. Influence line changes for reverse criticality type.

4.4. Increasing normal – decreasing neutral

Combined critical activities have at least three critical connections. Changing the duration of the activity can result in three or two critical connections. Increasing the activity duration causes the criticality type to become normal critical, with the critical path entering at the start of the activity and leaving at its end. Decreasing the activity duration makes no change in the project duration until the previously non-critical connection becomes critical, and the criticality type changes to neutral-reverse (then reverse). The effect of changing the activity duration can be inspected in figure 6., the influence line is shown in figure 12.

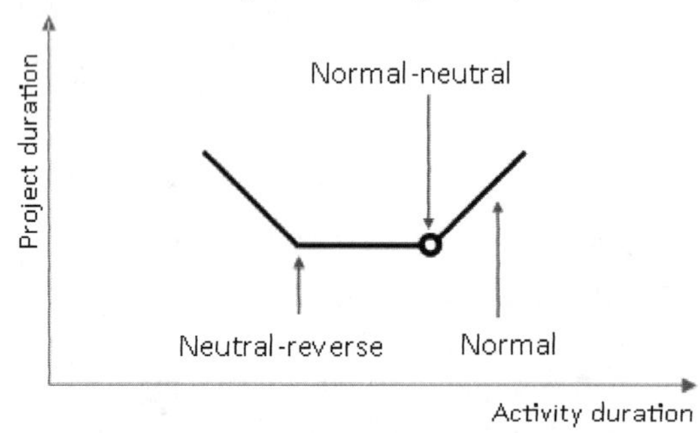

Figure 12. Influence line changes for normal-neutral criticality type

4.5. Increasing neutral – decreasing reverse

Increasing neutral – decreasing reverse activities act very similarly to the normal-neutral criticality, only contrariwise. In this case, increasing the activity duration does not affect the project duration until the critical path leaves the activity at the same point and after entering it. At this point, the activity is normal-neutral. Any further increase in the duration changes it to normal critical. Decreasing the activity duration eliminates one of the critical connections: the critical path enters at the finish of the activity and leaves at its start. The criticality type changes to reverse critical. This is presented in figure 7., the influence line is depicted in figure 13.

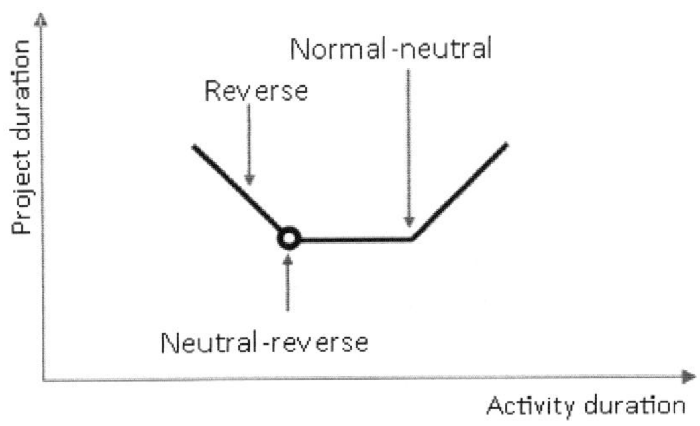

Figure 13. Influence line changes for neutral-reverse criticality type

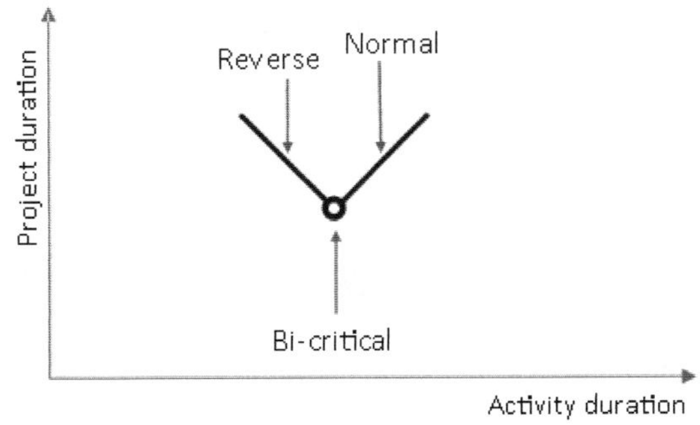

Figure 14. Influence line changes for bi-critical criticality type

4.6. Bi-critical

An activity that has the critical path entering and leaving both at the start and the finish points reacts to changing its duration with changing some of its critical relationships to uncritical. Increasing the activity duration results in a normal critical activity, which has the critical path entering at its start and leaving at its finish. In the case of normal critical activities, the project duration increases when the activity duration is increased. A decrement of the activity duration leads to reverse criticality, that again increases project duration. The criticality type with changing activity duration can be inspected in figure 8. The influence line is presented in figure 14.

5. Conclusions and further research

This research established that:

- influence lines are convex;
- they have zero, one, or a maximum of two break-points in the case of end-point relationships
- they are linear or piece-wise linear when activities are linear, non-splittable, non-stretchable, and relationships are the traditional end-point relationships.

Further research direction can be the description of the internal rules of influence lines where the generalizations listed in the 'Research objectives and limitations' section are allowed in the schedule network.

Acknowledgment

The research reported in this paper has been supported by the National Research, Development, and Innovation Fund (TUDFO/51757/2019-ITM, Thematic Excellence Program.)

References

[1] Hajdu, M. et al., 2016. How Many Types of Critical Activities Exist? A Conjecture in Need of Proof. Procedia Engineering, 164, pp.3–11. Available at: http://dx.doi.org/10.1016/j.proeng.2016.11.585.

[2] Wiest, J.D., 1981. Precedence diagramming method: Some unusual characteristics and their implications for project managers. Journal of Operations Management, 1(3), pp.121–130. Available at: http://dx.doi.org/10.1016/0272-6963(81)90015-2.

[3] Fondahl, J.W. 1961. A non-computer approach to the critical path method for the construction industry. Technical Report n°9 The Construction Institute, Department of Civil Engineering, Stanford University, Stanford California.

[4] Roy, G.B., 1962. Graphes et ordonnancement (Graphs and scheduling). Revue Française de Recherche Opérationnelle n° 25, 4e trimestre (1962), pp. 323-333 Available at: http://ekt.bme.hu/Cikkek/606-2021_Roy_1962.pdf

[5] IBM, 1964. Users' Manual for IBM 1440 Project Control System (PCS) Available at: http://ekt.bme.hu/Cikkek/12_27_1966_IBM.pdf

Analysing the Impact of Productivity in Indian Transport Infra Projects

Saurav Dixit

Associate Dean Research & Director International Collaboration, K. R. Mangalam University, Gurgaon, 122103, India.

*Corresponding author: Saurav.dixit@krmangalam.edu.in;

Abstract. The study's purpose is to analyze the impact of productivity in Indian transportation infrastructure projects. The study's methodology is to identify and analyse productivity attributes based on a literature review and expert focus group interviews. The identified attributes were identified and examined based on the respondents' relative weight. The data was gathered using a structured questionnaire across India, using both online and offline methods. SPSS 21 software was used to analyse the collected data. To analyse the collected data, quantitative research methodology was used, and the following tools and techniques were applied to the data: Reliability analysis was used to ensure the consistency of the data collected for the study, and the Relative importance index (RII) was used to set priorities and rank the attributes based on the weighted average score provided by the respondents. The study's findings show that planning and scheduling have the greatest impact on productivity in Indian transportation infrastructure projects, followed by construction methodology, storage area, poor construction method, and direction and coordination, with relative importance indexes (RII) of 0.79, 0.78, 0.77, 0.76, and 0.74, respectively. This study contributes to the current knowledge bank by identifying and analysing the impact of productivity in Indian transportation infrastructure projects. The study provides a solution for construction managers and project managers to apply the study's findings to their projects in order to control the issues of low productivity and project delays.

Keywords: Transportation; Infra Projects; Construction Productivity; Sustainability; Project Management.

1. Introduction

The Indian construction industry is diverse, fragmented, and appears to be un-organized. One of the most pressing issues for the industry has always been a skilled and qualified workforce. Because of the unique nature of the job, coordination, timely delivery, and quality have always been a source of concern. The construction industry in India is unorganized, and the majority of the workforce employed is uneducated and ignorant of sustainable practices, with only commercial viability as a motivator, with all other factors such as quality, sustainability, health and safety, and operability being ignored. The majority of construction projects in India are experiencing construction delays, budget overruns, poor quality, and low productivity. Low construction productivity is one of the major issues affecting the completion of construction projects on time and within budget.

The productivity could be measured at various levels, but there are three main measures of productivity are metronomic, case and pricing studies [1]. On the other hand, site-based productivity comparisons (case studies) were preferred because they help construction entities to find out the area of improvement and they can easily link them to their activities [2]–[4]. The financial wealth of nations is determined by their productivity growths. The nations experienced higher productivity growth translated into increases in the average wages of the workers, which contributes to the profits and tax revenue collection of the agencies. Although a few researchers try to understand the relationship between skill development and productivity in the construction industry. The trend is not consistent over time due to a number of reasons such as unplanned training sessions, consistency of skill development courses and the decrease in the number of participants [5].

With an average growth rate of 6-7 percent, the Indian economy is one of the fastest growing in the world. The construction industry in India is the second largest employer after agriculture and contributes about 10% to the economy on average. It will expand further as the rate of urbanization, followed by green and brownfield developments. To bridge the gap, the Indian government has implemented a number of policies, including the Smart Cities Mission, the Atal Mission for Rejuvenation and Urban Transformation (AMRUT), the Swachh Bharat Mission, Heritage City Development, and the Augmentation Yojana (HRIDAY). These missions will also help to increase productivity in Indian transportation infrastructure projects [6]–[11]. Construction productivity has been the area of interest for the research since the last 4-5 decades. A number of studies have been conducted in the field which includes: analysis of productivity, measurement techniques, causes of low productivity, factors affecting construction productivity and other studies.

2. Literature review

Table 1 previously identified attributes for productivity

References	Attributes of productivity
[12]–[16]	Planning and scheduling, Rework, Poor supervisor competency and Incomplete drawings
[12]–[20]	Tools and consumables, Coordination, Materials, Labour skills
[12]–[23]	Coordination, rework, supervision errors, drawing management, and construction equipment
[12]–[26]	Drawings, delay in response to information, truck availability, and material availability
[12]–[29]	Shop drawings, equipment's, motivation and support, and scheduling
[30]–[33]	Coordination and communication, timely feedback, owner's competence, favourable climatic condition, coordination among all team members, leadership, top management support, budget update, and the flow of funds

Productivity has been one of the most researched topics in the Indian construction industry in the last few decades. Factors affecting productivity may have a short-term or long-term effect on the project, some affect the productivity for a short duration but have a ripple effect on it. Productivity consists of various attributes like labour, finance, infrastructure, plant & machinery, facilities etc. Various studies in different countries have been carried out to identify the factor affecting labour productivity. Various methodologies and approaches have been adopted by researchers who have come with different schemes in the categorization of factors affecting productivity. [34] Proposed an analysis technique to measure capital productivity changes by the evaluation of factors influencing productivity levels in the construction industry. And applied it in the Australian construction industry. The research discoveries are relied upon to be useful for settling on strategy and key choices to enhance the capital productivity execution[35], [36].

3. Methodology

For this research paper, a quantitative research approach was used to identify the factors influencing the productivity of transportation infrastructure projects in India. The data was gathered using Question survey. A total of 115 valid questionnaires were received, yielding a response rate of 35%. All of those responded to the survey had prior experience working on transportation infrastructure projects.

3.1. Data collection

The data for the study were gathered through a structured questionnaire survey of 115 construction professionals and academicians from across India in order to attain the intended goal. The questionnaire consisted of 16 questions that attempt to cover the major factors that influence productivity. The factors were discovered with the assistance of various research on construction labour productivity. People with the necessary credentials and experience answered the questionnaires, so it can be safely assumed that the data obtained is accurate that can be used for the study because the respondents are knowledgeable associated with labour productivity and the factors that affect construction labour productivity.

3.2. Relative importance index

The received responses by the respondents were summarised in an Excel data sheet and the data analysed using SPSS software. Relative importance indices (RII) is performed to determine the priority of the significant factors and then followed by Reliability analysis performed to check the consistency of the data received.

$$Rii = \frac{\sum_{r=1}^{5} r^{*}n_r}{5N} \quad (1)$$

r is the rating on a Likert scale (1-5) as for the impact on construction efficiency for a specific element influencing construction profitability, nr is the number of respondents providing a specific Likert scale rating r, N is the aggregate number of respondents to a specific question [37]–[39]. The respondents were asked to rate the questions using a five-point scale '5' being the very high, '4' high, '3' moderate, '2' low, '1' very low impact on on-site productivity.

4. Results and discussion

4.1. RII

As per the study's findings, the most significant attributes influencing the productivity of transportation infrastructure projects in India are planning and scheduling, followed by method of construction, storage facility, poor workmanship technique, and direction and coordination, with relative importance indexes (RII) of 0.79, 0.78, 0.77, 0.76, and 0.74, respectively. The top five most important attributes have similar RII values. This finding needs to be validated, so a two-step cluster analysis was performed on the top five most important attributes to identify the main prediction models out of the five.

Table 2 Relative Importance Index (RII)

Rank	Total Responses	Total Score	RII	Attributes affecting the productivity of Indian transport Infra projects

1	115	455	0.79	Planning and scheduling
2	115	448	0.78	Type of Construction Methodology
3	115	441	0.77	Storage areas for materials
4	115	435	0.76	Poor construction method
5	115	428	0.74	Direction and coordination/communication
6	115	413	0.72	Adequate Crew and composition
7	115	409	0.71	Change in scope
8	115	406	0.71	Pep talk
9	115	403	0.70	Frequent change in drawings
10	115	392	0.68	Proper training provided prior to execution of work

5. Conclusion and recommendations

The construction industry's productivity has been studied and documented all over the world, and it advantages from over four decades of studies that has developed models, identified the factors affecting productivity, industry-level studies, the study of equipment and technology to improve productivity, and techniques to measure and improve construction productivity. As a result, construction and labour productivity are both increased [40], [41]. Several studies have been conducted to determine the factors that influence construction productivity, and the findings include the business culture, education level, tools and technologies used, values and ethics of the people and workers involved, politics, local laws and regulations governing the project, HR policies of the organization, importance given to the employee, religion of the peasant, and religion of the peasant. Smart cities and their parameters are being defined. The current study identifies the factors affecting the productivity of the Indian infra transport projects using a structured questionnaire survey approach, followed by the relative importance index. This study also highlights the importance of productivity in transport infra projects and provides a ranking of the most significant attributes impacting the productivity of transport infra projects in India.

6. Acknowledgement

The author would like to acknowledge the support and facility provided by the *K. R. Mangalam University Gurgaon* for conducting the research.

References

[1] B. Dyer, P. M. Goodrum, and K. Viele, "Effects of Omitted Variable Bias on Construction Real Output and Its Implications on Productivity Trends in the United States," *Journal of Construction Engineering and Management*, vol. 138, no. 4, pp. 558–566, 2012.

[2] M. R. Abdul Kadir, W. P. Lee, M. S. Jaafar, S. M. Sapuan, and A. A. A. Ali, "Factors affecting construction labour productivity for Malaysian residential projects," *Structural Survey*, vol. 23, no. 1, pp. 42–54, 2005.

[3] M. Chalker and M. Loosemore, "Trust and productivity in Australian construction projects: a subcontractor perspective," *Engineering, Construction and Architectural Management*, vol. 23, no. 2, pp. 192–210, 2016.

[4] S. Dixit, S. N. Mandal, J. V Thanikal, and K. Saurabh, "Evolution of studies in construction productivity : A systematic literature review (2006 – 2017)," *Ain Shams Engineering Journal*, vol. 10, no. 3, pp. 555–564, 2019.

[5] A. T. Gurmu and A. A. Aibinu, "Construction Equipment Management Practices for Improving Labor Productivity in Multistory Building Construction Projects," *Journal of Construction Engineering and Management*, vol. 143, no. 10, p. 04017081, 2017.

[6] S. Dixit, K. Sharma, and S. Singh, "Identifying and Analysing Key Factors Associated with Risks in Construction Projects," in *Emerging Trends in Civil Engineering, https://doi.org/10.1007/978-981-15-1404-3_3*, vol. 61, Springer, Singapore, 2020, pp. 25–32.
[7] S. Dixit, A. Stefańska, and A. Musiuk, "Architectural form finding in arboreal supporting structure optimisation," *Ain Shams Engineering Journal*, vol. 12, no. 2, pp. 2321–2329, 2021.
[8] M. N. Shah, S. Dixit, R. Kumar, R. Jain, and K. Anand, "Causes of delays in slum reconstruction projects in India," *International journal of construction management*, vol. 21, no. 5, pp. 452–467, 2021.
[9] S. Dixit, A. Stefańska, A. Musiuk, and P. Singh, "Study of enabling factors affecting the adoption of ICT in the Indian built environment sector," *Ain Shams Engineering Journal*, vol. 12, no. 2, pp. 2313–2319, 2021.
[10] S. Dixit, "Impact of management practices on construction productivity in Indian building construction projects: an empirical study," *Organization, Technology & Management in Construction*, vol. 13, no. 1, pp. 2383–2390, 2021.
[11] S. Dixit, "Study of factors affecting the performance of construction projects in AEC industry," *Organization, technology & management in construction: an international journal*, vol. 12, no. 1, pp. 2275–2282, 2020.
[12] N. Mani, K. P. Kisi, E. M. Rojas, and E. T. Foster, "Estimating Construction Labor Productivity Frontier: Pilot Study," *Journal of Construction Engineering and Management*, vol. 143, no. 10, p. 04017077, 2017.
[13] O. Moselhi and Z. Khan, "Significance ranking of parameters impacting construction labour productivity," *Construction Innovation*, vol. 12, no. 3, pp. 272–296, 2012.
[14] S. Mahmood, S. M. Ahmed, K. Panthi, and N. Ishaque Kureshi, "Determining the cost of poor quality and its impact on productivity and profitability," *Built Environment Project and Asset Management*, vol. 4, no. 3, pp. 296–311, 2014.
[15] S. Durdyev and S. Ismail, "On-site construction productivity in Malaysian infrastructure projects," *Structural Survey*, vol. 34, no. 4/5, pp. 446–462, 2016.
[16] S. Dixit, S. N. Mandal, J. V. Thanikal, and K. Saurabh, "Study of Significant Factors Affecting Construction Productivity Using Relative Importance Index in Indian Construction Industry," in *E3S Web of Conferences*, 2019, vol. 140.
[17] U. C. Gatti, G. C. Migliaccio, S. Schneider, and R. Fierro, "Assessing Physical Strain in Construction Workforce: A First Step for Improving Safety and Productivity Management," *Proceedings of the 27th International Symposium on Automation and Robotics in Construction (ISARC 2010)*, no. Isarc, pp. 255–264, 2010.
[18] K. P. Kisi, N. Mani, E. M. Rojas, and E. T. Foster, "Optimal Productivity in Labor-Intensive Construction Operations: Pilot Study," *Journal of Construction Engineering and Management*, vol. 143, no. 3, p. 04016107, 2017.
[19] R. Best, "Using purchasing power parity to assess construction productivity," vol. 10, pp. 1–10, 2010.
[20] M. A. Marhani, A. Jaapar, and N. A. A. Bari, "Lean Construction: Towards Enhancing Sustainable Construction in Malaysia," *Procedia - Social and Behavioral Sciences*, vol. 68, pp. 87–98, 2012.
[21] R. Navon, "Automated Productivity Control of Labor and Road Construction," *The 25th National Symposium on Automation and Robotics in Construction*, pp. 29–32, 2008.
[22] H. Rashid, "Construction Management : A Professional Approach of Factors Affecting the Labor Productivity," no. 1, pp. 283–287, 2015.
[23] P. F. Kaming, P. O. Olomolaiye, G. D. Holt, and F. C. Harris, "Factors influencing craftsmen's productivity in Indonesia," *International Journal of Project Management*, vol. 15, no. 1, pp. 21–30, 1997.
[24] S. Dixit, "Analysing Enabling Factors Affecting the On-site Productivity in Indian Construction Industry," *Periodica Polytechnica Architecture*, vol. 49, no. 2, pp. 185–193, Nov.

2018.
[25] M. Arashpour, R. Wakefield, N. Blismas, and E. W. M. Lee, "Analysis of Disruptions Caused by Construction Field Rework on Productivity in Residential Projects," *Journal of Construction Engineering and Management*, vol. 140, no. 2, p. 04013053, 2014.
[26] S. Dixit, K. Sharma, and S. Singh, *Identifying and Analysing Key Factors Associated with Risks in Construction Projects*, vol. 61. 2020.
[27] S. Kwon, G. Lee, D. Ahn, and H. Park, "A Modified-AHP Method of Productivity Analysis for Deployment of Innovative Construction Tools on Construction Site," pp. 45–50, 2014.
[28] A. T. Gurmu, "Tools for Measuring Construction Materials Management Practices and Predicting Labor Productivity in Multistory Building Projects," *Journal of Construction Engineering and Management*, vol. 145, no. 2, pp. 1–13, 2019.
[29] B. Vogl, "Measuring Construction: Prices, Output and Productivity," *Construction Management and Economics*, vol. 33, no. 9, pp. 775–777, 2015.
[30] Ministry of Business Innovation and Employment, "Construction sector productivity – the devil's in the detail," 2015.
[31] "PRODUCTIVITY MANAGEMENT OF WORKERS IN CONSTRUCTION INDUSTRY," vol. 2, no. Ii, p. 7074, 2010.
[32] F. C. Chia, M. Skitmore, J. Gray, and A. Bridge, "International comparisons of nominal and real construction labour productivity," *Engineering, Construction and Architectural Management*, vol. 25, no. 7, pp. 896–915, 2018.
[33] M. Loosemore, "Improving construction productivity: a subcontractor's perspective," *Engineering, Construction and Architectural Management*, vol. 21, no. 3, pp. 245–260, 2014.
[34] S. Dixit, A. K. Pandey, S. N. Mandal, and S. Bansal, "A study of enabling factors affecting construction productivity: Indian scnerio," *International Journal of Civil Engineering and Technology*, vol. 8, no. 6, 2017.
[35] S. Dixit, "Study of factors affecting the performance of construction projects in AEC industry," *Organization, Technology and Management in Construction*, vol. 12, no. 1, pp. 2275–2282, 2020.
[36] S. Dixit and K. Saurabh, "Impact of Construction Productivity Attributes Over Construction Project Performance in Indian Construction Projects," *Periodica Polytechnica Architecture*, vol. 50, no. 1, pp. 89–96, 2019.
[37] S. Singh, S. Dixit, and D. Varshney, "Sustainable construction management in education sector," vol. 7, no. 2, pp. 300–304, 2018.
[38] A. Sao, S. Singh, S. Dixit, A. K. Pandey, and S. Singh, "Quality, productivity and customer satisfaction in service operations: An empirical study," *International Journal of Mechanical Engineering and Technology*, vol. 8, no. 10, 2017.
[39] A. K. Pandey, S. Dixit, S. Bansal, S. Saproo, and S. N. Mandal, "Optimize the infrastructure design of hospital construction projects to manage hassle free services," *International Journal of Civil Engineering and Technology*, vol. 8, no. 10, pp. 87–98, 2017.
[40] M. N. Shah, S. Dixit, R. Kumar, R. Jain, and K. Anand, "Causes of delays in slum reconstruction projects in India," *International Journal of Construction Management*, pp. 1–16, Jan. 2019.
[41] S. Dixit, A. Stefańska, A. Musiuk, and P. Singh, "Study of enabling factors affecting the adoption of ICT in the Indian built environment sector," *Ain Shams Engineering Journal*, vol. 12, no. 2, pp. 2313–2319, Jun. 2021.

AUTHOR INDEX

Institute of Physics
Dirac House, Temple Back
Bristol BS1 6BE UK

ISSN: 1757-8981
ISBN 978-1-7138-4677-2

CURRAN ASSOCIATES INC.
proceedings.com